Global Crude Oil Business

Global Crude Oil Business

Dr. Sharad Goel

Pentagon Energy Press
AN IMPRINT OF PENTAGON PRESS

Global Crude Oil Business / Dr. Sharad Goel

ISBN 978-81-8274-496-7

First Published in 2011

Published by

PENTAGON ENERGY PRESS

An Imprint of

PENTAGON PRESS
206, Peacock Lane, Shahpur Jat,
New Delhi-110049
Phones: 011-64706243, 26491568
Telefax: 011-26490600
email: rajan@pentagonpress.in
website: www.pentagonpress.in

Printed at

New Elegant Printers, New Delhi.

This book is
dedicated
to my father
Late Shri Surender Kumar Goel

Contents

Preface

Discovery of crude oil and its products has been of great advantage to mankind. The human race has benefitted from crude products; gasoline or petrol, LPG, kerosene lubricants and asphalt. This natural resource has been of immense service to us. Crude oil trade is a booming business today. This is largely due to the ever-increasing costs of fuel. Many companies owning oil fields have decided to maximize profits from their upstream operations. Their main aim is to capture opportunities for growth. Further value creation is also their strategy. Many oil marketing companies are in contract with refiners to sell them physical crude oil for refining. Alternatively, they also have trading teams that enter into financial contracts to take charge of trading profits when time is ripe.

Hence there is a lot of manipulation in the market of crude oil price. The oil companies increase the rate of oil based on speculation. Demand and supply has nothing to do with it. This is mainly carried by the oil marketing and trading companies. Many a times the rate of crude rises and the market crashes. This happens because the traders had wrongly speculated that the prices of oil would go down. When this does not happen, the traders start buying oil contracts from the market. Apparently the traders and marketers are increasing their say how in the market. The government seems to be doing little about it. No wonder the price of gas is soaring!

Oil and gas are of paramount importance in economies worldwide. Oil, gas, hydro electricity, nuclear power and coal are the five constituents of primary oil. Oil and gas account for about 60 per cent of the total world's primary oil consumption. Crude oil is a mixture of hydrocarbons that exists in a liquid phase in natural underground reservoirs. Nations struggle to explore for oil, and import it at almost any cost. It is also an important contributor to the export realizations of many countries. In countries like Russia, nearly half the hard currency earnings come from crude oil exports. The figure rises to about 80% for Venezuela and 95% for Nigeria and Algeria. Since we have been tracking the data, the world has added more in proved reserves than it has used in oil production. In terms of years at current production rates, oil reserves are equivalent to 42, natural gas 60, and coal 133 years. Global proved oil reserves fell slightly in 2007, by 0.1%, and stand at 1.24 trillion barrels. But this is the result of reporting lags forcing us to roll over 2006 data. As is usually the case, the 2006 decline reported in last year's Statistical Review has now been revised up by 31 billion barrels, as new data came in.

Oil companies are increasingly exposed to price fluctuations in the volatile. Crude oil business since the turn off the new centaury. Since 1980s crude oil has seen a surge in its demand and prices worldwide. In the trade market, crude oil is sold in dollars and cents per barrel. Let us look at the current rate it enjoys in the market. If we take a stock of crude oil prices beginning from the turn of this century we see a rapid growth in the rates. This growth is of course directly proportional to the increase in its demand across nations, especially those with growing economies. But this year established a record set by shooting up crude prices to maximum. January saw crude being sold at $86.20 per barrel. In the month of February this price was increased by a dollar per barrel. March had oil sold at $97.46/barrel. But the steep graph did not stop here. April had crude oil priced at $104.31 and in May there was another leap of $13 thus placing it at $117.41 per barrel. Today a barrel of oil will cost you around $124.92. People have been paying for fuel through their noses. This is a trend that is being followed across nations all over.

In times like this it becomes necessary for us to understand the trading and marketing of crude oil. We need to know why we are paying what we are paying. Global analysts predict that the rate of crude oil is likely to go up to $150 per barrel. The developed nations are no longer building new refineries for production of oil. The need for new refineries is not being felt because natural resources are e fast depleting. It is no surprise that crude oil prices are sky rocketing. The world needs to be prepared for further increase in oil price in the coming times.

Crude oil prices are affected by numerous factors such as geographical and political environment, developments in legal system. OPEC decisions, speculative investments in the, market as well as demand and supply. The surge in global economic growth, as noted, stimulated demand for oil. Three generic groups of factors: Economic, Geopolitical and Technical has an influenced to a greater or lesser extent in the rise of oil prices and therefore other energy prices. We can better understand these in terms of how they gave momentum and durability to the price rise. In other words, how market agents perceived the first three sets of factors and interpreted or misinterpreted their importance influenced their actions in the market and determined market outcomes. Because these factors and events are generally well known and have been discussed and analyzed at length over the past few years, they are reviewed only briefly here.

Bulk amount of Crude Oil is transported from source to refineries by Crude Oil Tankers. India is the net importer of Crude Oil to the extent of about 75%. Bulk of the crude oil transportation for coastal refining centres as well as inland refineries is transported through Crude Oil Tankers. Crude oil tankers are classified on the basis of dead weight tonne (DWT) which is approximately equal to the carrying capacity.

Crude oil and petroleum products are traded globally 24 hours a day, every business day in both the physical and paper markets. With the daily physical consumption of oil of almost 84 million barrels and annual trade valued at over $1 trillion, the growth in paper energy trading seems assured, with new financial products evolving to meet the needs of producers, refiners, marketers, and consumers Global crude oil business is the guide to crude oil trading in the market and optimizing company's performance using correct financial derivatives, price risk strategies and tools.

Risk is the potential impact (positive or negative) to an asset or some characteristic of value that may arise from some present process or from some future event. In everyday usage risk is often used synonymously with probability and restricted to negative risk or threat. In professional risk assessments, risk combines the probability of an event occurring with the impact that event would be. Financial risk is often defined as the unexpected variability or volatility of returns, and thus includes both potential worse than expected as well as better than expected returns. In most financial markets there are a fairly small number of fundamental price drivers which can be easily translated into pricing and risk management models. In currency markets, for example, the commodity that has to be delivered is cash, a piece of paper which is easily stored, transferred and not sensitive to weather conditions.

Futures contracts have been used to manage cash market price risk for more than a century in the United States. Hedging allows a market participant to lock in prices and margins in advance and reduces the potential for unanticipated loss. Hedging reduces exposure to price risk by shifting that risk to those with opposite risk profiles or to investors who are willing to accept the risk in exchange for profit opportunity. Hedging with futures eliminates the risk of fluctuating prices, but also means limiting the opportunity for future profits should prices move favourably.

Global crude oil business is the guide to crude oil trading in the market and helping in optimizing company's performance using correct financial derivatives, price risk strategies and tools. This book provides the practical guidance in the effective trading of crude oil derivatives and their use as tools of risk management.

Dr. Sharad Goel

CHAPTER 1

Overview of Global Oil Business

1.1 INTRODUCTION TO THE OIL TRADE

The word petroleum is derived from the Latin words "Petra" meaning rock and "Oleum" meaning oil. It is also called mineral oil. In some parts of the world, a black, thin sticky liquids seeps out of the ground. This liquid is popularly known as *crude oil* or *petroleum.*

Petroleum is a complex mixture of hydrocarbons. As for the origin, Chinese used petroleum as a fuel as early as 200 BC. They were formed millions of years ago when large area of earth was covered by sea water. Tiny sea creatures called plankton died and sank to the sea bed. They mixed up with mud. The bacteria in mud began to bring about the decay of these tiny creatures. Process of decay took place slowly due to little oxygen dissolved in the depth of sea. The presence on the decaying material increased due to gradual deposition of mud and silt. Bacteria decay at high pressure with little oxygen changed the organic matter into crude oil. Petroleum and natural gas are found together, held in between the layers of non-porous rocks in the ground. These fuels are extracted by drilling a pipe through the rock.

Most geologists view crude oil and natural gas as the product of compression and heating of ancient organic materials over geological time. Oil is formed from the preserved remains of prehistoric zooplankton and algae which have been settled to the sea (or lake) bottom in large quantities under anoxic conditions. Terrestrial plants, on the other hand, tend to form coal. Over geological time this organic matter, mixed with mud, is buried under heavy layers of sediment. The resulting high levels of heat and pressure cause the organic matter to chemically change during digenesis, first into a waxy material known as kerogen which is found in various oil shale's around the world, and then with more heat into liquid and gaseous hydrocarbons in a process known as catagenesis. Because most hydrocarbons are lighter than rock or water, these sometimes migrate upward through adjacent rock layers until they become trapped beneath impermeable rocks, within porous rocks called reservoirs. Concentration of hydrocarbons in a trap forms an oil field, from which the liquid can be extracted by drilling and pumping. Geologists often refer to an "oil window" which is the temperature range that oil forms in – below the minimum temperature oil remains trapped in the form of kerogen, and above the maximum temperature the oil is converted to natural gas through the process of thermal cracking. Though this happens at different depths in different locations around the world, a 'typical' depth for the oil window might be 4–6 km. Note that even if oil is formed at extreme depths, it may be trapped at much shallower depths, even if it is not formed there (the Athabasca Oil Sands is one example). Three conditions must be present for oil reservoirs to form: first, a source rock rich in organic material buried deep enough for subterranean heat to cook it into oil; second, a porous and permeable reservoir rock for it to accumulate in; and last a cap rock (seal) that prevents it from escaping to the surface.

Crude or crude oil is a mixture of hydrocarbons that exists in a liquid phase in natural underground reservoirs. It remains liquid at atmospheric pressure after passing through surface separating facilities. Production volumes reported as crude oil include:

- Liquids technically defined as crude oil;
- Small amounts of hydrocarbons that exist in the gaseous phase in natural underground reservoirs, but which are liquid at atmospheric pressure after being recovered from oil well (casing head) gas in lease separators;
- Small amounts of non-hydrocarbons produced with the oil.

Basics of Crude Oil

(a) Crude oils are complex mixtures containing many different hydrocarbon compounds that vary in appearance and composition from one oil field to another. Crude oils range in consistency from water to tar-like solids, and in color from clear to black. An "average" crude oil contains about 84% carbon, 14% hydrogen, 1%-3% sulfur, and less than 1% each of nitrogen, oxygen, metals, and salts. Crude oils are generally classified as paraffinic, naphthenic, or aromatic, based on the predominant proportion of similar hydrocarbon molecules. Mixed-base crudes have varying amounts of each type of hydrocarbon. Refinery crude base stocks usually consist of mixtures of two or more different crude oils.

(b) Relatively simple crude oil assays are used to classify crude oils as paraffinic, naphthenic, aromatic, or mixed. One assay method (United States Bureau of Mines) is based on distillation, and another method (UOP "K" factor) is based on gravity and boiling points. More comprehensive crude assays determine the value of the crude (i.e., its yield and quality of useful products) and processing parameters. Crude oils are usually grouped according to yield structure.

(c) Crude oils are also defined in terms of API (American Petroleum Institute) gravity. The higher the API gravity, the lighter the crude. For example, light crude oils have high API gravities and low specific gravities. Crude oils with low carbon, high hydrogen, and high API gravity are usually rich in paraffins and tend to yield greater proportions of gasoline and light petroleum products; those with high carbon, low hydrogen, and low API gravities are usually rich in aromatics.

(d) Crude oils that contain appreciable quantities of hydrogen sulfide or other reactive sulfur compounds are called "sour." Those with less sulfur are called "sweet." Some exceptions to this rule are West Texas crudes, which are always considered "sour" regardless of their H2S content, and Arabian high sulfur crudes, which are not considered "sour" because their sulfur compounds are not highly reactive.

Crude Oil Types

Crude oil is arguably the world's most important and actively-traded commodity. Oil trades in a world market, and is bought and sold in relation to global prices. Because there are many different varieties and grades of crude oil, buyers and sellers have found it easier to refer to a limited number of reference, or benchmark, crude oils. Other varieties are then priced at a discount or premium, according to how their quality compares to that of the benchmark. The type of crude oil that is used as a benchmark in North America is West Texas Intermediate (WTI) oil, which is a light, sweet (low sulphur) crude. This is the price that is usually quoted in newspaper articles. Light sweet crude oils sell at higher prices than heavy sour (high sulphur) crudes, which are more difficult and expensive to refine and yield less of the more valuable oil products such as gasoline and jet fuel.

Typical approximate characteristics and properties and gasoline Potential of Various Crudes (Representative average numbers)

Crude source	Paraffins (% vol)	Aromatics (% vol)	Naphthenes (% vol)	Sulfur (% wt)	API gravity (Approx.)	Napht. Yield (% vol)	Octane no (typical)
Nigerian - Light	37	9	54	0.2	36	28	60
Saudi - Light	63	19	18	2	34	22	40
Saudi - Heavy	60	15	25	2.1	28	23	35
Venezuela - Heavy	35	12	53	2.3	30	2	60
Venezuela - Light	52	14	34	1.5	24	18	50
USA - Midcont Sweet	-	-	-	0.4	40	-	-
USA - W. Taxas Sour	46	22	32	1.9	32	33	55
North Sea - Brent	50	16	34	0.4	37	31	50

Source: US depts. of labour.

Varieties of Crude Oil

- Abu Bukhoosh
- Aktobe
- Al Shaheen
- Alaska North Slope
- Alba (Equ. Guinea)
- Alba (UK)
- Albian Heavy
- Algerian Condensate
- Al-Jurf
- Amenam Blend
- Ameriven-Hamaca
- Amna
- Antan Blend
- Arab Extra Light
- Arab Heavy
- Arab Light
- Arab Medium
- Arab Super Light
- Ardjuna
- Asgard
- Azeri (BTC)
- Azeri Light
- Bach Ho
- Balder
- Baobab
- Basrah Blend
- Bayu Undan
- BCF-17
- Belanak
- Belayim Blend
- Beryl
- Bintulu Condensate
- Bonga
- Bonny Light
- Boscan
- Bouri
- Bow River
- Brass River
- Brega
- Brent Blend
- Cabinda
- Cano Limon
- Captain
- Ceiba
- Cepu
- Cerro Negro
- Champion
- Chinguetti
- Cinta
- Clair
- Cold Lake
- Cossack
- CPC Blend
- Cusiana
- Dalia
- Daqing
- Dar Blend
- Djeno
- Doba
- Doroud
- Draugen
- Dubai
- DUC
- Dukhan
- Dulang
- Duri
- EA
- Ekofisk
- El Sharara
- Enfield
- Erha
- Es Sider
- Escravos
- Flotta
- Foinaven
- Forcados
- Foroozan Blend
- Forties
- Gippsland
- Girassol
- Grane
- Gullfaks
- Handil
- Harding
- Heavy Louisiana Sweet
- Heidrun
- Hibernia
- Hungo
- Iran Heavy
- Iran Light
- Isthmus
- Karachaganak Cond.
- Kashagan Khafji
- Kikeh
- Kirkuk
- Kissanje Blend
- Kole
- Kuito
- Kumkol
- Kutubu
- Kuwait
- Labuan
- Lavan Blend
- Light Louisiana Sweet
- Liuhua
- Liverpool Bay
- Mandji
- Marib
- Marlim
- Mars Blend
- Masila
- Maya
- Medanito
- Mesa-30
- Minas
- Miri
- Murban
- N'Kossa
- Nanhai Light
- Napo
- Nemba
- New Zafiro Blend
- NFC II
- Nile Blend
- Norne
- Northwest Shelf Cond.
- Nowruz/Soroush
- Olmeca
- Oman
- Oriente
- Oseberg
- Oso Condensate
- Palanca Blend
- Panyu
- Peng Lai
- Pennington
- Petrozuata
- Premium Albian
- Qatar Marine
- Qua Iboe
- Rabi Light
- Rang Dong
- Saharan Blend
- Sarir
- Schiehallion
- Senipah
- Seria Light
- Shah Deniz Condensate
- Shengli
- Siberian Light
- Sincor-Zuata
- Siri
- Sirri
- Sirtica
- Sleipner Condensate
- Sokol
- Souedieh
- South Arne
- Southern Green Canyon
- Statfjord
- Su Tu Den
- Suez Blend
- Syncrude Sweet Blend
- Syrian Light
- Tapis
- Terra Nova
- Tia Juana Light
- Triton
- Troll
- Turkmen Blend
- Umm Shaif
- Upper Zakum
- Urals
- Vityaz
- Wafra
- West Seno
- West Texas Intermediate
- West Texas Sour
- Western Canada Select
- White Rose
- Widuri
- Xikomba
- Yoho
- Zakum
- Zarzaitine

Products Derived from Crude

In India, following items are produced from crude oil:

- **Light distillates** (includes LPG, Mogas, Naphtha, LD i.e. Propylene, C-3, Propane, Hexane, Special Boiling Point Spirit, Benzene, Toluene, Petroleum Hydro Carbon Solvent, Natural Heptane, Methyl Tertiary Butyl Ether, Poly Isobutine, PBFS and MEKFS)
- **Middle distillates** (includes Kerosene, ATF, RTF, Jet A-1, HSD, LDO, MD i.e. Mineral Turpentine Oil, JP-5, Linear Alkyl Benzene Feed Stock, Aromex, Jute Batching Oil, Solvent 1425, Low Sulphur Heavy fuel HSD, DHCB and Special Kerosene)
- **Heavy ends** (includes Furnace oil, LSHS, HHS, RFO, Lube oils, Bitumen, Petroleum coke, Paraffin wax, other waxes etc.)There are in fact many products obtained from the processing of crude oil, and other hydrocarbon compounds. These include aviation gasoline, motor gasoline, naphtha, kerosene, jet fuel, distillate fuel oil, residual fuel oil, liquefied petroleum gas, lubricants, paraffin wax, petroleum coke, asphalt and other products. Following are some of the products:-

Gasoline: a mixture of relatively volatile hydrocarbons, with or without small quantities of additives, that have been blended to form a fuel suitable for use in internal combustion engines; includes gasoline used in aviation.

Kerosene: medium hydrocarbon distillates in the 150° to 280° C distillation range, and used as a heating fuel as well as for certain types of internal combustion engine; includes jet fuel,which is a fuel of naphtha, or of kerosene type, suitable for commercial or military purposes in aircraft turbine engines.

Distillates: middle distillate type of hydrocarbons. Included are products similar to number one and number two heating oils and diesel fuels. These products are used for space heating, diesel engine fuel and electrical power generation.

Residual fuel oil: these are fuels obtained as liquid still bottoms from the distillation of crude used alone or in blends with heavy liquids from other refinery process operations. It is used for the generation of electric power, space heating, vessel bunkering and various industrial purposes.

Oil units

1 Tonne = 7.33 Barrel = 1.165 Cubic Meters (kilolitres)
1 Barrel = 0.136 Tonnes = 0.159 Cubic Meters (kilolitres)
1 Cubic Metre = 0.858 Tonnes = 6.289 Barrels
1 Million Tonne = 1.111 Billion Cubic Meters Natural Gas
= 39.2 Billion Cubic Feet Natural Gas
= 0.805 Million Tonnes LNG
= 40.4 Trillion British Thermal Units

Over the last fifteen years oil has become the biggest commodity market in the world. During this period, oil trading has evolved from a primarily physical activity into a sophisticated financial market. In the process it has attracted the interest of a wide range of participants who now include banks and fund managers as well as the traditional oil majors, independents and physical oil traders.

As well as being the largest commodity market in the world, oil is also the most complicated. The physical oil market trades many different types of crude oil and refined products, and the relative values of each grade are continually shifting in response to changes in supply and demand on both a global and a local scale. The industry has therefore developed a complex set of interlocking markets not only establish prices across the entire spectrum of crude and products qualities, but also to enable buyers and sellers to accommodate changes in relative prices wherever they might occur.

The initial momentum for the expansion of the oil market came from the changing structure of the oil industry. Prior o 1973, oil trading was a marginal activity for most companies who only used the market to resolve any imbalances and demand that might emerge. Trading volumes were typically small and usually spot, and prices were much less transparent than they are today. And the industry was dominated by large integrated oil companies that have little use of external markets either as a means of obtaining supplies or as a basis for setting prices.

However the structure of the oil industry changed irreversibly in the 1970s with the nationalization of the upstream interest of the major oil companies in the Middle East and elsewhere, and trading become an essential component of any oil company's supply and marketing operations. Having lost access to large volumes of equity oil, the major oil companies were forced to buy at arm's length from their former host governments and the physical base of the international oil market expanded rapidly. With more oil being traded, external markets began to set the price for internal transfers as well as third party sales and companies started to buy and sell oil if better opportunities existed outside their own supply network fuelling the growth of the market.

But the driving force behind the rapid growth I oil trading is the huge variability in the price of oil. Daily price movements of $1/barrel are not uncommon and prices frequently change by up to 50 cents/barrel. Since there is no obvious upper or lower bound to oil prices, the value of a barrel of oil can double within the spaces of a few months. As a result, everyone involved in the industry is exposed to the risk of very large changes in the value of any oil that they are producing transporting, refining or purchasing and range of new markets have evolved in order to provide effective hedging instruments against the elaborate combination of absolute and relative price risks that characterizes the oil business. This has not only generated a very large volume of activity in its own right, but also attracted liquidity from other financial and commodity markets.

Trading Characteristics of Oil

Many of the characteristics features of the oil market are derived from the nature of oil itself. Despite the introduction of highly standardized paper trading instruments, oil remains a physical commodity. Like other primary commodity, the oil market is ultimately concerned with the transportation, processing and storage of an essential raw material as it travels from producer to consumer. However this is slow process, since crude oil may take several months to move from the well head through the refinery to the sales pump. As a result, prices often change because the right oil is not in the right place at the right time. This is very different from the financial markets, where assets can be moved instantaneously from one location to another if required.

Transportation, Processing and Storage

One of the most important characteristics of oil is that it is a liquid. As a liquid it requires specialized handling facilities for transportation, processing and storage. And it is these elements that provide the basic building blocks for the physical oil market.

(a) Transportation

Oil is transported either in ships or pipelines. In the international market, oil moves almost exclusively in ships and it is therefore the size of the ship that forms the basic trading unit. In the case of crude oil, quantities are typically large and usually depend on the capacity of the loading and discharge terminals, the length of the voyage and the relative cost of shipping. In the North Sea, which is the most active waterborne crude market in the world, 500,000 barrel cargoes are the norm. But for longer -haul crudes from West Africa or the Middle East, oil often moves in very large crude carriers (VLCCs) which can take up to 2 million barrels at a time. As a result, the scale of financial exposure associated with crude oil trading can be very large indeed.

Refined products, however, are usually traded in much smaller quantities. Long distance movements may involve shipments as large as 60,000 tones (about 500,000 barrels, depending on the type of product), but most of the international trade is conducted in smaller vessels holding 20-30,000 tones. And many of the most active product markets deal in much smaller, large sized quantities, of between 1,000 and 5,000 tones. Since products are usually traded ex-refinery and often sold to wholesale distributors who may not have the capacity to receive or store very large quantities, the basic trading units need to be much smaller than in the international crude market.

As well as fixing the size of trading unit the method of transport often determines the terms of trade. Crude oil is usually sold close to the point of production and title is transferred as oil flows from the loading terminal into the ship. Once loaded however, the oil can either be traded on the water or at the point of discharge. As a result, the same cargo of oil may be priced differently depending on the point of sale. Refined products are traded on a variety of terms depending on local circumstances, but it is important to realize that several markets can co -exist for the same product at the same location with prices that reflect different delivery arrangements or parcel sizes.

Oil is also transported and traded via pipelines. The most important pipeline markets are in the US where access is guaranteed in law to those who want to use them. In most cases oil is traded on ratable basis- a specified number of barrels per day over an agreed period such as a fortnight or a month- and the oil is sold free in pipeline (fip) at designated locations. But in the case of West Texas Intermediate, which forms the basis of the Nymex Light Sweet Crude Futures contract, oil is also sold in multiples of 1,000 barrels available from or delivered into storage facilities at Cushing, Oklahoma.

(b) Processing

Oil is a not normally used in its raw state. Crude oil must be processed through a refinery in order to turn it into marketable product such as gasoline, heating oil or fuel oil. The only exception is low sulphur crude oil which is sometimes burnt directly in power stations. Oil is therefore traded twice, first as a refinery feedstock, and secondly as a finished product. Although crude and product markets have rather different characteristics, they are inextricably linked by the technology and economics of refinery processes.

Crude oil markets operate between the producer and the refiner. The characteristics and the behavior of the crude oil market therefore depend on the preferences and needs of the refiner as well as the composition and nature of the supply. Because there are many different types of crude oil, relative value depends on the mix of products that can be obtained from them. In general, crudes that yield a higher proportion of the more valuable light products such as gasoline, naphtha, kerosene and heating oil can command a higher price than those which have a higher price than those which have a high yield of residual fuel oil. But there is no objective method of assessing the price of given crude since every refinery has a different configuration and its market value will depend on who is bidding at the time. And refineries in different regions may have very different views about the price they are prepared to pay.

Products markets operate between the refiner and the blender or wholesaler. They are usually much more localized than crude oil markets, since most refineries are positioned close to the end- user, and their process facilities are tailored to the needs of the local consumer. As a result, refined product prices can differ significantly from one market to another, reflecting the local structure of demand for the various petroleum product quality specifications.

(c) Storage

Oil must also be stored on its journey from the well head to the pump. As oil is a liquid, this requires the construction of specialized storage tanks at every stage in the supply chain. Stocks are necessary in any business that produces, manufactures and markets a physical commodity such as oil, and fluctuations in the level of stocks held at different points along the supply chain play an important role in determining the behavior of prices in the oil market. But holding stocks is also costly since it ties up cash and storage facilities are expensive to rent or build. Oil companies therefore try to keep their stocks as near to the minimum operating level as circumstances allow.

A surprisingly large amount of oil is required simply to fill the supply chain from well- head to customer. In addition, stocks are needed to keep system flowing since deliveries are usually made in discrete quantities and stocks are run down in the intervening period. Also companies need to hold extra stocks as an insurance policy against unexpected interruptions in supply or increases in demand from their customers. And finally companies often build up (or run down) stocks for purely speculative reasons either to profit from an upward price trend or to minimize the losses from downward price trend.

Crude Oil Supply Chain

The entire oil markets are divided into three phases on the basis of activities:

(1) The "upstream" comprises exploration and production;
(2) The "midstream" are the tankers and pipelines that carry crude oil to refineries, and;
(3) The "downstream", includes refining, marketing, and distribution, right down to the corner gasoline station or convenient store.

Any company that includes together significant upstream and downstream activities is said to be "integrated".

Crude Oil Value Chain – *from Wellhead to Gasoline Pump*

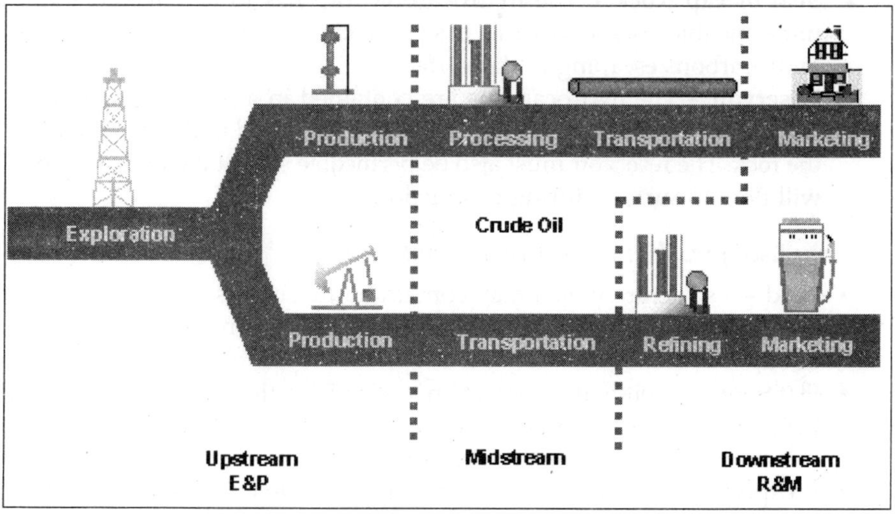

Exploration	**Production**	**Transportation**	**Refining**	**Marketing**
Using technology to find new oil resources	Bringing oil to the surface using natural and artificial methods	Moving oil to refineries and consumers with tankers, trucks and pipelines	Converting crude oil into finished products	Distributing and selling refined products

A. Upstream sector

The upstream sector includes the searching for potential underground or underwater oil and gas fields, drilling of exploratory wells, and subsequently operating the wells that recover and bring the crude oil and/or raw natural gas to the surface. The upstream activities can be related to following-

Petroleum Prospects

A prospect is a potential trap which geologists believe may contain hydrocarbons. A significant amount of geological, structural and seismic investigation is must to drill location from a lead to a prospect.

The five pre-requisites are –

1. **A source rock** – When organic-rich rock such as oil shale or coal is subjected to high pressure and temperature over an extended period of time, hydrocarbons form.
2. **Migration** – The Hydrocarbons are expelled from source rock by three density-related mechanisms: the newly-matured hydrocarbons are less dense than their precursors, which causes overpressure; the hydrocarbons are lighter medium, and so migrate upwards due to buoyancy, and the fluids expand as further burial causes increased heating. Most hydrocarbons migrate to the surface as oil seeps, but some will get trapped.
3. **Trap** – The hydrocarbons are buoyant and have to be trapped within a structural (e.g. Anticline, fault block) or stratigraphic trap.
4. **Seal or cap Rock** – The hydrocarbon trap has to be covered by an impermeable rock known as a seal or cap-rock in order to prevent hydrocarbons escaping to the surface.
5. **Reservoir** – The hydrocarbons are contained in a reservoir rock. This is a porous sandstone or limestone. The oil collects in the pores within the rock. The reservoir must also be permeable so that the hydrocarbons will flow to surface during production.

Terms used prevalent in upstream activities are –

- Lead – a structure which may contain hydrocarbons
- Dry Hole – Counter-intuitively, a formation that contains brine instead of oil.
- Flat Spot – An oil-water contact on a seismic section; flat due to gravity.
- Bright Spot – On a seismic section, coda that have high amplitudes due to a formation containing hydrocarbons.
- Prospect – a lead which has been fully evaluated and is ready to drill
- Chance of Success – An estimate of the chance of all the elements (see above) within a prospect working, described as a probability. High risk prospects have a less than 10% chance of working, medium risk prospects 10-20%, low risk prospects over 20%. Typically about 40% of wells recently drilled find commercial hydrocarbons.

- Hydrocarbon in Place – amount of hydrocarbon likely to be contained in the prospect. This is calculated using the volumetric equation – GRV × N/G × Porosity × Sh × FVF
- GRV – Gross Rock volume – amount of rock in the trap above the hydrocarbon water contact
- N/G – net/gross ratio – percentage of the GRV formed by the reservoir rock (range is 0 to 1)
- Porosity – percentage of the net reservoir rock occupied by pores (typically 5-35%)
- Sh – hydrocarbon saturation – some of the pore space is filled with water – this must be discounted
- FVF – formation volume factor – oil shrinks and gas expands when brought to the surface. The FVF converts volumes at reservoir conditions (high pressure and high temperature) to storage and sale conditions
- Recoverable hydrocarbons – amount of hydrocarbon likely to be recovered during production. This is typically 10-50% in an oil field and 50-80% in a gas field.

Reserves and Resources

Resources are hydrocarbons which may or may not be produced in the future. A resource number may be assigned to an undrilled prospect or an unappraised discovery. Appraisal by drilling additional delineation wells or acquiring extra seismic data will confirm the size of the field and lead to project sanction. The three categories of reserves generally used are proven –

1. **Proven reserves** – defined as oil and gas "Reasonably Certain" to be producible using current technology at current prices, with current commercial terms and government consent- also known in the industry as 1P. Some Industry specialists refer to this as P90 – i.e having a 90% certainty of being produced.
2. **Probable reserves** – defined as oil and gas "Reasonably Probable" of being produced using current or likely technology at current prices, with current commercial terms and government consent – Some Industry specialists refer to this as P50 – i.e having a 50% certainty of being produced. – This is also known in the industry as 2P or Proven plus probable.
3. **Possible reserves** – i.e "having a chance of being developed under favourable circumstances" – Some industry specialists refer to this as P10 – i.e having a 10% certainty of being produced. – This is also known in the industry as 3P or Proven plus probable plus possible.

Licensing and Concessionary Regimes

Petroleum resources are typically owned by the government of the host country. In most nations the government issues licences to explore, develop and produce its oil and gas resources, which are typically administered by the oil ministry. There are several different types of licenses. Typically oil companies operate in joint ventures to spread the risk, one of the companies in the partnership is designated the operator who actually supervises the work.

The terms and structures of upstream license agreements between companies and governments determine whether a specific oil and gas field can be developed on a commercial basis and how costs and profits will be divided over the field's life cycle.

Global comparisons of the economic performance of production sharing contracts (PSCs), tax and royalty, service, risk and joint venture contracts identify strengths, weaknesses, risks and opportunities associated with specific contracts. There are many issues, going beyond just the fiscal terms, that must be considered when negotiating licence agreements.

B. Midstream Sector

The midstream sector processes, stores, markets and transports commodities such as crude oil, natural gas and natural gas liquids (NGLs) such as ethane, propane and butane.

Marketing and Distribution

The business activities of marketing and distribution segment include purchasing refined oil products from the refining segment and third parities, and wholesale of refined oil products to domestic customers, direct sale and retail of the refined oil products through the retail network owned by this segment and provision of related services.

C. Downstream Sector

The downstream sector includes oil refineries, petrochemical plants, petroleum product distribution, retail outlets and natural gas distribution companies. The downstream industry touches consumers through thousands of products such as gasoline, diesel, jet fuel, heating oil, asphalt, lubricants, synthetic rubber, plastics, fertilizers, antifreeze, pesticides, pharmaceuticals, natural gas and propane.

Petroleum Refining Process

As mentioned previously, a barrel of crude oil has a mixture of all sorts of hydrocarbons in it. Oil refining separates everything into useful substances. Chemists use the following steps: The oldest and most common way to separate things into various components (called fractions), is to do it using the differences in boiling temperature. This process is called fractional distillation. You basically heat crude oil up, let it vaporize and then condense the vapor. Newer techniques use Chemical processing on some of the fractions to make others, in a process called conversion. Chemical processing, for example, can break longer chains into shorter ones. This allows a refinery to turn diesel fuel into gasoline depending on the demand for gasoline. Refineries must treat the fractions to remove impurities. Refineries combine the various fractions (processed, unprocessed) into mixtures to make desired products. For example, different mixtures of chains can create gasoline with different octane ratings. However, midstream operations are usually simply included in the downstream category.

REFINING PROCESSES
PRODUCTS AND THEIR USES

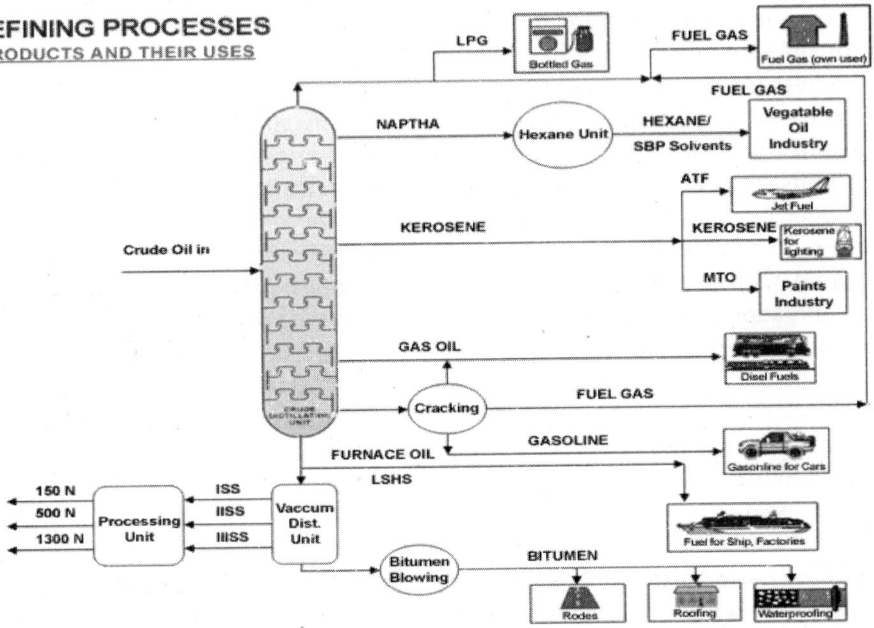

In the evolution of petroleum industry after crude oil was discovered the first process that got developed was fractional distillation of crude oil. This physical separation process since then still remains the key process. Simple distillation, cannot be applied in refining crude oil. Crude oil consists of many components with boiling points covering a great range of temperatures. Each finished product such as motor spirit or kerosene comprises a large number of components with different boiling points. Each product therefore has a boiling range having an initial boiling point (I.B.P.) and a final boiling point (F.B.P.). By simple distillation one cannot avoid considerable overlapping of the boiling ranges of the fractions. To obtain narrow fractions, fractional distillation using fractionating columns has to be employed. A mixture of two miscible liquids will start boiling at a temperature lying somewhere between the boiling points of the components, depending on their relative proportions in the mixture. Thus, for instance, by addition alcohol (boiling point 78°C) to the water, the boiling point of the lowerer the boiling point becomes. When such a mixture starts boiling, it is not only the more volatile of the two components which evaporates, the vapour also contains the less volatile component. When a liquid consists of two components, each of these contributes its own quotum to the total vapour pressure of the liquid, so that when boiling takes place the pressure in the still is made up of two parts, one exerted by the vapour of the first component, the other by the vapour of the second and, since the sum of these partial pressures of the components is equal to atmospheric, the combined vapours leave the liquid, that is to say, both of the components are able to evaporate at the same time. The Partial pressure of each component is principally dependent on the vapour pressure of the component in a pure state, and on its proportion in the liquid mixture, so that for the same proportion in the liquid the percentage in the vapour will be highest for the component with the highest vapour pressure. This can be expressed also by saying that the more valatile substance evaporates more readily, so that its percentage in the vapour is greater than in the liquid. Because the evaporated part of the liquid contains a higher percentage of the lighter component than the original liquid, the remainder will necessarily contain a smaller percentage. Consequently, the boiling point of the remaining liquid will be higher than that of the original liquid. This will be true as evaporation goes on, so that longer the evaportion proceeds the higher boiling point of the remainder of the liquid. Thus, for such a mixture there is no question of a constant boiling temperature. For petroleum products consisting of many components, with boiling points covering a great range of temperatures, this phenomenon will be still more pronounced. The boiling temperature will rise continuously from the temperature at which the product starts boiling i.e. the initial boiling point (I.B.P.), up to the temperature when the last remnant of liquid evaporates, i.e. the final boiling point (F.B.P.). This boiling range will, of course, be a function of the boiling points of all components present and their respective percentages in the mixture, but for a long series of components the initial boiling point will be principally determined by the components with the lower boiling points, while the final boiling point is more exclusively a function of the heavier hydrocarbons. Thus, the boiling range can be very wide.

Fractional Distillation

By simple distillation one cannot avoid considerable overlapping of the boiling ranges of the fractions, because the lightest components remain partly in the residue, while much heavier constituents are in the vapours. Consequently, the I.B.P. of the residue is well below the F.B.P. of the distillate; moreover, the boiling ranges of the products are by no means narrow. To obtain narrow fractions, a more complicated process is required, based principally on repeated distillation. Suppose a distillate is again partly evaporated: the more volatile components, which were already more concentrated in the distillate, than in the original, will again be more concentrated in the second distillate. By repeating this procedure several times, eventually a fraction can be produced consisting of the lightest constituents in a concentrated form. However, the efficiency of separation is very low; only a small part of the light fractions is obtained in this way, because at each evaporating stage, part of them is left behind in the liquid.

In order to get a quantitative separation between the desired fractions, it is necessary not only to effect evaporation in many successive stages but, moreover, to return the residue of each stage into the preceding one, in which it is again partly evaporated for purpose of driving out light fractions once more. As this goes on continually, the final residue will eventually be freed from light fractions. This process is called "rectification".

Fractionating Columns

The crude oil which is being pumped to the furnace first passes through heat exchangers, where it picks up heat from the outgoing vapours and residue. Before it reaches the furnac, its temperature has risen to, say, 180°C. In the furnace, the crude oil is heated to 340°-370°C, depending on the crude oil and on the residue needed. The heated oil leaving the pipe still passes to the fractionating columns, where the separation into fractions or "cuts" takes place. These columns are long cylindrical vessels set up on end and fitted with a large number of trays of a special type. The mixture of vapour and liquid leaving the pipe still is led in at a point near the bottom of one of these fractionating columns, where it rises gradually through the vapour uptakes shown. As it rises it cools and a certain amount condenses on each tray, until is kept just above the holes in the bubble-cap, so that all the vapour has to pass through liquid, and at the same time liquid head is not very large in order to avoid excessive pressure drops. It will be apparent that each tray is a little cooler than the one below it and that, therefore, lighter and lighter products will be present on each tray as the vapour passes up the column. As the vapours bubble through the liquid on these trays, that part of the vapour will condense which has the same boiling point as the liquid on the tray, and in addition any light fractions present in the liquid on the tray will be re-evaporated. The temperature throughout such a column is controlled at the bottom by the temperature of the crude oil leaving the pipe still, and at the top by pumping back a certain amount of the material which leave the top of the column after condensing it (this distillate pumped back is termed "reflux") In this way, by controlling the temperature at the top and at the bottom, the temperature variation throughout is kept under control, so that the temperature of each tray garies gradually from the bottom of the column to the top. The amount of liquid which is pumped back to the top of the column can be varied as required, in order to give the correct temperature at the top of the column, which, in turn, controls the final boiling point of the gasoline leaving at the top. In most cases, improvement in fractionating efficiency is obtained by supplying additional heat to the product in the bottom part of the fractionating column. This is done by pumping the oil at the bottom of the column either through a furnace or through a heat exchanger, where it is heated, and then back again over some stripping trays in the column. Such an arrangement is referred to as a "reboiler."

Side-Strippers

Although each tray has the tendency to collect liquid of a small range of boiling points, yet, on account of the complete range of vapours bubbling through any one tray, a certain amount of material will condense which has a lower boiling point than the bulk of the liquid on that tray. It is customary, therefore, to draw the liquid from a selected tray out of the main column into a small subsidiary column called a "side stripper" and allow it to flow down through several bubble trays. While passing over these trays, it meets an upward stream of steam, injected at the bottom of the stripper, which boils off the lightest components and thus narrows the boiling range. The vapours thus boiled off, together with the steam, re-enter the main column. Other means of supplying heat to side-strippers employ the roboiling principle, hot residue being used as heat medium. By using two or more of these side-strippers in conjunction with one main column, streams of white spirit, kerosene and gas oil can be obtained direct from the plant, without further distillation. This arrangement considerably simplifies the production of the various grades of distillates and places the refiner in a position to manufacture gasolines and the other products mentioned above of the correct distillation range, without the necessity of redistillation.

Vacuum and Pressure Distillation

The heaviest fractions of crude oil requires such high temperatures to evaporate them that cracking might occur. In order to avoid this, distillation is carried out under vacuum, thus lowering the boiling point.For the very light fractions, condensation of the vapours cannot be obtained at atmospheric pressure and temperature, so that distillation is carried out under pressure, which increases the boiling point and, consequently, the temperature at which condensation can be obtained.

A Typical Crude Distillation Unit (CDU)

Having discussed the process of fractional distillation and fractionating columns we now come to describe a typical crude distillation unit. Through the process of evolution over the years various developments and improvements in the equipment design have taken place. Bubble cap trays have been replaced by a variety of other types such as sieve trays and "Calming section trays" which is a proprietary shell device. Through application of "Calming section trays" Shell claim a reduction of 15 to 20% in the column area compared with conventional trays. Considerable improvements have been made in the design and construction of furnaces. The conventional furnaces have been superseded by vertical types with vastly improved combustion equipment such that excess air as low as 5% is now possible compared to 25-30% excess air in older furnaces. With escalating price of petroleum, waste heat recovery through improved heat exchange has assumed a great significance. In the latest design, therefore, has the heat transfer area as well as the configuration of heat exchanger trains. Last but not the least the advent of microprocessors has lead to process control and process optimization using online programmable logic controllers.

A typical crude distillation unit can have a variety of configurations depending upon the type of crude being processed, yield requirement etc. we give a Shell design of a 2 column crude distillation unit integrated wit a gas tail and high vacuum unit, to obtain a sharp separation, especially in the gas oil range a second column has been provided which works under a moderate vacuum. The two column set-up is claimed to give a gain in heavy gas oil yield of 1% wt. On crude compared to a one-column system, at virtually the same cost.

The crude oil is first processed through a desalter which is a standard feature in all crude units. Sodium and Magnesium Chlorides carried in the crude, dissociate as the temperature of the crude goes up there by releasing HCL in the system. This leads to corrosion. The crude is therefore, desalted. The crude after being preheated in heat exchangers goes to the preflash vassel. While the flashed vapours go directly to the primary crude splitter the liquid is heated in the furnace to a temperature around 250°C before entering this column. Apart from the top reflux a number of circulating refluxes have been provided which exchange heat with the crude oil in heat exchangers. The column tops are treated in a hydrotreater bottoms go to gasoline splitter and top to deethanizer and finally to C3–C4 splitter.

Kerosene and part of light gas oil is taken out as a side stream. The column bottom is passed through a furnace, raised to a temperature of around 350°C before entering the secondary crude splitter operates at slight vacuum. The top product is the balance of the light gas oil, the sidestream is the heavy gas oil and the bottom product is the long residue.

Long residue is heated to a temperature somewhat below 400°C before entering the high vacuum column. Top product is a gas oil stream, first side stream is a light waxy distillate which could be a feedstock for lube base and the second side stream is heavy waxy distillate for cracking. Amongst the physical separation processes, separation through fractional distillation depends upon the difference in volatility of different components. However when the volatility difference is too small other techniques would need to be adopted, some of these processes are described here after in brief.

Extractive Distillation

The difference in volatility of the components to be separated is enhanced by the addition of a solvent or an entrainer. The variation of distillation in which a solvent of low volatility is used to enhance separation is called Extractive Distillation. An example is the separation of butanes from butanes with furfural. When a solvent of high volatility is added as an entrainer, the operation is called azeotropic distillation is the production of high purity toluene using methyl ketone as the entrainer and also distillation to produce anhydrous alcohol.

Absorption

Absorption is generally used to separate a higher-boiling constituent from other components of a system of vapours and gases. The absorption medium is usually an oil in the range of gas oil. Absorption is widely employed in the recovery of natural gasoline from well gas and of vapours given off by storage tanks. Absorption also obtains light hydrocarbons from many refining processes (catalytic cracking, hydrocracking, coking etc.). The solvent oil may be heavy gasoline, kerosenes, or even heavier oils. The absorbed products are recovered by fractionating or steamstripping.

Adsorption

Adsorption is employed for about the same purpose as absorption; in the process just mentioned natural gasoline may be separated from natural gas by adsorption on charcoal. Adsorption is also used to remove undesirable colours from lubricating oils, usually employing activated clay. The use of molecular sieves in separating close boiling components deserves a special mention.

Filtration

Filtration after chilling is the usual method for removal of wax from waxy distillates. The mixture of wax and adhering oil obtained from the press is frozen and allowed to warm slowly so that the oil drains (sweats) from the cake thus further purifying the cake.

Crystallization

Crystallization ranks as one of the oldest physical separation operations. By means of crystallization. By means of crystallization, wax may be removed from lubricating oil to yield crystalline and microcystalline waxes of low oil content. Crystallization has also been applied in the separation of p-xylene from C8 aromatics in high purity in terephthalic acid manufacture.

Extraction

Extraction involves the removal of a component from a liquid by means of the selective solvent action of another liquid. In the Edeleanu process liquid sulphurdioxide is used to extract aromatics from kerosene pool, Liquid SO_2 selectively dissolves the aromatics leaving behind the low aromatic content kerosene as the finished products. The aromatic extract is then separated from SO_2 through rectification and SO_2 recirculated. Extraction is a widely used process in many application.

Lubricating Oil Base Stocks

An interesting application of some of these physical separation processes is in the production of lubricating oil base stocks. A variety of cuts is obtained from vacuum distillation of crude oil depending upon the grade of base stock required. These distitillate cuts are then subjected to the following process steps:

(a) The aromatic content in these cuts is required to be brought down for improvement of viscosity index. This is achieved through liquid-liquid extraction using phenol or furfural as a solvent for aromatics. The solvent is recovered in the solvent recovery section and recirculated.

(b) Next follows the step of solvent dewaxing. MEX or MIBK are generally used as solvents. Filtration after chilling through vacuum filters is the usual route to Wax removal.

(c) Finally the low aromatic content dewaxed oil is subjected to clay treatment or hydrotreatment, for colour improvement.

1.2 CONVERSION PROCESSES

When automobile industry was in its infancy, it became obvious that the production of large quantities of straight run gasoline from crude distillation would involve production of much larger quantities of residual fuels than the market could absorb. The demand for gasoline thus spurred on scientists to look for means of obtaining a higher yield of this product from crude oil. The cracking process brought the desired solution. Thermal cracking was begun on a commercial scale in the U.S.A. around the year 1913. Around the year 1946 Fluid Catalytic Cracking (F.C.C.) process was developed.

Cracking is essentially a high temperature treatment which breaks up heavier or large hydrocarbon molecules into small ones, often at the same time altering their internal structure. There is not sufficient hydrogen in the large molecules to supply the needs of the small molecules into which they crack. Cracked molecules are, therefore, mainly unsaturated and because of this they polymerize tosome extent as soon as they are formed the result is that heavier molecules are also formed at the same time. The total product of the cracking of a heavy oil (residue) therefore, contains gases, light hydrocarbons (gasoline) and hydrocarbons with large molecules, heavier than the original oil, as well as fractions boiling within the range of the oil which has been cracked.

Unlike the physical separation processes described earlier, in the conversion processes the basic structure of the hydrocarbon molecules gets converted into a different structure; hence the name. Numerous processes have been developed in petroleum refining and petrochemical industry to covert one hydrocarbon type into another. Some of the basic conversions steps have been described below:

- **Cracking or Pyrolysis:** Cracking Pyrolysis involves breaking down of large hydrocarbon molecules into smaller molecules by heat or catalytic action. Zeolite catalysts are commonly used.
- **Polymerization:** Linking of similar molecules; the joining together of light olefins.
- **Alkylation:** The union of an olefin with an aromatic for a paraffinic hydrocarbon.
- **Hydrogenation:** Addition of hydrogen to an olefin.
- **Dehydrogenation:** Removal of hydrogen from the molecule. Dehydrogenation of napthenes to produce aromatics is an important refining process for naphtha reforming.

- **Isomerization:** Alteration of the arrangement of the atoms in a molecule without changing the number of atoms of the numerous conversation processes applied in petroleum refining the following processes being important are described in detail:
  - Thermal cracking and Vis-breaking
  - Catalytic cracking
  - Hydrocracking
  - H-oil cracking and other processes to upgrade residual fuel oils
  - Naphtha reforming

Thermal Cracking

Thermal cracking is an attractive process for increasing the production of middle distillates. Nowadays it is mainly applied in two forms, which can be characterized as follows:

(a) In a simple refinery, where additional naphtha and gas oil fractions need to be produced, long residue (i.e. the fraction of crude boiling above about 350°C) will usually be the feed available for thermal cracking. All or part of this feed can be cracked to a varying degree of severity in a one or two-stage unit and with or without a high vacuum unit, depending on a particular situation. Often feedstock characteristics and product requirements are such that maximum severity of operation is desirable and possible. A varying degree of severity in a one or two-stage unit and with or without a high vacuum unit, depending on a particular situation. Often feedstock characteristics and product requirements are such that maximum severity of operation is desirable and possible. A varying degree of sophisticatication is conceivable for product work-up section. Usually a maximum recovery of gas oil is aimed at and therefore such units are named thermal gas oil units.

(b) In a complex refinery the straight run flashed distillate (boiling range between 350 and 500°C) is used as feedstock for the fluid catalytic cracker and/or hydrocracker. The remaining short residue (fraction boiling above approx. 500°C) which invariably constitutes the bulk of the residual fuel oil pool has a very high viscosity and consequently diluents (cutter stocks), such as gas oil are needed to blend the short residue to commercial grade fuel oils. The required amount of gas oil can be appreciable. It can be reduced to a significant extent by subjecting the short residue to the cracking, which lowers its viscosity. This operation is usually called visbreaking. It is cracked in the first furnace and flashed in a cyclone at a high temperature. The overhead of the cyclone goes to a fractionator. A side stream heavier than gas oil (roughly 350-450°C fraction) is recovered from the fractionator and cracked once again under more severe conditions in a distillate cracking furnace. The effluent of this furnace also goes to the fractionator. This lay out is referred to as two stage cracking. The feed is subjected to a high vacuum distillation before being cracked. The cracked residues are vacuum flashed to recover more feed for the cracking stage. A very high conversion can be achieved in this way, because each feed can be cracked under conditions yields are obtained while maintaining adequate residue stability. Gas oil yields of about 30% wt. On feed can be achieved. This scheme is particularly suitable for very light or waxy feedstocks. A soaker can be applied for the short residue cracking under this scheme.

Hydrocracking

(a) Introduction

Hydrogenation and hydrocracking are among the oldest catalytic processes used in petroleum refining. It was originally employed in Germany in 1927 for converting lignite into gasoline and later used to convert petroleum residues into distillable fractions. The first commercial hydrorefining installation in the United States was in the 1930's by Standard Oil Louisiana in Baton Rouge, Louisiana. There was no significant increase in hydrorefining of petroleum during the next 20 years. After that, hydrodesulfurization and mild hydrogenation processes experienced a tremendous growth in the 1950's, mostly because of large quantities of by-product hydrogen made available by catalytic reforming of low octane naphthas to produce high octane gasoline. Very little hydrocracking is effected in these operations. In 1960. UOP announced that a commercially feasible hydrocracking process was available to the petroleum industry, a process distinguished from hydrogen treating of the 1950's in that a substantial reduction in the molecular weight of the gross product is effected. It was immediately accepted and at present over 16 million MTA of capacity are licensed and considerable additional capacity is in the design and planning stages. UOP Inc.'s development, of course, started many years earlier. The first intensive work dating back to the early 50's included the conversion of residual oils into more valuable products.

(b) Applications

Hydrocracking is the most versatile of all petroleum refining processes - any fraction from naphtha to asphalt can be processed, producing any desired product with a molecular weight lower than that of the charge stock. At the same time as hydrocracking takes place, sulfur, nitrogen and oxygen are almost completely removed and olefins are saturated, thereby giving products which are a mixture of essentially pure paraffins, napthenes and aromatics.

(c) Reaction Requirements

Hydrogen is consumed in all hydrocracking reactions. Carbon-to-Carbon bonds in paraffins are broken and the fragments hydrogenated to form two or more lower molecular weight paraffins. Polycyclic aromatics are partially saturated and ring opening in one or more places follows-at least one ring is usually retained. Dual function catalysts containing both acid activity and hydrogenation-dehydrogenation activity are required. The patent literature is voluminous with disclosures of Group III metals composited with silicaalumina substrates to provide the necessary functions for hydrocracking. According to one theory, the hydrocracking reactions are initiated by partial dehydrogenation over the metallic components in the catalyst to form an olefin; of course, is in a very low concentration and is rapidly adsorbed on an acid site and converted to a carbonium ion cracks (usually in the B-position) to form two fragments may undergo further cracking, but at some critical molecular weight they become hydrogenated, and desorbed from the catalyst and constitute final product. It will be noted that this theory does not suggest a chain reaction. The metallic component in the catalyst effects the following:

(i) Formation of an olefin.
(ii) Hydrogenation of fragments.
(iii) Prevents condensation reactions which lead to catalyst fouling.

The hydrocracking reactions are carried out by pumping the charge stock into a catalyst bed - usually fixed - at elevated temperatures and pressures while recycling a hydrogen stream. The process conditions employed vary widely depending upon the nature of the charge stock and the products desired. However, pressures usually range between 500 and 3000 PSIG and temperatures between 260° and 427°C.

(d) Flow Schemes

UOP uses three basic flow schemes: (a) Single-stage, (b) Two-Stage, (c) Series-Flow; there are many modifications of each but pertinent features are as follows. The choice of flow scheme depends upon the charge stock, the desired products and the flexibility that is chosen for the process.

(i) Single Stage System

This system, fresh charge stock alongwith recycle hydrogen are heat exchanged and heated to the desired reaction temperature, then contacted with catalyst in the reactor. Hydrocracking reactions are exothermic; to control temperatures in the desired range, conversion per pass can be limited or cooling means such as quench recycle gas can be used. The reactor effluent is heat exchanged and passed to a high-pressure separator. Hydrogen-Highpressure separator liquid is flashed into a low-pressure drum where most of the dissolved gases are removed from the normal liquid product. The flash drum liquid is fed to a fractionator for separation into the desired products. Fractionator bottoms boiling ranage of the products can be recycled for complete conversion. Many units are on stream employing the single stage, scheme. Some convert naphtha into LPG. Others produce jet fuel diesel fuel and lube oil base stocks from vaccum gas oil and/or deasphalted oil.

(ii) Two Stage System

In this scheme, there are two complete reactor stages each containing a charge pump, recycle compressor, heat exchangers, heater, reactor, cooler and high-pressure fractionator. The first stages are separated by the stripper or fractionator. The first stage reactor decomposes nitrogen and sulphur compounds, saturates olefins and partially saturates polycyclic aromatics. The stripper or fractionator removed H_2S, NH_3 and other light gases, leaving a mixture of paraffins, napthenes and aromatics essentially free from impurtities. The second stage processes the "cleaned up" charge and converts it into the desired boiling range products. This system is the most versatile of the two-reactor schemes as regards variety of charge stocks and products which can be accommodated. A large number of such units are on stream, charging feeds as heavy as vacuum gas oil and propane deasphaled gas oils to produce naphtha, kerosene and distillates.

(iii) Series Flow System

The "Series flow" two-reactor system is similar to the one-reactor system except that there are two reactors in series, as shown, each containing a different catalyst. The first reactor performs the same function as the first reactor in the "two stage" scheme. The second reactor carries out the hydrocracking reactions in the presence of H_2S and NH_3. This scheme has the advantage of being more economical to build and operate in some instances. This scheme is being designed for a few cases, but only one is presently in operation.

(e) Catalysts

UOP has six different catalysts which used separately and in conjunction with each other in the three basic flow schemes: various combinations of catalysts and flow schemes provide wide flexibility of applications. Available catalyst types are as follows: DHC and HC catalysts, are used in HC Unibon only. (Unibon is the name given by UOP to its hydrocracking processes).

(i) DHC

These catalysts have the capacity of hydrocracking vacuum gas oils and deasphalted oils to maximize distillate, such as jet fuel and diesel fuel, and also to produce lube oil base stocks. There are two DHC catalysts: DHC-2 and DHC-5; selection depends upon the quality of charge stock and desired products.

(ii) HC-8

This catalyst is used in series flow units and has the capability of cracking heavy gas oil into gasoline in the presence of ammonia, H_2S and water. It can also be used to produce LPG from naphtha.

(iii) HC-9

This catalyst is primarily employed for the production of LPG from naphthas. It can also be used in the second stage of a two stage process wherein it is desired to saturate the aromatics, as well as hydrocrack naphthas and distillates.

(iv) RCD

This catalyst is employed for the direct desulfurization of reduced crudes wherein minimum hydrocracking is desired.

(v) BOC

This catalyst is employed when it is desired to process vacuum bottoms for the production of distillable hydrocarbons, such as vacuum gas oil.

Upgradation of Vacuum Residues

The world wide trends are a shift from residual fuels towards middle distillates. Increased Demand of jet fuel and Diesel oil for road transportation explains the growing market for middle distillates. While FCC yields high octane gasoline on the other, it does not produce any jet fuel or kerosene. FCC also produces heavy cycle oils and clarified oil which can only be blended into furnace oil pool.

Hydrocracking on the other hand is capable of giving much higher yields of distillates including jet fuel with enormous flexibility to produce more of one distillate or the other. This explains a relatively large percentage growth in the total Hydrocracking capacity, internationally. The capital cost of a Hydrocracker is higher than that of a FCC, for the same capacity and operating cost also higher particularly if by product hydrogen from catalytic reformers is not available. However, the choice between FCC and Hydrocracker will be determined by factors such as the demand for the various kinds of distillates, their prices and price differentials between those distillates and residual fuels. Both the Hydrocracker and the FCC use vacuum gas oil (V.G.O.) for feed stock with certain specified limits for metal contaminants i.e. Nickel and Vanadium and also Conradson Carbon Residue (CCR) - a property that determines the extent of Coke formation that will occur. The metal contaminants as well as the coke reduce the vacuum residue, therefore cannot be easily converted to distillates and thus goes to the residual fuel oil pool. The only processing routes available in the past, for the vacuum residues were Visbreaking and thermal cracking which would yield small quantities of distillates but basically reduce the viscosity thereby requiring lesser quantities of cutter stock for furnace oil blending.

Source: Extracted from the book 'Technology in Indian Petroleum Refining Industry by Department of Scientific & Industrial Research'.

1.3 OIL ECONOMICS: DEMAND, SUPPLY AND STOCKS

The price of oil has traditionally been determined by how closely supply and demand match each other. When there is more supply than consumers want, they can shop around for the best price leading to lower prices. If demand is higher than the amount available, consumers will compete with each other, bidding for the supplies they need driving prices up. However, it is not only the current supply/demand balance that determines market prices. Buyers and sellers also factor in what they expect will happen to prices in the future. If buyers think that supply might be lower in a few weeks or months and the price could go up, they will want to stockpile some oil now and might even be willing to pay a premium today to protect against a higher price in the future. Similarly, if buyers think that the supply of oil will increase in the future or that the price can be expected to decline soon, they will delay their purchases as long as possible or demand a discount on the price. The behavior of prices in any primary commodity market is strongly influenced by the fundamental forces of demand and supply. Although prices frequently change for other more ephemeral reasons, especially now that trading screens and on-line news services play such an important role of fundamentals in shaping the course of prices should not be forgotten.

Demand

The demand for oil, like other primary commodities, depends on mainly on the state of the global economy. Despite improvements in energy efficiency as a result of the price increase of the 1070s and early 1980s, oil demands remains closely linked to the growth in economic activity. Recently, oil demand has been growing fastest in the newly industrializing countries of the Asia - Pacific Rim where the economies are expanding very rapidly indeed. By contrast oil demands in industrialized countries are only just emerging from recession. And oil demand in the former Soviet Union has continued to fall as the economy contracts and the old, energy -intensive industries are no longer viable. However, the OECD countries still consume more than half the world's oil and it is in these countries that the oil markets are most developed and least constrained by government controls.

Oil demand is also highly seasonal. Peak demand for heating fuels such as kerosene, gas oil and residual fuel oil obviously comes in the winter, while peak demand for transport fuels such as gasoline and diesel comes in the summer. In addition, other products such as bitumen, which is used for road building, also display a clear seasonal pattern that can also affect oil price behavior at certain times of the year. Although the steady shift towards a greater share of transport fuels in the global demand barrel has reduced the annual variation in world oil consumption, there is still a difference of 3 to 4 million b/d between the winter peak and the summer trough in demand. Prices play an ambivalent in determining the oil demand. In short term, they appear to have very little impact on the level of oil consumption except in those markets such as electricity generation where there is direct competition with other fuels. In most markets, oil consumers can not easily react to price increases because this requires investment, either in a new car or new boiler. As a result, the impact of higher prices may take years to filter through. But in the longer-term, there is no doubt that the prices have a significant impact on the level of oil consumption.

The effect of prices on demand is clearly demonstrated by comparing the impact of consumer government taxes on the amount of oil consumed per head of the population in countries with similar levels of economic development. For example the US which still imposes very low taxes on oil products, consumes nearly twice as much oil per capita as the UK and France, which impose much higher taxes. And the fact oil products are either not taxed, or even subsidized, in many developing countries helps to explain the very high rates growth of oil demand achieved in recent years. We can list factors that affect the demand of crude oil in the international market as-

Determinants of price of oil and oil demand

The surge in global economic growth, as noted, stimulated demand for oil. Three generic groups of factors – Economic, Geopolitical and Technical – influenced to a greater or lesser extent the rise in oil prices and therefore other energy prices. We can better understand these in terms of how they gave momentum and durability to the price rise if we overlay them with a fourth; namely, 'Information'. In other words, how market agents perceived the first three sets of factors and interpreted or misinterpreted their importance influenced their actions in the market and determined market outcomes. Because these factors and events are generally well known and have been discussed and analyzed at length over the past three years, they are reviewed only briefly here. Table lists some of the events and factors that caused dislocations in the market. Key however was the disappearance of spare capacity.

Economic factors

It is generally held that this price 'shock' (it is debatable whether it can be called a 'shock') was demand-led and Asian-based. As noted earlier it is sometimes forgotten that since the late eighties most of the net growth in global oil demand occurred in the Asia Pacific region. This was perhaps understandably forgotten as the Asian Financial Crisis reversed this trend and, combined with the fallout from OPEC's internal differences at the time over production strategy, led to the price crash of (2000). Global economic growth in 2004 approached 5%, a level not experienced since the seventies, and world oil demand surged, led by China, North America, the rest of Asia and the Middle East. To remind us of some elements behind that growth;

- North America, principally the United States economy, came out of a post-9/11/2001 slump, lubricated by reelection year fiscal gifts in 2003/04 and historically low interest rates that in turn stimulated a mortgage boom, which together with a recovery in equity markets triggered a so-called wealth effect surge in household consumption (and ominously, record household debt).
- This surge in consumption saw increased U.S. imports of consumables and merchandise especially from China.
- China's expansion at or near double-digit levels was registered in a

sharp increase in oil demand, mostly diesel and LPG (Liquid Petroleum Gases, used as petrochemical feed stocks). Diesel demand was driven by increased transport by barge, rail and truck, certainly to move goods, but also to ship coal to power plants to alleviate congested rail transport. On top of that, fuel oil and diesel demand increased owing to the installation of independent oil-fired generation sets to make up for shortfalls in grid-based electricity supply.

- China's growth stimulated growth among its Asian neighbors including Japan.
- China and India, with a combined population approaching 2.5 billion people, experienced dramatic increases in incomes adding momentum to the global economy and therefore oil consumption.

As this new demand absorbed some spare capacity, political and technical developments removed supply; the result was the virtual disappearance of spare capacity. The carry over of years of underinvestment all along the delivery chain predisposed the system to geopolitical and technical stresses.

As the funds, cost and revenue all comprise of the economic factors. The major constituents of economic factors are –

(a) Production & consumption levels
(b) Storage availability and costs
(c) Insurance costs
(d) Interest rates
(e) Transportation costs
(f) Currency rates
(g) Other related markets

Geographical factors

The list of geopolitical developments that played a role in strengthening the rising price trend is long and not easily disentangled from their technical consequences. The invasion of Iraq and its impact on that country's oil supply not meeting analysts' expectations, the strikes in Venezuela and Nigeria, which removed supply, the Yukos Affair and the dampening of the rise in output from Russia, all either eroded the margin of spare capacity or sent signals that increased the nervousness of buyers. The thinner the spare capacity became, the greater was the psychological impact of each event on the market.

There is a tendency to view developments in the history of oil markets through a political lens. There may be good reasons for doing so and, while no additional evidence was needed of the continuing importance of geopolitics, since the last AEC in 2002, as noted earlier we have several significant examples of how it impacts oil and gas both here in the region and elsewhere, notably the former Soviet Union. While the political forces influencing supply and demand of hydrocarbons might well be the most interesting for political scientists and provide a fecund source of speculation regarding motives, interests and intrigue, we risk drawing the wrong conclusions about the future path of oil and gas if we ignore other factors, not the least of which are technical/scientific, social and economic. Pivotal is information about the industry and how that information is interpreted or misinterpreted in decision-making.

Much has been written and said about the role of 'speculators' through this period. As noted, the fact that spare capacity in crude production was whittled down to historic lows and that much of that crude was heavy sour grades unsuitable to the refining stock and product quality specs had a significant influence on market participants' actions. The relationship between forward days' inventory and prices has been examined in detail. But if spare production capacity - essentially a form of inventory - is taken into account along with actual inventories, it is not surprising that prices increased. In other words, at the end of the day tight supply conditions prevailed - these were fundamentals, which commodity traders took into account in their assessment of the market's direction, and prices were bid up.

These oil market developments since the last Arab Energy Conference underscore how economic, technical and political factors and events elsewhere in the world can impact on oil and gas producing countries everywhere.

The price of the key benchmark or reference crudes continued to rise, given further momentum by news and events. Just one example underscores the importance of information, in this case, information from governments. The statement by the US Vice President in early 2005 that the SPR would only be used in the event of a major oil supply cut of 5 or 6 mb/d gave market players some upside price comfort, while OPEC's credible defense of an implicit floor price provided protection on the downside. This was undone in the wake of hurricanes Katrina and Rita when IEA countries released strategic stocks. This merely underscores the importance of not getting locked into views of the future based on the past - the market and the forces acting on it are dynamic and evolving.

Analysts will be writing about the 2003-05 price surges for years to come. It is unlikely that tight capacity along the supply chain and the mismatch of crude with refining capacity will be dislodged as the most critical on the long list of factors that drove up the price. The roots of that tight capacity can be found in the distant history of the oil market, and reflect the considerable inertia in the energy delivery system.

Since crude oil is energy source driving the needs of the whole globe, geographical factors do play a mentionable role- -

(a) Pipeline/shipping availability
(b) Localized weather report
(c) Seasonality
(d) Locality

Political factors

Major all producing countries experiences political instability and the political state of the region thus affected by -

(a) Cartels/trade agreements

(b) Global events- war, embargoes etc
(c) Taxation/trade tariffs
(d) Regulation/legislation
(e) Environmental pressure

Technical factors

The oil industry is a complex and integrated technical system. When it has excess capacity along the chain from production, transportation, refining and distribution, and in the associated industries that supply and service various links in the chain, the system is generally capable of absorbing and offsetting upsets, fires or interruptions along the chain. But when the chain is tight, when capacity has no margin or swing, upsets can generate major dislocations in the market and thereby affect prices.

An early technical development that put pressure on oil prices had an ironic origin: a problem in the nuclear accidents of Japan's largest power utility, TEPCO. In May of 2003 TEPCO was required to shut all 17 of its nuclear reactors to verify if the problem was a generic fault. During the shut-down over the peak summer period, Tokyo faced electricity shortages and oil imports increased by 200 kb/d. The irony of this incident is that part of the rationale behind Japan's nuclear power program was to reduce its dependence on oil imports.

The erosion of spare capacity over 2003 to 2005 was of central importance to the market dislocations. Reduced to as little as 0.6 mb/d, and mostly heavy sour grades, the world spare crude production capacity was inadequate to make up for upsets.

Hurricanes Ivan in 2004 and Katrina and Rita in 2005 struck the US Gulf Coast, the world's largest single refining centre, at a critical time in the annual oil demand cycle. The 2005 storms removed over 1.5 mb/d of crude oil production and 75% of the region's gas production equal to 10% of U.S. gas supply, and shut in 20% of U.S. refining capacity. Their repercussions lasted for months.

Earlier in the summer, some observers blamed price increases on the tight US refinery capacity, but US refining capacity had been as tight in the mid nineties without increasing prices. But in 2004/05, global refining capacity was tight; specifically, capacity to process the marginal heavy barrel. This was the bottleneck: inadequate upgrading capacity to process the available sour crudes to produce 'sweet' products.

This 'mismatch' between the quality of available crude and the refining equipment available to process it to meet the product slate demanded in the market helped drive prices above $70.

Under tight market conditions even accidents to non-producing equipment can influence the market. When the Thunder Horse platform, nearing completion in the Gulf of Mexico, was damaged during hurricane Dennis earlier in summer 2005, the forward oil price increased because it was expected to start producing a quarter million b/d in the fourth quarter.

It is not a new insight and therefore not surprising that tight capacity or even the perception of tight capacity can compound price volatility. While political and technical upsets are inevitable they are not predictable. Reducing volatility then comes down to increasing capacity; increasing capacity raises many other issues including the long standing question of who should bear the responsibility and cost of doing so.

Comparative factors

When different alternatives are available to fulfill the energy needs a comparison do arises like –

(a) Preference shift
(b) Transaction costs
(c) Alternative commodities
(d) Product yields
(e) Relative quality

Demand for crude oil depends on the choices of many individual households and firms. However, owing to its importance for the economy and national security, the demand side is influenced by various private interest groups, the most influential of which being domestic oil refiners without foreign supply sources and governments, whose aim is to acquire sufficient quantities of petroleum from stable sources. Oil importing governments influence the petroleum market by means of fiscal instruments, antitrust policies, public funds for alternative energy research or petroleum exploration activities, political intervention in situations in which the interests of the nation are at stake, environmental regulations and strategic oil reserves.In addition to policy measures, governments in some oil importing countries, such as China, South Korea, Singapore, Taiwan, Thailand, Turkey and Brazil, own majority stakes in their countries' main oil companies. Nevertheless, the tendency over the last decade has been to privatise and deregulate the energy sector. As a result of its rapidly growing oil demand, China has begun to allow minority private ownership of its petroleum companies, as well as seeking to invest in foreign private companies. Among the consuming countries, the United States is the dominant player, being the world's largest producer, consumer and importer of petroleum. In 2004 this country imported over 10 million barrels of crude oil per day, over half of it coming from western hemisphere sources and another quarter coming from Atlantic basin sources, such as the North Sea, and north and west Africa. Saudi Arabia, the most important Middle Eastern source of crude oil for the United States, accounted for around 15% of total imports. In that year US petroleum consumption accounted for 25% of total world consumption. A distant second in terms of oil consumption was China with 7.6%, while the EU15 consumed 17.6% (BP, 2004). Because the United States is the most important market for oil, its West Texas Intermediate (WTI) grade of crude oil, though not traded internationally, is a leading benchmark for world petroleum prices.

Supply

The power of the producing countries is, in general, rooted in the characteristics of oil. Producers incur no storage costs, since petroleum is simply left in the ground, while consuming countries have to cover the technical costs of building storage facilities, interest on the value of oil stocks and various risks (e.g. environmental risks). In addition, oil production is not labour intensive and, therefore, the oil supply can be controlled easily by reducing depletion rates without affecting the labour market. Since there are no short term substitutes for petroleum, changes in supply are also effective. Moreover, demand for crude oil is highly insensitive to price changes (cf. Cooper, 2003). The most important player in the oil market – the Organization of the Petroleum Exporting Countries (OPEC), founded in 1960 in Baghdad – comprises a diverse group of developing nations, 4 highly dependent on oil exports and unified by their common interest in oil revenue maximisation. On average, petroleum exports represent over 68% of the total exports of these countries. Aiming to sustain world demand for oil (as opposed to replacing it with alternative energy sources), OPEC has to balance market share and profits.

The oil cartel's market power comes from the sheer size of its proven oil reserves (891 billion barrels) and exports (19.5 million barrels per day) – 78.3% and 48.7% respectively of the 2003 worldwide totals (OPEC, 2003). The Gulf countries also have the lowest production costs: USD 4.00 per barrel for Saudi Arabia or USD 4.50 for Iran, as compared, for example, with USD 9.85 for the North Sea and USD 12.50 for Brazil (Energy Intelligence, 2004). In addition, most OPEC oil is produced by 100% state-owned companies (as is the case in Algeria, Iran, Kuwait, Qatar, Saudi Arabia and Venezuela) or majority state-owned companies (Libya, Nigeria and United Arab Emirates).

Only in Indonesia is government participation in the oil sector very limited. The country with the largest weight among the oil exporting nations is Saudi Arabia. It has the world's largest proven petroleum reserves (one-quarter of the total) and some of the lowest production costs, and is the largest producer and net exporter of oil. As of May 2005, owing to the recent rapid increase in demand for petroleum, Saudi Arabia is the only country with any surplus production capacity: 900-1,400 thousand barrels per day, or some 13% of total capacity (EIA, 2005b). This enormous capacity has allowed Saudi Arabia to play the role of "swing" producer. The non-OPEC exporting countries, on the other hand, increased their international oil market share following the 1973-74 oil crisis, at the expense of OPEC. By contrast with the Middle Eastern countries, their oil production is characterised by technological difficulties (e.g. the North Sea) and high transportation costs (e.g. Alaska). The development of these oil reserves contributes to the geographical dispersion of petroleum production and the increased energy independence of some countries. While this would in principle imply a decline in international trade in crude oil, no significant decrease has yet been observed. Because of the different qualities of crude oil, it is more economical for some countries, whose refineries have not been adapted to process the newly discovered petroleum, to export their own grade of oil and to import the quality suitable for refining at home. Owing to the difference in construction materials used (related mainly to the quality of steel), the cost of building a refinery which processes heavy crude oil, for instance, can be six times that of a refinery with the same capacity built to process light grades of petroleum. A good example of oil trading conducted for petroleum quality reasons is the United Kingdom, which in 2003 exported 1,345 thousand barrels per day while importing 968 thousand barrels per day. Because North Sea oil is of a high quality, it is more suitable for processing at US refineries. Similarly, as is the case in Alaska, national oilfields may be located a long way from refineries and consumption locations and it may, therefore, be better to export the petroleum they produce (e.g. to Japan). In the end, these quality considerations contribute to the segmentation of the oil market.

In addition to the producing countries discussed above, the oil market includes both major oil companies and various smaller firms. The former are five very large, vertically integrated multinational corporations: British Petroleum (BP), ExxonMobil, Total, Royal Dutch/Shell and ChevronTexaco. Together, these produce 15.6% of the total annual world petroleum output (OPEC, 2003). Three of these companies have headquarters in the EU and account for their profits in euro or pounds sterling and use financial instruments to hedge against exchange rate risk.

The two American corporations also report some use of foreign currency derivatives. The smaller oil companies (the "independents") tend to operate in only one of the oil sector's technically segmented productive stages or to rely on a single country or region for their oil supplies.

Matching oil supply to demand has become much more difficult since the oil industry ceased to be properly integrated. Most oil producers simply maximize their output, subject to the technical constraints of the field, in order to get a quick return on the very large amounts of money they have invested in developing the oil field in the first place. And because their operating costs are typically much lower than the sunken capital costs, they will continue to produce until oil prices reach very low levels.

In the North-Sea, for example, most of the fields have operating costs of less than $5/barrel and are unlikely to be shut in unless prices fall below this level. As a result, the responsibility for restraining production below capacity lies with the twelve remaining members of the Organizations of Petroleum Exporting Countries (OPEC), who are committed to maintaining prices above their marginal cost of production in order to extract what they regard as a fair economic rent for oil. So far, they have succeeded, although competition for market share between OPEC members has frequently forced prices down until falling revenues have eventually restored a sense of discipline to the Organization. OPEC was particularly successful during the first half of the 1080s when Saudi Arabia was prepared to play the role of swing producer alone, but it has become more difficult to balance the market since the Saudis refused to continue cutting production at the end of 1985.

Two factors have made OPECs self appointed task more difficult. First there is the continued expansion of oil production outside OPEC. Although lower oil prices were initially expected to slow down the development of non- OPEC. Oil fields, this has not proved to be the case. By encouraging technological developments and forcing companies to cut costs, lower oil prices have actually made it easier to develop new oil fields outside the OPEC Countries. As a result the call on OPEC crude continues to grow more slowly than OPEC would like to see despite rising oil consumption.

Secondly, there is the inherent seasonality of oil demand. As the residual supplier to the world oil market, OPEC potentially faces large fluctuation in the level of production required by refiners at different times of the year. This not only makes it difficult to keep track of the underlying level of demand, but is also difficult to administer since OPEC. Members normally find it very hard to agree on how to allocate production between themselves. Recently, the problem has been solved by setting (and almost sticking to) a fixed production quota over a much longer period of time, thus leaving the market to handle the seasonal variation in the demand for crude.

Stocks

The level of stocks held by the world oil industry has fallen since the early 1980s. The reduction in partly due to the transfer of responsibility for strategic stocks from the oil companies to their governments, partly due to changes in the structure of oil demand, and partly due to improvements in the efficiency of company operations. And with renewed demand growth in recent years, the forward cover provided by OECD industry stocks has been sharply reduced/since the end of 1991, the seasonally adjusted level of OECD industry stocks has not increased while demand has grown by more than 2 million b/d. As a result the forward cover provided by total OECD industry stocks had fallen to only 64 days at the end of June 1995 compared with more than 80 days in the early 1980s, and dropped to only 59 days at the end of 1995 after supply disruptions and unexpectedly cold weather left companies holding very low stocks. However there is still some way to go before companies reach their historical minimum operating stock level of 55-days. A detailed study of the US oil industry published by Exxon showed that the US held a total of 89 days of oil stocks at the start of 1981 measured in terms of forward consumption. Out of this total, 7 days' worth was held by the government in the Strategic Petroleum Reserve (SPR), and 82 days' worth was held by companies. According to Exxon 58 of the 82 days' worth of company stocks were minimum operating stocks. More than a third of the minimum operating stocks (20 days' worth) occupied the pipelines and tankers that transport the oil filled the refineries that process it, and provided the "tank bottoms "for the storage facilities. The remaining 38days' worth of stocks represented the oil in transit through the system, of which a quarter (10 days' worth) was in the form of crude oil and three quarters (28-days' worth) was in the form of refined products.

Companies need to hold more stocks in the form of products than crude for two reasons. First the different types of refined product need to be kept separate from each other and move along their own distribution channels, which simply increases the amount of oil tied up in the supply chain once it has passed through the refinery. And secondly, refineries are not sufficiently flexible to vary their products yields in line with the seasonal variation in demand. For this reason, refiners are obliged to accumulate unwanted stocks of heating oil and residual fuel oil when they increase runs to meet peak gasoline demand during the summer, while the reverse occurs during the winter. It is this involuntary stock builds by refiners that create the characteristics seasonal pattern in the level of stocks held by the oil industry and influences the behavior of prices in the forward and futures markets.

Value Chain Economics

The following tables are estimates of the various cost components that make up the pump price of gasoline. The tables are based on a variety of sources and assume a crude oil price of $49/B and an average pump price of $2.00/gallon. The costs are based on data reported in the Oil & Gas Journal and Performance Profiles of Major Energy Producers 2007. The significant findings are that:

- Over 75% of an integrated oil company's net margins are in the upstream.
- Crude oil costs contribute nearly sixty percent to pump prices.
- Severance and gasoline taxes cost consumers nearly 52 cents/gallon or 26% of pump prices.
- Company net margins for refining and marketing are about 20 cents/gallon or 10% of pump price.

Oil and Gas Economics

Operation	Costs	Value	Gross Margin	Net Margin	Percent
Exploration	2.97	16.33	16.33	13.36	36%
Production	17.78	49.00	32.67	14.89	41%
Transportation	1.00	51.96	2.96	1.96	5%
Refining	3.70	60.46	8.50	4.80	13%
Distribution	1.90	63.69	3.23	1.33	4%
Marketing	0.80	64.85	1.16	0.36	1%
Pump Taxes	19.15	84.00	0.00	0.00	0%
				36.70	100%

Gasoline Cost Components

Component	$/B	Cents/Gallon	Percent of Pump Price
Crude Oil	49.00	116.67	58%
Operating Costs	4.90	11.67	6%
Taxes	21.65	51.55	26%
Company Net Margins	8.45	20.11	10%
Total	84.00	200.00	100%

1.4 WORLD OIL MARKET

Oil and gas are of paramount importance in economies worldwide. Oil, gas, hydro electricity, nuclear power and coal are the five constituents of primary oil. Oil and gas account for about 60 per cent of the total world's primary oil consumption. Crude oil is a mixture of hydrocarbons that exists in a liquid phase in natural underground reservoirs. Nations struggle to explore for oil, and import it at almost any cost. It is also an important contributor to the export realizations of many countries. In countries like Russia, nearly half the hard currency earnings come from crude oil exports. The figure rises to about 80% for Venezuela and 95% for Nigeria and Algeria.

Oil has many applications and it is hard to imagine the modern world without it. Almost all industries including agriculture are dependent on oil in one way or the other. Of the industries, oil & lubricants, transportation (including road, rail, sea and air), petrochemicals (some of the end products of petrochemicals include plastics, synthetic fibres, detergents and chemical fertilizers etc.), pesticides and insecticides, paints, perfumes, etc. are largely and directly affected by the oil prices as several products derived from crude oil are basic inputs in the production in these industries. The impact on these industries would result in spiraling effect on other industries and people. Without oil or its close associate natural gas, urban domestic life will become miserable. Oil light homes and streets and serves as a fuel for cooking. In cold countries, oil or gas is needed for heating homes. Metals are being progressively replaced by plastic, a product of oil and artificial fibers have made inroads into the domain of cotton. The indispensable ropes for agriculture and fishing, hitherto made from jute, are now being made from plastics. Polythene (plastic) bags, sheets and covers become indispensable in modern day's packaging and shopping. A wide range of chemicals, medicines and toiletry items is derived from oil.

There are in fact many products obtained from the processing of crude and other hydrocarbon compounds. These include aviation gasoline, motor gasoline, naphtha, kerosene, jet fuel, distillate fuel oil, residual fuel oil, liquefied petroleum gas, lubricants, paraffin wax, petroleum coke, asphalt and other products. The prices of crude are highly volatile. High oil prices lead to inflation that in turn increases input costs; reduces non-oil demand and lower investment in net oil importing countries. India, which is a net importer of oil, thus is often subject to the vagaries of price volatility in crude. Given this scenario, crude futures will come as a boon to everyone, ranging from the government and corporate to retail users. Crude oil is marketed principally in New York, London and Singapore. Futures are sold promising next-month delivery at agreed amount, price and location, in a minimum of 1000 bbl, and are settled daily. Oil is priced relative to certain standard kinds of crude. In London, it is Brent blend crude from the North Sea; about 2/3 of the world's crude oil is priced in terms of Brent. In New York, West Texas Intermediate light, sweet crude is the standard. OPEC prices its oil in terms of a basket of seven crudes: Saudi Arab light, Emirates Dubai crude, Nigerian Bonny light, Algerian Saharan blend, Indonesian Minas, Venezuelan Tia Juana, and Mexican Isthmus. Individual crudes are sold at a discount or premium, depending on quality and difficulty of transport. Crude oil is a very variable commodity.

Market Players

(a) Oil Producers

There is a great deal of concentration in the world oil industry: just ten companies control 68 percent of the world's proven oil reserves. Nine of the ten biggest oil reserve holders are state-owned National Oil Companies (NOCs). Many of these were formerly private sector companies that were nationalized in the 1970s. Eight of the ten largest oil producers in the world are NOCs. The others are large integrated private sector energy companies.

World's Top 10 Crude Oil Producers

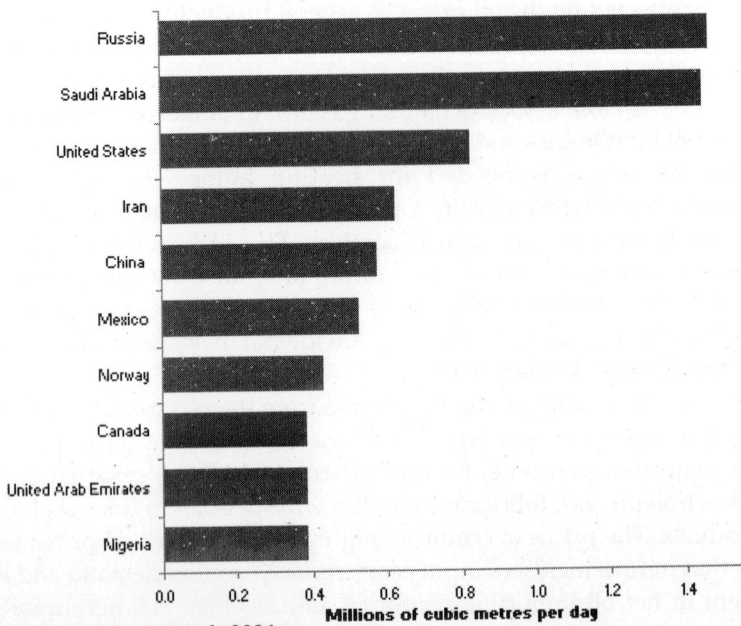

Source: Oil and Gas Journal, 2006.

Since 1960 the world oil market has been significantly influenced by the Organization of Petroleum Exporting Countries (OPEC). The goal of OPEC is to stabilize oil prices by adjusting their production levels and influencing the world's oil supply and demand balance. There are currently eleven members of OPEC, most of which are located in the Middle East and Africa. OPEC countries control close to 70 percent of the world's proven oil reserves and in 2005 accounted for about 41 percent of the world's supply of oil.

Canada holds the second largest oil reserves in the world, with over 178 billion barrels of oil. Over the next decade, Canada's importance as a leading oil producer is expected to increase, as oil sands production is projected to triple. Other key non-OPEC producers include: Russia, the United States, Mexico, China and Norway.

(b) Oil Consumers

Oil refineries are the primary users of oil. They convert crude oil into useable petroleum products such as gasoline, diesel, jet fuel and home heating oil. The refining industry's need for crude oil is driven by the demand for these products. The main consumers of oil continue to be the industrialized countries of the Organization for Economic Cooperation and Development (OECD), particularly the United States, Europe and Japan, which together consume about half of the world's annual oil output. However, consumption in emerging market regions is expanding at a faster pace (especially in China) as these markets grow rapidly and their use of energy in transportation, industry and residential sectors expands. The transportation sector accounts

World's Top 10 Crude Oil Reserve Holders

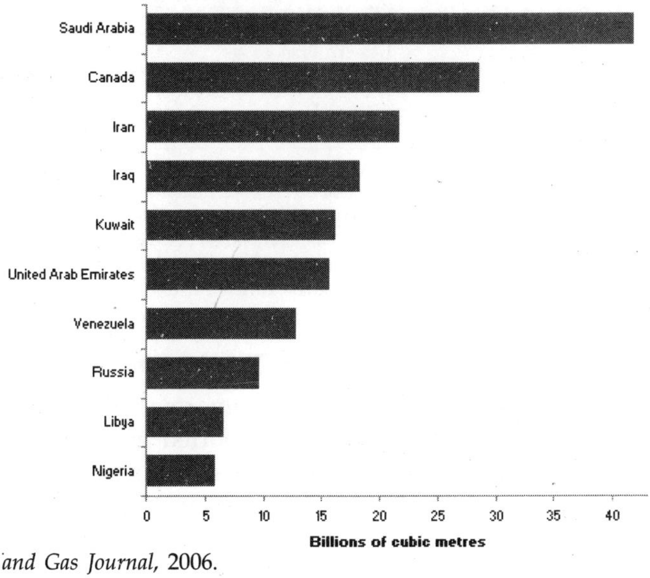

Source: Oil and Gas Journal, 2006.

for about two-thirds of the oil used in the world and for about half of the oil consumed in the United States.

(c) Oil Traders

Oil is a commodity that is widely traded around the world. Similar to other commodities, like coffee and soybeans, oil attracts investors who see an opportunity to make money by speculating on its price volatility. These traders are not generally involved in the actual production or use of oil – they buy and sell paper contracts, not actual oil – but can often have a significant influence on market prices.

Top World Oil Producers, Exporters, Consumers, and Importers, 2006

(millions of barrels per day)

Producers[1]	Total oil production	Exporters[2]	Net oil exports	Consumers[3]	Total oil consumption	Importers[4]	Net oil imports
1. *Saudi Arabia*	10.72	1. *Saudi Arabia*	8.65	1. United States	20.59	1. United States	12.22
2. Russia	9.67	2. Russia	6.57	2. China	7.27	2. Japan	5.10
3. United States	8.37	3. Norway	2.54	3. Japan	5.22	3. China	3.44
4. *Iran*	4.12	4. *Iran*	2.52	4. Russia	3.10	4. Germany	2.48
5. Mexico	3.71	5. *United Arab Emirates*	2.52	5. Germany	2.63	5. South Korea	2.15
6. China	3.84	6. *Venezuela*	2.20	6. India	2.53	6. France	1.89
7. Canada	3.23	7. *Kuwait*	2.15	7. Canada	2.22	7. India	1.69
8. United Arab Emirates	2.94	8. *Nigeria*	2.15	8. Brazil	2.12	8. Italy	1.56
9. *Venezuela*	2.81	9. Algeria	1.85	9. South Korea	2.12	9. Spain	1.56
10. *Norway*	2.79	10. *Mexico*	1.68	10. Saudi Arabia	2.07	10. Taiwan	0.94
11. *Kuwait*	2.67	11. Libya	1.52	11. Mexico	2.03		
12. *Nigeria*	2.44	12. *Iraq*	1.43	12. France	1.97		
13. Brazil	2.16	13. Angola	1.36	13. United Kingdom	1.82		
14. *Iraq*	2.01	14. Kazakhstan	1.11	14. Italy	1.71		

Note: OPEC members in italics.

1. Table includes all countries with total oil production exceeding 2 million barrels per day in 2006. Includes crude oil, natural gas liquids, condensate, refinery gain, and other liquids.
2. Includes all countries with net exports exceeding 1 million barrels per day in 2006.
3. Includes all countries that consumed more than 2 million barrels per day in 2006.
4. Includes all countries that imported more than 1 million barrels per day in 2006.

Source: Energy Information Administration (EIA). www.eia.doe.gov/emeu/cabs/ .

Crude Oil and Total Petroleum Imports Top 15 Countries

Monthly data on the origins of crude oil imports in March 2010 has been released and it shows that three countries exported more than 1.00 million barrels per day to the United States. The top five exporting countries accounted for 66 percent of United States crude oil imports in March while the top ten sources accounted for approximately 86 percent of all U.S. crude oil imports.

The top five sources of US crude oil imports for March were Canada (2.020 million barrels per day), Saudi Arabia (1.149 million barrels per day), Mexico (1.086 million barrels per day), Venezuela (0.984 million barrels per day), and Nigeria (0.939 million barrels per day). The rest of the top ten sources, in order, were Angola (0.490 million barrels per day), Iraq (0.475 million barrels per day), Brazil (0.299 million barrels per day), Algeria (0.276 million barrels per day), and Russia (0.248 million barrels per day). Total crude oil imports averaged 9.292 million barrels per day in March, which is an increase of 0.612 million barrels per day from February 2010. Canada remained the largest exporter of total petroleum in March, exporting 2.517 million barrels per day to the United States, which is an increase from last month (2.490 thousand barrels per day). The second largest exporter of total petroleum was Mexico with 1.265 million barrels per day.

Crude Oil Imports (Top 15 Countries)
(Thousand Barrels per Day)

Country	Mar-10	Feb-10	YTD 2010	Mar-09	YTD 2009
Canada	2,020	1,897	1,934	1,845	1,901
Saudi Arabia	1,149	881	1,000	944	1,128
Mexico	1,086	996	1,040	1,092	1,203
Venezuela	984	913	908	949	1,029
Nigeria	939	896	945	860	607
Angola	490	312	358	644	612
Iraq	475	540	506	587	570
Brazil	299	192	256	334	365
Algeria	276	282	296	215	242
Russia	248	214	199	219	173
Kuwait	218	228	169	181	218
Colombia	216	371	291	254	235
Ecuador	183	145	182	210	241
United Kingdom	142	260	177	85	56
Congo (Brazzaville)	124	76	81	56	44

Total Imports of Petroleum (Top 15 Countries)
(Thousand Barrels per Day)

Country	Mar-10	Feb-10	YTD 2010	Mar-09	YTD 2009
Canada	2,517	2,490	2,535	2,438	2,498
Mexico	1,265	1,134	1,178	1,199	1,330
Saudi Arabia	1,149	898	1,007	967	1,149
Venezuela	1,061	1,009	993	1,106	1,201
Nigeria	962	932	970	891	637
Angola	502	326	371	657	622
Russia	488	423	459	650	550
Iraq	475	540	506	587	570
Algeria	455	461	472	463	523
Brazil	302	226	296	338	390

(Contd.)

Country	Mar-10	Feb-10	YTD 2010	Mar-09	YTD 2009
United Kingdom	267	413	318	208	211
Colombia	251	386	318	283	265
Virgin Islands	228	187	243	264	321
Kuwait	218	228	173	181	224
Ecuador	183	152	184	215	245

Note: The data in the tables above exclude oil imports into the U.S. territories. *EIA-March 2010 Import Highlights: May 27, 2010.*

Pricing Differences among Various Types of Crude Oil

According to The International Crude Oil Market Handbook, 2001-2002 published by the Energy Intelligence Group, there are about 161 different internationally traded crude oils. They vary in terms of characteristics, quality, and market penetration.

Two crude oils which are either traded themselves or whose prices are reflected in other types of crude oil include West Texas Intermediate and Brent. Comparing these two crude oils with EIA's Imported Refiner Acquisition Cost (IRAC), the OPEC Basket, and NYMEX futures is important to understand the differences among the various types of crude oil that are often referred to in the press and by analysts. Generally, differences in the prices of these various crude oils are related to quality differences, but other factors can also influence the price relationships between each other.

(a) West Texas Intermediate

West Texas Intermediate (WTI) crude oil is of very high quality and is excellent for refining a larger portion of gasoline. Its API gravity is 39.6 degrees (making it a "light" crude oil), and it contains only about 0.24 percent of sulfur (making a "sweet" crude oil). This combination of characteristics, combined with its location, makes it an ideal crude oil to be refined in the United States, the largest gasoline consuming country in the world. Most WTI crude oil gets refined in the Midwest region of the country, with some more refined within the Gulf Coast region. Although the production of WTI crude oil is on the decline, it still is the major benchmark of crude oil in the Americas. WTI is generally priced at about a $2-per-barrel premium to the OPEC Basket price and about $1-per-barrel premium to Brent; although on a daily basis the pricing relationships between these can vary greatly.

(b) Brent

Brent Blend is actually a combination of crude oil from 15 different oil fields in the Brent and Ninian systems located in the North Sea. Its API gravity is 38.3 degrees (making it a "light" crude oil, but not quite as "light" as WTI), while it contains about 0.37 percent of sulfur (making it a "sweet" crude oil, but again slightly less "sweet" than WTI). Brent blend is ideal for making gasoline and middle distillates, both of which are consumed in large quantities in Northwest Europe, where Brent blend crude oil is typically refined. However, if the arbitrage between Brent and other crude oils, including WTI, is favorable for export, Brent has been known to be refined in the United States (typically the East Coast or the Gulf Coast) or the Mediterranean region.

Brent blend, like WTI, production is also on the decline, but it remains the major benchmark for other crude oils in Europe or Africa. For example, prices for other crude oils in these two continents are often priced as a differential to Brent, i.e., Brent minus $0.50. Brent blend is generally priced at about a $1-per-barrel premium to the OPEC Basket price or about a $1-per-barrel discount to WTI; although on a daily basis the pricing relationships can vary greatly.

(c) NYMEX Futures

The NYMEX futures price for crude oil, which is reported in almost every major newspaper in the United States, represents (on a per-barrel basis) the market determined value of a futures contract to either buy or sell 1,000 barrels of WTI or some other light, sweet crude oil at a specified time. Relatively few NYMEX crude oil contracts are actually executed for physical delivery. The NYMEX market, however, provides important price information to buyers and sellers of crude oil in the United States (and around the world), making WTI the benchmark for many different crude oils, especially in the Americas. Typically, the NYMEX futures prices tracks within pennies of the WTI spot price described above, although since the NYMEX futures contract for a given month expires 3 days before WTI spot trading for the same month ceases, there may be a few days in which the difference between the NYMEX futures price and the WTI spot price widens noticeably.

(d) OPEC Basket Price

For a discussion of crude oil pricing in general, and of the OPEC Basket price in particular, see EIA's OPEC Fact Sheet. OPEC collects pricing data on a "basket" of seven crude oils, including: Algeria's Saharan Blend, Indonesia's Minas, Nigeria's Bonny Light, Saudi Arabia's Arab Light, Dubai's Fateh, Venezuela's Tia Juana Light, and Mexico's Isthmus (a non-OPEC crude oil). OPEC uses the price of this basket to monitor world oil market conditions. As mentioned above, because WTI crude oil is a very light, sweet (low sulfur content) crude, it is generally more expensive than the OPEC basket, which is an average of light sweet crude oils such as Algeria's Saharan Blend and heavier sour crude oils (with high sulfur content) such as Dubai's Fateh. Brent is also lighter, sweeter, and more expensive than the OPEC basket, although less so than WTI. Since OPEC has (at least informally) tied its production management activity to the goal of maintaining the OPEC Basket price between $22 and $28 per barrel, market watchers now pay close attention to this oil price indicator.

(e) Imported Refiner Acquisition Cost

The Imported Refiner Acquisition Cost (IRAC) is a volume-weighted average price of all crude oils imported into the United States over a specified period. Because the United States imports more types of crude oil than any other country, it may represent the truest "world oil price" among all published crude oil prices. The IRAC is also usually similar to the OPEC Basket price, so it too is typically about $2 per barrel less than the WTI spot price and about $1 per barrel less than the Brent price. However, because the IRAC is not reported by EIA until nearly 2 months after the end of the month in question, i.e., the August IRAC average price would be reported sometime in late October, the IRAC is not a particularly timely measure of a "world oil price". Although EIA is generally the only organization that uses the IRAC, it is used by EIA as the "world oil price" in all of its forecast publications, including the Short-Term Energy Outlook, released monthly, as well as the Annual Energy Outlook and International Energy Outlook, both of which are released annually and provide an annual forecast looking out approximately 20 years in the future.

Market Developments in 2007-08

Global proved oil reserves fell slightly in 2007, by 0.1%, and stand at 1.24 trillion barrels. But this is the result of reporting lags forcing us to roll over 2006 data. As is usually the case, the 2006 decline reported in last year's Statistical Review has now been revised up by 31 billion barrels, as new data came in.

Oil reserves remain concentrated. The top ten reserves holders account for 81% of the total, OPEC members hold 76%, 10% are in the FSU. The OECD's share has edged up due to growth in Canadian proved reserves of oil sands, as the Albertan government doubled the estimate of reserves "under active development" to 21 billion barrels. Similarly, Venezuela increased official reserves by 7 billion barrels after an updated assessment of extra-heavy oil in the Orinoco. Brazil's reserves increased with greater deepwater activity; however, our figures exclude the recent Tupi discovery, which has not yet been recorded as a proved reserve. Qatari reserves were revised up by 12 billion barrels to update condensate and NGLs reserves in the North Field.

Global Oil Reserves (BP, 2008)

Global proved natural gas reserves rose by just over 1 Trillion cubic metres in 2007, or 0.6%, to stand at 177.4 Tcm. Reserves in the Middle East and FSU dwarf other regions, with 41% and 32% of global reserves, respectively. Growth last year came largely from the re-evaluation of earlier discoveries, led by Indonesia, Iran and China.

If there was a defining moment in oil markets in 2007, it was the re-emergence of OPEC in successfully managing its production. After prices had fallen rapidly in late 2006 and early 2007, when the OPEC price basket breached $50/bbl, two cuts were implemented, the latest and biggest one in February 2007.

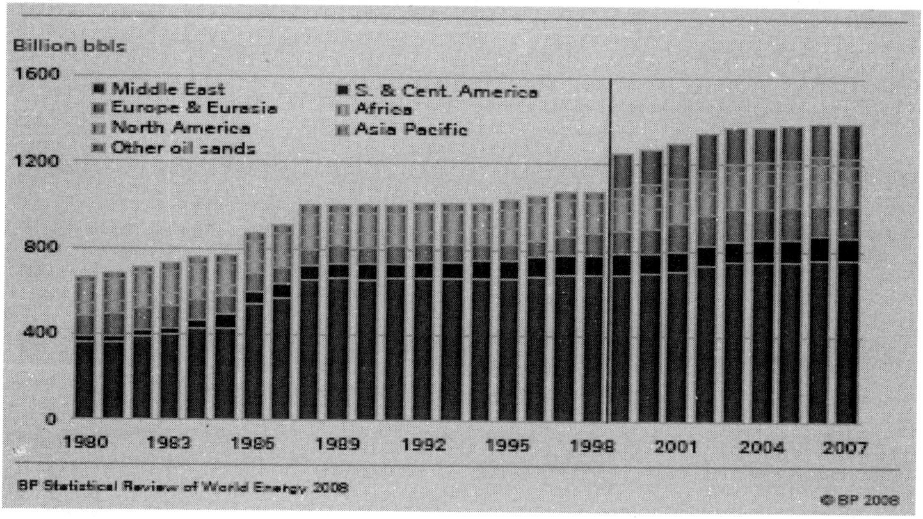

OECD Commercial Inventories (BP, 2008)

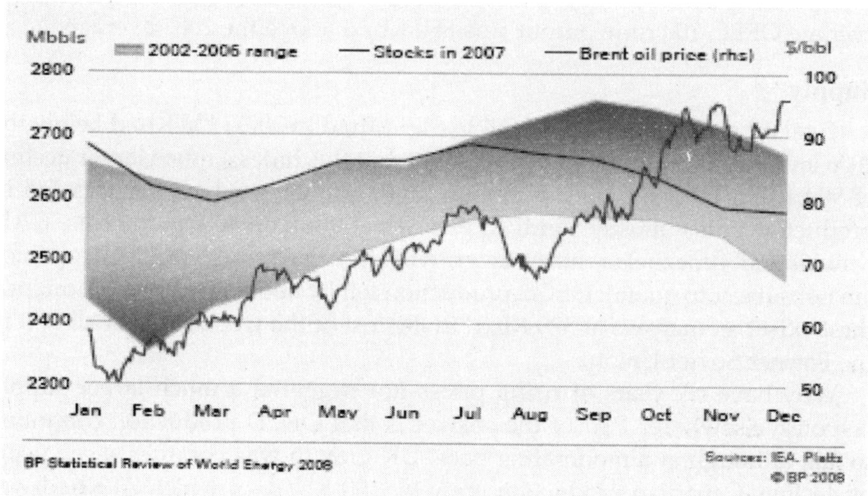

The decline in production drained OECD inventories over the second half of the year. By September 2007, inventories had fallen from eight year highs. Future prices moved into backwardation (with spot prices exceeding future prices). Ever since the second cut in February, crude prices have been on the rise.

World Oil Production Changes (BP, 2008)

In September, as prices continued to increase, OPEC announced a net 500 Kb/d increase in production effective November 2007. However, the immediate impact was neutralized by maintenance in Abu Dhabi, and this

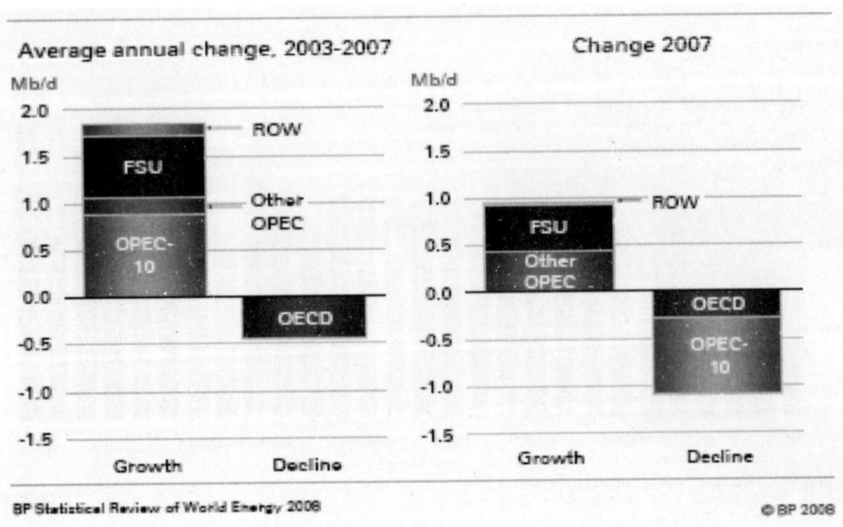

year partially offset by outages in Nigeria. For the first four months of 2008, average OPEC-10 crude output was 850 Kb/d above the 2007 average.

Supply

Total oil production in 2007 fell to 81.5 Mb/d in 2007, 130 Kb/d below the 2006 level. OPEC output fell by 360 Kb/d, but this hides a much larger decline of 900 Kb/d in crude output among those countries which participated in production cuts - mostly Saudi Arabia, which took on 40% of the cuts, UAE, Kuwait and Venezuela - offset by growth in Angola, Iraq, and NGLs (which are not subject to quota). OECD production fell by about 290 Kb/d. Combined these offset an increase of 520 Kb/d in the rest of the world, almost all of it in the Former Soviet Union.

Why have six years of rising prices not triggered a much larger supply response elsewhere? Part of the answer is that OECD production continued to fall, although at a moderating pace. UK growth was flat after seven years of decline. Canadian production grew by 100 Kb/d, of which 70 Kb/d was from oil sands, and US production increased for the first time since 1991 (40 Kb/d). But these could not outweigh accelerating decline in Mexico (210 Kb/d), and in mature fields in Norway (220 Kb/d).

2007 Oil Production: Largest Change (BP, 2008)

The only grouping where output increased substantially was the former Soviet Union. And here, a major shift is taking place. Russia's production, still up 210 Kb/d for the year, has since January 2008 been in year on year decline - mostly due to an unattractive tax system and lack of investment in new fields. The ramp-up of production in Azerbaijan by 210 Kb/d was able to mitigate Russia's slowdown. Bottlenecks in refining have in the past been blamed for the rise in crude prices. Has this changed?

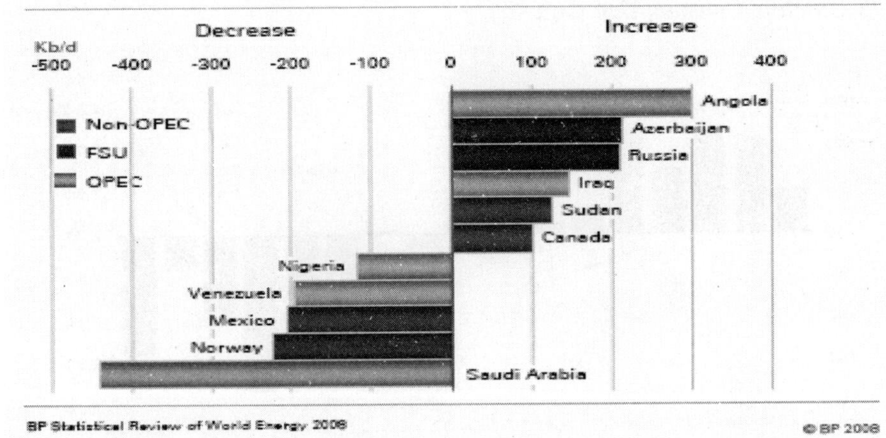

Russian Oil Production (BP, 2008)

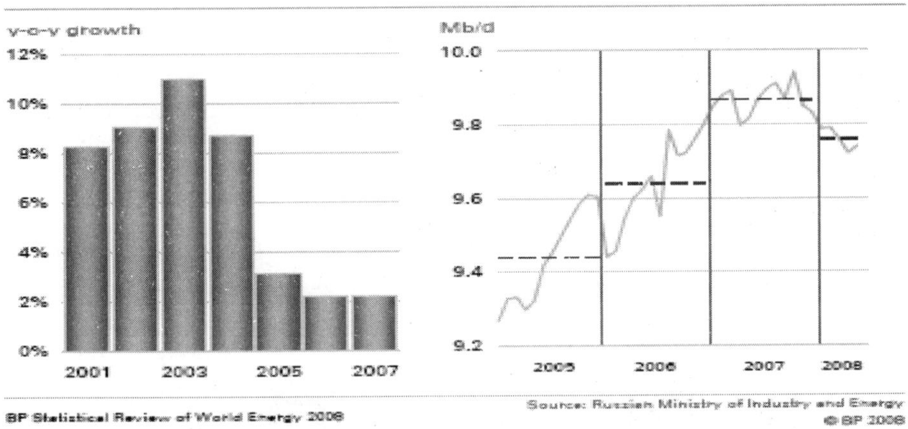

Oil Consumption Growth

Global oil consumption grew by 1 Mb/d, or 1.1%, in 2007, below the ten-year average of 1.4%. Two sources of rising consumption muted the demand response: Oil exporting countries and a number of fast growing non-OECD economies, including China. Viewed through a different lens, global consumption growth stayed positive in countries which subsidised oil products. Consumers in countries where prices are liberalized, and oil products are taxed, were first to be squeezed out of the global market place.

OECD consumption thus suffered its biggest decline since 1983. It fell for the second year, by 390 Kb/d or 0.9%. Non-OECD consumption growth, in contrast, accelerated for the second year running to 1.4 Mb/d, well above the ten year average.

US Gasoline Consumption Growth

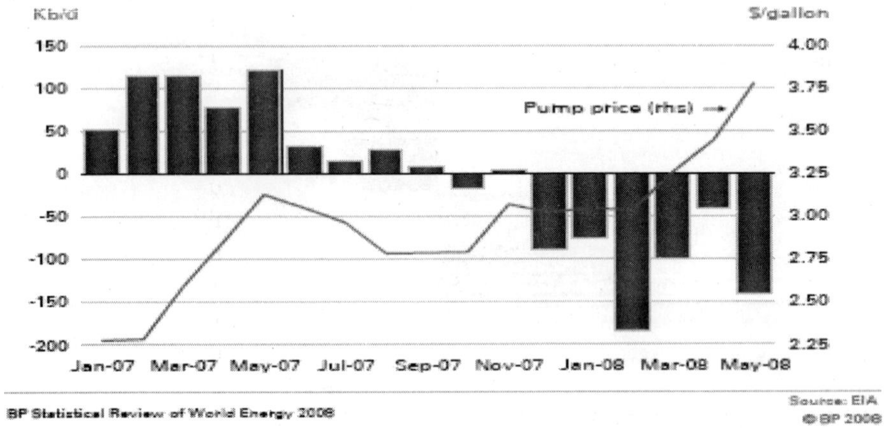

US consumption provides a good example for the wider picture in OECD markets. As pump prices rose from $2.30 per gallon at the beginning of the year to $3.00 at year-end, and then further into 2008, gasoline consumption contracted from growth of 100 Kb/d in early 2007 to a 140 Kb/d decline by May 2008.

Global Oil Consumption Growth

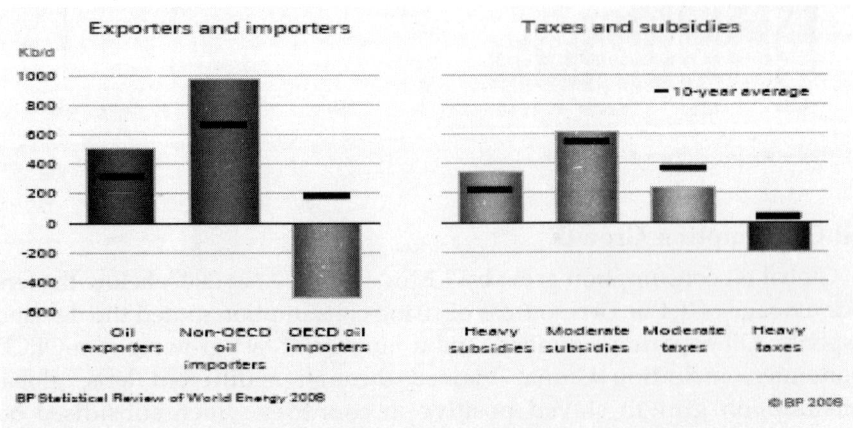

Last year's growth in non-OECD consumption had two components: fast growing emerging market economies, and oil exporters. With growth of 320 Kb/d China was the biggest source of non-OECD growth, although its growth rate slowed. Non-OECD consumption growth was led by Asia, with an increase of almost 600 Kb/d, more than half of which was in China and nearly one third was in India.

Worldwide, the consumption of oil exporters increased by 510 Kb/d or 3%

in 2007. It surpassed growth in importing countries, despite the fact that consumption in the main Former Soviet Union (FSU) exporters declined (by 83 Kb/d), because of an exceptionally warm winter.

Each region has its own story to tell, but they all fit into the general pattern. Latin America grew above average; it is a net exporting region where power shortages added to oil demand. In the Middle East, high economic growth met high subsidies; in Africa, economic growth finally caught on.

One quarter of global consumption was consumed at subsidized retail prices last year. In the subsidizing economies, consumption growth exceeded the 10 year average by 190 Kb/d; in taxing economies, it fell short by 360 Kb/d.

In many emerging economies, the fiscal burden of subsidies has put strain on public finances. In 2008, a whole range of countries - India, Thailand, Indonesia, and Egypt among them - have had to adjust. If high crude prices are sustained, this tendency will put further pressure on demand.

Consumer subsidies were a market distortion last year, which may help explain why oil demand in the non-OECD was less responsive to price changes than in the OECD. But to understand the impact on the oil price, one needs to turn to the supply side.

1.5 STRUCTURE OF WORLD OIL MARKETS

The most widely accepted theoretical approach to the economics of oil focuses on the prevailing oligopolistic market. According to Adelman (1993), the long-term marginal cost is a small fraction of the price of oil, even when making considerable allowances for the future values of the resources used up today ("user costs"). To support high price levels, the excess supply is restricted by a cartel. The market works in the following way: higher cost producers sell all they can produce, while low-cost producers satisfy the remainder of the demand at current prices and cut back production if needed. Econometric evidence on Saudi Arabia confirms the asymmetric behaviour of the low-cost petroleum suppliers: the country restricts production in reaction to negative demand shocks but does not expand production in response to positive ones, in order to sustain high prices (De Santis, 2000). The oligopolistic structure of the oil market or the dominant role of Saudi Arabia is supported in a number of other empirical studies (cf. Griffin, 1985; Alhajji and Huettner, 2000; and Dees et al., 2003).

Crude oil, as it dominates the international oil trade, because it is cheaper to transport. Oil should move to the nearest market first because of transportation costs. The "nearest first" pattern can, however, be distorted by refinery confi gurations, product demand mix, quality specifi cations and politics as discussed in previous sections. In addition, geographical proximity could be a source of short-term vulnerability (e.g. in the event of natural disasters such as hurricanes). This last point notwithstanding, the international oil trade is a relatively regional phenomenon. All panels show the same oil exporting countries ordered by time zone on the horizontal axis. The charts indicate oil exports to the United States, the euro area, Japan and four of the emerging economies of Asia for which there is bilateral data (i.e. China, India, Singapore and South Korea) as a percentage of a country's total oil exports. It is apparent that oil exports go predominantly to users that are geographically close.

The oil producing countries' imports of goods and services from the United States, the euro area, Japan and emerging Asia as a percentage of their total imports. Thus, not only do oil exports go predominantly to adjacent markets, but the oil exporting countries also generally import an equivalent percentage of goods and services from the countries to which they sell petroleum. Notably, the important oil exporting countries on the Arabian peninsula import most goods from emerging Asia. Apart from infl uencing trade in non-oil goods and services, oil export revenues also have an impact on fi nancial markets. Until a few years ago, oil earnings were deposited in international banks in US dollars. However, since 2002, OPEC countries have increasingly deposited their oil wealth in euro-denominated accounts (12% in the third quarter of 2001, against 25% in the second quarter of 2004) at the expense of US dollar-denominated accounts (75% in the third quarter of 2001, against 61.5% in the second quarter of 2004). These observations support the view that the oil exporting countries, as well as some oil importing countries, such as those within the euro area or in its vicinity, could potentially be willing to invoice or settle their oil contracts in euro or in another currency other than the US dollar. The next section discusses current practices as regards price setting in the oil market.

Despite such a substantial increase, the share of total petroleum exports that goes to the mature market economy nations in 2025 is projected to be almost 11 percentage points below their 2002 share, and the share of Persian Gulf exports going to the mature market economies is projected to fall by about 17 percent. The significant shift expected in the balance of OPEC export shares between the mature market economies and emerging economies is a direct result of the economic growth anticipated for the economically developing nations of the world, especially those of Asia. OPEC petroleum exports to emerging economies are expected to increase by 17 million barrels per day over the forecast period, with more than 70 percent of the increase going to the emerging economies of Asia. China, alone, is likely to import about 7.3 million barrels per day from OPEC in 2025, virtually all of which is expected to come from Persian Gulf producers.

North America's petroleum imports from the Persian Gulf in the reference case are expected to more than double over the forecast period. At the same time, over 46 percent of total North American imports in 2025 are expected to be from Atlantic Basin producers and refiners, with significant increases anticipated in crude oil imports from Latin American producers, including Venezuela, Brazil, Colombia, and Mexico. West African producers, including Nigeria and Angola, are also expected to increase their export volumes to

North America. Caribbean Basin refiners are expected to account for most of the increase in North America's imports of refined products.

With a moderate decline in North Sea production, Western Europe is expected to import increasing amounts from Persian Gulf producers and from OPEC member nations in both northern and western Africa. Substantial imports from the Caspian Basin are also expected. Mature market Asian nations are expected to increase their already heavy dependence on Persian Gulf oil. The emerging economies of the Pacific Rim are expected to more than double their total petroleum imports between 2002 and 2025.

Energy in the form of oil & gas is backbone of any economy worldwide. It accounts for over 60% of total world's primary consumption. Oil has multiple applications ranging from domestic to industrial uses. Metals are being progressively replaced by plastics, fibers a product and by product of oil. OPEC a group of oil producing and exporting countries plays a major role in oil business. Non- OPEC countries are also on the path of immense growth due to their exploration and production technology, aggressive cost reduction program, attractive physical policy and collaboration with other developing and developed nations. Demand is more than supply across world so the price sensitiveness of oil is a major concern for each country. Economic, geological, technical factors are focal area for oil price and its impact on economy.

International Oil and Gas Upstream Regimes and Contracts

Petroleum as a source of energy plays a vital role in the progress of economy of any country. This has led to evolution and development of different concession and contracting practices in the world over the years.

1. Contracting Regimes

Various forms and variants petroleum regimes are in vogue depending on priority and objectives sought to be achieved by a country. The practices followed in the international arena, can be divided into following four groups:

- (i) Concessionary arrangements;
- (ii) Contractual arrangements;
    - (a) Service Contracts,
    - (b) Production Sharing Contract (PSC),
    - (c) Risk Service Contracts/ Exploration Service Contracts,
    - (d) Risk Service Development Contracts,
- (iii) Joint Venture arrangements.

The contract regime and arrangements differ in terms of distribution of produce, management and control of operations, degree of intervention and participation of the State, risk sharing between the State and Contractor. The salient features, of these regimes are discussed below:

2. Concession Arrangements

Under concession arrangement, the State issues a license or grants concession to a Contractor granting the right to use an identified area for exploration and production of petroleum. The ownership of produce, if any, remains with concessionaire. Management and control of the operations, costs and risks associated with the operations remain entirely with the licensee. This arrangement is found in most of the developed countries like USA, UK, Canada, Italy and some of the developing countries like Somalia, etc.

The administration of a license arrangement is easy for the State as the State is assured of return in the form of royalty in case of discovery. Contractors like this system since there is no interference in the day to day operations from the government.

Up to early 1960 concession arrangements predominated the world scene. The earliest agreements consisted of only a royalty payment to the State. And later on taxes were added once governments gained more bargaining power. In the late 1970s and early 1980s, a number of governments created additional taxes to capture excess profits from unexpected high oil prices. Now there are numerous fiscal devices (like Signature and Production bonuses, rentals, Royalty, etc.), various types of taxation (like Corporate Income Tax, petroleum revenue tax, windfall profit tax, etc.), and sophisticated formulae for profit sharing found in concessionaire system.

3. Contractual Arrangements

There are a number of variants so far as contractual arrangements are concerned. It may be a pure service contract where payment is not linked to production or risk associated with oil business is not taken or shared by the contractor. On the other hand it may be linked to risk and production. A contract for exploration and development of petroleum through risk service contract refers as to how the produce is to be divided and shared among contractor and landowners, which is generally a State. The objective of private companies/international oil companies is to get a reasonable rate of return on investment keeping into account the risk of exploration and lead time required to produce from the earth, to gain long term access to new supply of petroleum through the right to export a significant part of production.

Under contractual arrangement, the State keeps the right of ownership of petroleum. For better understanding, these arrangements can be classified as follows:

(a) Pure Service Contracts;
(b) Exploration Service contracts;
(c) Development Service Contract;
(d) Production Sharing Contracts (PSC).

(a) Pure Service Contracts

This is when an oil company acts as a contractor for the host government, being paid to produce the hydrocarbons. In a pure service contract regime a contractor is required to carry out exploration and/or development work on behalf of the State and in turn receives a fee in cash and/or kind. All the risks of operation are borne by the State. Such contract may be for a specified activity or may encompass all the activities starting from exploration and culminating in production.

(b) Exploration Risk Service Contract

In an Exploration Risk Service contract, the Contractor is paid a fee for the exploration and/or development work and also has the right to participate in the profits made by charging a premium. The Contractor bears all the risk in so far as he is required to supply the necessary capital for exploring the petroleum. After a discovery, the field is taken over by the state for commercial exploitation. The reward of the Contractor is dependent on the discovery and exploitation of commercial reserves. An Exploration Risk Service contract is similar in nature to the production-sharing contract except that in case of former after development field is taken over by the National Oil Company and exploited by it. These are found in Iran, Iraq etc. Risk service contract is more beneficial to the government as Government keeps the ownership of produce but financing of operations and risk of exploration operations is borne by the Contractor.

(c) Development Service Contracts

Sometimes small discoveries are not developed by National Oil Company because of their small size. These are given on contract. Such contracts may be on the basis of getting all the production by National Oil Company and paying the charges for services provided by the Contractor or sharing of production by then two.

(d) Production Sharing Contracts (PSC)

In this arrangement a contract is entered into among State and/or National Oil Company (NOC) and the Contractor for exploration and production of oil and gas from an area. The Contractor bears all of the exploration and development costs, In case of discovery, the State shares production at predetermined rates as incorporated in the contract. The first Production sharing contract (PSC) was signed in 1966 in Indonesia. At least 50% of host governments now use production sharing contracts regime. PSCs are in use are North Korea, China, Indonesia, Philippines, Sudan, Libya, Oman, Qatar, India, Bahrain, etc. The acreage may be given at different stages to the contractor. It may be awarded at a stage when no exploration work has been done or after some exploration work. Such acreage or blocks are termed as Exploration Acreage or Blocks. There may be acreage in which discovery of oil or gas has been made such acreage is known as Discovered Fields or simply Fields. In India PSC are in operation for Exploration Blocks, Small and Marginal Fields and Medium Size Fields. The PSC contains provision for deciding as to appraisal program, development program and overall sharing of oil or gas on the basis of a formulae given in the PSC.

Some Important Features of Contractual Regimes

Fiscal Regime

Fiscal regime deals with the principles according to which the produce from the contract area is distributed among the State and contractor and levy of taxes, duties etc. In essence it is concerned with the share of Government take and Contractor's take. There are various fiscal tools for fulfillment of conflicting objectives of the State and Contractor under varying circumstances. The characteristics, of some commonly used tools are as follows:

(a) Signature and Production Bonuses

Bonus payments are common elements of the State's receipt. The two most common forms of bonuses are (i) Signature Bonus and (ii) Production Bonus. In India these are generally confined to and taken in case of discovered fields. Signature bonus is the cash payment to the State at the time of signing of the contract. It is a payment to the State for awarding right to explore in the licensed area. In some cases, it can be a payment for technical information obtained from State. These are generally small and token payments.

Production bonuses are cash payments made to the State in the event of a commercial discovery and after commencement of production at various stage s of production. They have no particular purpose other than receipt of certain unquestionable revenues by the State at points in time defined in the contract. There are several tiers of such payments with fresh sums being paid as production continues to rise (i.e. production bonuses linked to barrel of oil per day produced) or it reaches pre-specified levels (i.e. linked to cumulative production achieved). These are triggered, particularly at a high level of production. They are unrelated to profit and could produce disincentive effects if payments are set at high levels. They may prompt companies to avoid reaching threshold level of production to reduce and avoid financial burden of paying production bonuses.

(b) Profit Oil Sharing

Profit oil sharing by State means a portion of oil, which is left after meeting all costs of operating, exploration and development of a field, and therefore called profit oil, is directly shared by State with contractor. This is one of the main features of production sharing contracts.

(c) Cost Oil Recovery Limit

The contractors are asked to bid the percentage of cost oil recovery. This limit is closely associated to profit oil sharing. It lays down what volume of production will be available to company to recover its cost. Cost oil recovery limit has effect of delaying recovery of cost to company and state starts sharing the profits at an early stage of the project.

(d) Rentals

Rentals levied during exploration phase, are payments made per year usually prior to the first day of new license or contract year. They could be a lump sum payment as in Abu Dhabi; a constant payment per measured area in Alberta, payment that increases over a time period e.g. in India rental varies per year for year 1 to year 5.

(e) Royalty

Royalty is one of the most commonly used fiscal tools for the State's revenue. It is linked to top gross value and/or amount of production. Royalty can be a fixed amount or percentage or alternatively linked to a sliding scale. In India royalty is payable even it follows a production sharing regime. Companies would pay a royalty on any oil produced, together with a profits tax (which can have expenditure offset against it). In some cases there are also various bonuses and ground rents (license fees) payable to the government - for example a signature bonus payable at the start of the licence. Licences are awarded in competitive bid rounds on the basis of either the size of the work programme (number of wells, seismic etc) or size of the signature bonus.

A PSA is more complex than a Royalty system - The companies bid on the percentage of the production that the host government receives (this may be variable with the oil price), There is often also participation by the Government owned National Oil Company (NOC). There are also various bonuses to be paid. Development expenditure is offset against production revenue.

(f) Corporate Income Tax

All countries in the world have some form of Corporate Income Tax (CIT). Petroleum exploration activities would normally be subject to provisions of this tax legislation. The rate of Income Tax applicable to petroleum operations is generally same as that applicable to other business activities in the country.

(g) Hydrocarbon Tax (HT)

Hydrocarbon tax was first introduced in Norway and Denmark. It is calculated in the same manner as Corporate Income Tax except that an additional uplift on investment is allowed. HT may or may not be deductible for CIT and vice-a-versa. In India it can be compared with cess.

(h) Additional Profit Tax (APT)

Additional profit tax is a profit sharing, fixed or sliding scale, which is linked to rate of return (ROR), earned by the project on a given date. It is also to avoid and take care of windfall profit. Certain contracts also contain separate provisions for windfall profits.

Major Oil Trade Markets

Traditionally, markets where in the oil flows internationally has been classified as the North Sea, Mediterranean, Far East, West Africa, United states, Middle East Gulf. Some of these major markets are briefly discussed below:

United States of America has oil reserves but suffers from economic inviability of exploiting those resources. The market comprises of East coast, Gulf Coast and Mid- Continent. East cost is dependent on crude imports from North Sea,West Africa,Middle east and Latin America on FOB basis. Gulf coast is the biggest refining centre in the US. Key domestic crude in the US are WTI, Light Louisiana Sweet (LLS), Heavy Louisana Sweet (HLS), West Texas Sour (WTS), Alaskan North Slope (ANS). Active Spot markets for domestic & imported Crude,which in turn have supported the growth of very large forward, future and Derivative markets.

Canada is a net exporter of oil, natural gas, coal, and electricity. It is one of the most important sources for U.S. energy imports. Canada has considerable natural resources and is one of the world's largest producers and exporters of energy. In 2004, Canada produced 18.6 quadrillion British Thermal Units (Btu) of total energy, the fifth-largest amount in the world. Canada is consistently among the top sources for U.S. oil imports, and it is the largest source of U.S. natural gas and electricity imports. In total for 2006, Canada exported to the United States 2.3 million barrels per day (bbl/d) of oil and petroleum products (11 percent of U.S. supply), 3.6 trillion cubic feet of natural gas (16 percent of U.S. supply), and 41.2 billion kilowatt hours of electricity (1 percent of U.S. supply).

Mexico is a major non-OPEC oil producer, with one of the world's largest oil companies, Pemex. In 2006, Mexico was the sixth-largest oil producer in the world, and the second largest in the Western Hemisphere (behind the United States). State-owned Petroleos Mexicanos (Pemex) holds a monopoly on oil production in the country and is one of the largest oil companies in the world. However, oil production in the country has begun to decrease, as production at the giant Cantarell field declines. The oil sector is a crucial component of Mexico's economy: while its relative importance to the general Mexican economy has declined, the oil sector still generates over 10 percent of the country's export earnings. More importantly, the government relies upon earnings from the oil industry (including taxes and direct payments from Pemex) for one-third of total government revenues. Therefore, any decline in production at Pemex has a direct effect upon the country's overall fiscal balance.

Venenzuela contains some of the largest oil and natural gas reserves in the world. It consistently ranks as one the top suppliers of U.S. oil imports and is among the top ten crude oil producers in the world. Venezuela is one of the world's largest exporters of crude oil and the largest in the Western Hemisphere. In 2006, the country was the sixth-largest net oil exporter in the world. The oil sector is of central importance to the Venezuelan economy: it accounts for more than three-quarters of total Venezuelan export revenues, about half of total government revenues, and around one-third of total gross domestic product (GDP). Venezuela was a founding member of the Organization of the Petroleum Exporting Countries (OPEC).

South and Central America has limited energy resources, it is important to world energy markets as a transit center for oil via the Panama Canal and as a potential energy transit center between North and South America. With hardly any domestic hydrocarbon reserves, Central American countries rely heavily on imported oil for their energy needs. The countries of Central America, including Belize, Costa Rica, El Salvador, Guatemala, Honduras, Nicaragua, and Panama, have traditionally been dependent upon agricultural exports for a large portion of their economic activity. the countries have been especially affected by high world oil prices in recent years. Partially offsetting this, many have been able to secure preferential pricing for oil from Venezuela and Mexico. Despite the lack of sizable oil reserves, Central America remains an important transit center for oil via the Panama Canal and as a potential energy transit center between North and South America. Europe & Eurasia or North Sea market comprises of United Kingdom, Norway, Denmark and Netherland. Market witnesses physical trade of the commodity Brent. For the purpose forward paper of 15- day Brent at the IPE. Brent is also traded through swap contracts and also through Brent CFD's i.e. Contract for Differences.

United Kingdom *with* its significant North Sea reserves, the UK is a major European oil and natural gas producer. The United Kingdom (UK) is an important political and economic power in Europe and the world. It has the second-largest economy in the European Union (EU) with a nominal 2005 gross domestic product (GDP) of $2.2 trillion. The UK is the largest producer of oil and natural gas in the EU. However, after years of being a net exporter of both fuels, the UK became a net importer of natural gas in 2004. Government estimates also predict that the country will become a net importer of oil by the end of the decade. Due to decline in the production of oil by the reason of mismatching of discovery and maturity of oil fields, UK government is aiming at exploiting marginal fields.

Norway is a major non-OPEC source of oil. It is a major world supplier of oil and natural gas, especially to the European Union. Norway's economy is highly dependent on its offshore oil and natural gas sector, which provides the government with its largest single source of revenue and the largest contribution to GDP. In recent years, high oil prices have made for government budget and current account surpluses, and rising disposable income. Norway's dependence upon oil and gas revenues present long-term challenges for the country, especially because many industry analysts believe that North Sea oil and gas production has already reached or passed its peak. Norway's dependence upon oil and gas revenues present long-term challenges for the country, especially because many industry analysts believe that North Sea oil and gas production has already reached or passed its peak.

Russia holds the world's largest natural gas reserves, the second largest coal reserves, and the eighth largest oil reserves. Russia is also the world's largest exporter of natural gas, the second largest oil exporter and the third largest energy consumer. Russia's economic growth over the past seven years has been driven primarily by energy exports, given the increase in Russian oil production and relatively high world oil prices during the period. The share of energy use from coal and nuclear has stayed constant, while energy use from oil has decreased from 27 percent to around 19 percent. Russia's economy is heavily dependent on oil and natural gas exports, making it vulnerable to fluctuations in world oil prices. According to an International Monetary Fund (IMF) a $1 per barrel increase in Urals blend oil prices for a year is estimated to raise federal budget revenues by 0.35 percent of GDP, or $3.4 billion. In order to manage windfall oil receipts, the government established a stabilization fund in 2004 worth. Raw materials, such as oil, natural gas, and metals, dominate merchandise exports and account for over two-thirds of all Russian export revenues.

Middle East or Mediterranean markets mainly comprises of Algeria, Libya, Egypt, Syria, Tunisia and Russia. Crude mainly flows from former Soviet Union and Iran on FOB basis via Sidi Kerir. Also the oil is shipped via Black Sea port of Novorossiysk and Ukrainian. Middle East Gulf basket of crude oil basket is the richest in all other markets. It comprises of Saudi Arabia, Iran, Iraq, Kuwait, Oman, Qatar, UAE and Yemen. Other then Oman all is OPEC members. Most Gulf crude is sold on a term basis, partly because of the time required to ship oil to refineries in the US, Europe, Far East and partly because major OPEC Producers such as Saudi Arabia & Kuwait are reluctant to see their Crude being traded on the spot market and term contracts usually prohibit the secondary trading of Gulf Crudes.

Syria is the only significant oil producer in the Eastern Mediterranean, with crude oil production of an estimated 365,000 barrels per day (bbl/d) and total liquids production of 416,000 bbl/d in 2005. Israel and the Palestinian Authority, Jordan, and Lebanon import almost all of their petroleum requirements. After the initiative enacted by the Cypriot government to proceed in handing out research and drilling rights for expected oil reserves deep under the sea, estimated to be worth some 450 billion USD at current prices.

Iran, one of OPEC's founding members, holds the world's third-largest proven oil reserves, and the world's second-largest natural gas reserves. and ranks amongst the world's top three holders of proven oil and natural gas reserves. Iran is OPEC's second-largest exporter after Saudi Arabia, and is the fourth-largest exporter of crude oil globally after Saudi Arabia, Russia, and Norway. Natural gas accounts for half of Iran's total domestic energy consumption, while the remaining half is predominately oil consumption. The continued exploration and production of the offshore South Pars natural gas field in the Persian Gulf is a key part of in Iran's energy sector development plan.

Iraq Security issues continue to hold back rehabilitation of Iraq energy sector. Iraq has the world's third largest proven petroleum reserves and some of the lowest extraction costs, although just a fraction of its known fields are in development. in 2006, crude oil export revenues represented around 60 percent of GDP and 89 percent of government revenues. Iraq was the world's 15th biggest oil producer and Iraq meets approximately 94 percent of its energy needs with petroleum. Iraq's use of abundant natural gas resources and hydropower is limited. the stabilization of Iraq is highly correlated with Iraq's economic success or failure, which in the medium-term is highly dependent on its hydrocarbons industry.

Kuwait contains 101.5 billion barrels of proven oil reserves (including its share of the Neutral Zone), or roughly 8 percent of the world's total oil reserves. Kuwait's economy is heavily dependent on oil export revenues. Current high oil prices are producing a surge in oil export revenues for Kuwait, with 2005 the best year for oil export revenues in the past decade. Non-oil sectors of the Kuwaiti economy, particularly services, have also experienced strong growth fueled by the inflow of oil revenues. Kuwait would like to diversify its economy away from near-complete dependence on oil revenues. Currently, the country relies on oil revenues for around 90-95 percent of total export earnings and around two-fifths of GDP. Kuwait channels around 10 percent of its oil revenues into the "Future Generations Fund" for the day when oil income runs out. The bulk of this reserve is invested in the United States, Germany, the United Kingdom, France, Japan, and Southeast Asia. In order of importance, foreign assets are believed to be invested in stocks and bonds, fixed yield instruments (mostly short term), and real estate. Kuwait follows a generally conservative investment policy.

Oman is a significant non-OPEC oil exporter, although production has declined in recent years. Oman's economy is heavily reliant on oil revenues, which account for about 75 percent of the country's export earnings and 40 percent of its gross domestic product (GDP). The slowdown in economic growth occurred largely as a result of declining oil production in Oman. To help offset these declines, the government has devoted considerable resources to new exploration and production activities, enhanced oil recovery projects, and introduced policies aimed at diversifying the country's economy away from the oil sector. The development of natural gas reserves in Oman is a central part of this effort, and natural gas production is likely to expand considerably during the next several years. All of Oman's domestic energy consumption is supplied by natural gas and oil, reflecting the country's relative abundance of oil and natural gas reserves.

Qatar holds the world's third largest natural gas reserves and is the single largest supplier of liquefied natural gas. Qatar is also a member of OPEC and exports considerable amounts of oil. Qatar is a member of the Organization of the Oil Exporting Countries (OPEC) and is a significant oil producer. Domestically, the vast majority of Qatar's total energy consumption comes from natural gas (79 percent), while the balance is supplied by oil.

Saudi Arabia is the biggest oil producer in the Organization of the Petroleum Exporting Countries (OPEC). With one-fifth of the world's proven oil reserves, some of the lowest production costs, and an aggressive energy sector investment initiative, Saudi Arabia is likely to remain the world's largest net oil exporter. From January-November 2006, Saudi Arabia supplied the United States with 1.4 million barrels per day of crude oil, or approximately 14 percent, of U.S. crude oil imports. Between mid-2003 and mid-2006, Saudi Arabia showed strong economic performance due to high oil prices, increasing oil production and export earnings, paired with structural reforms, economic diversification, and stable macroeconomic policymaking. Saudi Arabia remains heavily dependent on oil and petroleum-related industries, including petrochemicals and petroleum refining. Saudi Arabia is increasing its oil production capacity to 12 million barrels per day (bbl/d), by 2009. Angola, Nigeria, Gabon, Congo, Cameroon are the countries that formed the crude basket of Africa. Out of these Nigeria and Gabon are members of the OPEC and yield sweet crude for the European and American markets needs.

Angola, a significant Sub-Saharan Africa oil producer, joined the Organization of Petroleum Exporting Countries (OPEC) in January 2007. On January 1, 2007, Angola became the 12th member of the OPEC. As an OPEC member, Angola will have to pay $2 million per year in membership fees and might be restricted by OPEC quotas on oil production. Angola is the second largest oil producer in Sub-Saharan Africa behind Nigeria. Angola is set to experience oil production increases over the next five year period as new offshore projects come online. Angola exports crude oil primarily to China, the United States, Europe and Latin America. The majority of natural gas produced in Angola is either flared or used in oil recovery. To help reduce flaring, Chevron and Sonangol are planning to build a five-million-ton liquefied natural gas (LNG) plant, which could be operational in 2010.

Nigeria is rich in mineral wealth, with petroleum and natural gas being the country's major mineral products. Nigeria's economic growth primarily comes from the country's oil sector. On April 21, 2007, Nigeria held presidential elections, first time in Nigeria's history that the country passed control from one civilian government to another. During the 16 months preceding the election, militant activity in the Niger Delta (especially near Warri and Port Harcourt) has severely impacted Nigeria's oil production potential by shutting-in an estimated 20 percent of total production. The Nigerian economy is heavily dependent on the oil sector, which accounts for 95 percent of the country's total export revenues.

Gabon's economic growth is highly dependent on oil production. As a result, Gabon is seeking to simultaneously increase investment in oil and boost non-oil sector growth for future stability. Gabon's economy is highly dependent on oil production, with the country's oil exports accounting for 51 percent of GDP and 63 percent of government revenues. Gabon hopes to increase growth in non-oil sectors as it is currently facing significant declines in both oil production and reserves. The country's other main exports are timber, uranium and manganese but oil still represents 70 percent of the total exports. In recent years, Gabon has taken steps to market itself as a high end ecotourism destination. Nonetheless, the country is expected to continue relying on oil production in the near future. Gabon has been in a dispute with Equatorial Guinea over three islands in the Corsico Bay that could potentially contain oil deposits. Gabon is also a signatory to the Extractive Industries Transparency Initiative that supports governance and revenue transparency from the oil, gas and mining sectors. In 1996, Gabon retracted its membership in the OPEC, citing the organization's high annual dues as the reason.

Algeria's hydrocarbon sector accounts for 98 percent of export revenues. Oil and natural gas exports, which made up 98 percent of Algerian exports (by value) in 2006, are the main driver of Algerian economic growth. With continuing investments being made in Algerian oil and gas development, both sectors have potential for increasing production capacity over the next few years. The Economic Community of West African States (ECOWAS) was formed to promote economic and development growth in West Africa. Major exports from the region include energy products, minerals and agricultural products. The Gambia, Ghana, Guinea, Guinea-Bissau, Liberia, Mali, Niger, Nigeria, Senegal, Sierra Leone, and Togo. The leaders established ECOWAS to promote regional integration and economic growth in West Africa, as well as to create a monetary union in the region. However, ECOWAS has encountered problems in the process of regional integration including: political instability and lack of good governance that has plagued many member countries, the insufficient diversification of national economies, the absence of reliable infrastructure, and the multiplicity of organizations for regional integration with the same objectives

Egypt is a significant oil producer and a rapidly growing natural gas producer. The country's first liquefied natural gas (LNG) export terminal began operation in January 2005. The Suez Canal and Sumed Pipeline are strategic routes for Persian Gulf oil shipments, making Egypt an important transit corridor. Though Egypt's net exports of crude oil and petroleum products have declined in recent years, higher prices on world markets have pushed Egypt's oil revenues upward. Additionally, Egypt's economy is continuing its gradual recovery from the declining growth rates it experienced in 2001 and 2002, but with a growth rate still far below what was achieved in the 1990s.

Asia Pacific or Far- East markets includes no major oil producer. India, Australia, Brunei, China, Indonesia, Malaysia, Papua New Guinea, Vietnam constitutes this markets. Majority of Far East Crude are traded on the basis of the Weekly Asian Petroleum Price Index (APPI).

Malaysia is a net exporter of oil and natural gas. Malaysia is a significant Southeast Asian producer of oil and natural gas, although proven oil reserves have declined in recent years. Malaysia's western coast runs alongside the Strait of Malacca, an important route for seaborne energy trade that links the Indian and Pacific Oceans.

Australia is rich in energy resources and leads the world in total coal exports. However, in 2006, severe drought, increasing inflation, and high oil prices all contributed to a deceleration of economic growth. ustralia is rich in natural resources with significant petroleum, natural gas and coal reserves. Petroleum accounts for a large share of energy consumption, but due to declining output, Australia is facing a growing dependence on petroleum imports. Australia is one of the few countries belonging to the OECD that is a significant net energy exporter.

China is the world's most populous country and the second largest energy consumer behind the United States. Rising oil demand and imports have made China a significant factor in world oil markets. Together with strong economic growth, China's demand for energy is surging rapidly. EIA forecasts that China's oil consumption will increase by almost half a million barrels per day in 2006, or 38 percent of the total growth in world oil demand. China is the world's third-largest net importer of oil behind the United States and Japan, an important factor in world oil markets.

India boasts a growing economy, and is increasingly a significant consumer of oil and natural gas. Together with the country's impressive growth, India has also become a significant consumer of energy resources. According to EIA estimates, India was the fifth largest consumer of oil in the world during 2006.

Oil Trade Movements

Trade Flows worldwide (millions tonnes)

The data illustrates the flow of crude oil and products between sources of production and the regions of consumption. The intra-area movements of oil

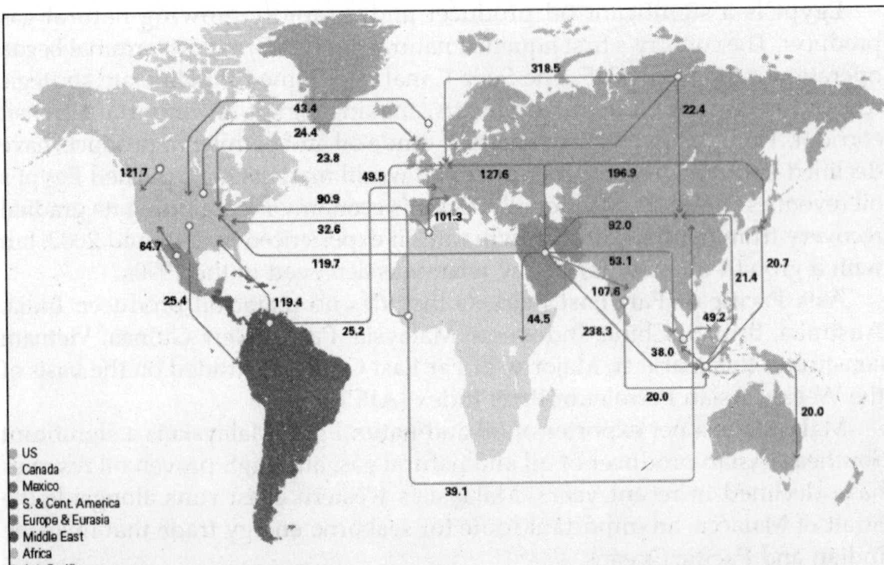

Source: BP Statistical Review 2008.

Trade Movements

Thousand barrels daily	1998	1999	2000	2001	2002	2003	2004	2005	2006	2007	2008	Change 2008 over 2007	2008 share of total
US	10382	10550	11092	11618	11357	12254	12898	13525	13612	13632	12872	-5.6%	23.6%
Europe	11017	10670	11070	11531	11895	11993	12538	13261	13461	13953	13751	-1.4%	25.2%
Japan	5259	5346	5329	5202	5070	5314	5203	5225	5201	5032	4925	-2.1%	9.0%
Rest of World*	14438	15050	15880	16436	16291	17191	18651	19172	20287	22937	23078	0.6%	42.2%
Total World	41096	41616	43371	44787	44613	46752	49290	51182	52561	55554	54626	-1.7%	100.0%
Exports US	1011	956	890	910	904	921	991	1129	1317	1439	1967	36.7%	3.6%
Canada	1603	1520	1703	1804	1959	2096	2148	2201	2330	2457	2498	1.6%	4.6%
Mexico	1770	1739	1814	1882	1966	2115	2070	2065	2102	1975	1609	-18.5%	2.9%
S. & Cent. America	3240	3145	3079	3143	2965	2942	3233	3528	3681	3570	3616	1.3%	6.6%
Europe	1344	1851	1967	1947	2234	2066	1993	2149	2173	2273	2023	-11.0%	3.7%
Former Soviet Union	3569	4019	4273	4679	5370	6003	6440	7076	7155	8334	8184	-1.8%	15.0%
Middle East	18702	18341	18944	19098	18062	18943	19630	19821	20204	19680	20128	2.3%	36.8%
North Africa	2712	2726	2732	2724	2620	2715	2917	3070	3225	3336	3260	-2.3%	6.0%
West Africa	3094	2985	3293	3182	3134	3612	4048	4358	4704	4830	4587	-5.0%	8.4%
Asia Pacific +	3496	3543	3736	3914	3848	3978	4189	4243	4312	6004	5392	-10.2%	9.9%
Rest of World*	556	791	940	1505	1551	1351	1631	1542	1359	1656	1363	-17.7%	2.5%
Total World	41057	41016	43371	44789	44613	46752	49290	51182	52561	55554	54626	-1.7%	100.0%

" Includes unidentified trade.

t Excludes Japan. Excludes trade between other Asia Pacific countries and India prior to 2007.

Note: Annual changes and shares of total are calculated using thousand barrels daily figures.

Imports and Exports 2008

	Million tonnes				Thousand barrels daily			
	Crude imports	Product imports	Crude exports	Product exports	Crude imports	Products imports	Crude exports	Product exports
US	487.2	149 5	6 9	87.7	9756	3116	139	1829
Canada	34.6	13.8	96.4	27.2	693	288	1931	566
Mexico	2.0	24.0	72.8	7.2	40	500	1459	150
S. & Cent. America	30.4	41.4	123.8	54.5	610	864	2479	1136
Europe	542.2	138.7	14.0	83.5	10859	2892	281	1741
Former Soviet Union	t	7.1	311.3	93.5	1	149	6235	1948
Middle East	11.0	9.7	895.0	105.7	220	202	17925	2203
North Africa	16.4	8.2	130.7	30 8	329	170	2618	642
West Africa	0.9	13.5	223.0	5.8	18	281	4466	121
East & Southern Africa	25.4	9.0	17.0	0.8	509	189	340	17
Australasia	24.4	19.2	13.7	1.8	489	400	274	37
China	178.8	39.0	3.7	15.0	3581	812	75	313
India	127.7	22.0	+	34.4	2557	458	+	718
Japan	203.2	41.0	-	17.1	4069	856	-	356
Singapore	54.5	76.4	2.0	74.0	1092	1593	40	1542
Other Asia Pacific	231.0	114.6	46.5	70.1	4627	2390	931	1460
Unidentified*	-	0.8	12.8	18.9	-	17	257	394
Total World	1969.9	727.9	1969.9	727 9	39451	15176	39451	16175

* Includes changes in the quantity of oil in transit, movements not otherwise shown, unidentified military use. etc

+ Less than 0.05.

Note: Bunkers are not included as exports. Intra-area movements (for example, between countries in Europe) are excluded

Inter-area Movements, 2008

Million tonnes From	US	Canada	Mexico	S&Cent. America	Europe	Africa	Austra-lasia	China	India	Japan	Singa-pore	Other Asia Pacific	Rest of World	Total
US	-	13.1	17.0	25 4	24 4	1.8	0.8	0.8	0.7	3.6	4.3	14	1.6	94.6
Canada	121.7	-	-	0.1	1.6	-	-	t	t	0.1	t	t	t	123.6
Mexico	64.7	1.4	-	4.3	7.7	-	-	-	.1.9	-	0.1		t	80.0
S. & Cent. America	119.4	1.0	1.3	-	25.2	1.1	-	16.5	5.8	0.1	7.8	0.1	0.1	178.3
Europe	43.4	8.3	5.7	4.8	-	16.9	t	0.2	0.5	1.4	5.3	1.4	9.7	97.6
Former Soviet Union	23.8	1.6	0.2	3.0	318.5	1.1	0.6	22.4	2.0	8 2	5.0	6.6	11.8	404.8
Middle East	119.7	6.3	0.6	5.8	127.6	44.5	5.4	92.0	107.6	196.9	53.1	238.3	3.1	1000.7
North Africa	32.6	8.9	0.2	5.1	101.3	1.0	0 3	4.2	4.3	0.4	0.1	3.1	-	161.5
West Africa	90.9	5.2	t	15.0	49.5	4.5	-	39.1	16.6	1.1	0.1	6.7	0.1	228.8
East & Southern Africa	-	-	-	t	0.6	-	-	10.6	0.8	4.9	0.2	0.6	-	17.8
Australasia	1.8	t	t	-	t	-	-	0.9	0.2	2 6	3.5	6.6	t	15.5
China	0.8	t	-	4.0	0.3	0.4	0.2	-	0.1	1.3	2.1	9.2	0.3	18.8
India	0.3	-	-	1.6	3.3	-	-	0.2	-	14	6.8	20.0	0.8	34.4
Japan	-	-	0.5	0.1	1.2	-	2.6	4.9	0.6	-	4 6	2.4	0.1	17.1
Singapore	-	t	t	1.0	24	1.6	12 1	4.5	2.9	1.4	-	49.2	0.8	76.0
Other Asia Pacific	5.3	t	0.6	1.5	2 5	0.5	20.0	21.4	5.8	20.7	38.0	-	0.2	116.5
Unidentified*	12.3	2.5		~	14.9	-	1.7	-	-	0.2	-	-	-	31.7
Total Imports	636.6	48.4	26.0	71.9	680.9	73.4	43.6	217.8	149.7	244.2	130.9	346.7	28.8	2697.8

(Contd.)

Table *(Contd.)*

Million tonnes From	US	Canada	Mexico	S&Cent. America	Europe	Africa	Australasia	China	India	Japan	Singapore	Other Asia Pacific	Rest of World	Total
Thousand barrels daily form														
US	-	270	352	529	508	37	16	16	14	75	89	29	34	**1967**
Canada	2459	-	-	3	34	-	-	t	t	1	t	t	t	**2498**
Mexico	1299	28	-	88	155	-	-	-	38	-	1		t	**1609**
S. & Cent. America	2418	21	26	-	509	23	-	334	116	1	163	1	2	**3616**
Europe	900	168	118	100	-	352	1	5	11	29	110	28	202	**2023**
Former Soviet Union	492	33	4	61	6428	22	12	453	40	165	105	133	236	**8184**
Middle East	2398	125	12	116	2564	899	108	1844	2167	3960	1072	4800	64	**20128**
North Africa	665	179	4	103	2039	20	5	83	87	9	2	63	–	**3260**
West Africa	1825	105	1	301	992	90	-	783	332	21	1	134	3	**4587**
East & Southern Africa	-	-	-	t	12	-	-	213	16	99	5	12	–	**357**
Australasia	35	t	t	-	1	-	-	19	3	52	70	132	t	**312**
China	15	t	-	83	6	9	5	-	1	27	43	191	6	**388**
India	6	-	-	32	70	-	-	5	-	30	142	417	17	**718**
Japan	-	-	11	2	24	-	54	103	12	-	97	51	2	**356**
Singapore	-	t	1	22	50	33	250	93	61	29	-	1026	17	**1583**
Other Asia Pacific	108	1	12	32	52	11	403	442	117	422	786	–	5	**2391**
Unidentified*	252	51	-	-	307	**307**	35	-	-	5	-	–	–	**650**
Total imports	12872	981	541	1473	13751	1496	889	4393	3016	4925	2685	7017	588	54625

*Includes changes in the quantity of oil in transit, movements not otherwise shown, unidentified military use, etc.

\+ Less than 0.05

(for example, crude oil and products moving between countries within Europe). Also shown is a map of major oil trade movements.

Crude Oil Prices

Crude oil prices behave much as any other commodity with wide price swings in times of shortage or oversupply. The crude oil price cycle may extend over several years responding to changes in demand as well as OPEC and non-OPEC supply. The U.S. petroleum industry's price was heavily regulated through production or price controls throughout much of the twentieth century. In the post World War II era U.S. oil prices at the wellhead averaged $26.64 per barrel adjusted for inflation to 2008 dollars. In the absence of price controls, the U.S. price would have tracked the world price averaging $28.68. Over the same post war period the median for the domestic and the adjusted world price of crude oil was $19.60 in 2008 prices. That means that only fifty percent of the time from 1947 to 2008 have oil prices exceeded $19.60 per barrel. Until the March 28, 2000 adoption of the $22-$28 price band for the OPEC basket of crude, oil prices only exceeded $24.00 per barrel in response to war or conflict in the Middle East. With limited spare production capacity, OPEC abandoned its price band in 2005 and was powerless to stem the surge in oil prices, which was reminiscent of the late 1970s.

Crude Oil Prices 1947 – August, 2009 (Dollars)

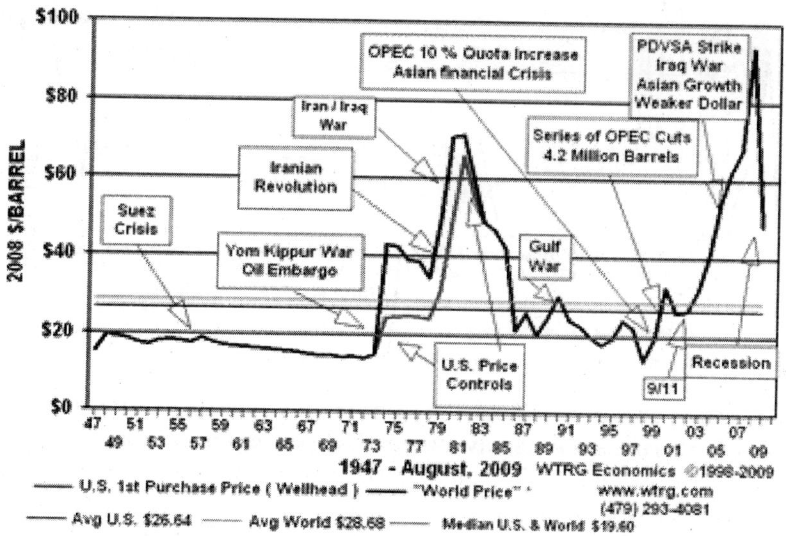

*World Price – The only very long-term price series that exists is the U.S. average wellhead or first purchase price of crude. When discussing long-term price behavior this presents a problem since the U.S. imposed price controls on domestic production from late 1973 to January 1981. In order to present a consistent series and also reflect the difference between international prices and U.S. prices we created a world oil price series that was consistent with the U.S. wellhead price adjusting the wellhead price by adding the difference between the refiners acquisition price of imported crude and the refiners average acquisition price of domestic crude.

The Very Long-term View

The very long-term view is much the same. Since 1869, US crude oil prices adjusted for inflation have averaged $22.52 per barrel in 2008 dollars compared to $23.42 for world oil prices. Fifty percent of the time prices U.S. and world prices were below the median oil price of $16.71 per barrel. If long-term history is a guide, those in the upstream segment of the crude oil Industry should structure their business to be able to operate with a profit, below $17.65 per barrel half of the time. The very long-term data and the post World War II data suggest a "normal" price far below the current price.

Crude Oil Prices 1869-2009

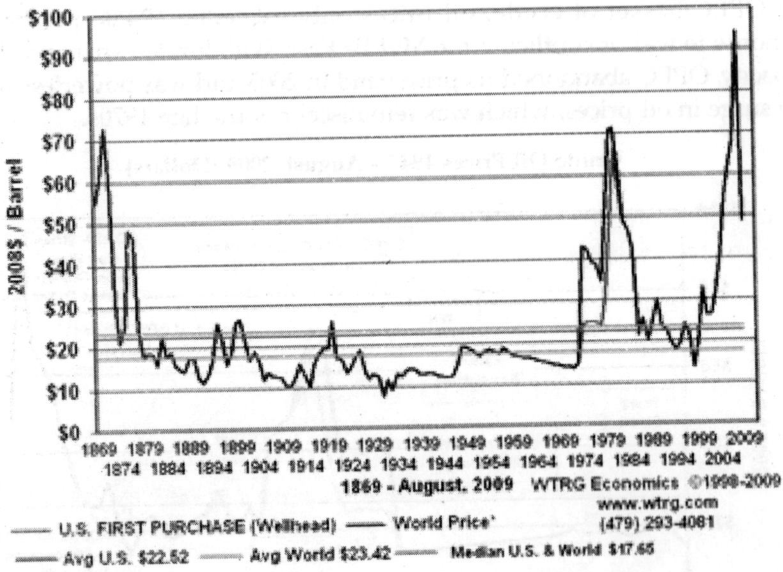

The results are dramatically different if only post-1970 data are used. In that case, U.S. crude oil prices average $32.36 per barrel and the more relevant world oil price averages $35.50 per barrel. The median oil price for that period is $30.04 per barrel. If oil prices revert to the mean this period is likely the most appropriate for today's analyst. It follows the peak in U.S. oil production eliminating the effects of the Texas Railroad Commission and is a period when the Seven Sisters were no longer able to dominate oil production and prices. It is an era of far more influence by OPEC oil producers than they had in the past. As we will see in the details below influence over oil prices is not equivalent to control.

Crude Oil Prices 1970-2008 (Dollars)

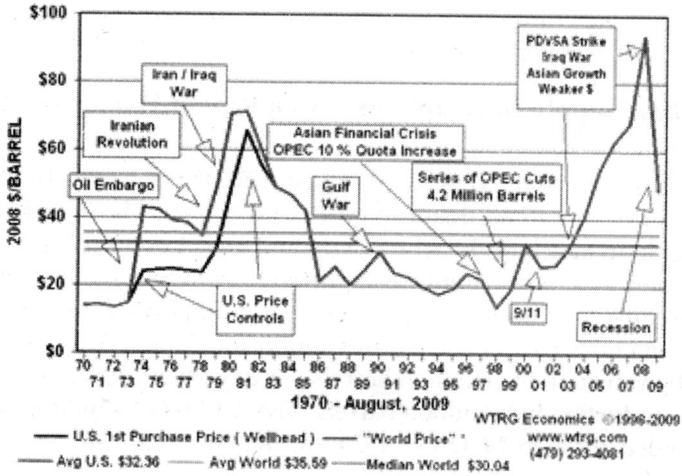

Post-World War II – Pre-Embargo Period

Crude Oil prices ranged between $2.50 and $3.00 from 1948 through the end of the 1960s. The price oil rose from $2.50 in 1948 to about $3.00 in 1957. When viewed in 2008 dollars an entirely different story emerges with crude oil prices fluctuating between $17 and $19 during most of the period. The apparent 20% price increase in nominal prices just kept up with inflation. From 1958 to 1970, prices were stable near $3.00 per barrel, but in real terms the price of crude oil declined from above $19 to $14 per barrel. The decline in the price of crude when adjusted for inflation for the international producer suffered the additional effect in 1971 and 1972 of a weaker US dollar. Established in 1960 OPEC, with five founding members Iran, Iraq, Kuwait, Saudi Arabia and Venezuela, took over a decade to establish its influence in the world market. Two of the representatives at the initial meetings had studied the the Texas Railroad Commission's methods of influencing price through limitations on production. By the end of 1971, six other nations had joined the group: Qatar, Indonesia, Libya, United Arab Emirates, Algeria and Nigeria. From the foundation of the Organization of Petroleum Exporting Countries through 1972 member countries experienced steady decline in the purchasing power of a barrel of oil.

Throughout the post war period exporting countries found increasing demand for their crude oil but a 40% decline in the purchasing power of a barrel of oil. In March 1971, the balance of power shifted. That month the Texas Railroad Commission set proration at 100 percent for the first time. This meant that Texas producers were no longer limited in the volume of oil that they could produce. More importantly, it meant that the power to control crude oil prices shifted from the United States (Texas, Oklahoma and Louisiana) to OPEC. Another way to say it is that there was no more spare capacity in the

U.S. and therefore no tool to put an upper limit on prices. A little over two years later OPEC, through the unintended consequence of war, obtained a glimpse of the extent of its power to influence prices.

Middle East Supply Interruptions - Yom Kippur War - Arab Oil Embargo

In 1972, the price of crude oil was about $3.00 per barrel. By the end of 1974, the price of oil had quadrupled to over $12.00. The Yom Kippur War started with an attack on Israel by Syria and Egypt on October 5, 1973. The United States and many countries in the western world showed support for Israel. Because of this support, several Arab exporting nations and Iran imposed an embargo on the countries supporting Israel. While these nations curtailed production by 5 million barrels per day other countries were able to increase production by a million barrels. The net loss of 4 million barrels per day extended through March of 1974 and represented 7 percent of the free world production.

Any doubt the ability to control crude oil prices had passed from the United States to OPEC was removed during the Arab Oil Embargo. The extreme sensitivity of prices to supply shortages became all too apparent when prices increased 400 percent in six short months. From 1974 to 1978, the world crude oil price was relatively flat ranging from $12.21 per barrel to $13.55 per barrel. When adjusted for inflation world oil prices were in a period of moderate decline.

OPEC Oil Production 1973-2009 (Mbbl/d)

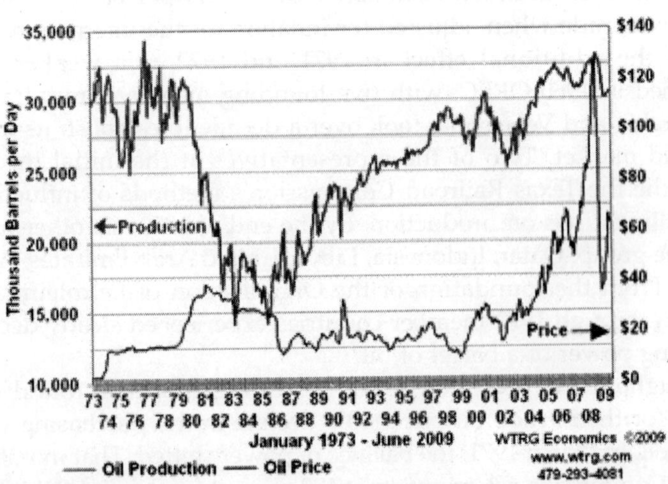

Crises in Iran and Iraq

In 1979 and 1980, events in Iran and Iraq led to another round of crude oil price increases. The Iranian revolution resulted in the loss of 2 to 2.5 million barrels per day of oil production between November 1978 and June 1979. At one point production almost halted. The Iranian revolution was the proximate cause of what would become the highest price in post-WWII history. However, its impact on prices would have been limited and of relatively short duration had it not been for subsequent events. Shortly after the revolution, production was up to 4 million barrels per day. In September 1980, Iran already weakened by the revolution was invaded by Iraq. By November, the combined production of both countries was only a million barrels per day and 6.5 million barrels per day less than a year before. Consequently worldwide crude oil production was 10 percent lower than in 1979. The combination of the Iranian revolution and the Iraq-Iran War cause crude oil prices to more than double increasing from $14 in 1978 to $35 per barrel in 1981. Three decades later Iran's production is only two-thirds of the level reached under the government of Reza Pahlavi, the former Shah of Iran. Iraq's production remains a million barrels below its peak before the Iraq-Iran War.

Iran Oil production 1973-2009 (Mbbl/d)

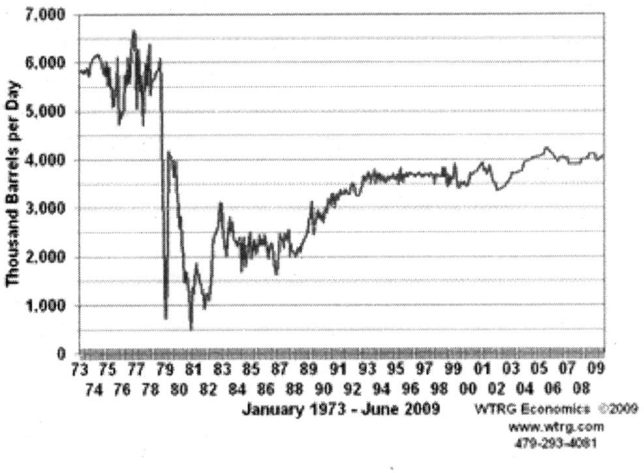

Iraq Oil production 1973-2009 (Mbbl/d)

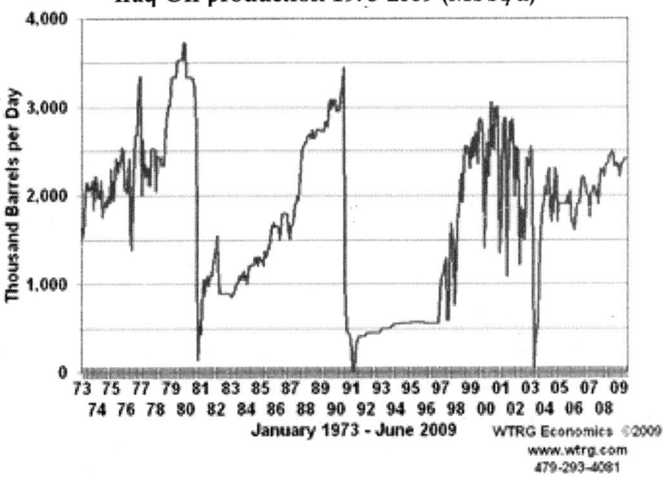

US Oil Price Controls

The rapid increase in crude prices from 1973 to 1981 would have been much less were it not for United States energy policy during the post Embargo period. The US imposed price controls on domestically produced oil. The obvious result of the price controls was that U.S. consumers of crude oil paid about 50 percent more for imports than domestic production and U.S. producers received less than world market price. In effect, the domestic petroleum industry was subsidizing the U.S. consumer. Did the policy achieve its goal? In the short-term, the recession induced by the 1973-74 crude oil price spike was somewhat less severe because U.S. consumers faced lower prices than the rest of the world. However, it had other effects as well. In the absence of price controls, U.S. exploration and production would certainly have been significantly greater. Higher petroleum prices faced by consumers would have resulted in lower rates of consumption: automobiles would have achieved higher miles per gallon sooner, homes and commercial buildings would have has better insulated and improvements in industrial energy efficiency would have been greater than they were during this period. Consequently, the United States would have been less dependent on imports in 1979-1980 and the price increase in response to Iranian and Iraqi supply interruptions would have been significantly less.

US Oil Price Controls 1973-1981

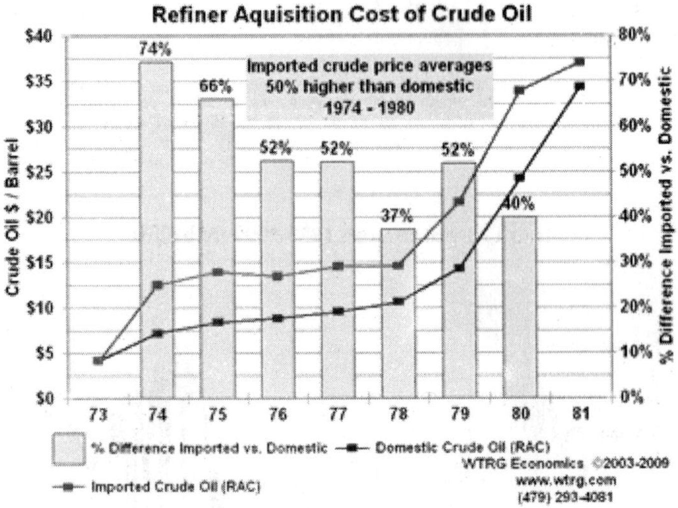

OPEC Control of Crude Oil Prices

OPEC has seldom been effective at controlling prices. While often referred to as a cartel, OPEC does not satisfy the definition. One of the primary requirements is a mechanism to enforce member quotas. The old joke went something like this. What is the difference between OPEC and the Texas Railroad Commission? OPEC doesn't have any Texas Rangers! The only enforcement mechanism that has ever existed in OPEC was Saudi spare capacity.

With enough spare capacity at times to be able to increase production sufficiently to offset the impact of lower prices on its own revenue, Saudi Arabia could enforce discipline by threatening to increase production enough to crash prices. In reality even this was not an OPEC enforcement mechanism unless OPEC's goals coincided with those of Saudi Arabia.

During the 1979-80 period of rapidly increasing prices, Saudi Arabia's oil minister Ahmed Yamani repeatedly warned other members of OPEC that high prices would lead to a reduction in demand. His warnings fell on deaf ears.

Surging prices caused several reactions among consumers: better insulation in new homes, increased insulation in many older homes, more energy efficiency in industrial processes, and automobiles with higher efficiency. These factors along with a global recession caused a reduction in demand which led to falling crude prices. Unfortunately for OPEC only the global recession was temporary. Nobody rushed to remove insulation from their homes or to replace energy efficient plants and equipment - much of the reaction to the oil price increase of the end of the decade was permanent and would never respond to lower prices with increased consumption of oil.

Higher prices also resulted in increased exploration and production outside of OPEC. From 1980 to 1986 non-OPEC production increased 10 million barrels per day. OPEC was faced with lower demand and higher supply from outside the organization.

From 1982 to 1985, OPEC attempted to set production quotas low enough to stabilize prices. These attempts met with repeated failure as various members of OPEC produced beyond their quotas. During most of this period Saudi Arabia acted as the swing producer cutting its production in an attempt to stem the free fall in prices. In August of 1985, the Saudis tired of this role. They linked their oil price to the spot market for crude and by early 1986 increased production from 2 MMBPD to 5 MMBPD. Crude oil prices plummeted below $10 per barrel by mid-1986. Despite the fall in prices Saudi revenue remained about the same with higher volumes compensating for lower prices.

A December 1986 OPEC price accord set to target $18 per barrel bit it was already breaking down by January of 1987and prices remained weak.

The price of crude oil spiked in 1990 with the lower production and uncertainty associated with the Iraqi invasion of Kuwait and the ensuing Gulf War. The world and particularly the Middle East had a much harsher view of Saddam Hussein invading Arab Kuwait than they did Persian Iran. The proximity to the world's largest oil producer helped to shape the reaction.

Following what became known as the Gulf War to liberate Kuwait crude oil prices entered a period of steady decline until in 1994 inflation adjusted prices attained their lowest level since 1973.

The price cycle then turned up. The United States economy was strong and the Asian Pacific region was booming. From 1990 to 1997 world oil consumption increased 6.2 million barrels per day. Asian consumption accounted for all but 300,000 barrels per day of that gain and contributed to a price recovery that extended into 1997. Declining Russian production contributed to the price recovery. Between 1990 and 1996 Russian production declined over 5 million barrels per day.

Non-OPEC Production and Crude Oil Prices (Mbbl/d)

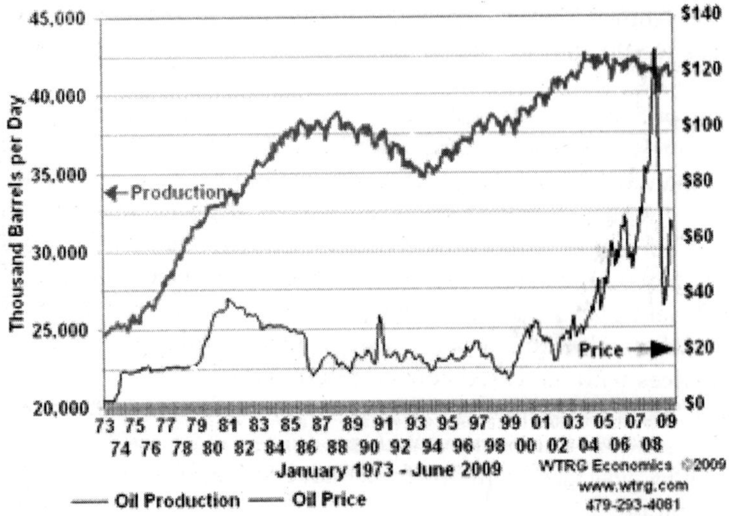

OPEC Production and Crude Oil Prices (Mbbl/d)

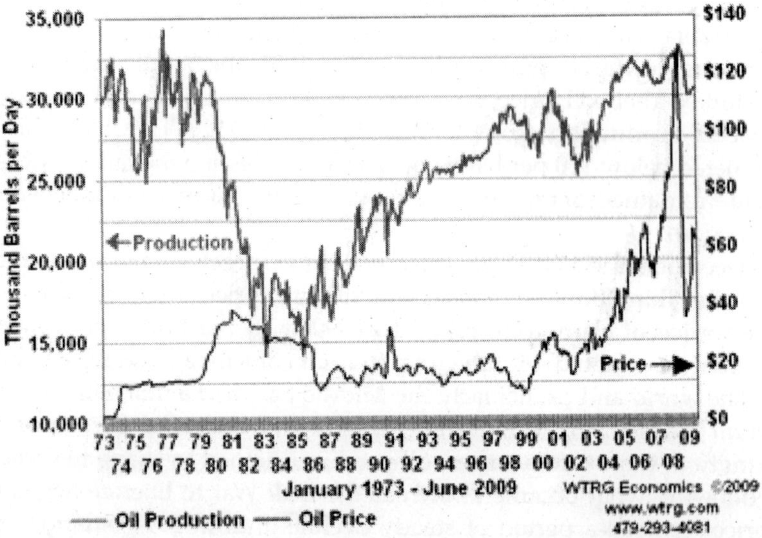

Russian Crude Oil Production

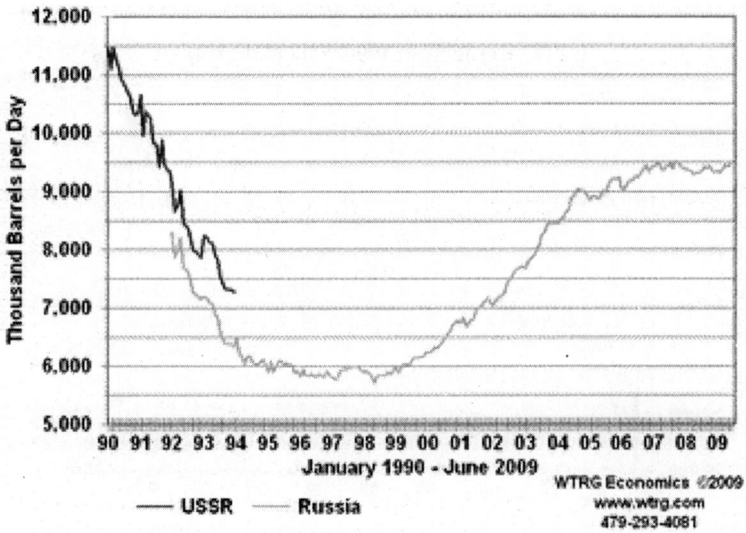

OPEC continued to have mixed success in controlling prices. There were mistakes in timing of quota changes as well as the usual problems in maintaining production discipline among its member countries. The price increases came to a rapid end in 1997 and 1998 when the impact of the economic crisis in Asia was either ignored or severely underestimated by OPEC. In December, 1997 OPEC increased its quota by 2.5 million barrels per day (10 percent) to 27.5 MMBPD effective January 1, 1998. The rapid growth in Asian economies had come to a halt. In 1998 Asian Pacific oil consumption declined for the first time since 1982. The combination of lower consumption and higher OPEC production sent prices into a downward spiral. In response, OPEC cut quotas by 1.25 million b/d in April and another 1.335 million in July. Price continued down through December 1998. Prices began to recover in early 1999 and OPEC reduced production another 1.719 million barrels in April. As usual not all of the quotas were observed but between early 1998 and the middle of 1999 OPEC production dropped by about 3 million barrels per day and was sufficient to move prices above $25 per barrel. With minimal Y2K problems and growing US and world economies the price continued to rise throughout 2000 to a post 1981 high. Between April and October, 2000 three successive OPEC quota increases totaling 3.2 million barrels per day were not able to stem the price increases. Prices finally started down following another quota increase of 500,000 effective November 1, 2000.

Russian production increases dominated non-OPEC production growth from 2000 forward and was responsible for most of the non-OPEC increase since the turn of the century. Once again it appeared that OPEC overshot the mark. In 2001, a weakened US economy and increases in non-OPEC production put downward pressure on prices. In response OPEC once again entered into a series of reductions in member quotas cutting 3.5 million barrels by September 1, 2001. In the absence of the September 11, 2001 terrorist attack this would have been sufficient to moderate or even reverse the trend.

OPEC Production 1990-2007 (Mbbl/d)

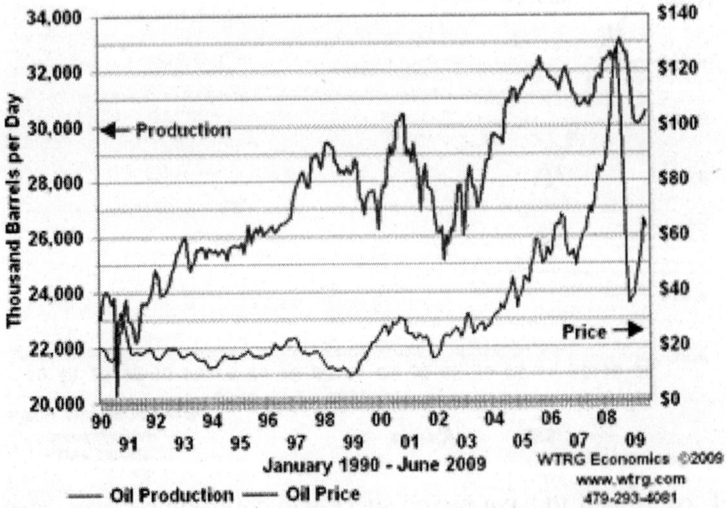

In the wake of the attack crude oil prices plummeted. Spot prices for the U.S. benchmark West Texas Intermediate were down 35 percent by the middle of November. Under normal circumstances a drop in price of this magnitude would have resulted an another round of quota reductions but given the political climate OPEC delayed additional cuts until January 2002. It then reduced its quota by 1.5 million barrels per day and was joined by several non-OPEC producers including Russia who promised combined production cuts of an additional 462,500 barrels. This had the desired effect with oil prices moving into the $25 range by March, 2002. By mid-year the non-OPEC members were restoring their production cuts but prices continued to rise and U.S. inventories reached a 20-year low later in the year.

By year end oversupply was not a problem. Problems in Venezuela led to a strike at PDVSA causing Venezuelan production to plummet. In the wake of the strike Venezuela was never able to restore capacity to its previous level and is still about 900,000 barrels per day below its peak capacity of 3.5 million barrels per day. OPEC increased quotas by 2.8 million barrels per day in January and February, 2003. On March 19, 2003, just as some Venezuelan production was beginning to return, military action commenced in Iraq. Meanwhile, inventories remained low in the U.S. and other OECD countries. With an improving economy U.S. demand was increasing and Asian demand for crude oil was growing at a rapid pace.

The loss of production capacity in Iraq and Venezuela combined with increased OPEC production to meet growing international demand led to the erosion of excess oil production capacity. In mid 2002, there was over 6 million barrels per day of excess production capacity and by mid-2003 the excess was below 2 million. During much of 2004 and 2005 the spare capacity to produce oil was under a million barrels per day. A million barrels per day is not enough spare capacity to cover an interruption of supply from most OPEC producers.

In a world that consumes over 80 million barrels per day of petroleum products that added a significant risk premium to crude oil price and is largely responsible for prices in excess of $40-$50 per barrel. Other major factors contributing to the current level of prices include a weak dollar and the continued rapid growth in Asian economies and their petroleum consumption. The 2005 hurricanes and U.S. refinery problems associated with the conversion from MTBE as an additive to ethanol have contributed to higher prices.

Venezuelan Oil Production (Mbbl/d)

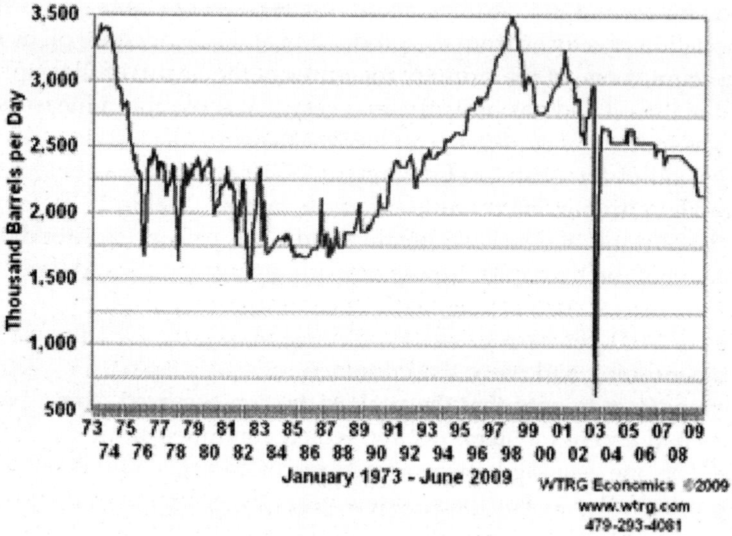

Excess Crude Oil Production Capacity (Mbbl/d)

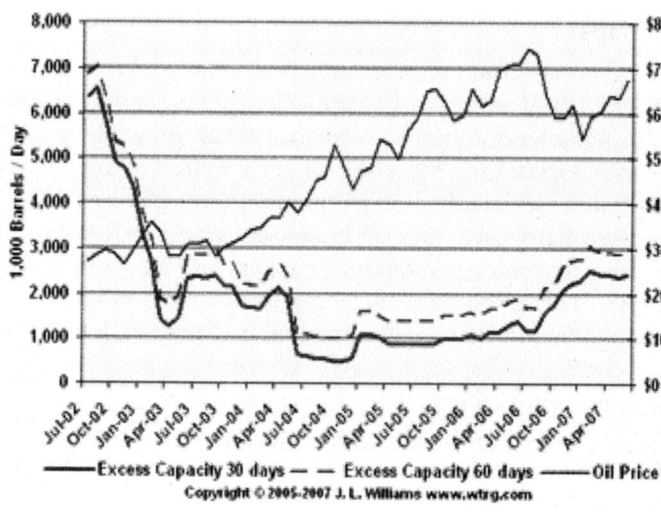

One of the most important factors supporting a high price is the level of petroleum inventories in the U.S. and other consuming countries. Until spare capacity became an issue inventory levels provided an excellent tool for short-term price forecasts. Although not well publicized OPEC has for several years depended on a policy that amounts to world inventory management. Its primary reason for cutting back on production in November, 2006 and again in February, 2007 was concern about growing OECD inventories. Their focus is on total petroleum inventories including crude oil and petroleum products, which are a better indicator of prices that oil inventories alone.

Impact of Prices on Industry Segments

The Rotary Rig Count is the average number of drilling rigs actively exploring for oil and gas. Drilling an oil or gas well is a capital investment in the expectation of returns from the production and sale of crude oil or natural gas. Rig count is one of the primary measures of the health of the exploration segment of the oil and gas industry. In a very real sense it is a measure of the oil and gas industry's confidence in its own future.

At the end of the Arab Oil Embargo in 1974 rig count was below 1500. It rose steadily with regulated crude oil prices to over 2000 in 1979. From 1978 to the beginning of 1981 domestic crude oil prices exploded from a combination of the the rapid growth in world energy prices and deregulation of domestic prices. At that time high prices and forecasts of crude oil prices in excess of $100 per barrel fueled a drilling frenzy. By 1982 the number of rotary rigs running had more than doubled.

It is important to note that the peak in drilling occurred over a year after oil prices had entered a steep decline which continued until the 1986 price collapse. The one year lag between crude prices and rig count disappeared in the 1986 price collapse. For the next few years the economy of the towns and cities in the oil patch was characterized by bankruptcy, bank failures and high unemployment.

After the Collapse

Several trends established were established in the wake of the collapse in crude prices. The lag of over a year for drilling to respond to crude prices is now reduced to a matter of months. (Note that the graph on the right is limited to rigs involved in exploration for crude oil as compared to the previous graph which also included rigs involved in gas exploration.) Like any other industry that goes through hard times the oil business emerged smarter, leaner and more conservative. Industry participants, bankers and investors were far more aware of the risk of price movements. Companies long familiar with accessing geologic, production and management risk added price risk to their decision criteria.

Technological improvements were incorporated:

- Increased use of 3-D seismic data reduced drilling risk.

- Directional and horizontal drilling led to improved production in many reservoirs.
- Financial instruments were used to limit exposure to price movements.
- Increased use of CO_2 floods and improved recovery methods to improve production in existing wells.

In spite of all of these efforts the percentage of rigs employed in drilling for crude oil decreased from over 60 percent of total rigs at the beginning of 1988 to under 15 percent until a recent resurgence.

U.S. Rotary Rig Count 1974-2005
Crude Oil and Natural Gas Drilling

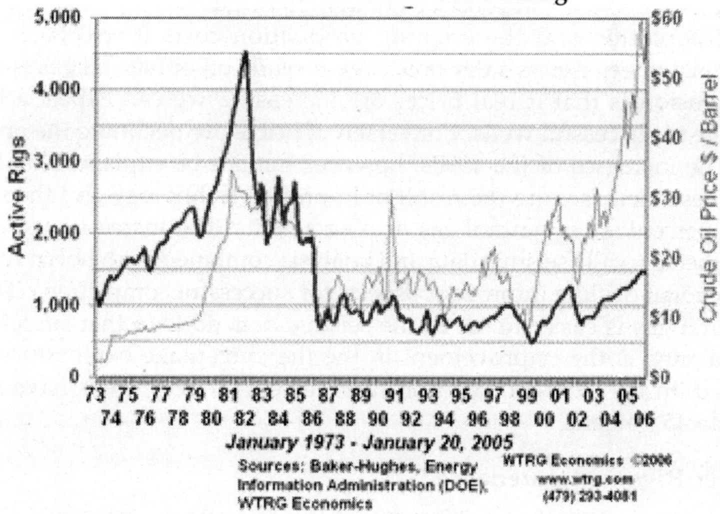

Well Completions – A measure of success?

Rig count does not tell the whole story of oil and gas exploration and development. It is certainly a good measure of activity, but it is not a measure of success. After a well is drilled it is either classified as an oil well, natural gas well or dry hole. The percentage of wells completed as oil or gas wells is frequently used as a measure of success. In fact, this percentage is often referred to as the success rate. Immediately after World War II 65 percent of the wells drilled were completed as oil or gas wells. This percentage declined to about 57 percent by the end of the 1960s. It rose steadily during the 1970s to reach 70 percent at the end of that decade. This was followed by a plateau or modest decline through most of the 1980s. Beginning in 1990 shortly after the harsh lessons of the price collapse completion rates increased dramatically to 77 percent. What was the reason for the dramatic increase? For that matter, what was the cause of the steady drop in the 1950s and 1960s or the reversal in the 1970s? Since the percentage completion rates are much lower for the more risky exploratory wells, a shift in emphasis away from development would result in lower overall completion rates. This, however, was not the case. An examination of completion rates for development and exploratory wells shows the same general pattern. The decline was price related as we will explain later. Some would argue that the periods of decline were a result of the fact that every year there is less oil to find. If the industry does not develop better technology and expertise every year, oil and gas completion rates should decline. However, this does will not explain the periods of increase.

The increases of the seventies were more related to price than technology. When a well is drilled, the·fact that oil or gas is found does not mean that the well will be completed as a producing well. The determining factor is economics. If the well can produce enough oil or gas to cover the additional cost of completion and the ongoing production costs it will be put into production. Otherwise, its a dry hole even if crude oil or natural gas is found. The conclusion is that if real prices are increasing we can expect a higher percentage of successful wells. Conversely if prices are declining the opposite is true. The increases of the 1990s, however, cannot be explained by higher prices. These increases are the result of improved technology and the shift to a higher percentage of natural gas drilling activity. The increased use of and improvements to 3-D seismic data and analysis combined with horizontal and and directional drilling improve prospects for successful completions. The fact that natural gas is easier to see in the seismic data adds to that success rate. Most dramatic is the improvement in the the percentage exploratory wells completed. In the 1990s completion rates for exploratory wells have soared from 25 to 45 percent.

Workover Rigs - Maintenance

Workover rig count is a measure of the industry's investment in the maintenance of oil and gas wells. The Baker-Hughes workover rig count includes rigs involved in pulling production tubing, sucker rods and pumps from a well that is 1,500 feet or more in depth. Workover rig count is another measure of the health of the oil and gas industry. A disproportionate percentage of workovers are associated with oil wells. Workover rigs are used to pull tubing for repair or replacement of rods, pumps and tubular goods which are subject to wear and corrosion. A low level of workover activity is particularly worrisome because it is indicative of deferred maintenance. The situation is similar to the aging apartment building that no longer justifies major renovations and is milked as long as it produces a positive cash flow. When operators are in a weak cash position workovers are delayed as long as possible. Workover activity impacts manufacturers of tubing, rods and pumps. Service companies coating pipe and other tubular goods are heavily affected.

Crude Oil Prices in 2008

Oil prices roses again in 2008 with dated Brent averaging $97.26 for the year 2008 was another record breaking year for crude, making it the seventh consecutive year that prices have risen. Dated Brent daily oil prices peaked at $144.22/bbl on July 3rd in 2008 and bottomed out at $33.66/bbl on the 24th of December. The key crudes quoted are Brent, West Texas Intermediate (WTI), Nigerian Focados and Dubai in US$ per barrel.

EIA World Nominal Oil Price Chronology: 1970-2008

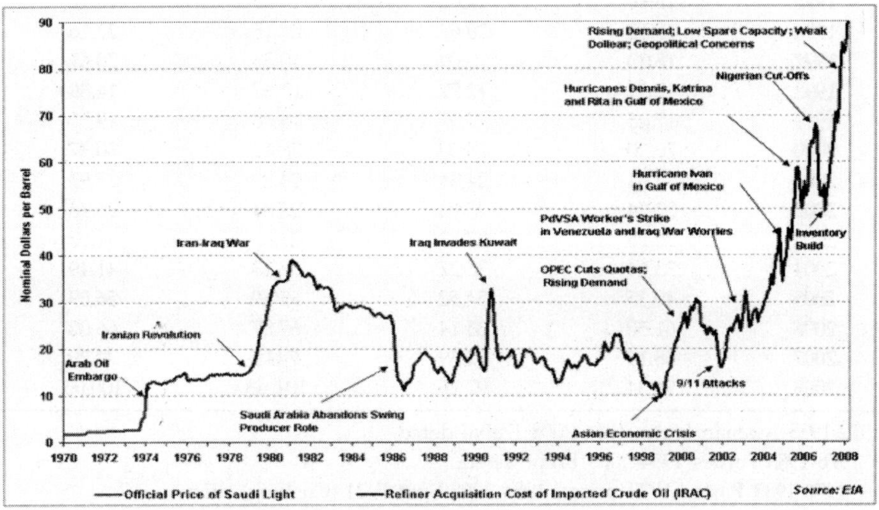

Spot Crude Prices (Source: Platts)

US dollars per barrel	Dubai S/bbl-	Brent s/bbl+	Nigerian Forcados $/bbl	WTI $/bbit
1972	1.90	-	-	-
1973	2.83	-	-	-
1974	10.41	-	-	-
1975	10.70	-	-	-
1976	11.63	12.80	12.87	12.23
1977	12.38	13.92	14.21	14.22
1978	13.03	14.02	13.65	14.55
1979	29.75	31.61	29.25	25.08
1980	35.69	36.83	36.98	37.96
1981	34.32	35.93	36.18	36.08
1982	31.80	32.97	33.29	33.65
1983	28.78	29.55	29.54	30.30
1984	28.06	28.78	28.14	29.39
1985	27.53	27.56	27.75	27.98
1986	13.10	14.43	14.46	15.10
1987	16.95	18.44	18.39	19.18
1988	13.27	14.92	15.00	15.97
1989	15.62	18.23	18.30	19.68
1990	20.45	23.73	23.85	24.50
1991	16.63	20.00	20.11	21.54
1992	17.17	19.32	19.61	20.57

(Contd.)

US dollars per barrel	Dubai S/bbl-	Brent s/bbl+	Nigerian Forcados $/bbl	WTI $/bbit
1993	14.93	16.97	17.41	18.45
1994	14.74	15.82	16.25	17.21
1995	16.10	17.02	17.26	18.42
1996	18.52	20.67	21.16	22.16
1997	18.23	19.09	19.33	20.61
1998	12.21	12.72	12 62	14.39
1999	17.25	17.97	18.00	19.31
2000	26.20	28.50	28.42	30.37
2001	22.81	24.44	24.23	25.93
2002	23.74	25.02	25.04	26.16
2003	26.78	28.83	28.66	31.07
2004	33.64	38.27	38.13	41.49
2005	49.35	54.52	55 69	56.59
2006	61.50	65.14	67.07	66.02
2007	68.19	72.39	7448	72.20
2008	94.34	97.26	101 43	100.05

1972-1985 Arabian Light, 1986-2008 Dubai dated.
+ 1976-1983 Forties 1984-2008 Brent deted.
++ 1976-1983 Posted WTI prices, 1984-2008 Spot WTI (Cushing) price.

Global Petroleum Trade Mechanism

This section looks at petroleum trade and transportation. The main oil producing areas are not the same as the main consuming areas. Hence, oil must be moved from regions where supply is greater than demand - exporting regions – to regions where demand is greater than supply -- importing regions. The "have/have not" balance results in oil flows from one international region to another, from one country to another, and from one region within a country to another. These flows, dictated by economics, logistics, and temporary imbalances in supply and demand, are central to the efficient operation of the oil market.

(a) Regional Importers and Exporters

As one might expect, the world's three largest consuming regions – North America, Europe, and Asia-Pacific – are all importers. All the other regions are exporters. The Middle East still exports vastly more oil than any other region, despite the strong growth in production in other areas in recent years. This global dependence on Middle East oil makes the geo-political importance of the Middle East readily understandable.

There is less variation among the importing regions. In the decade preceding its 1997-98 financial crisis, Asia-Pacific's economic boom propelled it into the Number 1 spot, with import growth more than double that of any other region's. Even though the United States is the largest individual importer, both net and gross, North America as a region ranks third; because Canada and Mexico are two of the United States' three top oil suppliers, their exports to the United States offset U.S. imports from these neighbors in the regional calculation. (Net imports are gross imports minus gross exports). This has kept North America's import dependency down to not much over 30 percent, half that of the Asia-Pacific region at its 1997 peak. The latter's import dependency, and even its import volume, have declined since then, but not by enough to threaten its position as the world's top importing region.

(b) Oil Trade: Highest Volume, Highest Value

There is more trade internationally in oil than in anything else. This is true whether one measures trade by how much of a good is moved (volume), by its value, or by the carrying capacity needed to move it. All measures are important and for different reasons. Volume provides insights about whether markets are over- or under-supplied and whether the infrastructure is adequate to accommodate the required flow. Value allows governments and economists to assess patterns of international trade and balance of trade and balance of payments. Carrying capacity allows the shipping industry to assess how many tankers are required and on what routes. Transportation and storage play a critical additional role here. They are not just the physical link between the importers and the exporters and, therefore, between producers and refiners, refiners and marketers, and marketers and consumers; their associated costs are a primary factor in determining the pattern of world trade.

(c) Distance: The Nearest Market First

Generally, crude oil and petroleum products flow to the markets that provide the highest value to the supplier. Everything else being equal, oil moves to the nearest market first, because that has the lowest transportation cost and therefore provides the supplier with the highest net revenue, or in oil market terminology, the highest *netback*. If this market cannot absorb all the oil, the balance moves to the next closest one, and the next and so on, incurring progressively higher transportation costs, until all the oil is placed. The recent growth in United States dependence on its Western Hemisphere neighbors is an illustration of this "nearer-is-better" syndrome. For instance, Western Hemisphere sources now supply over half the United States import volume, much of it on voyages of less than a week. Another quarter comes from elsewhere in the area called the Atlantic Basin, those countries on both sides of the Atlantic Ocean. This oil, coming especially that which comes from the North Sea and Africa, and takes just 2-3 weeks to reach the United States, boosts the so-called *short-haul* share of U.S. imports to over three-quarters. Most North Sea and North and West African crude oils stay in the Atlantic Basin, moving to Europe or North America on routes that rarely take over 20 days. In contrast, voyage times to Asia for just the nearest of these, the West African crude oils, would be over 30 days to Singapore, rising to nearly 40 for Japan. Not surprisingly, therefore, most of Asia's oil comes instead from the Middle East, only 20-30 days away.

Mexico and Venezuela have consciously helped the trend toward short-haul shipments. They pro-actively took the strategic decision to make as large and as profitable a market as possible for poor quality crudes, since their reserves are unusually biased toward those hard-to-place grades. Both countries therefore targeted their nearest markets, the U.S. Gulf Coast and the Caribbean, for joint venture refinery investments. They began with refineries that had traditionally run their crudes, and then with refineries that might be upgraded to do so. This policy has turned poor quality crudes into the preferred crude at these sites, significantly increasing the crude oil self-sufficiency of the Western Hemisphere.

A change in trade flow patterns can also be of critical importance to the shipping industry. For example, the Suez crisis of 1957 forced tanker owners back to using the much longer route around the Cape of Good Hope, and resulted in the development of Very Large Crude Carriers (VLCC's) to reduce that voyage's higher costs. The shift to short-haul routes in the 1990's was also critical. Using the growth in world trade volumes as a proxy for demand, tanker owners had been expecting a return to a strong tanker market. But the combination of the surge in short haul imports in the Atlantic Basin and the shift of Middle East exports from the longer United States to shorter Asian voyages led to a sharp decline in average voyage length. This decline was accelerated by the return of Iraqi crude exports, many of which move on the extremely short route from the Black Sea end of the Iraq-Turkey pipeline to the Mediterranean. The tanker owners' outlook was thus fading even before world trade volumes were undermined by the Asian crisis.

(d) Quality, Industry Structure and Governments

In practice, trade flows do not always follow the simple "nearest first" pattern. Refinery configurations, product demand mix, product quality specifications, all three of which tie into quality and politics can all change the rankings. Different markets frequently place different values on particular grades of oil. Thus, a low sulfur diesel is worth more in the United States, where the maximum allowable sulfur is 0.05 percent by weight, than in Africa, where the maximum can be 10 to 20 times higher. Similarly, African crudes low in sulfur are worth relatively more in Asia, where they may allow a refiner to meet tighter sulfur limits in the region without investing in refinery upgrades. Such differences in valuing quality can be sufficient to overcome transportation cost disadvantages, as the relatively recent establishment of a significant trade in long-distance African crudes to Asia shows. The cost of moving oil into a particular market can be further distorted from the principle of nearest first by government policies such as tariff.

In addition, both buyers and sellers may impose restrictions. For instance, the United States prohibits the importation of Iranian and Libyan oil, and the United Nations allows only limited sales of Iraqi oil. On the seller's side, Mexico formerly limited sales to the United States to 50 percent of its exports, reflecting concerns about dependence on the United States specifically and about dependence on one geographic market in general. Saudi Arabia's national security concerns, on the other hand, dictate that it maintain a very high profile as a supplier to the United States market, even at the cost of lower netbacks. Indeed, it was the top United States crude supplier in 1999.

(e) Crude versus Products

Crude oil dominates the world oil trade. The risk-weighted economics clearly favor siting refineries close to consumers rather than close to the wellhead. This siting policy takes maximum advantage of the economies of scale of large ships, especially as local quality specifications are increasingly fragmenting the product market. It maximizes the refiner's ability to tailor the product output to the market's short-term surges such as those caused by weather, equipment outages, etc. In addition, this policy also guards against the very real risk that governments will impose selective import restrictions to protect their domestic refining sector.As noted in the section on Refining, there are a limited number of refining centers that are at odds with this general rule, having been developed to serve particular export markets. These export refining centers - Singapore, the Caribbean, and the Middle East - give rise to some regular inter-regional product moves, but they are the exception. The inter-regional products trade is largely a temporary market-balancing function. Some inter-regional flows are extremely short lived, as when extremely cold weather in Europe causes the United States to export heating oil there. A longer-lived example arose when a large proportion of European drivers opted for diesel cars, leaving the region in the late 1990's with surplus capacity to produce gasoline for export to the United States.

(f) Tankers and Pipelines

There are two modes of transportation for inter-regional trade: tankers and pipelines. Tankers have made global (intercontinental) transport of oil possible, and they are low cost, efficient, and extremely flexible. Pipelines, on the other hand, are the mode of choice for transcontinental oil movements. Not all tanker trade routes use the same size ship. Each route usually has one size that is the clear economic winner, based on voyage length, port and canal constraints and volume. Thus, crude exports from the Middle East - high volumes that travel long distances - are moved mainly by Very Large Crude Carriers (VLCC's) typically carrying over 2 million barrels of oil on every voyage. The VLCC's economies of scale outweigh the constraints imposed: they are too large for all the ports in the United States except the Louisiana Offshore Oil Port (LOOP). Thus, they must have some or all of their cargo transferred to smaller vessels, either at sea (lightering) or at an offshore port (transshipment). In contrast, ships out of the Caribbean and South America are routinely smaller and enter ports in the United States directly. Because of such ship size differences, a long voyage can sometimes be cheaper on a per barrel basis than a short one.

Pipelines are critical for landlocked crudes and also complement tankers at certain key locations by relieving bottlenecks or providing shortcuts. The only inter-regional trade that currently relies solely on pipelines is crude from Russia to Europe. Export pipelines are also needed for production from the Caspian Sea region, where the protracted commercial and political debate illustrates the greatest negative for pipelines crossing national boundaries: their political vulnerability.

Pipelines come into their own in intra-regional trade. They are the primary option for transcontinental transportation, because they are at least an order of magnitude cheaper than any alternative such as rail, barge, or road, and because political vulnerability is a small or non-existent issue within a nation's border or between neighbors such as the United States and Canada. (Pipelines are also an important oil transport mode in mainland Europe, although the system is much smaller, matching the shorter distances.)

The development of large diameter pipelines during World War II allowed the development of the vast pipeline network in North America that moves crude oil and product within Canada, from Canada into the United States, and within the United States. Domestically, the 200,000 miles of pipelines account for about two-thirds of all the oil shipments, when adjusted for volume and distance. They move domestic crudes from producing areas like California, the Rockies, and West Texas, and imported crudes from the receiving ports, and transport them to the refining centers. The United States also relies heavily on pipelines to transport petroleum products from refining centers, such as the Gulf Coast, to consuming regions, like the East Coast.

Fungibility is an important factor in transportation economics. Because the oil is broadly interchangeable (fungible), it can be mixed without a significant diminution in value. As environmental mandates have required different regional and seasonal qualities of gasoline, the required batching for transport and segregation for storage has increased substantially. Thus the logistics flexibility inherent in a product's fungibility - the ability to substitute one shipment for another, to exchange between regions, for instance - has disappeared. While this is invisible to consumers during normal times, it contributes to market upheavals and price spikes in times of surprises in demand or supply, as during the early driving season in 2000.

(g) Import Dependency

As already noted, the United States continues to sit at the top of the national rankings of importers, now accounting for around a quarter of total world imports. Yet its import dependency, the percentage of demand met by imports, is significantly lower, at about 50 percent, than that of its international partners. Industrialized countries such as Japan and Germany have import dependency levels of 90-100 percent.

The Middle East has long been regarded as politically unstable and its oil supply, therefore, subject to disruptions. U.S. policy makers have viewed its increased dependence on Western Hemisphere supplies and its decreased dependence on the Middle East as a welcome development. (Again, an international contrast: the Asia-Pacific region now relies on the Middle East for almost 90 percent of its imports and hence 50 percent of its consumption.) As the Middle East holds the majority of the world's oil reserves, increasing U.S. dependence on this region had been viewed as the inevitable partner of import growth. In light of the shift toward Western Hemisphere, short-haul import sources, Saudi Arabia is the only significant Middle East supplier left, and then only because of its willingness to trade security for revenue as noted above. Yet, although U.S. dependence on the long-haul Middle East has fallen sharply, this has not made U.S. prices less vulnerable to a disruption in Middle East supplies. Since oil is a global market, the relevant measure for that vulnerability is not U.S. dependence, but world dependence on Middle East oil – and that has not shrunk.

While interdependence in the Western Hemisphere has been a major United States policy objective, it is important to realize that this combination of size and geographic closeness creates its own short-term vulnerability. Take for instance the case of Hurricane Roxanne, which severely damaged a large part of Mexico's production facilities in the Gulf of Mexico in late 1995. Some 40 million barrels of Mexico's production was eliminated, the vast majority earmarked for refineries along the U.S. Gulf Coast. These refiners, less than a week's sailing time away, had little time to compensate for this sudden hole in their planned supplies.

(h) Crude Oil and Products Mix

Crude oil dominates U.S. imports just as it dominates world trade and for much the same reasons. Therefore, all of the U.S. leading suppliers are major crude producers. Imports of crude oil, having grown to replace declining domestic production and to meet growing demand, now account for around 80 percent of the total. Product import volumes have stayed relatively stable. In spite of the seeming stability in product imports, there have been significant structural shifts over the last couple of decades in the mix of products that the United States imports. Residual fuel oil, for instance, formerly accounted for the majority of all product imports, but its share has shriveled into insignificance as utilities and industrial users have switched to other fuels, particularly nuclear and natural gas. In its place, the United States now imports a much higher proportion of petroleum products that are reprocessed or blended by the oil industry, such as the unfinished gasoline and gasoline blending components that are central to reformulated gasoline supply. Canada is the one country that delivers oil to the United States by pipeline. (Only its relatively new offshore Eastern Canadian production, from fields like Hibernia, depends on tankers.) The vast majority of Canada's crudes are landlocked and rely almost exclusively on trunk-lines from Western Canada that tie into the U.S. transcontinental network to reach their main export markets, which lie all across the Northern Tier of the United States. As domestic production has declined, these Canadian crudes have had a greater reach into the United States.

(i) U.S. Exports

Since the United States is the world's largest importer, it may seem surprising that it also exports around 1 million barrels a day of oil, predominantly petroleum products. Due to various logistical, regulatory, and quality considerations, it turns out that exporting some barrels and replacing them with additional imports is the most economic way to meet the market's needs. For example, the Gulf Coast may export lower quality gasoline to Latin America while the East Coast imports higher quality gasolines from Europe. Exports in the 1990's have been at record highs, the efficiency of the oil market has been increased, and consumers everywhere have benefited.

(j) U.S. Regional Trade

There are significant differences between different parts of the United States in terms of their involvement in and dependence on international trade. Most of these differences are the direct result of the uneven distribution of both production and refining across the United States. The East Coast imports over half of all the products that come to the United States, because it is the largest consuming area in the United States but, for historical reasons, it has only enough capacity to meet around 1/3 of those needs from its own refining. It fills the product gap with supplies from other parts of the United States, particularly the Gulf Coast, and with imports. Its limited volume of refining capacity also keeps it a distant third as a crude importer. However, because its local production is so insignificant, its crude import dependency is the highest of all, at almost 100 percent. The only other region that imports significant amounts of products is the Gulf Coast. Its focus is not, like the East Coast's, on products that can be supplied directly to the consumer, but on refinery feedstock and blend stocks, to support its role as the main U.S. refining and petrochemical center. That role, plus the need for all the Midwest's non-Canadian crude imports to move through the Gulf Coast's ports and pipelines too, has also led to the Gulf Coast being by far the most important crude oil importing region in the United States, accounting for nearly two-thirds of the total.

The trade among regions of the United States is focused on the eastern half of the country. The Midwest and East Coast account for 90 percent of the inter-regional flow, the flow between Petroleum Administration for Defense Districts. The Gulf Coast is by far the largest supplier, accounting for more than 80% of the inter-PADD flow. In contrast, the Rockies and the West Coast are isolated, in petroleum logistics terms, from the rest of the country. The easy flow of petroleum from the Gulf Coast to the Midwest and the East Coast mean that incremental supply is more readily available to those markets in the event of a demand surge or supply drop. In contrast, the West Coast, and the California market in particular, cannot so readily attract incremental supplies. Thus, the California refinery outages that occurred in the Spring of 1999, resulted in a large price increase as market players scrambled for additional supply, none of which was available close at hand, or cheaply. The California market's isolation is more than just geographic: the State imposes unique and stringent quality restrictions on both its gasoline and its off-highway diesel, making what otherwise might be available to augment California product supplies unsuitable.

1.6 STATISTICAL OVERVIEW OF WORLD OIL MARKETS

Oil Production by Region

Global oil production increased by 380,000 b/d in 2008, or 0.4%, to 81.8 mb/d in 2008. OPEC production increased by 990,000 b/d despite production cuts instituted late in the year, Saudi Arabia saw the largest production increase, with output rising by 400,000 b/d. Russian production fell by 90,000 b/d, the first decline since 1998. OECD production fell by 750,000 b/d, with Mexico registering the world's largest decline (310,000 b/d).

Crude oil production data includes crude oil, shale oil, oil sands and NGLs (natural gas liquids – the liquid content of natural gas where this is recovered separately). It excludes liquid fuels from other sources such as biomass and coal derivatives.

Reserves-to-production (R/P) ratios are available for the world and by region and feature in the Energy charting tool. R/P ratios represent the length of time that those remaining reserves would last if production were to continue at the previous year's rate. It is calculated by dividing remaining reserves at the end of the year by the production in that year. World oil production tables are in both thousand barrels daily and million tonnes.

Global oil production increased by 865,000 b/d in February from 83.82 to 84.69 million barrels per day according to the latest oil market report of the International Energy Agency. Most of which was interestingly not produced by OPEC but by the Non-OPEC countries Canada, USA, United Kingdom, China, Brazil, and Kazakhstan. OPEC only contributed 200,000 b/d of the total February production increase.

Since OPEC cut its production in 2008 the cartel has not brought any of its spare capacity on the oil market. OPEC production in the first two months of 2010 was only 1.14 million barrels per day higher than in 2009, at a level of 34.27 million versus 33.13 million b/d. In the same period total production capacity, measured as production + surplus production capacity as documented by the International Energy Agency, increased by 1.85 million b/d from 37.9 to 39.8 million b/d.

The lack of a response from OPEC to oil prices which have been consistently rising since early February, now at a level near 80 dollars per barrel, is telling of the conservative nature of the cartel. In the summary of OPEC's recent meeting in Vienna on 17 March it was concluded that "*although world oil demand is projected to increase marginally during the year, this rise will be more than offset by the expected increase in non-OPEC supply...in light of the foregoing, the Conference again decided to maintain the current oil production ceiling unchanged**." Surprisingly these statements are contradicted by OPEC's own

Production (Million tones)

Country	1998	1999	2000	2001	2002	2003	2004	2005	2006	2007	2008	Change 2008 over 2007	2008 share of total
US	368.1	352.6	352.6	349.2	346.8	338.4	329.2	313.3	310.2	309.8	305.1	-1.8%	7.8%
Canada	125.1	121.0	126.9	126.1	135.0	142.6	147.6	144,9	153.4	159.5	156.7	-2.0%	4.0%
Mexico	173.5	165.2	171.2	176.6	178.4	188.8	190.7	187.1	183.1	172.7	157.4	-9.1%	4.0%
Total North America	**666.7**	**638.8**	**650.8**	**651.8**	**660.2**	**669.8**	**667.4**	**645.3**	**646.7**	**642.0**	**619.2**	**-3.8%**	**15.8%**
Argentina	44.0	41.8	40.4	41.5	40 9	40.2	37.8	36 2	35 8	34.9	34.1	-2.6%	0.9%
Brazil	49.8	56.3	63.2	66.3	74.4	77.0	76.5	84.6	89.2	90.4	93.9	3.6%	2.4%
Colombia	38.5	41.6	35.3	31 0	29 7	27.9	27.3	27.3	27.5	27.6	30.5	10.3%	0.8%
Ecuador	19.6	19.5	20.9	21.2	20.4	21.7	27.3	27.6	27.7	26.5	26.2	-1.2%	0.7%
Peru	5.7	5.2	4.9	4.8	4 8	4.5	4.4	5.0	5 1	5 1	5.3	4.0%	0.1%
Trinidad & Tobago	6.7	6.9	6.8	6.5	7.5	7.9	7.3	8.3	8.3	7.2	6.9	-5.0%	0.2%
Venezuela	179.6	160.9	167.3	161.6	148.8	131.4	150.0	151.0	144.2	133.9	131.6	-1.9%	3.4%
Other S & Cent. America	6 2	6.2	6.6	6.9	7.8	7 8	7.3	7.2	7.1	7.2	7.0	-2 7%	0.2%
Total S & Oat America	350.0	338.4	345.3	33.9.9	334.2	318.3	337.9	347.1	345.0	332.7	335.6	0.6%	0.6%
Azerbaijan	11.4	13 9	14.1	15.0	15.4	15 5	15.6	22 4	32 5	42.8	44.7	4.2%	1.1%
Denmark	11.6	14.6	17.7	17.0	18.1	17.9	19.1	18.4	16.7	15.2	14.0	-7.7%	0.4%
Italy	5.6	5.0	4.6	4.1	5.5	5.6	5.5	6.1	5.8	5.9	5.2	-10.9%	0.1%
Kazakhstan	25.9	30.1	35.3	40.1	48.2	52.4	60.6	62.6	66.1	68.4	72.0	5.1%	1.8%
Norway	149.6	149.7	160.2	162.0	157.3	153.0	149.9	138.2	128.7	118.8	114.2	-4.1%	2.9%
Romania	6.6	6.4	6.3	6.2	6.1	5.9	5.7	5.4	5.0	4.7	4.7	-0.4%	0.1%
Russian Federation	304.3	304.8	323.3	348.1	379.6	421.4	458.8	470.0	480.5	491.3	488.5	-0.8%	12.4%
Turkmenistan	6.4	7.1	7.2	8.0	9.0	10.0	9.6	9.5	9.2	9.8	10.2	3.8%	0.3%
United Kingdom	132.6	137.4	126.2	116.7	115.9	106.1	95.4	84.7	76.6	76.8	72.2	-6.3%	1.8%
Uzbekistan	8.2	8.1	7.5	7.2	7.2	7.1	6 6	5.4	5.4	4 9	4.8	-2.7%	0.1%

Other Europe & Eurasia	24 2	22 7	22.4	22 2	23 6	24 0	23.4	21.9	21.6	21 4	20.3	-5.3%	0.5%
Total Europe & Eurasia	686.5	699.6	724.7	74.6	786.0	818.9	850.1	844.8	848.0	860.0	851.0	-1.3%	21.7%
Iran	190.8	178.1	189.4	186.5	172.7	203.7	207.9	206.2	208.0	209.7	209.8	-0.2%	5.3%
Iraq	104.2	128.3	128.8	123.9	104.0	66.1	100.0	90.0	98.1	105.3	119.3	13.0%	3.0%
Kuwait	110.0	102.6	109.1	105.8	98.2	114.8	122.3	129.3	132.7	129.9	137.3	5.3%	3.5%
Oman	44.7	45.0	47.6	47.5	44.5	40.7	38.9	38.6	36.9	34.6	36.0	3.7%	0.9%
Qatar	33.6	34.3	36.1	35.7	35.2	40.8	46.0	47.3	50.9	53.6	60.8	13.2%	1.5%
Saudi Arabia	455.7	423.6	456.3	440.6	425.3	485.1	506.0	526.8	514.3	494.2	515.3	4.0%	13.1%
Syria	28.6	28.8	27.3	28 9	27.2	26 2	24.7	22.4	21.6	20.6	19.8	-4.1%	0.5%
United Arab Emirates	123.5	117.4	123.1	118.2	108.4	122.2	124.7	129.0	139.0	136.4	139.5	2.0%	3.6%
Yemen	18.0	19.2	21.3	21.5	21.5	21.1	19.9	19.6	17.9	16.3	14.4	-11.6%	0.4%
Other Middle East	2.3	2.2	2.2	2.2	2.2	2.2	2.2	1.6	1.4	1.6	1.5	-5 7%	-
Total Middle East	**1111.3**	**1079.4**	**1141.2**	**1110.8**	**1039.4**	**1123.0**	**1192.7**	**1210.8**	**1221.0**	**1202.2**	**1253.7**	**4.0%**	**31.9%**
Algeria	61 8	63 9	66.8	65.8	70.9	79.0	83.6	86 4	86.2	85 5	85.6	-1.3%	2.2%
Angola	36.0	36.7	36.9	36.6	44.6	42.5	48.2	61.2	69.7	84.3	92.2	9.1%	2.3%
Cameroon	5.3	4.8	4.5	4.1	3.6	3.4	4.5	4.2	4.4	4.2	4.3	2.3%	0.1%
Chad	-	-	-	-	-	1.2	8.8	9.1	8.0	7.5	6.7	-11.5%	0.2%
Republic of Congo (Brazzaville)	13.6	13.7	13.1	12.1	11.9	11.1	11.2	12.7	13.5	11.5	12.9	12.3%	0.3%
Egypt	43.0	41.4	38.8	37.3	37 0	36.8	35.4	33.9	33.7	34.1	34.6	1.3%	0.9%
Equatorial Guinea	4.1	5.0	4.5	8.8	10.1	12.0	17.1	18 5	17.7	18.2	17.9	-2.1%	0.5%
Gabon	16.8	17.0	16.4	15.0	14.7	12.0	11.8	11.7	11.7	11.5	11.8	2.2%	0.3%
Libya	69.6	67.0	69.5	67.1	64.6	69.8	76.6	82.1	85.6	86.0	86.2	-0.1%	2.2%
Nigeria	106.0	100.8	105.4	110.8	102.3	110.3	121.9	125.4	120.0	114.2	105.3	-8.0%	2 7%
Sudan	0.6	3.1	8.6	10.7	11.9	13.1	14.9	15.0	16.3	23.1	23.7	2.6%	0.6%
Tunisia	4.0	4.0	3.7	3.4	3.5	3.2	3.4	3.4	3.3	4.6	4.2	-8.9%	0.1%
Other Africa	3.0	2.7	2.7	2.5	3.0	3.5	3.6	3.5	3 2	2.9	2.7	-8.5%	0 1 %
Total Africa	363.9	360.0	370.9	374.5	318.1	397.6	440.9	467.2	473.3	488.5	488.1	-0.4%	12.4%

(*Contd.*)

Table (*Contd.*)

Australia	27.4	26 7	35.3	31.8	31.5	26.6	24.8	24.5	23 4	24.1	23.8	-1.5%	0.6%
Brunei	7.6	8.9	9.4	9.9	10.2	10.5	10.3	10.1	10.8	9.5	8.5	-10.1%	0.2%
China	160.2	160.2	162.6	164.8	166.9	169.6	174.1	180.8	183.7	186.7	189.7	1.4%	4.8%
India	34.7	34.6	34.2	34.1	35 2	35.4	36.3	34.6	35.8	36.2	36.1	-0.5%	0.9%
Indonesia	74.2	68.6	71.5	68.0	63.0	57.7	55.2	53.0	49.9	47 4	49.1	3.2%	1.2%
Malaysia	35.7	33.9	33 7	32.9	34.5	35.6	36.5	33 9	32 5	33 6	34.3	1.8%	0.9%
Thailand	5.0	5.4	7.0	7.5	8 2	9.6	9.1	10.8	11.8	12.7	13.4	5.5%	0 3%
Vietnam	12.1	14.6	16 2	17.1	17.3	17.7	20.8	19.4	17.8	16.4	15.4	-6.0%	0.4%
Other Asia Pacific	10.1	10.2	9.4	9.1	9.0	9.1	8.7	9.3	9.3	10.5	10.9	3.1%	0.3%
Total Asia pacific	367.1	363.1	379.2	375.2	375.7	371.8	375.8	376.4	375.0	377.0	381.2	0.9%	9.7%
Total World	3545.4	3479.3	3612.1	3598.4	3573.6	3699.6	3864.8	3891.6	3908.8	3902.3	3902.3	0.4%	100.9%
of which:													
European Union	168.5	174.8	166.3	155.6	158.2	148.2	137.7	125.7	114 6	113 1	105.9	-6.6%	2.7%
OECD	1011.5	988.9	1011.1	999.5	1005.3	995.6	977.9	931.8	912.0	898.0 :	864.1	-4.0%	22.0%
OPEC	1564.4	1501.5	1581.0	1541.6	1458.4	1545.1	1669.7	1715.4	1726.3	1707.8	1758.3	2.7%	44.8%
Non-OPECJ	1618.5	1607.8	1637.7	1632.2	1649.0	1640.9	1636.6	1599 0	1581.7	1570.3	1543.6	-2.0%	39.3%
Former Soviet Union	362.5	370.0	393.4	424.6	466.2	513 6	558.5	577 1	600.8	624.2	627.0	o.2'v:	16.0%

*Includes crude oil, shale oil, oil sands and NGLs (the liquid content of natural gas where this is recovered separately) Excludes liquid fuels from other sources such as biomass and coal denvatives

++ Excludes Former Soviet Union ¨Less than 0.05%.

Note: Growth rates are adjusted for leap years.

Production* (Thousand barrels daily)

Country	1998	1999	2000	2001	2002	2003	2004	2005	2006	2007	2008	Change 2008 over 2007	2008 share of total
US	8011	7731	7733	7669	7626	7400	7228	6895	6841	6847	6736	-1.8%	7.8%
Canada	2672	2604	2721	2677	2858	3004	3085	3041	3208	3320	3238	-2.0%	4.0%
Mexico	3499	3343	3450	3560	3585	3789	3824	3760	3683	3471	3157	-9.1%	4.0%
Total North America	14162	13678	13904	13906	14069	14193	14137	13696	13732	13638	13131	-3.8%	15.8%
Argentina	890	847	819	830	818	806	754	725	716	699	682	-2.6%	0.9%
Brazil	1003	1133	1268	1337	1499	1555	1542	1716	1809	1833	1899	3.6%	2.4%
Colombia	775	838	711	627	601	564	551	554	559	561	618	10.3%	0.8%
Ecuador	385	383	409	416	401	427	535	541	545	520	514	-1.2%	0.7%
Peru	116	107	100	98	98	92	94	111	116	114	120	4.0%	0.1%
Trinidad & Tobago	134	141	138	135	155	164	152	171	174	154	149	-5.0%	0.2%
Venezuela	3480	3126	3239	3142	2895	2554	2907	2937	2808	2613	2566	-1.9%	3.4%
Other S. & Cent. America	125	124	130	137	152	153	144	143	141	143	138	-2.7%	0.2%
Total S. & Cent America	6908	6699	6813	6722	6619	6314	6680	6899	6866	6636	6686	0.6%	8.5%
Azerbaijan	231	279	282	301	31-!	313	315	452	654	869	914	4.2%	1.1%
Denmark	238	299	363	348	371	368	390	377	342	311	287	-7.7%	0.4%
Italy	117	104	95	86	115	116	113	127	120	122	108	-10.9%	0.1%
Kazakhstan	537	631	744	836	1018	1111	1297	1356	1426	1484	1554	5.1%	1.8%
Norway	3138	3139	3346	3418	3333	3264	3189	2969	2779	2556	2455	-4.1%	2.9%
Romania	137	133	131	130	127	123	119	114	105	99	99	-0.4%	0.1%
Russian Federation	6169	6178	6536	7056	7698	8544	9287	9552	9769	9978	9886	-0.8%	12.4%
Turkmenistan	129	143	144	162	182	202	193	192	186	198	205	3.8%	0.3%

(*Contd.*)

Table (*Contd.*)

United Kingdom	2807	2909	2667	2476	2463	2257	2028	1809	1636	1638	1544	-6.3%	1.8%
Uzbekistan	191	191	177	171	171	166	152	126	125	114	111	-2.7%	0.1%
Other Europe & Eurasia	506	474	465	465	50"	509	496	468	457	451	427	-5.3%	0.5%
Total Europe & Eurasia	14199	14480	14950	15450	16289	16973	17579	17541	17598	17819	17591	-1.3%	21.7%
Iran	3855	3603	3818	3794	3543	4183	4248	4233	4282	4322	4325	-0.2%	5.3%
Iraq	2121	2610	2614	2523	2116	1344	2030	1833	1999	2144	2423	13.0%	3.0%
Kuwait	2232	2085	2206	2148	1995	2329	2475	2618	2690	2636	2784	5.3%	3.5%
Oman	905	911	959	961	900	824	785	782	747	701	728	3.7%	0.9%
Qatar	701	723	757	754	764	879	992	1028	1110	1197	1378	13.2%	1.5%
Saudi Arabia	9502	8853	9491	9209	8928	10164	10638	11114	10853	10449	10846	4.0%	13.1%
Syria	576	579	548	581	548	527	495	450	435	415	398	-4.1%	0.5%
United Arab Emirates	2643	2511	2626	2534	2324	2611	2656	2753	2971	2925	2980	2.0%	3.6%
Yemen	380	405	450	455	457	448	420	416	380	345	305	-11.6%	0.4%
Other Middle East	49	48	48	47	48	48	48	34	32	35	33	-5.7%	?
Total Middle East	22984	22328	23516	23006	21623	23357	24788	25262	25499	25168	26200	4.0%	31 9%
Algeria	1461	1515	1578	1562	1680	1852	1946	2015	2003	2016	1993	-1.3%	2.2%
Angola	731	745	746	742	905	862	976	1246	1421	1720	1875	9.1%	2.3%
Cameroon	105	95	88	81	72	67	89	82	87	82	84	2.3%	0.1%
Chad	-	-	-	-	-	24	168	173	153	144	127	-11.5%	0.2%
Republic of Congo (Brazzaville)	264	266	254	234	231	215	216	246	262	222	249	12.3%	0.3%
Egypt	857	827	781	758	751	749	721	696	697	710	722	1.3%	0.9%
Equatorial Guinea	83	100	91	177	204	242	345	373	358	368	361	-2.1%	0.5%
Gabon	337	340	327	301	295	240	235	234	235	230	235	2.2%	0.3%
Libya	1480	1425	1475	1427	1375	1485	1624	1751	1834	1848	1846	-0.1%	2.2%
Nigeria	2167	2066	2155	2274	2103	2263	2502	2580	2474	2356	2170	-8.0%	2.7%
Sudan	12	63	174	217	241	265	301	305	331	468	480	2.6%	0.6%

Tunisia	85	84	78	71	74	68	71	73	70	97	89	-8.9%	0.1%
Other Africa	63	56	56	53	63	71	75	72	66	59	54	-8.5%	0.1%
Total Africa	7644	7563	7804	7697	7994	S402	9268	9846	9992	10320	10285	-0.4%	12.4%
Australia	644	625	809	733	730	624	582	580	554	567	556	-1.5%	0.6%
Brunei	157	182	193	203	210	214	210	206	221	194	175	-10.1%	0.2%
China	3212	3213	3252	3306	3346	3401	3481	3627	3684	3743	3795	1.4%	4.8%
India	737	736	726	727	753	756	773	738	762	770	766	-0.5%	0.9%
Indonesia	1520	1408	1456	1389	1289	1183	1129	1087	1017	969	1004	3.2%	1.2%
Malaysia	779	737	735	719	757	776	793	744	717	743	754	1.8%	0.9%
Thailand	130	140	176	191	204	236	223	265	286	309	325	5.5%	0.3%
Vietnam	245	296	328	350	354	364	427	398	367	337	317	-6.0%	0.4%
Other Asia Pacific	217	218	200	195	193	195	186	201	203	229	237	3.1%	0.3%
Total Asia Pacific	764 I	7556	7874	7813	7336	7750	7804	7845	7810	7862	7928	0 9%	9.7%
Total World	73538	72325	74861	74794	74431	76990	80256	81089	81497	81443	81820	0.4%	100.0%
of which:													
European Union	3553	3684	3493	3285	3339	3128	2902	2659	2422	2388	2239	-6.6%	2.7%
OECD	21500	21103	21521	21303	21430	21165	20766	19861	19458	19148	18400	-4.0%	22.0%
OPEC	32277	31054	32569	31914	30318	32136	34658	35736	36007	35714	36705	2.7%	44.8%
Non-OPECJ	33870	33719	34278	34220	34580	34355	34191	33513	33171	32930	32295	-2.0%	39.3%
Former Soviet Union	7391	7552	8014	8660	9533	10499	11407	11839	12318	12799	12821	0.2%	16.0%

Includes crude oil, shale oil, oil sands and NGLs (the liquid content of natural gas where this is recovered separately] Excludes liquid fuels from other sources such as biomass and coal derivatives.

+ Excludes Former Soviet Union. ? Less than 0.05%.

Notes: Annual changes and shares of total are calculated using million tonnes per annum figures. Growth rates are adjusted for leap years.

Source: BP Statistical Review 2009.

figures. In the OPEC March monthly oil market report a total global demand increase of 0.88 million b/d is forecast from 84.36 in 2009 to 85.24 million b/d in 2010. In the same report Non-OPEC supply is expected to increase by 0.41 million b/d from 51.02 in 2009 to 51.43 million b/d in 2010. Showing that non-OPEC is unable to meet increasing demand according to OPEC's own analysts.

In light of these numbers from the OPEC oil market report, the overstatement of non-OPEC supply at the OPEC members meeting can be seen as an excuse to stave off adding additional supply to the market for as long as possible, as OPEC prefers to keep prices at a level of at least 75 dollars per barrel.

- **World oil production capacity:** Total oil production capacity in February 2010 increased by 785,000 b/d from January 2010, from 89.36 to 90.15 million b/d. World production capacity is measured here as the sum of world liquids production excluding biofuels plus total OPEC spare capacity excluding Iraq, Venezuela and Nigeria.
- **World crude oil production:** Latest figures from the Energy Information Administration (EIA) show that crude oil production including lease condensates decreased by 184,000 b/d from November to December 2009. Resulting in total production of crude oil including lease condensates of 73.04 million b/d.
- **OPEC crude oil production:** Total crude oil production excluding lease condensates of the OPEC cartel increased by 200,000 b/d to a level of 29.24 million b/d, from January to February 2010, according to the latest available estimate of the IEA. Average crude oil production in 2009 was 28.7 million b/d, versus 31.43 and 30.37 million b/d in respectively 2008 and 2007.
- **Non-OPEC crude oil production:** Total Non-OPEC crude oil production excluding lease condensates decreased by 79,000 b/d to a level of 42.19 million b/d, from November to December 2009, according to the latest available estimate of the EIA. Crude oil production for November 2009 was revised upwards in the EIA International Petroleum Monthly of March from 41.93 to 42.27 million b/d. Average crude oil production in 2009 was 41.62 million b/d, versus 41.32 and 41.80 million b/d in respectively 2008 and 2007.

Oil Consumption by Region

As ever, the world economy is the key driver of energy consumption. It is easy to forget that until the middle of 2008, the economy continued to grow. In retrospect, last year represented the end of one of the strongest periods of economic growth ever recorded. However, the economy had already started to slow, most likely not unrelated to the high price of energy, and the financial crisis in September then triggered a sharp recession – with critical implications for global energy consumption. Energy prices followed these economic headlines, making for a year of very different parts. Oil prices increased steadily early in the year, exceeding $140 per barrel in early July - a record even on an inflation-adjusted basis. But then prices collapsed, falling by more than 70% by the end of the year. Market prices for natural gas and coal followed similar trajectories. Over the whole of 2008, average prices for all forms of primary energy increased significantly, with annual oil prices rising for a seventh consecutive year, a first in the nearly 150-year history of the oil industry. Primary energy consumption growth slowed in 2008, as did growth for each of the fossil fuels. All the net growth in energy consumption came from the rapidly industrializing non-OECD economies, with China alone accounting for nearly three-quarters of global growth. For the first time, non-OECD energy consumption surpassed OECD consumption. For a sixth consecutive year, coal was the fastest-growing fuel with obvious implications for global CO_2 emissions. The use of renewable fuels again rose rapidly, often benefi tting from government support. Although renewable energy continues to play only a small role in the world's energy mix, the share is rising rapidly in some countries and there are the beginnings of a material impact. In 2008 the world was no longer supply-constrained, as production growth exceeded that of consumption for all fossil fuels, particularly later in the year. Expanded OPEC production drove increases in world oil supply, even as consumption declined. The cost-effective development of unconventional gas, enabled by technological innovation, drove the largest-ever increase in US natural gas supply, and for coal, strong growth in China was once again a key driver. Seen in this context, fundamental market forces help to explain the downward pressure seen on energy prices later in the year. BP data confirms that the world has enough proved reserves of oil, natural gas and coal to meet the world's needs for decades to come. The challenges the world faces in growing supplies to meet future demand are not below ground, they are above ground. They are human, not geological.

Global energy consumption growth slowed in 2008, with slower economic growth and higher average prices breaking a string of five consecutive years of above-average growth. Production growth exceeded that of consumption for all fossil fuels. For the year as a whole, prices for all forms of traded energy rose substantially despite sharp declines late in the year. Oil prices reached inflation-adjusted record highs, rising for a seventh consecutive year. Internationally traded coal prices rose more steeply than other fuels. Natural gas prices in Europe and for Asia-Pacific liquefied natural gas (LNG) rose in line with or more rapidly than oil prices, while North American price increases were more moderate. Non-OECD primary energy consumption exceeded OECD consumption for the first time. Primary energy consumption in china is +7.2% and observed –2.8% in US as per BP Data.

World total primary energy consumption is grew by 1.4% in 2008, the slowest growth since 2001. Non-OECD primary energy consumption exceeded OECD consumption for the first time. The Asia-Pacific region accounted for 87% of the world's energy consumption growth. For the third consecutive year, coal accounted for the majority of primary energy consumption growth.

Chinese consumption growth slowed for the fifth consecutive year, yet China accounted for nearly three-quarters of global growth. Energy consumption among exporting regions remained robust, with above-average growth in the Middle East and Africa. Consumption in the US fell by 2.8%, the largest decline since 1982.

(In millions barrels per day)

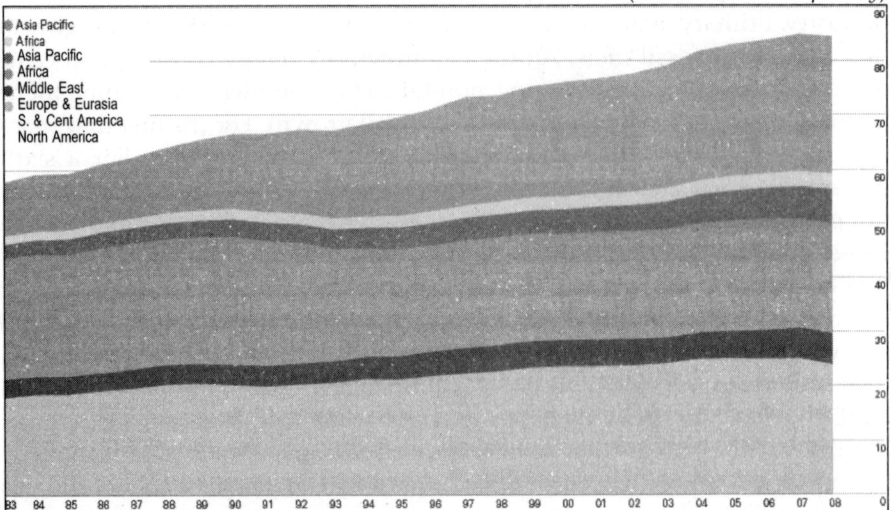

World oil consumption fell by 420,000b/d, the largest decline since 1982. OECD consumption fell by 1.5 million b/d, driven by a decline of nearly 1.3 million b/d in the US. China again recorded the world's largest incremental growth, rising by 260,000b/d. Consumption growth was above the 10-year average in the exporting regions of the Middle East, South and Central America, Africa and the Former Soviet Union.

Source: BP Statistical Review 2009.

World oil consumption fell by 420,000 b/d, the largest decline since 1982.OECD consumption fell by 1.5 million b/d, driven by a decline of nearly 1.3 million b/d in the US. China again recorded the world's largest incremental growth, rising by 260,000 b/d. Consumption growth was above the 10-year average in the exporting regions of the Middle East, South and Central America, Africa, and the FSU. Oil consumption data includes inland demand plus international aviation, marine bunkers and oil products consumed in the refining process. Consumption of fuel ethanol and biodiesel is also included. Annual changes and shares of total are calculated using million tonnes per annum figures rather than thousand barrels daily.

Dated Brent averaged $97.26 per barrel in 2008, an increase of 34%. Other benchmark crudes registered similar gains. Prices began the year just below $100 and rose sharply through mid-year, peaking above $144 in early July. But with strong OPEC production growth and the world economy - and oil consumption - weakening, inventories increased and prices declined sharply thereafter, ending the year below $40. An easing of refinery bottlenecks led to a narrowing of crude quality differentials in the second half of the year. Global oil consumption declined by 0.6%, or by 420,000 barrels per day (b/d), the fi rst decline since 1993 and the largest decline since 1982. Consumption in OECD countries fell by 3.2% (1.5 million b/d), a third consecutive year of decline led by a drop of 6.4%, or nearly 1.3 million b/d, in the US. Outside the OECD, consumption growth slowed to 1.1 million b/d, with continued strong growth among oil exporting countries partly offset by slower growth among Asia-Pacific importers. Oil production outside OPEC fell by 1.4% or 610,000b/d, the largest decline since 1992.While oil consumption declined, global oil production increased by 0.4%, or 380,000b/d. Although OPEC began cutting production late in the year, average annual production rose by 2.7%, or 990,000b/d. Middle Eastern OPEC members accounted for all of the net increase, with Saudi Arabian production rising by nearly 400,000b/d and Iraqi output rising by 280,000b/d. Oil production outside OPEC fell by 1.4% or 610,000b/d, the largest decline since 1992.

OECD production fell by 4%, or 750,000b/d, with declines in North America and Europe; Mexican production fell by 310,000b/d, the world's largest decline. Russian production fell by 90,000b/d, the first decline since 1998. These declines were partly offset by continued growth in Brazilian output of 70,000b/d. Refining capacity additions in 2008 totaled some 830,000b/d, with most of the additions concentrated in the Asia-Pacific region, especially China where capacity grew by 2.9% (220,000b/d). However, global crude runs fell by 250,000b/d because of lower oil consumption. The biggest throughput changes were a fall of 3.4% (510,000b/d) in the US, partially offset by growth of 4.4% (290,000b/d) in China. The combination of increasing refining capacity and lower crude runs meant that global refinery utilization in 2008 fell for the third year running to 84.8%, the lowest level since 2003. Global oil trade fell by 1.7% or 930,000b/d, the first decline since 2002 and the largest since 1987, driven by declining consumption among importing countries (especially in the OECD) and rising domestic consumption among the world's oil exporters. Global oil consumption first time decline by -0.6% since 1993, as per BP Data.

- **OPEC oil consumption** – Oil consumption in all OPEC oil producers combined decreased by 48,000 b/d from November to December 2009. Resulting in a consumption level of 5.77 million b/d. Average OPEC oil consumption in 2009 was 5.94 million b/d, versus 5.76 and 5.30 million b/d in respectively 2008 and 2007.
- **Iran oil consumption** – Oil consumption in Iran increased by 11,000 b/d from November to December 2009 to a level of 1.54 million b/d. Average Iranian oil consumption in 2009 was 1.52 million b/d, versus 1.64 and 1.52 million b/d in respectively 2008 and 2007.
- **Saudi Arabia oil consumption** – Oil consumption in Saudi Arabia decreased by 129,000 b/d from November to December 2009 to a level of 1.59 million b/d. Average Saudi Arabian oil consumption in 2009 was 1.82 million b/d, versus 1.65 and 1.52 million b/d in respectively 2008 and 2007.
- **OECD oil consumption** – Oil consumption in OECD countries increased by 1.67 million b/d from November to December 2009. Resulting in a consumption level of 45.27 million b/d. Average OECD oil consumption in 2009 was 43.92 million b/d, versus 46.10 and 47.68

Consumption* (Thousand barrels daily)

Country	1998	1999	2000	2001	2002	2003	2004	2005	2006	2007	2008	Change 2008 over 2007	2008 share of total
US	18917	19519	19701	19649	19761	20033	20732	20802	20687	20680	19419	-6.4%	22.5%
Canada	1913	1926	1937	2023	2067	2132	2248	2247	2246	2323	2295	-1.0%	2.6%
Mexico	1844	1842	1910	1899	1837	1885	1918	1974	1970	2027	2039	0.5%	2.3%
Total North America	22674	23286	23548	23571	23665	24050	24898	25023	24904	25030	23753	-5.4%	27.4%
Argentina	467	445	431	405	364	372	388	414	432	487"	508	4.1% '	0.6%
Brazil	2034	2114	2056	2082	2063	1985	1999	2048	2102	2274	2397	5.3%	2.7%
Chile	247	249	236	230	228	228	240	254	264	346	358	3.8%	0.4%
Colombia	266	238	232	245	222	222	223	230	240	234	234	0.3%	0.3%
Ecuador	145	131	129	132	131	137	141	168	182	196	204	4.0%	0.2%
Peru	155	159	155	148	147	140	153	152	147	154	172	10.8%	0.2%
Venezuela	475	474	496	545	594	479	523	576	607	665	719	8.1%	0.8%
Other S. & Cent. America	1152	1166	1173	1223	1228	1267	1271	1288	1323	1324	1308	-1.3%	1.6%
Total S & Cent America	4942	4976	4908	5009+	4977	4830	4938	5129	5297	5681	5901	3.7%	6.9%
Austria	255	250	244	265	271	293	285	294	294	278	278	-0.1%	0.3%
Azerbaijan	116	111	123	81	74	86	92	108	99	92	71	-25.9%	0.1%
Belarus	175	154	143	149	145	148	153	146	165	157	159	0.9%	0.2%
Belgium & Luxembourg	656	670	702	669	691	748	785	815	839	832	836	0,4%	1.1%
Bulgaria	100	93	84	87	98	115	105	109	116	117	120	2.1%	0.1%
Czech Republic	174	174	169	179	174	185	203	211	208	206	210	1,4%	0.3%
Denmark	223	222	215	205	200	193	189	195	197	196	189	-4.0%	0.2%
Finland	221	224	224	222	226	239	224	233	225	226	225	-1.2%	0.3%
France	2016	2044	2007	2023	1967	1965	1978	1960	1956	1921	1930	0.7%	2.3%

Germany	2915	2824	2763	2804	2714	2664	2634	2605	2624	2393	2505	4.9%	3.0%
Greece	375	384	407	412	416	405	438	436	454	445	438	-1.5%	0.5%
Hungary	157	151	145	142	140	138	142	163	169	169	169	-0.2%	0.2%
Iceland	18	18	19	18	19	18	20	21	20	21	19	-9.1%	?
Republic of Ireland	152	172	170	185	182	178	185	196	195	198	190	-4.4%	0.2%
Italy	1974	1980	1956	1946	1943	1927	1873	1819	1813	1759	1691	-3.9%	2.1%
Kazakhstan	176	147	158	186	193	183	187	207	227	244	229	-6.4%	0.3%
Lithuania	76	63	49	56	53	51	55	58	59	59	64	9.4%	0.1%
Netherlands	854	880	897	942	952	962	1003	1070	1043	962	982	0.8%	1.2%
Norway	215	216	201	213	208	219	210	212	217	222	210	-4.4%	0.2%
Poland	424	431	427	415	420	435	460	479	516	535	545	2.6%	0.6%
Portugal	322	330	324	327	338	317	322	331	300	302	286	-5.1%	0.3%
Romania	242	195	203	217	226	199	230	223	219	223	227	2.2%	0.3%
Russian Federation	2554	2625	2583	2566	2606	2622	2619	2601	2709	2706	2797	3.1%	3.3%
Slovakia	80	73	73	68	76	71	68	81	82	86	89	3.2%	0.1%
Spain	1381	1423	1452	1508	1526	1559	1593	1619	1602	1617	1574	-2.5%	2.0%
Sweden	338	337	318	318	317	332	319	315	322	308	302	-1.5%	0.4%
Switzerland	279	271	263	281	267	259	258	262	269	243	258	6.4%	0.3%
Turkey	640	638	677	645	656 t	662	667	656	635	656	690	5.8%	0.8%
Turkmenistan	75	80	79	83	86	95	95	100	118	121	123	1.7%	0.1%
Ukraine	302	272	255	273	278	286	293	294	318	325	326	0.5%	0.4%
United Kingdom	1741	1721	1697	1697	1693	1717	1764	1802	1785	1714	1704	-1.0%	2.0%
Uzbekistan	146	143	138	135	130	148	134	112	105	111	113	1.5%	0.1%
Other Europe & Eurasia	462	428	404	428	454	496	512	549	562	5.85	610	4.3%	0.7%
Total Europe & Eurasia	19831	19745	19568	19747	19738	19915	20092	20285	20463	20031	20158	0.6%	24.3%

(Contd.)

Table (*Contd.*)

Iran	1194	1221	1301	1314	1413	1498	1558	1620	1693	1693	1730	2.1%	2.1%
Kuwait	180	202	202	206	222	238	266	295	267	269	300	11.7%	0.4%
Qatar	34	33	36	40	54	48	53	62	75	90	104	16.5%	0.1%
Saudi Arabia	1262	1302	1426	1471	1505	1616	1670	1756	1841	2054	2224	8.1%	2.7%
United Arab Emirates	283	271	255	292	320	333	355	376	402	425	467	10.6%	0.6%
Other Middle East	1304	1338	1382	1431	1433	1404	1467	1508	1539	1554	1599	2.7%	2.0%
Total Middle East	4258	4367	4602	4754	4948	5138	5370	5617	5817	6084	6423	5.5%	7.8%
Algeria	194	187	192	200	222	231	240	251	260	288	311	8.3%	0.4%
Egypt	559	573	564	548	534	550	567	629	610	650	693	6.4%	0.8%
South Africa	450	461	475	486	499	512	523	526	537	549	558	1.7%	0.7%
Other Africa	1161	1216	1200	1218	1228	1244	1269	1322	1290	1288	1319	2.4%	1.6%
Total Africa	2364	2438	2432	2452	2484	2537	2600	2728	2696	2776	2881	3.8%	3.4%
Australia	825	843	837	845'	846″	851	856	886	918	925	936	1.5%	1.1%
Bangladesh	76	68	66	80	80	83	83	94	93	94	95	1.9%	0.1%
China	4228	4477	4772	4872	5288	5803	6772	6984	7382	7742	7999	3.3%	9.6%
China Hong Kong SAR	185	194	202	244	268	270	316	287	305	324	294	-9.4%	0.4%
India	1963	2134	2254	2284	2374	2420	2573	2569	2580	2748	2882	4.8%	3.4%
Indonesia	914	980	1064	1086	1138	1141	1225	1231	1173	1201	1217	1.5%	1.5%
Japan	5504	5598	5557	5422	5347	5440	5269	5343	5213	5039	4845	-3.5%	5.6%
Malaysia	404	435	435	442	482	473	485	469	459	481	475	-1.4%	0.6%
New Zealand	127	129	131	134	139	149	150	154	156	156	158	1.2%	0.2%
Pakistan	350	363	373	366	357	321	325	312	356	388	389	0.2%	0.5%>
Philippines	392	375	348	347	331	331	338	315	284	300	288	-4.0%	0.3%
Singapore	651	619	654	716	699	668	748	794	853	916	958	5.0%	1.3%
South Korea	2030	2178	2229	2235	2282	2300	2283	2308	2317	2389	2291	-4.9%	2.6%
Taiwan	786	964	1003	991	999	1069	1084	1090	1097	1123	1074	-4.8%	1.3%

Thailand	691	698	669	656	705	746	824	852	834	823	797	-4.1%	0.9%
Other Asia Pacific	432	447	479	506	528	537	568	596	601	629	641	1.8%	0.8%
Total Asia Pacific	19559	20501	21073	21225	21863	22601	23899	24283	24620	25277	25339	0.2%	30.1%
Total World	73628	75313	76132	76758	77675	79071	81796	830S5	83797	84878	84455	-0.6%	100.0%
of which:													
European Union	14849	14813	14692	14861	14797	14867	15032	15205	15211	14746	14765	0.1%	17.9%
OECD	46569	47467	47651	47692	47675	48276	49073	49490	49274	48830	47303	-3.2%	55.5%
Former Soviet Union	3741	3700	3623	3627	3667	3745	3760	3772	3948	3973	4045	1.5%	4.8%
Other EMEs	23319	24146	24859	25440	26333	27050	28963	29804	30575	32075	33108	3.1%	39.7%.

*Inland demand plus international aviation and marine bunkers and refinery fuel and loss. Consumption of fuel ethanol and biodiesel is also included. ?Less than 0.05%.

Notes: Annual changes and shares are calculated using million tonnes per annum figures. Growth rates are adjusted for leap years. Differences between these world consumption figures and world production statistics are accounted for by stock changes, consumption of non-petroleum additives and substitute fuels, and unavoidable disparities in the definition, measurement or conversion of oil supply and demand data.

million b/d in respectively 2008 and 2007.

- **North America oil consumption** – Oil consumption in North America increased by 726,000 b/d from November to December 2009. Resulting in a consumption level of 23.23 million b/d. Average oil consumption in North America in 2009 was 22.51 million b/d, versus 23.50 and 24.72 million b/d in respectively 2008 and 2007.
- **European Union oil consumption** – Oil consumption in the European Union increased by 230,000 b/d from November to December 2009. Resulting in a consumption level of 13.43 million b/d according to JODI statistics. Average consumption in the European Union in 2009 was 13.6 million b/ d, versus 14.25 and 14.32 million b/d in respectively 2008 and 2007.
- **United States oil consumption** – Oil consumption in the US increasde by 560,000 b/d from November to December 2009. Resulting in a consumption level of 19.11 million b/d. Average consumption of oil in the US in 2009 was 18.68 million b/d, versus 19.50 and 20.70 million b/d in respectively 2008 and 2007.
- **Mexico oil consumption** – Oil consumption in Mexico increased by 187,000 b/d from November to December 2009. Resulting in a consumption level of 2.02 million b/d. Average oil consumption in Mexico in 2009 was 1.87 million b/d, versus 1.95 and 1.94 million b/d in respectively 2008 and 2007.
- **Canada oil consumption** – Oil consumption in Canada decreased by 22,000 b/d from November to December 2009. Resulting in a consumption level of 2.09 million b/d. Average consumption in Canada in 2009 was 1.96 million b/d, versus 2.06 and 2.08 million b/d in respectively 2008 and 2007.
- **France oil consumption** – Oil consumption in France increased by 246,000 b/d from November to December 2009. Resulting in a consumption level of 1.96 million b/d. Average consumption of oil in France in 2009 was 1.84 million b/d, versus 1.94 and 1.94 million b/d in respectively 2008 and 2007.
- **Germany oil consumption** – Oil consumption in Germany decreased by 36,000 b/d from November to December 2009. Resulting in a consumption level of 2.31 million b/d. Average oil consumption in Germany in 2009 was 2.42 million b/d, versus 2.56 and 2.47 million b/ d in respectively 2008 and 2007.
- **Italy oil consumption** – Oil consumption in Italy increased by 22,000 b/d from November to December 2009. Resulting in a consumption level of 1.52 million. Average consumption in Italy in 2009 was 1.53 million b/d, versus 1.60 and 1.67 million b/d in respectively 2008 and 2007.
- **Spain oil consumption** – Oil consumption in Spain increased by 10,000 b/d from Nov. to Dec. 2009. Resulting in a consumption level of 1.44 million b/d. Average oil consumption in Spain in 2009 was 1.46 million b/d, versus 1.54 and 1.59 million b/d in respectively 2008 and 2007.
- **United Kingdom oil consumption** – Oil consumption in the United

Kingdom decreased by 50,000 b/d from November to December 2009. Resulting in a consumption level of 1.5 million b/d. Average oil consumption in the United Kingdom in 2009 was 1.57 million b/d, versus 1.61 and 1.67 million b/d in respectively 2008 and 2007.

- **Poland oil consumption** – Oil consumption in Poland decreased by 36,000 from November to December 2009. Resulting in a consumption level of 517,000 b/d. Average consumption in Poland in 2009 was 530,000 b/d, versus 527,000 and 507,000 b/d in respectively 2008 and 2007.
- **Netherlands oil consumption** – Oil consumption in the Netherlands increased by 66,000 b/d from November to December 2009 to a consumption level of 959,000 b/d. Average oil consumption in the Netherlands in 2009 was 905,000 b/d, versus 945,000 and 920,0000 b/d in respectively 2008 and 2007.
- **Japan oil consumption** – Oil consumption in Japan increased by 66,000 b/d from November to December 2009. Resulting in a consumption level of 5.08 million b/d. Average oil consumption in Japan in 2009 was 4.43 million b/d, versus 4.92 and 5.13 million b/d in respectively 2008 and 2007.
- **South Korea oil consumption** – Oil consumption in South Korea increased by 133,000 b/d from November to December 2009. Resulting in a consumption level of 2.43 million b/d. Average consumption in South Korea in 2009 was 2.25 million b/d, versus 2.21 and 2.29 million b/d in respectively 2008 and 2007.
- **China oil consumption** – Oil consumption in China increased by 125,000 b/d from November to December 2009. Resulting in a consumption level of 8.27 million b/d according to JODI statistics. Average oil consumption in China in 2009 was 8.05 million b/d, versus 6.92 and 7.29 million b/d in respectively 2008 and 2007.
- **India oil consumption** – Oil consumption in India increased by 45,000 b/d from Nov. to Dec. 2009. Resulting in a consumption level of 2.92 million b/d. Average oil consumption in India in 2009 was 2.85 million b/d, versus 2.60 and 2.43 million b/d in respectively 2008 and 2007.
- **Taiwan oil consumption** – Oil consumption in Taiwan increased by 90,000 b/d from Nov. to Dec. 2009. Resulting in a consumption level of 1.04 million b/d. Average consumption in Taiwan in 2009 was 976,000 b/d, versus 978,000 and 958,000 b/d in respectively 2008 and 2007.

Oil Product Consumption by Region

Regional consumption of oil products classified in thousand barrels per day as: light distillates, middle distillates, fuel oil and others. Light distillates consist of aviation and motor gasolines and light distillate feedstock. Middle distillates consist of jet, heating kerosenes, gas and diesel oils (including marine bunkers). Fuel oil includes marine bunkers and crude oil used directly as fuel. Others consist of refinery gas, LPG, solvents, petroleum coke, lubricants, bitumen, wax, other refined products and refinery fuel and loss.

(In thousand barrels daily)

Regional Consumption by product group	1998	1999	2000	2001	2002	2003	2004	2005	2006	2007	2008	Change 2008 over 2007	2008 share of total
North America													
Light distillates	9849	9998	10106	10211	10523	10675	10979	11020	11131	11205	**10857**	-3.1%	45.7%
Middle distillates	6450	6628	6811	6812	6655	6861	7133	7249	7297	7318	**6924**	-5.4%	29.2%
Fuel oil	1506	1415	1518	1411	1209	1271	1340	1396	1055	1082	**946**	-12.5%	4.0%
Others	4869	5245	5112	5137	5278	5243	5446	5359	5420	5425	**5026**	-7.4%	21.2%
Total North America	22674	23286	23548	23571	23665	24050	24898	25023	24904	25030	**23753**	-5.1%	100.0%
of which: US													
Light distillates	8579	8716	8813	8890	9167	9275	9518	9548	9599	9597	**9220**	-3.9%	47.5%
Middle distillates	5545	5700	5852	5884	5735	5886	6116	6198	6226	6199	**5790**	-6.6%	29.8%
Fuel oil	869	814	893	794	686	763	859	914	683	718	**615**	-14.4%	3.2%
Others	3924	4290	4143	4082	4172	4109	4239	4143	4178	4166	**3794**	-8.9%	19.5%
Total US	18917	19519	19701	19649	19761	20033	20732	20802	20687	20680	**19419**	-6.1%	100.0%
S. & Cent. America													
Light distillates	1553	1668	1540	1517	1506	1496	1452	1551	1637	1746	**1801**	3.2%	30.5%
Middle distillates	1641	1625	1617	1657	1653	1655	1776	1833	1873	2040	**2126**	4.2%	36.0%
Fuel oil	866	761	801	827	789	754	740	744	766	810	**814**	0.6%	13.8%
Others	882	922	950	1009	1029	925	971	1002	1021	1086	**1160**	6.8%	19.7%
Total S. & Cent. America	4942	4976	4908	5009	4977	4830	4938	5129	5297	5681	**5901**	3.9%	100.0%
Europe													
Light distillates	4315	4314	4194	4086	4025	3923	3875	3764	3615	3491	**3365**	-3.6%	20.9%
Middle distillates	6630	6690	6734	6991	6934	7137	7347	7595	7853	7763	**7937**	2.2%	49.3%
Fuel oil	2195	2093	1967	1986	1995	1952	1888	1855	1773	1667	**1633**	-2.0%	10.1%
Others	2950	2948	3050	3056	3117	3158	3221	3298	3273	3137	**3179**	1.4%	19.7%
Total Europe	16090	16045	15946	16120	16071	16170	16331	16513	16514	16058	**16114**	0.3%	100.0%

Former Soviet Union													
Light distillates	893	858	797	851	882	937	918	934	993	1053	**1096**	4.1%	27.1%
Middle distillates	1026	1022	1008	1021	1048	1085	1108	1148	1208	1274	**1324**	4.0%	32.7%
Fuel oil	1046	937	765	693	670	643	578	558	615	511	**455**	-11.0%	11.2%
Others	775	882	1053	1062	1066	1080	1157	1132	1133	1136	**1170**	3.0%	28.9%
Total Former Soviet Union	3741	3700	3623	3627	3667	3745	3760	3772	3948	3973	**4045**	1.8%	100.0%
Middle East													
Light distillates	849	859	910	949	1011	1059	1177	1241	1325	1326	**1379**	4.0%	21.5%
Middle distillates	1471	1502	1567	1618	1666	1723	1767	1849	1950	2013	**2124**	5.5%	33.1%
Fuel oil	1240	1295	1292	1319	1329	1317	1390	1463	1450	1532	**1644**	7.3%	25.6%
Others	698	712	832	868	941	1039	1036	1064	1092	1213	**1276**	5.2%	19.9%
Total Middle East	4258	4367	4602	4754	4948	5138	5370	5617	5817	6084	**6423**	5.6%	100.0%
Africa													
Light distillates	563	570	574	585	580	605	632	637	617	634	**666**	5.1%	23.1%
Middle distillates	926	971	999	1013	1039	1066	1070	1119	1137	1181	**1232**	4.3%	42.8%
Fuel oil	498	501	459	436	425	380	390	437	417	416	**429**	3.1%	14.9%
Others	378	396	401	419	440	486	509	534	525	545	**554**	1.7%	19.2%
Total Africa	2364	2438	2432	2452	2484	2537	2600	2728	2696	2776	**2881**	3.8%	100.0%
Asia Pacific													
Light distillates	5201	5486	5636	5771	6021	6283	6644	6781	6826	7147	**7196**	0.7%	28.4%
Middle distillates	7234	7678	7813	7968	8108	8246	8774	9030	9089	9151	**9496**	3.8%	37.5%
Fuel oil	3657	3713	3675	3481	3402	3547	3558	3552	3632	3614	**3403**	-5.9%	13.4%
Others	3467	3624	3949	4005	4331	4526	4923	4921	5073	5364	**5244**	-2.2%	20.7%
Total Asia Pacific	19559	20501	21073	21225	21863	22601	23899	24283	24620	25277	**25339**	0.2%	100.0%
of which: China													
Light distillates	1280	1249	1270	1256	1352	1489	1673	1776	1853	1983	**2175**	9.7%	27.2%
Middle distillates	1241	1467	1590	1669	1791	1945	2285	2528	2723	2879	**3178**	10.4%	39.7%

(Contd.)

Regional Consumption by product group	1998	1999	2000	2001	2002	2003	2004	2005	2006	2007	2008	Change 2008 over 2007	2008 share of total
Fuel oil	741	771	775	780	792	880	956	904	938	892	**716**	-19.8%	8.9%
Others	967	990	1136	1167	1352	1489	1858	1777	1867	1988	**1930**	-2.9%	24.1%
Total China	4228	4477	4772	4872	5288	5803	6772	6984	7382	7742	**7999**	3.3%	100.0%
of which: Japan													
Light distillates	1611	1702	1735	1720	1759	1788	1793	1822	1813	1790	**1675**	-6.4%	34.6%
Middle distillates	1949	1978	1958	1957	1930	1904	1856	1876	1758	1619	**1500**	-7.4%	31.0%
Fuel oil	858	840	785	678	637	755	660	683	664	664	**712**	7.2%	14.7%
Others	1086	1078	1079	1067	1021	993	960	963	979	966	**959**	-0.7%	19.8%
Total Japan	5504	5598	5557	5422	5347	5440	5269	5343	5213	5039	**4845**	-3.8%	100.0%
World													
Light distillates	23224	23754	23758	23969	24549	24976	25676	25929	26145	26601	**26359**	-0.9%	31.2%
Middle distillates	25378	26115	26550	27080	27104	27773	28974	29823	30407	30740	**31163**	1.4%	36.9%
Fuel oil	11007	10714	10477	10154	9818	9865	9884	10004	9708	9631	**9324**	-3.2%	11.0%
Others	14020	14730	15348	15555	16203	16457	17263	17308	17537	17905	**17609**	-1.7%	20.8%
Total World	73628	75313	76132	76758	77675	79071	81796	83065	83797	84878	84455	-0.5%	100.0%
European Union													
Light distillates	4006	4025	3923	3827	3763	3671	3619	3520	3367	3255	**3139**	-3.5%	21.3%
Middle distillates	6103	6170	6234	6463	6404	6592	6795	6996	7221	7109	**7234**	1.8%	49.0%
Fuel oil	2060	1936	1785	1796	1792	1743	1703	1705	1651	1553	**1516**	-2.4%	10.3%
Others	2681	2683	2750	2775	2838	2861	2916	2985	2972	2829	**2876**	1.6%	19.5%
Total European Union	14849	14813	14692	14861	14797	14867	15032	15205	15211	14746	**14765**	0.1%	100.0%
OECD													
Light distillates	16728	17027	17069	17064	17389	17495	17747	17753	17739	17764	**17149**	-3.5%	36.3%
Middle distillates	15749	16107	16276	16529	16295	16684	17095	17483	17676	17459	**17088**	-2.1%	36.1%

Fuel oil	4809	4650	4622	4403	4163	4294	4205	4237	3784	3692	**3513**	-4.8%	7.4%
Others	9284	9684	9683	9696	9828	9803	10026	10015	10074	9915	**9552**	-3.7%	20.2%
Total OECD	46569	47467	47651	47692	47675	48276	49073	49490	49274	48830	**47303**	-3.1%	100.0%
Other EMEs													
Light distillates	5603	5869	5892	6055	6277	6544	7011	7241	7412	7785	**8115**	4.2%	24.5%
Middle distillates	8603	8986	9265	9530	9761	10004	10771	11192	11523	12007	**12751**	6.2%	38.5%
Fuel oil	5152	5128	5090	5057	4986	4928	5102	5209	5309	5428	**5356**	-1.3%	16.2%
Others	3961	4163	4611	4797	5309	5574	6079	6161	6330	6855	**6886**	0.5%	20.8%
Total Other EMEs	23319	24146	24859	25440	26333	27050	28963	29804	30575	32075	**33108**	3.2%	100.0%

Excludes Former Soviet Union.

Notes: Annual changes and shares of total are calculated using thousand barrels daily figures.

'Light distillates' consists of aviation and motor gasolines and light distillate feedstock.

'Middle distillates' consists of jet and heating kerosenes, and gas and diesel oils (including marine bunkers).

'Fuel oil' includes marine bunkers and crude oil used directly as fuel.

'Others' consists of refinery gas, LPG, solvents, petroleum coke, lubricants, bitumen, wax, other refined products and refinery fuel and loss.

Distribution of Proved Oil Reserved in 1988, 1998 and 2008

Global proved oil reserves in 2008 fell by 3 billion barrels to 1,258 billion barrels, with an R/P ratio of 42 years. Declines in Russia, Norway, China and other countries offset increases in Vietnam, India and Egypt.

The 2007 figure has been revised higher by 23.1 billion barrels, with the largest upward revisions in Venezuela and Angola. Proved reserves of oil are generally taken to be those quantities that geological and engineering information indicates with reasonable certainty can be recovered in the future from known reservoirs under existing economic and geological conditions.

Proved oil reserves at end 2008	At end 1988 Thousand million barrels	At end 1998 Thousand million barrels	At end 2007 Thousand million barrels	Thousand million tonnes	At end 2008 Thousand million barrels	Share of total	R/P ratio
1	2	3	4	5	6	7	8
US	35.1	28.6	30.5	3.7	30.5	2.4%	12.4
Canada	11.9	15.1	28.6	4.4	28.6	2.3%	24.1
Mexico	53.0	21.6	12.2	1.6	11.9	0.9%	10.3
Total North America	100.0	65.3	71.3	9.7	70.9	5.6%	14.8
Argentina	2.3	2.8	2 6	0.4	2.6	0.2%	10 5
Brazil	2.8	7.4	12.6	1.7	12.6	1.0%	18.2
Colombia	2.1	2.5	1.5	0.2	1.4	0.1%	6.0
Ecuador	1.5	4.1	4.0	0.5	3.8	0.3%	20.3
Peru	0.9	0.9	1.1	0.2	1.1	0.1%	25.5
Trinidad & Tobago	0.6	0 7	0.9	0.1	0.8	0.1%	15.2
Venezuela	58.5	76 1	99.4	14.3	99.4	7.9%	*
Other S. & Cent America	0.6	1.1	1.4	0.2	1.4	0.1%	27.7
Total S. & Cent. America	69.2	95.6	123.5	17.6	123.2	9.8%	50.3
Azerbaijan	n/a	n/a	7.0	1.0	7.0	0.6%	20.9
Denmark	0.5	0.9	1.1	0.1	0.8	0.1%	7.7
Italy	0.8	0.8	0.9	0.1	0.8	0.1%	21 1
Kazakhstan	n/a	n/a	39.8	5.3	39.8	3.2%	70.0
Norway	7.3	11.7	8.2	0.9	7.5	0.6%	8.3
Romania	1.2	1.2	0.5	0.1	0.5	†	13.3
Russian Federation	n/a	n/a	80.4	10.8	79.0	6.3%	21.8
Turkmenistan	n/a	n/a	0.6	0.1	0.6	†	8.0
United Kingdom	4.3	5.1	3.4	0.5	3.4	0.3%	6.0
Uzbekistan	n/a	n/a	0.6	0.1	0.6	†	14.6
Other Europe & Eurasia	63.2	2.1	2.1	0.3	2.1	0.2%	13.4
Total Europe & Eurasia	77 3	104.9	144.6	19.2	142.2	11.3%	22.1
Iran	92 9	93 7	138.2	18.9	137.6	10.9%	86.9
Iraq	100.0	112.5	115.0	15.5	115.0	9.1%	*
Kuwait	94.5	96.5	101.5	14.0	101.5	8.1%	99.6
Oman	4.1	5.4	5.6	0.8	5.6	0.4%	20.9
Qatar	4.5	12.5	27.4	2.9	27.3	2.2%	54.1
Saudi Arabia	255.0	261.5	264.2	36.3	264.1	21.0%	66.5
Syria	1.8	2.3	2.5	0.3	2.5	0.2%	17.2
United Arab Emirates	98.1	97.8	97.8	13.0	97.8	7.8%	89.7

1	2	3	4	5	6	7	8
Yemen	2.0	1.9	2.7	0.3	2.7	0.2%	23.9
Other Middle East	0.1	0.2	0.1	‡	0.1	†	10.6
Total Middle East	653.0	684.3	755.0	102.0	754.1	59.9%	78.6
Algeria	9.2	11.3	12.2	1.5	12.2	1 0%	16 7
Angola	2.0	4.0	13.5	1.8	13.5	1.1%	19.7
Chad	-	-	0.9	0.1	0.9	0.1%	19.4
Republic of Congo (Brazzaville)	0.8	1.7	1.9	0.3	1.9	0.2%	21.3
Egypt	4.3	3.8	4.1	0.6	4.3	0.3%	16.4
Equatorial Guinea	-	0.6	1.7	0.2	1.7	0.1%	12.9
Gabon	0.9	2.6	3.2	0.4	3.2	0.3%	37.0
Libya	22.8	29.5	43.7	5.7	43.7	3.5%	64.6
Nigeria	16.0	22.5	36.2	4.9	36.2	2.9%	45.6
Sudan	0.3	0.3	6.7	0.9	6.7	0.5%	38.1
Tunisia	1.8	0.3	0.6	0.1	0.6	†	18.5
Other Africa	1.0	0.7	0 6	0.1	0.6	†	12.0
Total Africa	89 0	77.2	125.3	16.6	125.6	10.0%	33.4
Australia	3 4	4.1	4.2	0.5	4.2	0.3%	204
Brunei	1.2	1.0	1.1	0.1	1.1	0.1%	16.9
China	17.3	17.4	16.1	2.1	15.5	1.2%	11.1
India	4.5	5 4	5.5	0.8	5.8	0.5%	207
Indonesia	9.0	5.1	4.0	0.5	3.7	0.3%	10 2
Malaysia	3.4	4.7	5.5	0.7	5.5	0.4%	19.8
Thailand	0.1	0.4	0.5	0.1	0.5	†	3.9
Vietnam	0.1	1.9	3.4	0.6	4.7	0.4%	40.8
Other Asia Pacific	1.0	1.3	1.1	0.1	1.1	0.1%	12.8
Total Asia Pacific	39.8	41.3	41.3	5.6	42.0	3.3%	14.5
Total World	998.4	1068.5	1261.0	170.8	1258.0	100.0%	42.0
of which: European Union	8.3	8.9	6.7	0.8	6.3	0.5%	7.7
OECD	118.3	89.2	90.3	12.0	88.9	7.1%	13.2
OPEC	764.0	827.2	957.1	129.8	955.8	76.0%	71.1
Non-OPEC+	173.5	157.6	174.7	23.6	174.4	13.9%	14 8
Former Soviet Union	60.9	83 8	129.2	17.4	127.8	10.2%	27 2
Canadian oil sands*	n/a	n/a	150.7	24.5	150.7		
Proved reserves and oil sands	n/a	n/a	1411.7	195.3	1408.7		

*=More than 100 years. ‡=Less than 0.05. †=Less than 0.05%. +Excludés Former Soviet Union. *'Remaining established reserves', less reserves under active development'.

Notes: Proved reserves of oil – Generally taken to be those quantities that geological and engineering information indicates with reasonable certainty can be recovered in the future from known reservoirs under existing economic and operating conditions. Reserves-to-production (R/P) ratio – If the reserves remaining at the end of any year are divided by the production in that year, the result is the length of lime that those remaining reserves would last if production were to continue at that rate.source of data – The estimates in this table have been compiled using a combination of primary official sources, third-party data from the OPEC Secretariat, World Oil. Oil & Gas Journal and an independent estimate of Russian reserves based on information in the public domain Canadian proved reserves include an official estimate of 22 0 billion barrels for oil sands 'under active development'. Reserves include gas condensate and natural gas liquids (NGLs) as well as crude oil. Annual changes and shares of total are calculated using thousand million barrels figures.

- **OECD crude oil stocks –** Industrial inventories of crude oil in the OECD in January 2010 decreased to 979 million from 994 million barrels in November according to the latest IEA statistics. Current OECD crude oil stocks are 13 million barrels higher than the five year average of 964 million barrels. In the February Oil Market Report of the IEA a total stock level of 979 million barrels was tabulated for December which has been revised downwards to 968 million barrels in the March'10. Industrial product stocks in the OECD in January 2010 increased to 1447 million from 1431 million barrels in December according to the latest IEA Statistics. Current OECD product stocks are 42 million barrels higher than the five year average of 1447 million barrels. In the February Oil Market Report of the IEA a total stock level of 1420 million barrels was tabulated for December which has been revised upwards to 1447 million barrels in the March.'10
- **Europe crude oil stocks** – Industrial inventories of crude oil in OECD Europe in January 2010 increased to 341 from 338 million barrels in December according to the latest IEA statistics. Current OECD Europe crude oil stocks are 5 million barrels higher than the five year average of 336 million barrels. In the February Oil Market Report of the IEA a total stock level of 337 million barrels was tabulated for December which has been revised upwards to 338 million barrels in the February'10. Industrial product stocks in OECD Europe in January 2010 increased to 590 million from 573 million barrels in December 2009 according to the latest IEA statistics. Current OECD Europe product stocks are 20 million barrels higher than the five year average of 553 million barrels.
- **Pacific crude oil stocks** – Industrial inventories of crude oil in OECD Pacific in January 2010 increased to a level of 162 million from 161 million barrels in December 2009 according to the latest IEA statistics. Current OECD Pacific crude oil stocks are 8 million barrels higher than the five year average of 169 million barrels. In the February Oil Market Report of the IEA a total stock level of 170 million barrels was tabulated for December which has been revised downwards to 161 million barrels in the March'10. Industrial product stocks in OECD Pacific in January 2010 increased to 168 million from 161 million barrels in December according to the latest IEA Statistics. Current OECD Pacific product stocks are 13 million barrels lower than the five year average of 180 million barrels.
- **North America crude oil stocks** – Industrial inventories of crude oil in OECD North America in January 2010 decreased to 474 million from 469 million barrels in December 2009 according to the latest IEA statistics. Current OECD North America crude oil stocks are 14 million barrels higher than the five year average of 460 million barrels. In the February Oil Market Report of the IEA a total stock level of 472 million barrels was tabulated for December which has been revised downwards to 469 million barrels in the March'10. Industrial product stocks in North

America in January 2010 decreased to 690 million from 696 million barrels in December 2009 according to the latest IEA Statistics. Current North American product stocks are 17 million barrels higher than the five year average of 673 million barrels. In the February Oil Market Report of the IEA a total stock level of 691 million barrels was tabulated for December which has been revised upwards to 696 million barrels in the March '10.

- **US gasoline stocks** – Gasoline stocks in the United States in February 2010 increased to 231 million from 226 million barrels in December according to the latest EIA Statistics. Current Gasoline stocks are 22 million barrels higher than the five year average of 209 million barrels.

Refinery Capacities. Margines and R/P Ratio

The refinery capacity data presented in this Review represents the sum of reported atmospheric crude distillation and condensate splitting capacity. Capacity should comprise the amount of input that a distillation facility can process under usual operating conditions, taking into account scheduled downtime. Figures are in thousand barrels per calendar day ie the capacity taking into account normal scheduled downtime. (See Table on pp. 112-13)

Global refining throughputs fell last year, for the first time since 2002. The biggest fall was in the US, where crude runs were down by 510 Kb/d. Chinese throughputs grew by 290 Kb/d and there were small increases in S&C America, Africa, the Middle East and elsewhere in Asia-Pacific. Refinery throughputs are based on the quantity of crude and condensate processed in atmospheric distillation units and condensate splitters. Figures are in thousands of barrels per day. Also shown is a chart of refinery utilization by region. (See Table on p. 114)

Refining Margins

Refining margins in 2008 were the lowest since 2004 but still above the 10-year average. The 2008 average of \$6.52/bbl (2009 basis) was nearly \$3.50/bbl lower than 2007's record of \$9.96/bbl (2009 basis) with all of the margin adjustment occurring in the US. Margins in other regions grew year-on-year. In each case, refinery margins are based on a single crude oil appropriate for that region and have optimized product yields based on a generic refinery configuration (cracking, hydrocracking or coking), again appropriate for that region. The margins are on a semi-variable basis, i.e. the margin after all variable costs and fixed energy costs have been deducted.

Reserves-to-production (R/P) ratios are available by country and feature in the table of oil reserves and the Energy charting tool. R/P ratios represent the length of time that those remaining reserves would last if production were to continue at the previous year's rate. It is calculated by dividing remaining reserves at the end of the year by the production in that year. There is a time series of crude oil reserves, which can be found in the table shown above. Data are measured in thousand million barrels.

Refinery Capacities (Thousand barrels daily')	1998	1999	2000	2001	2002	2003	2004	2005	2006	2007	2008	Change 2008 over 2007	2008 share of total
US	16261	16512	16595	16785	16757	16894	17125	17339	17443	17594	17621	0.2%	19.9%
Canada	1844	1861	1861	1917	1923	1959	1915	1896	1914	1907	1951	2.3%	2.2%
Mexico	1449	144S	1481	1481	1463	1463	1463	1463	1463	1463	1463	-	1.7%
Total North America	19664	19922	19937	20183	20143	20316	20503	20693	20821	20904	21035	0.3%	23.7%
Argentina	648	643	624	612	609	609	610	612	614	625	625	-	0 7%
Brazil	1768	1796	1849	1849	1854	1915	1915	1916	1916	1935	1942	0.4%	2.2%
Netherlands Antilles & Aruba	545	545	545	545	545	545	545	545	545	585	595	1.7%	0.7%
Venezuela	1215	1248	1291	1291	1291	1291	1306	1311	1311	1311	1311	-	1.5%
Other S. & Cent. America	1938	1976	1999	1988	2028	2017	2023	2041	2061	2070	2115	2.2%	2.4%
Total S. & Cent America	**6114**	**6208**	**5308**	**6285**	**6327**	**6377**	**6399**	**6425**	**6447**	**6526**	**6588**	**1.0%**	**7.4%**
Belgium	732	736	770	785	803	805	782	778	774	745	745	-	0.8%
France	1918	1933	1984	1961	1987	1967	1982	1978	1959	1962	1962	-	2.2%
Germany	2206	2240	2262	2274	2286	2304	2320	2322	2390	2390	2366	-1.0%	2.7%
Greece	403	403	403	412	412	412	412	418	425	425	425	-	0.5%
Italy	2385	2485	2485	2485	2485	2485	2497	2515	2526	2497	2486	-0.4%	2.8%
Netherlands	1266	1282	1282	1283	1287	1287	1289	1279	1279	1266	1261	-0.5%	1.4%
Norway	310	323	318	307	310	310	310	310	310	310	310	-	0.3%
Russian Federation	5572	5439	5361	5299	5357	5371	5371	5421	5473	5522	5547	0.5%	6.3%
Spain	1247	1247	1247	1247	1333	1333	1358	1377	1377	1377	1377	-	1.6%
Sweden	422	422	422	422	422	422	422	422	422	422	422	-	0.5%
Turkey	713	713	713	713	713	713	693	613	613	613	613	-	0.7%
United Kingdom	1848	1777	1778	1769	1785	1813	1848	1819	1836	1819	1821	0.1%	2.1%
Other Europe & Eurasia	6239	6145	5981	5930	5844	5841	5877	5814	5675	5711	5750	0.7%	6.5%
Total Europe & Eurasia	2526	20145		24337	25024	25063	25161	25067	25059	25060	*25086*	0.1%	28.3%

Iran	1505	1607	1607	1607	1607	1517	1652	"552	1737	1822	1832	0.5%"	2.1%
Iraq	634	634	639	644	644	644	644	649	665	674	681	1.0%	0.8%
Kuwait	921	921	740	759	809	909	931	931	931	931	931	-	1.1%
Saudi Arabia	1762	1823	1806	1806	1810	1890	2075	2100	2100	2100	2100	-	2.4%
United Arab Emirates	235	290	440	674	711	645	620	620	620	625	673	7.7%	0.8%
Other Middle East	1145	1158	1158	1160	1238	1238	1238	1238	1268	1369	1376	0.4%	1.6%
Total Middle East	6202	6433	6390	6650	6819	6943	7160	7190	7321	7592	7592	0.9%	8.6%
Total Africa	2846	2950	2879	3140	3222	3171	3110	3228	3066	3052	3228	5.8%	3.6%
Australia	810	828	828	815	829	756	763	711	694	733	734	0 1 %	0.8%
China	4592	5401	5407	5643	5479	5487	6289	6587	7029	7511	7732	2.9%	8.7%
India	1356	2190	2219	2261	2303	2293	2558	2558	2872	2983	2992	0.3%	3.4%
Indonesia	1096	1119	1127	1127	1092	1057	1057	1057	1127	1157	1157	-	1.3%
Japan	5144	5087	5010	4705	4721	4683	4567	4529	4542	4598	4650	1.1%	5.2%
Singapore	1246	1246	1255	1255	1255	1255	1255	1255	1255	1255	1255	-	1.4%
South Korea	2598	2598	2598	2598	2598	2598	2598	2598	2633	2671	2712	1.5%	3.1%
Taiwan	732	732	732	874	1159	1159	1159	1159	1140	1197	1197	-	1.4%
Thailand	890	914	1062	1062	1062	1062	1062	1065	1092	1112	1187	6.7%	1.3%
Other Asia Pacific	1258	1344	1403	1512	1487	1416	1410	1438	1445	1454	1482	19%	1.7%
Total Asia Pacific	19722	21459	21641	21852	21984	21766	22716	22957	23829	24671	25098	1.7%	28.3%
Total World	79699	82017	82161	82997	83518	83635	85051	85565	86542	87794	88627	0.9%	100.0%
of which: European Union	15262	15304	15373	15455	15692	15713	15787	15829	15875	15827	15788	-0.3%	17.8%
OECD	43688	44041	44197	44141	44413	44500	44645	44692	44916	45145	45271	0.3%	51.1%
Former Soviet Union	8578	8447	8279	8124	8021	8035	8075	8019	7965	8014	8079	0.8%	9.1%
Other EMEs	27433	29529	29685	30732	31084	31100	32331	32854	33661	34635	35277	1 9%	39.8%

Atmospheric distillation capacity on a calendar-day basis. Source: Includes data from Parpinelli Tecnon.

Note: Annual changes and shares of total are calculated using thousand barrels daily figures.

Refinery throughputs (Thousand barrels daily)*	1998	1999	2000	2001	2002	2003	2004	2005	2006	2007	2008	Change 2008 over 2007	2008 share of total
US	14889	14804	15067	15128	14947	15304	15475	15220	15242	15156	14645	-3.4%	19.5%
Canada	1709	1714	1765	1812	1862	1880	1957	1882	1835	1834	1796	-2.1%	2.4%
Mexico	1446	1378	1363	1398	1390	1436	1436	1415	1407	1395	1392	-0.2%	1.9%
S. & Cent. America	5120	5191	5337	5332	4941	5040	5430	5402	5356	5470	5479	0.2%	7.3%
Europe & Eurasia	19592	18997	19305	19457	19488	19991	20454	20845	20902	20705	20661	-0.2%	27.5%
Middle East	5639	5632	5430	5406	5482	5566	5805	6007	6276	6366	6403	0.6%	8.5%
Africa	2386	2343	2200	2440	2377	2301	2304	2477	2367	2470	2495	1.0%	3.3%
Australasia	866	880	851	847	855	823	820	757	749	767	756	-1.4%	1.0%
China	3060	3686	4218	4215	4395	4823	5382	5916	6155	6563	6851	4.4%	9.1%
Japan	4212	4149	4145	4107	3986	4118	4038	4136	4026	3995	3946	-1.2%	5.2%
Other Asia Pacific	7974	8269	8884	9020	8796	9263	9916	10035	10392	10711	10754	0.4%	14 3%
Total Work													100.0%
of which:													
European Union	14092	13519	13767	13698	13477	13731	14089	14172	14002	13601	13549	-0.4%	180%
OECD	39886	39345	39754	39762	39083	39888	40456	40269	40072	39578	38787	-2.0%	51.6%
Former Soviet Union	4519	4471	4583	4797	5060	5276	5413	5713	5900	6073	6173	1.7%	8.2%
Other EMEs	22490	23226	24229	24602	24376	25381	27149	28111	28735	29781	30219	1.5%	40.2%

Input to primary distillation units only. Source: Includes data from Energy Security Analysis. Inc (ESAI)

Note: Annual changes and shares of total are calculated using thousand barrels daily figures.

Chart of Regional Refining Margins

(In US dollars per barrels)

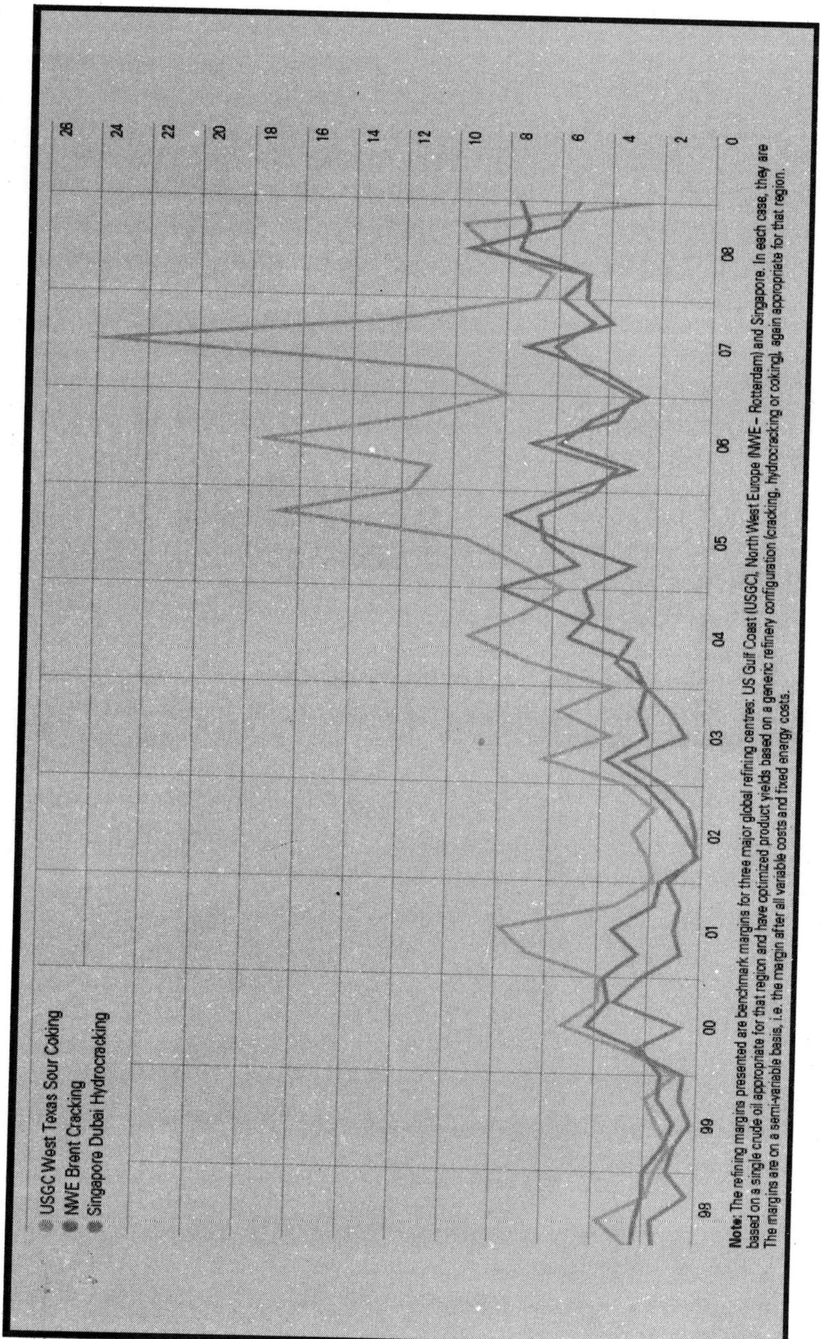

Note: The refining margins presented are benchmark margins for three major global refining centres: US Gulf Coast (USGC), North West Europe (NWE – Rotterdam) and Singapore. In each case, they are based on a single crude oil appropriate for that region and have optimized product yields based on a generic refinery configuration (cracking, hydrocracking or coking), again appropriate for that region. The margins are on a semi-variable basis, i.e. the margin after all variable costs and fixed energy costs.

Oil Reserve to Production Ratio

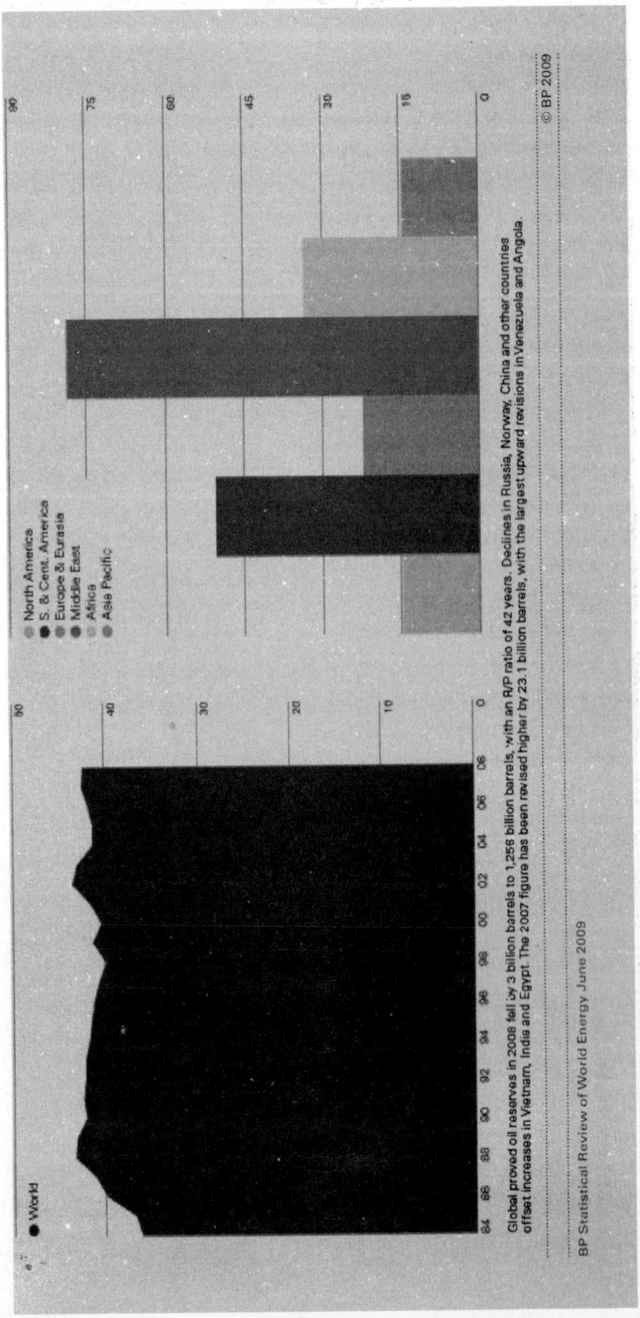

Global proved oil reserves in 2008 fell by 3 billion barrels to 1,258 billion barrels, with an R/P ratio of 42 years. Declines in Russia, Norway, China and other countries offset increases in Vietnam, India and Egypt. The 2007 figure has been revised higher by 23.1 billion barrels, with the largest upward revisions in Venezuela and Angola.

© BP 2009

BP Statistical Review of World Energy June 2009

CHAPTER 2

The Role of Shipping and Freight in Oil Trading

2.1 INTRODUCTION TO FREIGHT AND WORLD SCALE

Bulk amount of Crude Oil is transported from source to refineries by Crude Oil Tankers. India is the net importer of Crude Oil to the extent of about 75%. Bulk of the crude oil transportation for coastal refining centers as well as inland refineries is transported through Crude Oil Tankers. Crude oil tankers are classified on the basis of dead weight tonne (DWT) which is approximately equal to the carrying capacity.

Crude Oil transportation is mainly done through crude vessels in international market. Pipeline is cheaper medium but involves much complexity. Selection and type of crude carrier are important in transportation as it incur various cost, time and capacity to load and move from origin to destination. Major costs are freight cost, ocean loss, insurance, wharfage etc and crude carrier are Panamax tankers, Aframax tankers, Suezmax tankers, VLCC and ULCC. Charter party contract, delivery and redelivery surveys, bunkers, condition of vessels, cargo spaces, delivery certificates, damage notification and tanker characteristic are some of the important parameters for shipping of crude oil. There are various methods of chartering merchants ships that encounters in the world freight market are voyage chartering, time chartering, trip time chartering, bareboat chartering and contracts of affreightment. Demurrage and its calculation is area of concern in shipping because it acts some cost to both buyer and seller.

Cross market linkages and spillover effects between Forward Freight Agreements (FFAs) and futures contracts on the commodities transported by Panamax vessels can aid decision making in the very volatile freight markets. Results indicate that there are significant spillover effects between freight and commodity derivatives markets. These relationships run stronger from the commodity futures markets to forward freight markets. Following these results, market participants may monitor changes in the futures markets of the commodities transported by Panamax vessels to enhance their decisions in FFA and spot markets.

Selection of Crude Carriers

The vessel size for crude oil transportation gets determined by the parcel size, which is typically constrained by the tankage of receiving refinery and the port restriction at both the load port and discharge port Suezmax crude carriers are those which are less than 1,50,000 DWT in full laden condition and can be pass through the Suez Canal. Aframax Crude carriers are flat bottom carriers which can voyage short hauls in the Caribbean, Mediterranean, North Sea and Far East. They have flexibility of loading and unloading at almost every terminal. Many ships in this category meet the Panamax size restriction of 60,000-70,000 DWT - the max size that can pass through the Panama Canal.

Supply of imported as well as Bombay High Crude Oil to coastally located Indian refineries such as BPCL Mumbai refinery, HPCL Mumbai, Vizag, KRL, and CPCL refineries is done through crude tankers based on load port and disport limitation.

Freight Cost

Freight cost refers to the cost of transportation of crude oil from the load port to discharge port. Per unit freight cost depends on the size of vessels utilized for transporting Crude oil. The shipping industry publishes the tanker rates for various DWT vessels as per the World Scale (WS). During the year AFRA, (Average freight rate assessment) on a monthly basis publishes the prevailing percentage of world scale for different size of crude vessels.

The percentage WS for various sizes of vessels is broadly taken as under. It is however never fixed. It reflects the prevailing freight rate in the shipping market. World Scale revised every year.Average freight rate assessment published monthly.

ULCC	40-50 ($/MT)
VLCC	50-70 ($/MT)
LRII	70-100 ($/MT)
LRI	100-120 ($/MT)
MR	120-150 ($/MT)

Panamax tankers: Ships in the 55-70,000 DWT size range, 70,000 DWT being the maximum size tanker able to transit the Panama Canal. The need to pass through a series of Canal locks dictates a maximum length of 274.3 meters and maximum breadth of 32.3 meters. In the Atlantic Basin trades Panamax vessels have a competitive advantage over larger tankers due to physical trading and local port depth restrictions. North American imports of crude and fuel oil comprise the bulk of Panamax tanker business. Typical double hull ship of 60,000 DWT. 228.6 m length overall × 32.2 m breadth × 12.6m draft – Lightship. 11,000 tons of steel.

Aframax tankers: Tankers in the 75,000-120,000 DWT size range. AFRA is Average Freight Rate Assessment. At one time Aframax was used to refer to ships up to 79,999 DWT, the upper limit of one of six deadweight groups for which the AFRA rate is assessed. Aframax has since become a general term for ships in this overall size range. Aframax ships are traditionally employed on a wide variety of short and medium-haul crude oil trades. The biggest tanker that can be accommodated fully laden in the ports of the US – the world's largest importer of oil – is 100,000 DWT, and this only at a limited number of ports. Many of the more modern ships in the Aframax size range are built as long-haul product tankers, with epoxy-coated tanks. Typical double hull ship of 100,000 DWT. 253.0 m lengths overall × 44.2 m breadth x11.6 m draft. Lightship. 14.850 tons of steel.

Suezmax tankers: Suezmax tankers are ships in the 120,000-200,000 DWT size range and are generally identified as those capable of lifting one million barrel cargoes. The name was originally bestowed on such ships because from 1980, when a development project which deepened the waterway to 16.1 meters was completed, the largest tankers able to transit the Suez Canal fully laden were those of 140,000-150,000 DWT. This association will effectively become redundant later this year when a project to deepen the Canal to 18.9 meters is completed. The Canal may be further deepened to 20.1 meters by 2005 and 22.0 meters by 2010.Typical double hull ship of 150,000 DWT. 274.0 m length overall × 50.0 m breadth × 14.5 m draft Lightship. 20,000 tons of steel.

Very large crude carriers (VLCCs): VLCCs are tankers in the 200,000-320,000 DWT size range. Ships of this size were prompted by the rapid growth in global oil consumption during the 1960s and, in 1967, closure of the Suez Canal, necessitating voyages around the Cape of Good Hope. Today, VLCCs are the most effective way of transporting large volumes of oil, including 2 million barrel cargoes, to customers over relatively long distances. Relatively simple ships, VLCCs are subdivided into a number of cargo tanks by two longitudinal and several transverse bulkheads. Typical double hull ship of 280,000 DWT. 335.0 m length overall × 57.0 m breadth × 21.0 m draft-Lightship. 35,000 tons of steel.

Ultra large crude carriers (ULCCs): Tankers in excess of 320,000 DWT. Most ships of this type were built in the mid to late 1970s and are now approaching 25 years of age. Ordered to take advantage of the economies of scale in a buoyant market, they were delivered as oil prices skyrocketed and demand collapsed. There are now fewer than 40 of these ships remaining. Rather inflexible and limited to serving a few deepwater ports. ULCCs never achieved their full potential. In early 2000 a tanker owner ordered two 440,000 DWT ULCCs, with 2 options, the first ULCC order in 20 years. Typical double hull ship of 410,000 DWT 377 0 m length overall × 68 0 m breadth × 23.0 m draft

Lightship. 45,000 tons of steel. Other tankers include Eiffel Tower, Empire State Building, Queen Elizabeth 2, and Tanker Queen Elizabeth 2.

Ocean Loss

Ocean loss comprises of the loading loss at the load port, transit loss during voyage and unloading loss at the disport. Ocean loss normally taken as 0.2% to 0.3% of the Bill of Lading (BL) quantity for arriving at the landed cost.

Insurance

Insurance is the charge incurred for covering the voyage risk. It is normally taken as 0.06%-0.1% on FOB crude cost but could be higher in case of war or other eventualities prevailing at a particular time in the international arena.

Wharfage

Refers to port related charge such as piloting, water tax, port usage charges etc.

2.2 CHARTER PARTY CONTRACTS AND NEGOTIATION

A Time charter commences with the delivery of a vessel to her Time charterers, and it ends with the corresponding redelivery back to that vessel's Owners. In between those occasions, with the exception of certain 'off hire' periods, hire money is payable by the Chatterers the Owners and the Master is to follow the orders of the Time charterers with regard to the prosecution of his ship's voyages, unless, of course, those Charterers instruct him contrary to the terms and conditions of the Charter party. Scrutiny of the terms of a Time charter party will reveal that the Owners will be required to deliver the ship to her Time charterers at some prearranged geographical location, that location may be a safe berth in a named port or, perhaps, at any one safe port out of a whole number of alternatives in a particular range, say 'Antwerp/Hamburg Range, Inclusive'; which port is normally left to the Charterers to nominate some time after the fixing of the business and, usually, at an agreed time before delivery is to take place.

Delivery and Redelivery Surveys

The Time charter Delivery and Redelivery Clause(s) will generally specify that a survey is to be conducted on each occasion, often stating that such surveys are to be performed by jointly approved surveyors half of whose fees are to be reimbursed by the Chatterer, and half by the Owner. Furthermore, it will probably be agreed that the time spent in performing the Delivery Survey is to be for the Owner's expense, whilst the time spent over the Redelivery Survey operation is to be for the account of the Time charter – i.e. the vessel remaining on hire.

The Delivery and Redelivery Surveys set out to establish three main points, which are:

(i) The quantity of bunkers (fuel for the main engine and auxiliary equipment) on board (and also, where applicable, the quantity of any boiler water).
(ii) The condition of the ship – particularly the cargo carrying areas.
(iii) That the holds and/or tanks are fit for the carriage of cargo.

Logically, in order to accurately assess the quantity if delivery is made on picking up an inbound pilot, the expenses for towages, pilot age etc. become a matter for the Time chatterer's account of bunkers onboard, the handover of the vessel wither on to or off her Time charter should occur where a survey can physically take place, 'at a dock or wharf as the New York Produce says, or even at a convenient anchorage. However, a certain amount of negotiation into this aspect often arises when fixing the Time charter, concerning where exactly in a port or location, the delivery/redelivery operation will take place; which decision may depend on the relative strengths of the freight market and whether either Owners or Charterers are in a position to dictate terms. One might read from a Time charter text, that 'the vessel is to deliver upon taking the inbound pilot' (T.I.P.) or, perhaps, 'redelivery upon dropping the outbound pilot' (D.O.P.), which points may be some distance away form the dock or wharf alongside which the vessel is consigned. Since it is unlikely to be possible for a survey to be performed at the pilot station one may be excused for wondering why such a delivery/redelivery position is agreed.

The advantage to Owners of this somewhat 'illogical' arrangement, however, is that after arrival at the delivery port, the expenses of towages, pilot age, line-handling, etcetera, involved in the passage inwards from the delivery point at the pilot station to the berth (or in the reverse direction upon redelivery), become a matter for the Time charterer's account, and it may happen that the savings thus made by the Owner can be considerable, and worth several day's hire money. Of course, the surveyor needs to adjust the eventual tank-soundings accordingly, to reflect the actual bunker consumption of the inward passage, or an estimate for the outward journey from the cargo berth.

For example, in the case of delivery at 'taking inward pilot Hooghly River', followed eventually by a survey on arrival at a berth in, say, Haldia, the bunker consumption difference between these two places could be substantial. At this juncture it should be noted that a distinction is sometimes made in the chartering arrangements surrounding delivery at a pilot station. Should delivery be scheduled to take place 'upon taking inward pilot' (T.I.P.), it could be argued that in the event of a suspension of the pilot age service (e.g. because of adverse weather/strike action, etcetera) that the vessel, although arrived at the correction position, has not physically delivered on Time charter as the pilot has not boarded. Thus during negotiations Owners may be expected to attempt to word the charter party delivery clause to read 'upon arrival pilot station' (A.P.S.), so as to avoid that risk. There is also, of course, occasions when a vessel puts to sea 'unfixed', her Owners instructing the Master to head the ship in a certain direction, whilst they attempt to arrange suitable employment. If that employment happens to be a Time charter, both parties to the contract will be forced to agree upon some arbitrary delivery position, without being able to arrange an on-hire survey. Thus the vessel may deliver on Time charter, for example, 'on passing Gibraltar', or even retroactively, perhaps 'on dropping the outward pilot' at, say, Geneva – the previous port. Assuming that agreement could be reached on the date and time of delivery and also on the quantity of bunkers remaining aboard at that time, it would still be prudent to arrange a 'delivery' survey at the first opportunity, if only to assess the vessel's condition at the time of the survey, since this statement may prevent a future dispute on the occasion of the vessel's redelivery. Thus, with regard to the three aims of the on and off hire surveys, it will be of value to examine these in more detail; since the statements of the surveyor in these respects may be the catalyst of subsequent, avoidable disputes.

Bunkers

The reason for the first objective – that of assessing the quantity of bunkers onboard the vessel – is simply to enable the Chatterer to purchase from the Owner the bunkers remaining onboard the vessel at the time of delivery, and for the reverse process to take effect on the occasion of her redelivery. In order to proceed with this element of the time charter accounting procedure it is obviously necessary to ascertain the quantity of bunkers involved. Most time charters, furthermore, contain a clause ' specifying the price per ton of bunkers sold and purchased in this way, either entering an actual figure or perhaps referring to the 'current market price' or utilizing some similar phrase. The clause is likely to continue, to describe the quantity to be onboard either at the time of delivery or redelivery, adding a phrase such as, '... but always sufficient to reach the nearest main bunkering port', or, alternatively, 'expected about... tons'. Occasions may arise, however, where it is possible for one party or the other, to secure considerable profits in connection with the bunker reimbursement operation. For example, shortly before his vessel's delivery on time charter the Owner may manage to secure a large supply of bunkers at a cost below the price entered in the charter. Once he has been paid by the time charterer in accordance with the contract, and reimbursed the bunker suppliers, any balance is his. Other than shrewdness on the part of the Time charterer when negotiating the business, by including a relevant price per ton in the charter party text, or some limiting phrase such as 'not to exceed... tons', there appears little to legally prevent the Owner so doing. However, at the time of redelivery – under English Law at least – the Charterer is not in a similar position to the fortunate Owner. Even if the charter party contains no controlling limitation as to maximum or minimum quantity remaining onboard on redelivery, it is taken that a Charterer is not allowed to profiteer in this way. He must act reasonably and return the vessel with at least sufficient quantity so as to enable the ship to safely reach the nearest port where bunkers are freely available, but cannot legally take the opportunity to 'refill' the bunker tanks at, say, the port of final discharge, when such an action would not be so as to perform a further voyage but, patently, to make a profit by playing on differences in the bunker prices.

Since the basic intentions of both parties to a time charter contract are – or should be – far removed from bunker price profiteering, it is perhaps best, to remove all temptation by specifying in the charter party approximate tonnages which the vessel is expected to have onboard at the times of delivery and redelivery and the applicable price(s), and this approach is commonly utilized. The Inter tank time, in fact, avoids mentioning any fixed price for bunkers by utilizing the sentence – 'The price for the bunkers to be the lowest listed prices published by a major oil company prevailing at the times of delivery/redelivery at the respective ports'. This clause thus circumvents the problem sometimes encountered, where the actual ports of delivery and/or redelivery are unknown at the time of fixing and of drawing up the charter party. A price inserted in the contract at that time could have but little relevance to the actual prices ruling at the appropriate ports and so, by inserting this clause, balance and equity is achieved. Perhaps the fairest clause of all is provided by the New York Produce which states 'That the Charterers, at the port of delivery, and the Owners, at the port of redelivery, shall take over and pay for all fuel remaining onboard the vessel at the current prices in the respective ports, the vessel to be delivered with no less than... tons and not more than... tons, and to be redelivered with not less than... tons and not more than.....tons'.

Condition of the Vessel

The second objective of Delivery and Redelivery surveys is that of assessing the condition of the ship at those times, and which leads particularly to the matter of damages caused to the vessel during the currency of the time charter. Most time charters specify that the vessel is to be redelivered to Owners in substantially the same condition as on her delivery fair wear and tear excepted, and this second objective of the surveys sets out to establish whether the vessel is in substantially the same condition, or whether her condition has deteriorated beyond what can be described as fair wear and tear. In the unfortunate case that excessive deterioration has occurred, it may be due to lack of proper maintenance and therefore unlikely to be the fault of the Charterer; but occasionally a vessel has suffered damages over and above any connection with either fair wear and tear or with poor maintenance, and the survey is designed to list these defects so that the Owner can be adequately reimbursed for the cost of their repair. Most, if not all damages normally incurred on a time charter result from stevedore damages occurring during loading and discharging operations, and because of the difficulties inherent in collecting appropriate reimbursement from Stevedoring companies as a group, most Charterers enter a clause in the charter party detailing the action to be taken by the Master when dealing with such matters at the time the damages are actually caused. Normally such a clause will require the Master to officially notify the Stevedores in writing of any damages they have allegedly caused within, say, twenty-four hours of the incident, or certainly prior to departure from the port concerned. Some Owners or Charterers issue a special form for this purpose, in which the damage is given its exact location – e.g. 'No. 3 Hold, hopper plate, starboard side, forward part' – and description – e.g. 'one indent approximately 3 cm in length, 8 mm in depth'. An example of such a form is included in this Chapter and illustrates what is necessary. The Charterer, of course, in his capacity of Disponent Owner, is entitled to claim against any Stevedores causing damage, for appropriate reimbursement for those damage repair costs. Any funds he may collect from these sources can be held over for possible use in settling eventual end-of-hire claims that the Owner may present, issued on the basis of the Redelivery survey report. Stevedores sometimes immediately repair damages themselves on their own account, whilst others can be expected to refuse to pay any claims unless they are properly advised of damages allegedly caused, and held responsible for same by the Master (or one of his officers) either at the time of the occurrence of the damages or shortly afterwards, thus being given an opportunity to inspect same. The prudent Charterer will ensure that standing instructions are issued to the Master, as well as being included in the time charter party, to promptly claim against Stevedores suspected of causing damage – if possible obtaining their admission of liability. This will enable the Charterer to thereafter submit a further copy of the damage notification form to the Stevedores, together with an estimate of the cost, of repair of the damage (often a quotation issued by a reputable repair company), as well as an invoice requesting appropriate reimbursement. Failure by the Master to comply with these time charter requirements can involve the Charterer in loss, in which case the latter can reasonably withhold equivalent payment to the Owner, when such damage is identified by the Redelivery Survey. Of course, minor stevedore damage can often be described` as wear and tear, and again the subject have to be examined in the light of the vessel's trade.

An Owner (or a Disponent Owner for that matter) can hardly charter a vessel for the extensive carriage of scrap metal, for example, and not anticipate a certain amount of cargo-handling damage during the loading or unloading operations. Naturally, excessive damage incurred even during scrap trading is claimable, but what may be fair wear and tear or excessive in such circumstances is debatable. However, providing the Redelivery survey accurately illustrates the damages caused during the currency of the charter, and the Master efficiently and conscientiously claims against Stevedores and such like when necessary, at least the evidence will have been properly compiled and it will need to be left to good sense and fair interpretation between the Owner and Charterer to reach an equitable solution to any disagreements that may arise. Ideally any repairs in respect of damages incurred during a period charter should for convenience be repaired during the currency of the time charter, on the occasion of interim repairs and/or dry dockings or, alternatively, at the end of the time charter as a prerequisite before redelivery is affected. If the latter course of action is taken, then logically hire should continue whilst the repairs are being affected, but it is often more convenient to both parties to reach an agreement for the Charterer to reimburse the Owner for the reasonable cost of repairs and for the Owner to put these repairs in hand on the occasion of the vessel's next dry docking, or on some other suitable occasion. It is not always possible, however, to easily obtain agreement on the cost of repairs at the termination of a time charter, although it would perhaps be possible for the Charterer and Owner to agree to accept the expressed opinion of their jointly-appointed surveyor in this regard.

Cargo Spaces

The vessel's cargo-spaces are not only inspected with regard to their condition in respect of damages and wear and tear, but also, as the third objective of the Delivery and Redelivery surveys, in regard to their fitness to carry cargo. Most, if not all, dry cargo time charters require the vessel to present 'with holds clean-swept and ready to receive cargo'. Since the vessel is almost certainly to be redelivered 'in alike condition' it is only reasonable that the Owners expect her to be returned similarly, with clean-swept holds at the termination of the charter. Tanker charters are not so explicit and one is left to rely on what is reasonable. For instance, a tanker delivering in Rotterdam with the next loading port in North Africa, could deliver on time charter and proceed in ballast to the loading port, washing tanks and gas-freeing as necessary enroute, without interfering with the time charter arrangements apart from the small extra fuel consumption involved in the cleaning operation. Providing the Owners accept similar treatment on redelivering this would seem an equitable solution.

Delivery Certificate

DATE: ________________ PORT__________________

This is to certify that the Liberian flag vessel m.v. 'UNICORN' of 9607 net and of 14516 gross registered ton, has been delivered by her Owners, Messrs. UNICORN CORPORATION to her Time charterers, WHEATCHART LIMITED, at the port of SEVILLE on the 9th MAY, 1979 at 2300 hours local time, according to the terms, conditions and exceptions of the time charter-party dated 15th April, 1979. At the time of delivery, quantities of bunkers and fresh water remaining on board were calculated and certified to be:

FUEL OIL 211.00 Metric Tons

DIESEL OIL 911.30 Metric Tons

FRESHWATER 210.00 Metric Tons

The vessel delivered with clean swept holds in accordance with the Time charter requirement.

The on-hire condition survey report dated 9th May. 1979, is attached herein, being duly signed by the Master and Joint Surveyor.

Delivery and Redelivery Certificates

Having carried out the required on or off-hire survey, the surveyor, or the agent, is left to draw up and issue the Delivery and Redelivery Certificate, detailing the date and time of handover, the quantity of bunkers (and perhaps boiler water) onboard at the time, and confirming, where necessary the holds/ tanks condition, attaching at the same time the condition survey report. This certificate being signed and agreed by all parties becomes a vital document in assessing hire payments due, and also of the commencement and termination of various liabilities of the Owner and Charterer.

Stevedore's Damage Notification

Port: ______________________ Date: ______________________

To STEVEDORES Name ________________________________

Address ______________________________

____

____

The following damages occurred on hoard m.v.

____

____

During the loading/discharging operations at: ______________________

____

Description and location repaired satisfactorily thereof damage paid unsatisfactorily not repaired

Master Stevedores

IMPORTANT: Master to ensure that notification of damages given to Stevedores within time specified in Charter party.

Tanker Characteristics (Board)

The amount of information required to be entered in a time charter party concerning a ship's description depends very largely on the nature and whereabouts of the intended trade upon which the vessel is to be employed. In some cases, particularly those involving a straightforward trip-charter, very limited information will probably suffice. For more involved and lengthy period employment, far more information may be required, and it is quite normal for extensive details to be provided and guaranteed as to accuracy by the Owners, such information being included in the contract in the form of a separate side-clause. An example of such detailed information is particularly provided by the tanker charter parties, both the Shell time and Inter tank time having appropriate addenda (a document entitled the 'Form B' in the case of Shell time, and Part III for the Inter tank time), which parts themselves amount to several pages of comprehensive description of the time chartered vessel.

In addition to information entered in a time charter party, it is quite commonplace for a Charterer to request and to receive copies of a vessel's General Arrangement and/ or Capacity Plan, from which a considerable amount of what may appear superficial but is perhaps valuable information, e.g. the overhang of decks above holds - will be obtainable, enabling the Charter to exercise more fully his responsibilities as a Disponent Owner, without his continually having to contact the actual Owners for such incidental information. In fact, both the Shell time and Inter tank time

Charters require the following plans to be supplied by the Owners

(1) General Arrangement Plan, including Loading Scale.
(2) Detailed Cargo Manifold Arrangement Drawing.
(3) Pumping Arrangement Plan.
(4) Plan of Cargo Tank Ventilating System.
(5) Manufacturer's Characteristic Curves of Pumps, if Centrifugal Pumps installed

The completion of a vessel's description clause should not be dealt with carelessly, for various items of information contained in that clause may have a profound effect upon other parts of the charter. For example, a vessel's described speed and bunker consumption is vitally important in gauging whether or not the ship is performing efficiently and whether the Charterer is entitled to claim for off-hire time, for alleged under-performance. Even where there may be no claim for poor performance, such information provides the basis for measuring deviation time and bunker consumption on the occasions when a vessel is withdrawn from service for routine dry docking, or for some similar event. Equally important, a ship's constant weights - a tonnage allowance in respect of water, stores, provisions, spare parts, etcetera - may figure prominently when a Charterer is calculating the cargo quantity he expects the vessel to carry on any particular voyage. Even the gross and net tonnages are of relative importance, and are not mere window-dressing, since on one or other of these figures are based various port charges, light-dues, tonnage-dues, etcetera. Errors in one figure or other could possible incur the Charter in unanticipated extra port-expenses; whilst a vessel's country of registry is one more factor that deeply concerns many modern Charterers, especially in these days of arbitrary action taken by various groups of shore workers, on the basis of whether these groups, individuals therein, feel any real or imaginary animosity towards the flag of visiting ships. Obviously, a vessel's dimensions - its draft, cubic capacity, length overall, beam and air-draft, for example - are all of varying importance whatever a vessel's trade or period of charter, for a Charter must satisfy himself that the ship is able physically to perform the voyage(s) on which he intends to employ her. For more specific trades, more specific information will be required. Thus the tanker which is to trade out of Indonesia carrying waxy crude oil must have efficient heating coils which should be described as such in the charter party, whilst the bunker to be would best be fitted with a CO_2 system. However, a brief and general list of details to facilitate the efficient handling of a 'period' time charter would perhaps be as follows; additional information being supplied as necessary depending on the envisaged trade:

(1) Vessel's name
(2) Year of build
(3) Name and location of Owners
(4) Flag
(5) Class
(6) Call letters
(7) GRT/NRT
(8) Summer Deadweight & Draft
(9) Fresh Water Allowance
(10) Cubic Capacity
(11) Speed and Consumption
(12) Constant weights

Having accurately entered details of a vessel's dimensions, tonnages and fittings, however, does not end an Owner's obligations in this matter. A fundamental alteration in the described details - say a change of flag - should not be undertaken without prior assent and approval of the Charterers, whilst undertakings that the vessel is in good condition, e.g. 'tight, staunch, strong, in good order and condition, in every way fit for the service' are expected to be reasonably applied throughout the period of time charter. Indeed, the Shell time states that the Owners 'undertake to use their best endeavors so to maintain the vessel during the period of her service', going on to say that Owners 'bind themselves to take every possible precaution to maintain the tanks, valves and pipelines. Whilst the majority of dry cargo forms are not so specific in defining an Owner's duty to maintain his vessel during the currency of the time charter, to that standard to which the ship aspired upon her delivery on the charter, nevertheless, there is an implication towards this effect if nothing else, and the fact that the vessel must maintain her described class - e.g., Lloyds 100 AI, or similar - places an onus upon the Owner to keep his ship in good condition at least in that respect. It stimulates initiative.

If there are any essential movements of supplies to be transported, liners which have their scheduled services and, therefore, commitments on fixed routes, cannot be easily diverted for transportation of essential supplies. During the last two world wars, the major part of the world fleet was employed for strategic purposes. It was extremely difficult for the neutral countries to charter tonnage on the world market for the carriage of vital foodstuffs and raw materials. In such times of emergency the Government can employ the country's own tramp fleet for carriage of vital cargoes at fixed minimum rates as was done in all shipping countries during the Second World War In times of war, ships of belligerent countries are operated as tramps. Familiarity with this type of operation, therefore, is a great asset in times of trouble.

Finally it can be stated that the tramp fleet provides great flexibility to transport capacity in the changing patterns of world trade and also seasonal or sudden and unexpected variations in seaborne trade, which is a regular feature of world trade. Wars or threats of war, famines, bumper harvests, underproduction in some countries and overproduction in other countries regularly engender movements of cargo, which cannot be foreseen and provided for. Since liners are tied to their fixed routes and trades, it is only tramp ships, which can effectively cater to such movements of cargo and smoothen the flow of international trade. With the growth in world trade, goods and commodities are moving in larger quantities in international trade. If on some routes these goods and commodities of a homogeneous nature move in very large quantities, shipping develops to carry these commodities economically.

For example, ships with extensive refrigerator capacity have been built. Even within the category of refrigerated ships there has been further specialization such as fruit carriers, meat carriers or carriers of dairy produce. Apart from the phenomenal growth of the specialized tanker fleet to carry oil after the first world war, ships have now been built or converted not only for the bulk carriage of ore but also for the bulk carriage of rock phosphate, liquid gases, wine-even orange juice. Highly specialized vessels are closely connected to the trade for which they were built or converted and, therefore, carry out voyages on these routes and resemble liners, in that they operate on fixed routes. 'But the more specialized ships become then, paradoxically the less like liners they are in terms of cargo composition and freighting arrangement. The practical decision whether such vessels are liners or tramps will then depend on the conditions under which the cargo is carried.

If the shipper buys all the cargo space for one voyage or for a series of voyages or a period of years, then there is little practical reason to think of the vessel concerned as a liner. If the ship's space is not hired as a whole, but a charge is made for the actual cargo carried, for practical purposes the ship is a liner. Where the goods, although homogeneous in nature are dispatched by numerous shippers each sending small consignments, then a liner-type operation is required. In recent years, especially in the last decade there has been spectacular growth of specialized bulk carrier fleet for carriage of ore, food grains, coal, etc. Since these bulk carriers are hired or chartered by single shippers they are functionally tramps. Except those industry-owned and industry-tied bulk carriers such as a good proportion of specialized ore carriers, a very large part of the bulk carrier fleet is chartered by shippers as in the case of traditional tramps. There is no reason to distinguish such bulk carriers from tramps for purposes of our study.

Charter Negotiation

Before examining aspects of tanker shipping in any depth it is firstly desirable to touch upon the various alternative methods of chartering merchant-ships that one encounters in the world's freight markets. Thus will the reader be able to place into perspective the inherent variations of each of these methods, and focus attention on the specialized features of period employment, which will be covered in the subsequent chapters.

Voyage Chartering

As the name implies, voyage chartering covers a situation where a ship is employed to carry a particular cargo from a certain port or area, to a designation i.e. for a voyage, the Ship-Owner being reimbursed by freight money paid either on a lump-sum basis, or in relation to the actual cargo quantity carried. The Owner is responsible for the returning expenses of his ship, such as employment of crew, purchase of stores and provisions, etcetera, as well as the incidental voyage expenses incurred, covering port charges, light-dues, special voyage insurance, bunker fuel supply, and so on – in short, dealing with the host of incidental matters necessary to efficiently operate a modern merchant-ship. The risks of sudden and alarming rises in the price of bunker fuel are his, and he must provide the funds with which to pay port expenses enroute, even though, under the particular terms of the contract, he may not receive his freight money until after the cargo has been discharged.

The Ship-Owner also runs the risk of the duration of the voyage being other than contemplated when 'fixing' the business. He may well discover that his vessel will be unexpectedly delayed for, perhaps weeks, awaiting the availability of a berth at which, say, to discharge his ship. Most voyage charters, however, include a clause by which the Ship-Owner may be entitled to compensation in the form of liquidated damages (demurrage) for any port time spent in excess of a total, stipulated time (lay time) laid down in the charter party. On a voyage charter, the risks of sudden and alarming rises in the price of bunker fuel are the Owners. It can also be specifically agreed that either the Charterer or the Ship-Owner assumes responsibility for cargo-handling expenses involved in loading and/or discharging the vessel, the freight rate being adjusted to reflect any such extra expense the Ship-Owner may thus incur. A multitude of voyage charter forms exits, covering the entire spectrum of ships and cargoes. Some are specifically drawn up for an individual trade, whilst others, of a more general character, being capable of adaptation to suit almost any imaginable commodity or circumstance.

Consecutive Voyages

On occasions a vessel may be chartered for a whole series of voyage, each voyage being considered to be separate venture, with its own freight settlement and individual establishment of lay time used. Such series of voyages are termed Consecutive Voyages, and that phrase adequately serves to describe this method of chartering.

Time Chartering

Time chartering, as its name implies, arises from an instance where a Charterer hires a ship for a period of time. The Owner is still required to operate his ship, but instead of freight, he receives previously agreed sums of hire money, in advance and at regular intervals; whilst the responsibilities of arranging the vessel's employment and bunker fuel purchase pass on to the Charterer, as does added responsibility for payment of expenses such as port dues, canal tolls, bunkers and the like, incurred on the vessel's voyages. Whilst running expenses remain very much the responsibility of the Ship-Owner, the Charterer may utilize the ship to carry his own cargoes, in which case it is possible that he will receive no freight monies, instead saving himself the task of voyage chartering various ships, and the expense of paying freight money to their Owners.

However, should the time charterer contract the vessel to perform a voyage, he is still entitled in his capacity of Disponent Owner to receive any applicable freight and/or demurrage money and, indeed, he may have time chartered the vessel as a pure speculation, judging that freight market rates were about to rise, and hoping to capitalize by voyage-chartering out and earning more from freights than he is committed to pay in hire. Alternatively, some Charterers employ a vessel on a period basis in order to secure carriage for their goods at a predetermined and relatively static cost per ton or unit, and in so doing overcome the only too well-known freight-rate vagaries that beset international shipping. Or it may be that they have sufficient employment of their own for only a part of the period of charter, and thus need to 'sublet' the vessel for occasional 'fill-in' voyages or, perhaps, for another period time charter.

Indeed, it is quite normal for a Charterer having a contractual commitment to move a given quantity of cargo between two ports, to time charter a suitable ship for reasons of security against unknown freight liabilities, whilst seeking return cargoes for the empty, or ballast, leg of each trip in his role of Disponent Owner of the time chartered vessel; thus producing an overall saving in this aspect of his transportation of the commodity.

It is important to note that cargo-handling expenses arising when a ship is fixed on at time charter will always remain the Charterer's responsibility, the Owner having no expenses in this respect, and no involvement in the appointment of stevedores. Compared with voyage-chartering, there are relatively few time charter forms, and of those that there are, some provide merely a basis for the contract, with many amendments being made to the printed text, and additional clauses being added; although others, particularly tanker-charters, rely more fully on the printed words. Nevertheless, all time charter parties fulfill the role of providing a firm skeleton on which to build the flesh of a contractual agreement.

General cargo vessels and bulk-carriers are fixed mainly on either the New York Produce Exchange Time Charter Form (1946); the Baltime (1939); or the Linertime (1968) documents; whilst tankers have more choice at their disposal since most major oil companies have their own Company form, with titles such as Shelltime, Beepeetime, Texacotime, et al; they being, in fact, similar in layout and wording. There are, however, tanker time charter parties other than those produced by the oil companies and forms such as 'Intertank time' are available, published by Intertank -an association of independent tanker owners. For ease of reference, copies of the blank forms of each of the BP Tank Voyage and Shell Time 4 charters (as adapted by Indian Charterers and Oil Companies) are included in the Appendix, and the reader will find reference is made to relevant sections of these forms in subsequent pages. Finally, it should be noted that time charters as described above are best considered as period time charters, in order to distinguish them from trip charters which are described immediately hereunder.

Trip Time Chartering

During recent years 'trip-chartering' has become a somewhat prominent feature of chartering, being a method of employment comparable both with voyage chartering (in that such a trip represents a single voyage or a 'round trip'), and with time chartering (since the contracting partners assume the usual roles and responsibilities associated with period employment). To all intents and purpose, however, trip charters can be considered simply a shorter variant of time chartering. The rate of hire applicable to trip charters is generally related to current 'spot-market' voyage freight-rates and not to the somewhat lower freight-rate levels that are normally associated with period time charters, which lower rates result from the security obtained by Owners and their reduction of risk expectantly provided by such longer period of employment. Such charters are identical to time charter period employment, and all that is subsequently written in this book on time chartering can be assumed to similarly apply to trip-chartering.

Bareboat (Demise) Chartering

A further variation of period charter employment applicable to merchant vessels is that termed bareboat chartering or, alternatively, chartering by demise. This form of chartering is, perhaps, not as common today as it once was, although it is still to be encountered on the world's shipping scene, particularly in the European Coastal and Short-Sea trades. In such a contract the Ship-Owner virtually relinquishes all responsibility and rights in respect of his vessel for a specified period of time, in return for a prearranged, regular amount of 'hire'. The Charterer becomes a 'Disponent Owner' in the strictest sense of the word, operating, crewing and chartering the vessel as though he were, in fact, the Owner.

Contracts of Affreightment

Finally, and for the sake of completeness, Contracts of Affreightment arise where Charterers (or Owners in some cases) undertake to transport a given quantity of some specific commodity or general goods, from place to place over an agreed period. How that person manages the task is left to his own choice, and the contract simply provides for the movement of the goods, the contractor being remunerated accordingly on, say, a tonnage or volume basis and, in turn, arranging employment of whatever ships are required for his own account and in his own way.

The position of period employment in the spectrum of various methods of chartering merchant ships having been established.

2.3 DEMURRAGE

In layman's language, demurrage is detention of a ship, freight car, or other cargo conveyance during loading or unloading beyond the scheduled time of departure. Compensation is usually paid for such a detention. Separate freight charge, in addition to ordinary shipping costs, which is imposed according to the terms of a carriage contract upon the person responsible for unreasonable delays in loading or unloading cargo. In maritime law, demurrage is the amount identified in a charter contract as damages_payable to a ship owner as compensation for the detention of a ship beyond the time specified by a charter party for loading and unloading or for sailing. Demurrage is intended to serve the public interest by facilitating the flow of commerce through the prompt loading and unloading of cargo. In general, the person liable for demurrage is the one who assumed the duty to unload or load the cargo but failed to fulfill it. A consignee who agrees to unload a shipment but unreasonably delays in doing so is liable for the charge. Payment of demurrage is excused only if the delay was unavoidable, such as a delay caused by a natural disaster or the fault of the carrier. Reciprocal demurrage may be imposed upon a carrier who unreasonably delays in providing transportation to customers. The practical effect of reciprocal demurrage is a reduction in the customer's shipping charges unless the contractual amount exceeds that figure. If a person against whom demurrage is imposed fails to pay, the carrier might have a right to keep the goods until payment is made. This is known as demurrage lien, enforceable only if authorized by statute, contract, or custom.

Demurrage Calculation

STATEMENT OF FACTS	*Lord Port*
VSL ARRIVED	30/0130 HRS
ANCHORED	30/0700 HRS
NOR TENDERED	04/1600 HRS
PILOT ONBOARD	13/1645 HRS
COMMENCED MOORING	13/1930 HRS
COMPLETED MOORING	13/2000 HRS
COMMENCED HOSES CONNECTION	13/2000 HRS
COMMENCED DEBALLASTING	13/2000 HRS
COMPLETED HOSES CONNECTION	13/2205 HRS
COMPLETED DEBALLASTING	14/0145 HRS
COMMENCED LOADING	14/0230 HRS
COMPLETED LOADING	15/0900 HRS
COMMENCED HOSES DISCONNECTION	15/0900 HRS
COMPLETED HOSES DISCONNECTION	15/1000 HRS
COMMENCED UNMOORING	15/1000 HRS
COMPLETED UNMOORING	15/1030 HRS

DOCUMENTS ONBOARD	15/1615 HRS
PILOT OFF	15/1645 HRS
VSL SAILED	15/1715 HRS
NEXT PORT	INDIA
ETA NEXT PORT	08/08/O.M.
BUNKERS ON ARRIVAL	BUNKERS ON DEPARTURE
FO: 2G70 MTONS	FO: 2235 MTONS
DO: 108 MTONS	DO: 89 MTONS
FW: 300 MTONS	FW: 300 MTONS
DRAFT ON ARRIVAL	DRAFT ON DEPARTURE
FWO: 05 M 30 CM	FWO: 16 M 48 CM
AFT: 08 M 30 CM	AFT: 16 M 48 CM
MEAN: 06 M 60 CM	MEAN: 16 M 48 CM

BILL OF LADING FIGURES ARE AS FOLLOWS:

GROSS/NET BBLS	962815/962333
CROSS/NET MTONS	129910/129833
GROSS/NET TONS	127958/127782
GROSS/NET API	34.8J/ 34.87
GROSS/NET SPEC GRAY	0.8506/0.8505
BS/W %	0.05 X
SHIPMENT NBR	SHELL PET. OEV. CO. LTD.
CONSIGNEE	STASCO
B/L DATE	15/07/2001.
HBR/SEI/B/I COPIES	I SEI, 3 ORIGINALS - 4

DELAYS AND STOPPAGES:

DATE AND TIME	*REASONS*	
FM 04/1600 ARRIVED BEFORE	HRS TO	12/0001 HRS - LAY DAYS.
FM 12/0001 OPR	HRS TO CLEARENCE.	13/1645 HRS - AWTG
FM 13/2000 OEBALLASTING.	HRS TO	14/0145 HRS -
FM 15/1030 SHIFTING.	HRS TO	15/1130 HRS -
FM 15/1220 SHORE	HRS TO FIGURE.	15/1230 HRS - AWIG
FM 15/1230 CARGO	HRS TO	15/1615 HRS - AWIG

DOCUMENTS.

PROTEST FROM MASTER TO TERMINAL FOR:

1. FREE WATER FOUND IN CARGO.
2. OIFF. 8/W SHIP/SHORE FIGURE.

PORT : Vizag

DATE : 07/06/06

CARGO : NIGERIAN GRADC A.PI. B/L 962.815 GR,

BBL8

LIGHT C.O.

TIMESHEET DISCHARGE PORT

	DATE	*TIME*
END OF SEA PASSAGE	12/08/01	1000
N.O.R. TENDERED	12/08/01	1115
N.O.R. ACCEPTED RECEIVED	12/08/01	1712
FREE PRATIQUE	12/08/01	1615
PILOT ON BOARD	12/08/01	1210
TUGS SECURED (5)	12/08/01	1245
COMMENCED MOORING TO ORE BERTH NO. 1	12/08/01	1315
COMPLETED MOORING, PILOT OFF TUGS OFF	12/08/01	1415
COMPLETED HOSES CONNECTION (2 × 10 T)	12/08/01	1425
COMMENCED TANKS ULLAGING	12/08/01	1525
COMPLETED TANKS ULLAGING	12/08/01	1625
COMPLETED CALCULATION	12/08/01	1712
M/T SATYAMURTI ALONGSIDE	12/08/01	1820
HOSES CONNECTED FM SATYAMURTI	12/08/01	1840
COMMENCED DISCHARGING (REQUESTED RATE 1500MT/HR)	12/08/01	1000
COMPLETED DISCHARGING	13/08/01	1005
COMMENCED TANKS ULLAGING	13/08/01	1035
COMPLETED TANKS ULLAGING	13/08/01	1130
COMPLETED CALCULATION	13/08/01	1100
HOSES DISCONNECTED FM SATYAMURTI	13/08/01	1350
MT SATYAMURTI CAST OFF	13/08/01	1200
CARGO DOCUMENTS COMPLETED	13/08/01	1416
POB FOR SHIFTING TO MPB BERTH	13/08/01	1416
FENDERS OFF COMMENCED UNMOORING	13/08/01	1500
COMPLETED UNMOORING	13/08/01	1520

World Shipping

To understand the importance of shipping in our lives, we only need to look at the diverse types of goods we use everyday. From the mouse to the computer screen, the shirt on your back and shoes on your feet to the fuel that powers our vehicles and factories around the world every day, what we now call basic necessities of modern life are brought to us by world trade. Some 90% of world trade is transported by sea. In 2006, seaborne trade reached over 30 trillion ton-miles (an increase of 49% compared to trade in 1996).

The three main types of goods transported by sea are dry bulk, oil and containerised cargo. Dry bulk accounted for 38% of the world's seaborne trade in 2006. Oil trade formed the second largest type of cargo accounting for 36% and containerised cargo contributed 15%. These trades are carried by more than 20,000 merchant ships to various ports around the world, leading to a global cargo throughput of 14.8 billion tonnes and container throughput of 440 million TEUs. The operation of these ships generates an estimated annual income of about US$630 billion (S$1.05 trillion) in freight rates within the global economy. That's about 5 percent of total world trade.

Looking Ahead

In recent years, the world has seen the centre of maritime gravity moving to Asia. Asian countries' share of world seaborne trade increased to 39% in 2006 (compared with 31% in 1996) and where nearly half of the world's merchant fleet is domiciled in Asia[5]. 13 of the world's top 20 ports are also located in Asia, handling around 36% of the world's containers. Three of the world's largest shipbuilding nations are in Asia accounting for 90% of global market share.

Besides the growth of the world's port operations, shipping and shipbuilding have also contributed significantly to the global economy through its demand for supporting services such as ship financing, marine insurance, maritime legal and arbitration services, maritime education and research & development (R&D) etc. The global shipping portfolio of commercial banks is estimated to be about US$ 272 billion in 2006. Currently, the market capitalisation of the 182 public shipping companies listed on 35 exchanges is about US$ 359 billion and major Asian exchanges (comprising, Singapore and Tokyo) account for as much as 34% of the global value. Risk management tools such as trading in forward freight agreements (FFAs) has also grown significantly with notional value of contracts rising from US$ 7 billion in 2002 to US$ 56 billion in 2006.

With Singapore strategically positioned within Asia, we are well placed to leverage on these growth opportunities. To this end, Singapore will continue working towards developing well-integrated business infrastructure and conducive business environment, skilled manpower and advanced maritime R&D and technology in support of the continued growth of the Singapore Maritime Cluster that serves global needs.

SHEET 1

Demurrage Calculations

Vessel Name						
Loading at BONNY				**Time Elapsed**		
Event	**Time Start**	**Time Ends**	**Time Elapsed**	**Days**	**Hours**	**Mins**
TimeUsed	05/07/01	15/07/01 16:30	10/01/00 16:29	10	10	29
Less						
Lay time Allowance						
Nor Allowance						
Shifting	13/07/01 16:45	13/07/01 20:00	00/01/00 03:15	0	3	15
Bad Weather			00/01/00 00	0	0	0
Deballasting	13/07/01 21.40	14/07/01 01.20	00/01/00 03:40	0	3	40
Documents on Board	15/07/01 10:00	15/07/01 13:00	00/01/00 03:00	0	3	0
			Total Deductions	2	2	55

(Contd.)

Discharge at: Vizag				Time Elasped		
Event	**Time Start**	**Time Ends**	**Time Elapsed**	**Days**	**Hours**	**Mins**
Time Used	12/08/01 14:15	16/08/01 09:05	03/01/00	3	18	50
Less						
Lay time Allowance				1	12	
Not Allowance					0	
Shifting			00/01/00 00.00	0	0	0
Over Pumping			00/01/00 00.00	0	0	0
			00/01/00 00.00			
			Total Deductions	0	12	0
Dem Rate	5	23,000.00	Total Times Used both Ports	14	11	19
Dem at Loadport	5	196,043.06	Total Deductions at both Ports	0	15	55
Dem at Disport	5	52,548.61	Time User	13	19	24
Total Demurrage	5	248,291.67	Less Laytime			
			3 days	3		
			Time on Demurrage	10	19	24
			Days	10.80833		

2.4 THE FORWARD FREIGHT AGREEMENT (FFA) MARKET

An independent association of FFA broking Baltic Exchange members formed in 1997 and is serviced by the Baltic.

Aims

- Promote the trading of forward freight agreements (FFAs).
- Promote high standards of conduct amongst market participants.
- Liaise with the Baltic Exchange to ensure production of high quality indices for use by the freight futures industry.
- Provide a forum for brokers and principals to resolve problems as they arise.
- Develop and promote use of standard contracts.
- Develop the use of other 'Over the Counter' and exchange traded derivative products for freight risk management.

Baltic Exchange

The Baltic Exchange is an independent, London-based association of shipbrokers, shipowners, charterers, financial institutions, maritime lawyers, educators, insurers and related associations involved in the shipping industry. The Baltic Exchange is the world's only independent source of maritime market information for the trading and settlement of physical and derivative contracts. Its international community of over 550 members encompasses the majority of world shipping interests and commits to a code of business conduct overseen by the Baltic.

World's premier ocean freight market for mostly dry bulk cargo (such as coal, grains, and iron ore) where ship owners and cargo owners are matched by the exchange's members called ship brokers. Situated in London, it is also the market for buying and selling ships (accounting for about half the world's new and second-hand tonnage), and where forward freight agreements are traded. The exchange started informally in 1744 in the Virginia & Maryland Coffee House at Threadneedle Street frequented by merchants and ships' captains, and its full name is Baltic International Freight Futures Exchange.

Baltic Exchange members are responsible for a large proportion of all dry cargo and tanker fixtures as well as the sale and purchase of merchant vessels. The exchange provides daily freight market prices and maritime shipping cost indices which are used to settle freight futures (known as *Forward Freight Agreements* or FFAs). Originally operating a trading floor, the exchange's members' transactions are today done solely over the telephone.

In the very traditional maritime business, a widening array of financial instruments for speculation and investment has brought about a new awareness of screen systems for risk management and for trading (which includes pure speculation as well as more conservative arbitrages). Both exchange traded and over the counter derivatives markets for crude oil, natural gas and refined products have grown enormously in the twenty + years since

NYMEX's 1983 launch of WTI futures on crude oil. In the freight markets, the inchoate futures markets of the 1980s evolved into the Forward Freight Agreements (FFA's or just "paper freight") between parties - usually an owner and a charterer. In the post Enron era, freight traders now are benefiting from a new paradigm-financial clearing, which is basically the credit guarantee from a financial entity that could step into to support an FFA commitment, if necessary.

The dynamics of the ocean freight market are not unlike those of the better known crude and oil products markets; tanker shipping people have seen the successes of their customers - where futures and derivatives markets enable users to manage extreme volatility. On the liquid side, seaborne movements of crude oil and petroleum products are closely intertwined with the vagaries of the underlying raw materials markets. Rates for VLCCs - crude oil tankers of 300,000 MT deadweight (with a carrying capacity slightly in excess of 2 Million Barrels) were providing a return to owners averaging $95,000/day in 2004 (rates levels rivaling those of the early 1970s - a time when some of the great shipping fortunes were made). In 2005, hires averaging $59,000/day were seen on VLCC voyages from the Persian Gulf to Japan, according to London based Drewry Shipping Consultants. In April 2006, ships owners were netting under $25,000/ day for the same ships on the same voyages - in line with average rates for 2002. By end May 2006, rates on similar vessels had firmed to levels in excess of $40,000/day; August 2006 saw spot rates approaching $90,000 day for such large tankers, and forward rates approaching $120,000/ day for November/December 2006 forward slots. With fears of Alaskan oil shortages, Middle Eastern tensions and supply disruptions in both West Africa and the Caribbean, spot tanker rates were at historical highs. As was seen in the Summer of 2006, huge amounts of hedge able risk in tanker markets are tied to the uncertainty premiums surrounding oil supply, and, in turn, oil and product prices. As oil's vagaries evaporated, tanker rates calmed down. When December 2006 actually came around, spot VLCC's were fetching under $30,000/day. Financial risk management techniques are now plying their way to the dry cargo side, where China-induced waves of iron ore consumption of iron ore, and bulk raw materials such as metallurgical coal, have brought about quick and usually unpredictable shifts in demand. Consider that Singapore listed Noble Group uses its major presence in the dry bulk shipping arena to manage freight risk for its customers. Noble is working as a logistics partner with raw material producers in Australia (and elsewhere) to market their output into the seemingly re-igniting Chinese raw material markets. In this role, it uses FFA markets to protect margins. Dry cargo rates are at record levels, and, typical of these market dynamics, forward fixing by charterers and freight operators (who then transport iron ore, coal and other commodities) has led a huge uptick in period time charter (forward) activity. Spot rates for Cape size bulk carriers are averaging more than $80,000/day over a worldwide average of trips, with FFA rates for 2007 above $60,000/day out though end-2007. Pana maxes (similar to, but not identical with the Kams armaxes in the fleet of Quintana Maritime) have been hired at $45,000/day over a composite of geographical runs.

FFA levels are in the mid/upper $30,000s for 2007. Screen trading has played a growing role in the freight markets, by the Oslo based International Maritime Exchange (Imarex), where both tanker and dry routes are traded, with a seamless link into the clearing capabilities of the Norwegian Futures and Options marketplace (NOS). Clearing of over the counter transactions in the freight rates for deep sea tanker shipments of crude oil and products were launched on NYMEX's Clear Port in the Summer of 2005, and around the same time on London's on LCH Clearnet. In 2006, the Singapore Exchange (SGX), launched its "Asia Clear" service. The impetus of these clearing activities has been the financial investors – whose ISDA swap contracts demand the creditworthiness offered by the clearing approach. The backing of these financial clearers supplements the maritime industry's historical "principal to principal" approach to managing credit and performance risk generally and is invaluable where business partners in deals are new or unknown counterparties. The Imarex screen trading system is fully integrated with the Norwegian Futures and Options Exchange's clearing role. Freight and commodity interactions are enhanced by the ability to trade FFA's on a screen. In the dry cargo markets, over the counter coal swap markets (which can be traded through electronic arenas such as Global coal) have given rise to arbitrages between swap prices in Rotterdam (delivered) and Richards Bay (origin).

Freight traders, in turn, have traded differentials between these freight synthetics and actual voyages booked between South Africa and North Europe. Increasingly, rates in the physical market are keyed off the levels seen in the FFA arena. Some owners, including Dry Ships, Quintana and Diana Shipping have actually indexed vessel hires to the freight quotes underlying the FFA markets. Growth has come on the financial as well as raw material fronts. At the same time that cargo interests are integrating freight risk management into their programs, bankers were seeking ways to bring in a new set of counterparties other than the usual sometimes over-extended suspects-credit worthy entities that could "buy" freight from their clients. Credit worthiness of counterparties is crucial to banks and financial brokers, who are regulated entities; not surprisingly, they view the clearing aspect as critical in supporting the continued growth of the freight derivatives markets. The FFA markets are big – although the preponderance of private deals prevents a precise measurement of their size. According to one study, by Boston based consultants Celent, tanker derivative trades (mainly done over the counter) comprised $7.5 Billion of an overall $30 Billion freight swap (notional value) marketplace for 2004, with dry cargo freight (including coal markets) accounting for the balance.

The physical tanker market exceeds the size of the FFA market and has been estimated variously to be in the magnitude of $30–$50 Billion annually. Data for 2006 suggests that cleared contracts represent roughly 15% of the FFA market. As consolidation waves have spilled over the maritime markets (more in tankers than dry) the deals, measured in the hundreds of $Millions, have brought about huge risks to shipping companies themselves – and to two other vital groups of stakeholders – financiers (usually big banks that specialize in shipping) and investors (who might own stocks such as OMM, OSG, Torm and others). "Forward earnings visibility", which refers to future freight earning capacity with known revenues now from credit-worthy sources, is critical. Rating agency Standard & Poors, in its 2006 publication "Key Rating Factors for Shipping Companies" says: "A high level of long-term contract cover with creditworthy customers is the best way to mitigate the high volatility in spot rates, which could be a key rating factor." Its discussion of revenue visibility, where time charters and leases are discussed, goes on to comment: "Shipping companies are also increasingly using forward freight agreements (FFA) to manage its spot exposure. Freight derivatives provide a means of hedging exposure to freight market risk through the trading of specified time-charter and voyage rates for forward positions." Indeed, just like Ratings Agencies, banks analyzing shipping company credits must look at commitments under freight derivatives similarly to their view of leases and charters. In the pre-clearing environment, major shipping banks including Fortis and Royal Bank of Scotland have both reportedly worked closely with their shipowner clients in utilizing freight swaps (under "belt and suspender" type ISDA contracts) as a means of securing forward revenue, instead of the traditional "time charter" – an operating lease extending out typically two or three years. Rates are tied to indices of individual routes mainly from London's Baltic Exchange (which produces daily broker quotes for dozens of shipping routes) and several from Platts. Settle prices for forward maturities are also gathered by the Baltic Exchange; recently Imarex became an information provider feeding the Baltic. In describing how freight swaps are used to manage risk, a representative of leading inter-dealer broker GFI, says "we are doing strip trades, mostly spread over a number of months during the calendar year." Often, such trades are designed to replicate the coverage afforded by the time charter.

Bankers have seen increasing mandates to infuse risk management with derivatives into their formidable ship finance activities, ie using trades like those described by the GFI representative to provide a part of the comfort for secured loans. NYSE listed shipping companies, like OMI, OSG and General Maritime (GMR) are now employing full time traders to talk to principals and brokers in the derivatives markets; these traders may also handle energy fuels hedging. In the dry markets, NYSE traded Navios Corporation views freight as a portfolio, where risk management in the forward paper markets is combined with physical freighting activities such as time charter operation of vessels. Over the counter dry bulk players such as Dry Ships (DRYS) have also been actively using the "paper markets" to manage their freight risks. Anyone doubting that financial shipping has melded with physical need look no further than the 3Q 2006 acquisition by financial powerhouse Morgan

Stanley of privately held Heidenreich Marine - a user of the FFA markets in managing and optimizing revenues for more than 90 tankers in a range of sizes. It takes two sides to make a market, however.

Other parts of the shipping and finance communities are engaged in speculative trading in risk, contrasted with risk mitigating. The past two years have seen an influx in interest from financial players, including hedge funds. In conference presentations, speakers from the financial sector have estimated that more than 100 participants from the financial fold had entered the freight derivatives markets. The broker community, a sign of market health, now includes entrants from the financial sector such as GFI (which works through joint ventures with physical market brokers ACM), Tradition Financial, Icapp (tied in with London broker J Hyde) and Carnegie (in the Nordic markets). Major London and Oslo brokers including Clarksons, SSY, Bassoe, Lorentzen Stemoco and Platou have been able to combine intermediation in physical shipping activities with brokerage in the FFA markets.

A handful of fund managers, mainly in the London markets, have sprouted up in response to institutional interest in the freight sector. Though a great deal of secrecy is involved with the world of hedge funds, evidence suggests that these pools of money have played a role in the sector's growth. Typical trades for these actors include arbitrage activity-where an offsetting position is taken against a privately held or listed shipping equity. Imagine the gains in "shorting" the tanker market in August 2006 when chartering managers at leading Scandinavian tanker owners were playing it spot. Hedge funds with energy complex expertise have also been tied to some of the coal freight route arbitrage trading mentioned earlier. Broker GFI, mentioned earlier in the article, recently opened an office in South Africa to handle additional coal related activities. Software vendors have developed systems to support management of freight positions as well as price sensitive exposures to derivatives in fuel.

The vendor roster includes specialists Triple Point Technology and Sol Arc-two companies offering platforms that integrate information on commodities and logistics. Triple Point, offering "Commodity XL" boasts a client list that reads like a Who's Who of the shipping and commodities world, including Noble Group, General Maritime, Glencore (a big oil trader) and Morgan Stanley. Sol Arc has achieved notable successes among airlines, and oil traders - with a client roster including names like Koch Industries and Chevron. But, oil companies also own and charter tankers and barges, and SolArc, offering its "Right Angle" software, has now teamed up with Veson Nautical, a leader in the ship management software area (estimating voyage profitability and then comparing against actual results). Veson Nautical, based in Boston, has now introduced a module specifically for linking FFA trade management with its industry standard setting voyage calculations. As maritime company CFO's are realizing that, in effect, they are running portfolios of commodities (by virtue of their freight and fuel hedging and trading that must be "marked to the market" daily), such a tie-in is probably just what the risk doctor ordered.

FFAs and the Baltic Exchange

The market for Forward Freight Agreements (FFAs) continues to gain the interest of a wide range of companies including banks, commodity traders and ship-owners. The latest statistics show that 12,758 tanker FFAs were transacted in 2006, equating to 329,434,250 metric tonnes of oil transported by sea with an estimated freight value of USD 5.5bn. On the dry side in the third quarter of 2006 alone, 511,105 lots were traded, with a lot representing one day's hire of a vessel or 1000 tonnes of freight. Compiled by the London headquartered Baltic Exchange, the provider of settlement data for the shipping markets, the statistics are based on confidential returns provided by 19 different FFA broking companies. The Baltic Exchange says that it believes that the publishing of trading volume data will assure market newcomers of the depth of liquidity now available across the FFA market. Commenting on the results, Baltic Exchange chief executive Jeremy Penn said: "Freight is a readily tradable commodity. The FFA market has performed strongly throughout 2006 and we expect to see continued growth of volumes in 2007."

As a neutral organisation made up of ship brokers, ship-owners, charterers and traders, the Baltic Exchange underpins the growing freight derivatives market. Not only does it publish independent daily dollar assessments covering vessel hire rates for the capesize, panamax, supramax and handy size markets as well as World scale assessments for the tanker markets which are used to settle the vast majority of FFA contracts, but it also publishes benchmark forward curve data for all the most heavily traded FFAs. The recent launch of the Baltic's handy size index covering 28,000 dwt bulk carriers, means that owners and charterers of smaller vessels have the opportunity to trade paper for the first time. The Baltic Exchange is actively involved in promoting growth in the FFA market, holding workshops, forums and training courses in shipping centers around the world.

The organisation, whose history stretches back over 250 years, is increasingly focusing its efforts on Asia and has just opened a representative office in Singapore. The Baltic sees a clear need for reporting on new routes for the Asian market and the potential for publishing to take place in local time for certain routes to aid the development of the derivatives market in Asia. Several new routes have already been proposed and are currently under consideration. "We see our role as a servant of the shipping markets," comments Jeremy Penn. "As an organisation with over 550 member companies active in the shipping business all our services are geared towards providing the freight markets with the information and support they need." The Baltic Exchange's indices are widely regarded as being the market's most reliable and independent assessments available. "Key to our success is the involvement of our international panel of independent shipbrokers which provides us with data on a daily basis," says the Baltic's chief executive. "We are able to reflect freight market movements with real accuracy. With input from our numerous consultative groups made up of member firms, we ensure that our vessel descriptions and routes assessed reflect what's really happening in the market."

The Baltic Exchange is a unique organization whose members include the world's leading ship broking, owning and chartering companies as well as banks and commodity houses active in the freight markets. Governed by its members, the Baltic also provides an important role in maintaining standards in the freight market. All Baltic Exchange members are expected to adhere to its famous motto "our word our bond" and the Baltic provides its members with a dispute resolution service and helps its members recover monies owed to them.

What are FFAs and how do they Work?

FFAs are derivatives. Derivatives are risk management tools, the value of which is derived from the value of an underlying asset. In the shipping industry, the underlying asset is the freight rate for a specific physical trade route which receives a daily assessment on one of the Baltic Exchange Indices. While not all routes receive such an assessment, as demand for FFAs increases, the Baltic Exchange continually evaluates additional trade routes for inclusion in the indices. Freight derivatives serve as a means of hedging exposure to freight market risk by providing for the purchase and sale of a freight rate (the "contract rate") along a named voyage route (the "contract route") over a specified period of time (the "contract period").

Contracts are cash settled, and there is no physical delivery. On settlement, if the contract rate is less than the average of the rates for the contract route over the contract period (the "settlement rate"), as determined by reference to the relevant index, the seller of the FFA is required to pay the buyer an amount equal to the difference between the contract rate and the settlement rate multiplied by the number of days specified in the contract (the "settlement sum"). Conversely, if the contract rate is greater than the settlement rate the buyer is required to pay the seller the settlement sum.

The freight derivatives market began with the trading of voyage rates for certain "dry" cargo routes in the early 1990's and later was expanded to include "wet" tanker routes. Until recently FFAs were used almost exclusively by participants in the shipping industry, such as shipowners and charterers, to hedge against fluctuations in freight rates. As the basic forms and means of trading FFAs have evolved, new participants such as investment banks and financial institutions have entered the market for the purpose of speculation. The conventional wisdom is that the paper market for FFAs will soon come to surpass the underlying physical market in terms of dollar value.

Types of FFAs: Swaps vs. Futures

There are two categories of freight derivatives. One is privately-negotiated derivatives known as "over the counter" ("OTC") derivatives or "swaps". The other is standardized, exchange-traded derivatives known as "futures" or cleared contracts. One of the major differences between swaps and futures is the assumption of counter party risk; that is, with OTC derivatives the risk of default by either party to the FFA is assumed by the other party to the agreement (the "counter party"), whereas with exchange-traded derivatives the risk of default by either of the parties is assumed by a clearing house, and the parties to the agreement only assume exposure to default by the clearing house. The assumption of counter party risk by a clearing house makes it almost certain that the holder of an in-the-money futures contract will be able to collect and has the potential to make this type of derivative more liquid than the OTC variety. This liquidity, combined with the ability to mask the true identities of the counterparties when trading on an exchange (as opposed to the transparency that comes with negotiating FFAs at arms-length, as is the case with swaps) generally makes futures more appealing to speculators. The liquidity of freight futures has been further enhanced by the entry of new clearing facilities to the marketplace.

The International Maritime Exchange ("Imarex") in is the principal exchange dedicated to FFAs and uses the Norwegian Futures and Options Clearinghouse ("NOS") to clear its transactions. Until recently, Imarex/NOS was the exclusive forum for clearing freight futures trades. However, in 2005, due to increasing demand from FFA traders, the New York Mercantile Exchange (NYMEX) and LCH. Clearnet also began clearing trades. Later in 2006, Forward Freight Agreement Brokers Association ("FFABA") members Clarksons, Ifchor and Freight Investor Services plan to launch a multi-user screen based trading system which they have jointly developed as part of an effort to generate more FFA futures trading.

While swaps may lack the liquidity of their exchange-traded counterparts, the fact that they can be individually negotiated makes them a useful hedging tool for participants in the physical marketplace who are seeking to offset specific risks to which they find themselves exposed through contracts of affreightment and/or time charters. Parties to swap agreements generally use either the FFABA form contract or the International Swaps and Derivatives Association ("ISDA") Master Agreement and Schedule.

ISDA is a trade organization whose primary purpose is to encourage the prudent and efficient development of the privately-negotiated derivatives business. The ISDA prescribed form agreement for swap transactions includes a Master Agreement which sets forth the terms of the ongoing legal and credit relationships between the parties. Any amendments to the Master Agreement are set forth in an attached schedule. One of the major issues agreed to in the schedule is the choice of law provision, which typically provides that either New York law or English law govern the contract. From a legal perspective, the advantage of the ISDA form is that it has been used as the standard form of derivative contract for the great majority of traded commodities and, as such, it has been subject to judicial review and found to be enforceable in a number of international jurisdictions. As a result, the legal risk of uncertainty surrounding the interpretation of the terms of the ISDA Master Agreement is greatly reduced. This has not always been the case with the FFABA form agreement.

Before the 2005 amendments to the FFABA form, provisions setting forth events of default, termination and close-out netting rights were noticeably absent, and an aggrieved party litigating any such matter would have been forced to rely on English common law, which governs both the original and the revised FFABA contract, to show that its counterparty had violated the agreement. The FFABA has sought to bridge the gap between its form and the ISDA form by incorporating ISDA's approach in listing specific and detailed events of default and in providing express termination and close-out netting provisions. While there is still no universally accepted form of OTC contract, the harmonization of the two most widely used forms and the movement towards a standard OTC contract with clearly definable rights and obligations should have the effect of enhancing the liquidity of swaps and bringing more participants into this segment of the marketplace.

The FFA market is still very much a work in progress. While there have undoubtedly been significant advancements in both the swaps and futures components of the market in terms of enhancing liquidity and reducing participants' exposure to uncertainty and credit risk, these measures have as yet largely gone untested.

Dry Freight or Dry-Bulk FFAs

The Baltic Exchange, Baltic Dry Index which measures the cost for shipping goods like iron ore and grains, doubled over the past 12 months and has risen more than fourfold since 2006.

The trading volume of dry freight derivatives, a market estimated to be worth about $200 billion in 2007, grew as those needing ships attempted to contain their risks and investment banks and hedge funds looked to make profits from speculating on price movements. At the close of the 2007 financial year, the number of traded lots on dry FFAs doubled the derived physical product.

The Baltic Dry Index (BDI) is a number issued daily by the London-based Baltic Exchange. Not restricted to Baltic Sea countries, the Index tracks worldwide international shipping prices of various dry bulk cargoes. The index provides "an assessment of the price of moving the major raw materials by sea. Taking in 26 shipping routes measured on a timecharter and voyage basis, the index covers Handymax, Panamax, and Capesize dry bulk carriers carrying a range of commodities including coal, iron ore and grain."

Freight derivative trading members of the Baltic Exchange have signed an agreement to boost confidence in the growing freight derivatives market. The agreement formalises existing arrangements between brokers and principals and includes an undertaking to ensure the continued integrity of the Baltic's daily indices and route assessments.

The Baltic Exchange, the Freight Market Information Users Group (FMIUG) and the Forward Freight Agreement Brokers' Association (FFABA) are working in support of the for the following joint objectives:

- To increase the liquidity of the freight derivatives market
- To encourage membership of the Baltic Exchange.
- To increase awareness of, and confidence in the reliability and robustness of Baltic Exchange freight market information.

These are to be achieved through:

1. Maintaining the integrity and consistency of the daily indices and route assessments by close adherence to the *Manual for Panellists* by the reporting panellists.
2. Maintaining as close as possible a correlation between the route assessments and the physical/cash markets. The widest possible industry participation in this process will continue to be encouraged through the Freight Market and Consultative Group (FMFCG).
3. Permanent production and dissemination of freight market information (via balticexchange.com and the news vendors) by the Baltic Exchange.
4. Endeavouring to ensure the creditworthiness of market participants. · Best price dissemination by FFABA brokers in line with accepted market practices.
5. Worldwide marketing of Baltic freight market information and the benefits of membership of the Exchange by the FMIUG, FFABA and the Exchange itself.
6. The development and use of one standard contract, updated according to the latest developments in the market, by close cooperation between the principals and the FFABA.

Forward Freight Agreements (FFA) - The standard contracts FFABA 2005

The role of freight derivatives as a financial tool is on the increase. Their emergence is a symptom of both the growing sophistication of freight and commodities markets and the appetite of financial institutions for increasing diversification of their investment portfolios. Freight derivatives in the form of Forward Freight Agreements or FFAs have bee around for more than twenty years. FFAs are bespoke swap contracts traded "over the counter" (OTC) between counterparties on a principal to principal basis as a means of hedging exposure to freight market risk. They used to be the preserve of the biggest players in the industry and attracted little external interest due to a lack of liquidity and obscure pricing. Now, however, a number of factors mean that barriers to entry are getting lower, there is an improved level of liquidity and as a consequence interest in freight derivatives is increasing rapidly.

This note explains what freight derivatives are and how they can be used to manage risk and examines the issues thrown up by the recently introduced FFABA 2005 standard contract.

The Basics

A straight forward swap freight derivative is essentially a "bet" on whether the market freight for a certain type of voyage or a certain period of time over a designated route or combination of routes will be higher or lower than the price struck as the basis of the "bet". The winning party receives the difference between these two prices, multiplied by the agreed contract quantity. Managing Freight Price Buying the fixed price side of the bet means that the other party takes the risk of the market rising. Conversely, the risk of buying the fixed price is that the market might fall. The opportunity to lock in a fixed price for freight over a particular route is obviously attractive to commodities traders who wish to mitigate the risk of an increase in their chartering costs which can impact heavily on price and profit margins.

The ability to hedge the cost of freight permits traders to price their business with a greater degree of certainty and long term stability. Why have Freight Derivatives become more Popular? While the volatility of the freight market is something that commodity traders wish to hedge against, it is a can also be a selling point for those wishing to speculate. Large movements in price within short periods also mean the potential for large profits.

Recent developments in OTC and exchange traded freight derivatives are both adapting to and driving the development of the market. In response to the increasing sophistication of the freight derivatives market and a demand for uniformity, the Freight Forward Agreement Brokers Association (FFABA) has produced a new 2005 contract that brings the terms on which freight derivatives are traded more in line with other derivative products.

Exchange Trading

As an alternative to the traditional FFAs outlined above, exchange trading and clearing have increased liquidity in the market, which in turn has increased interest. Smaller units of freight can be bought and sold thus providing greater flexibility for larger users and greater access to companies previously priced out of the market.

Freight futures and options traded exchanges such as IMAREX provide a relatively simple method of participating in the market. The contracts traded through IMAREX are cleared by the Norwegian Futures and Options clearing house (NOS). The advent of clearing provides advantages for markets participants. First and most importantly, the risk of default by a counterparty is virtually eliminated. The clearing house effectively acts as counterparty to both buyer and the seller of the contract. Second, this process can be completed instantly at the click of a mouse. Cleared contracts are settled on a daily basis and traders pay or obtain the difference between the price of their contract and the market index.

All of the signs are that screen market derivative trading is increasing and it will be interesting to see whether the traditional FFA market decreases as a result.

The New FFABA 2005 Form – Important Changes and Potential Pitfalls As outlined above, FFAs are OTC products which are generally traded on a standard form and agreed between principals with a broker acting as intermediary. The new 2005 FFABA form was created with the objective of introducing an improved contract acceptable to the industry but which would eliminate the risk related deficiencies of the old form.

The key to the operation of the 2005 document is its incorporation of the 1992 ISDA Master Agreement 1992 (Multi currency cross-border) without schedule. The ISDA Master Agreement is the most widely used contract that exists for derivatives transactions. Although there is a 2002 version, the 1992 version is still widely used and it is this version which forms part of the new FFABA form.

Generally, the ISDA Master Agreement is agreed between counter parties through the negotiation of a customised schedule that produces a bespoke umbrella agreement under which all derivatives deals between the two parties are executed. The 2005 FFABA contract incorporates by reference the ISDA Master Agreement with a set of standard elections and with no separate schedule. The objective is to create a standard form document that ensures market conformity and maintains liquidity.

In addition to the incorporation of the ISDA Master Agreement in relation to individual trades, the 2005 FFABA contract also puts in place a master agreement structure for existing and future transactions.

We highlight below some of the important changes in the new form.

Default/Termination

The default provisions of the old version of the FFABA contract were rather vague and it was a matter of judgement as to when a party was entitled to terminate a particular contract. This uncertainty is removed by the incorporation of the Events of the Default clause of the ISDA master agreement.

Netting

The netting provisions are an improvement on the old FFABA contract as it means that a single balancing payment can be made between parties to cover all contracts made in the same currency in respect of which payment must be made on the same day. This has the added effect of avoiding difficulties in jurisdictions where insolvency law does not already provide for setting off of debts owed to an insolvent company, so long as the relevant jurisdiction recognises netting provisions.

Exclusive Jurisdiction is Conferred on the English High Court

The new jurisdiction clause is a potential concern for anyone who trades with overseas counterparties. Any dispute must be settled in the English courts and in some countries it can be difficult to enforce such a judgment. Although this change brings the form in line with ISDA 1992, the possibility of refusal to honour a judgment introduces an important new factor in determining counter party risk.

Invoice-based Payment Terms

The new invoice based payment terms are designed to eliminate problems caused by counter parties who fail to issue invoices. A party who does not produce an invoice has no right to demand payment by a particular date.

Warranties

Under the new form, the warranties set out in clause 11 only need to be correct as at the date of the contract whereas under the corresponding clause in the 2000 form warranties were continuing. They were, however, limited to good standing, capacity and consents. By ISDA section 3 there are separate continuing representations and warranties covering acceptance of binding obligations, absence of conflicts, as well as absence of any event of default termination. Taken together, the warranties provided by the 2005 form are wider.

Other Issues

We understand that uptake of the new FFABA form is excellent and that it is being used without amendment by traders. Yet the new form does raise some issues of potential concern, particularly for traders who, unlike banks or financial institutions, will probably not be negotiating full ISDA Master Agreements with counterparties. In the 2005 FFABA contract, the ISDA Master Agreement is incorporated without schedule and accordingly there are certain provisions (such as the cross default clause or the provisions relating to events of automatic early termination or additional termination events) that will not apply unless amendments are made. In addition, clauses 20 and 21 aim to apply the ISDA Master Agreement to all existing and future trades. However, the wording of the clauses is not clear regarding incorporation of the other new provisions of the FFABA into future and past dealings. An amendment to the terms could remove this uncertainty.

Exchange trading and clearing offer a reduction in counter party risk and a boost to liquidity. However, FFAs still provide flexibility in terms of contract routes, quantities and pricing structure.

Whilst it is obviously necessary in terms of liquidity to maintain a standard form across the industry, it is also a good idea for traders to review the effect of new documentation particularly if they are unfamiliar with the 1992 ISDA Master Agreement and the way it is intended to work. The 2005 FFABA contract is in effect a hybrid document the nature of which remains untested. It is therefore particularly important for banks and traders to understand the potential issues that could arise even if it might not be possible or desirable to amend the contract.

2.5 FREIGHT DERIVATIVES

Freight derivatives are financial instruments to be used by organisations or individuals as part of a risk management strategy to counter the negative economic affect of freight price alterations that harm the financial prospect of a company. The company can trade in future levels of freight rates, primarily for dry bulk carriers and tankers. Such instruments include exchange traded futures contracts and options on futures contracts, plus OTC (over-the-counter) freight forward contracts like FFAs (Forward Freight Agreements) swaps and swaptions. Cleared freight futures contracts are traded on the Oslo based exchange Imarex (International Maritime Exchange ASA), and in the past used to be traded on BIFFEX and INTEX. Freight derivatives are primarily used by shipowners and operators, oil companies, trading companies and grain houses as tools for managing freight market risk. A freight derivative is a financial contract between two parties, which sets an agreed future price for carrying commodities at sea. The contract does not involve any actual freight or any actual ships. It is purely a financial agreement

Given the current volatility of the freight markets, managing freight market risk is a significant issue for the shipping industry. One method of managing this risk is through the purchase and sale of forward freight agreements (FFAs).

Freight Derivatives, which includes Forward Freight Agreement (FFA) and options based on these, are financial instruments for trading in future levels of freight rates, for dry bulk carriers and tankers. These instruments are settled against various freight rate indices published by the Baltic Exchange (for Dry and most Wet contracts) & Platt's (Asian Wet contracts). FFAs are often traded over-the-counter (through broker members of the Forward Freight Agreement Brokers Association - FFABA - such as Clarkson's Securities, SSY - Simpson, Spence and Young, Ifchor, FIS - Freight Investor Services, BGC Partners, GFI Group Inc, ACM Shipping Ltd, BRS, Tradition-Platou and ICAPHYDE); but screen-based trading is becoming more popular, through various screens. Trades can be given up for clearing by the broker to one of the clearing houses that support such trades. There are four clearing houses for freight: NOS Clearing, LCH.Clearnet. NYMEX (NY Mercantile Exchange) and Singapore Stock Exchange (Singapore). Freight derivatives are primarily used by shipowners and operators, oil companies, trading companies and grain houses as tools for managing freight rate risk. Recently with Commodities now standing at the forefront of international economics; the large financial trading houses, including banks and hedge funds have entered the market. A financial instrument's value that is derived on the future levels of freight rates, such as "dry bulk" carrying rates and oil tanker rates. Freight derivatives are used most often by end users (such as ship owners and grain-houses) and by suppliers (such as integrated oil companies and international trading corporations) to mitigate risk and hedge against price spikes in the supply chain.

As with all derivatives, market speculators, like hedge funds and individual traders, participate in both the buying and selling of these contracts providing for a new, more liquid, marketplace. Freight derivatives now include exchange-traded futures (ETFs), swaps futures and the older "Forward Freight Agreements", which were sold over-the-counter. Freight derivatives are most often used to hedge risks against large swings in price, a model popularized in the agriculture and commodities industries.

Freight derivatives are a relatively new product in the global marketplace, but the advent of clearing services has brought increased safety, and with it liquidity, into the business. The freight derivatives market has been hitting record highs. Whilst many markets are suffering from the liquidity crunch, freight derivatives have seen a huge increase in business. The massive increase in the demand from China for goods and raw materials has been one of the main driving forces behind growth.

In recent months a tide of new participants has entered the market. Investment banks, hedge funds and commodity traders are looking to take advantage of the attractive volatility levels and increased liquidity levels the FFA market offers.

Indeed, while many other markets are dwindling, investors are reportedly enjoying spectacular returns from freight derivatives. However, it is widely acknowledged that a thorough understanding of the fundamentals driving the global freight and commodity markets and a sound risk management strategy are essential for any market participant.

The Baltic exchange is a major exchange for freight derivatives. The derivative can be used by a company to reduce the risk for rising oil prices or for political situations.

There are basically two types of freight derivatives namely: (a) Freight Futures (also called ***cleared contracts***), *and* (b) OTC forward freight agreements (FFAs).

Both types of freight derivatives are cash settled. Freight derivatives usually reference various freight rate indices published by the Baltic Exchange (for "dry" and "wet" contracts) and Platt's (e.g. Asian Wet contracts).

Freight derivatives evolved as the need of ship owners and charterers (ship hirrs) to hedge against the volatility of freight rates (which is a factor of price thus profit). They allow the users to hedge against risk of a certain type of voyage over a certain period of time and a specific route (or combination of routes). Increasingly, they also allow investors to speculate on the freight markets. Investment banks, financial trading houses and hedge funds have trading desks specialise in freight derivatives.

The history of freight derivatives can be traced to the establishment of the BIFFEX and the creation of the BFI on 1 January 1985. The BFI was a basket of freight routes and recorded the movements in dry cargo freight rates. OTC swaps on individual routes began to be traded in early 1990fs as users looked to hedge against risks relating to specific routes.

The Baltic Tanker and Dry Bulk Indices

The 'wet' tanker routes, TDI and the TCI were added in February 1998. The BFI was split in 1999 into the BCI and the BPI and the BHMI was added in November 1999. In January 2003, the BLPG was introduced.

Forward Freight Agreements (FFAs)

FFAs are cash settled privately negotiated bespoke financial contracts between two parties whereby one party agrees to pay the other party an amount equal to the difference between the contract price of the underlying index (e.g. one of those discussed above) or the contract rate of a specified route and the settlement price of the index or the route.

Freight Derivatives Trading

Freight derivatives are often traded over-the-counter through FFABA broker members but screen-based trading is possible and becoming popular. Freight derivatives trades can be "given up" for clearing to one of the clearing houses that support such freight derivatives trades.

FFA Master Agreements

The most widely used OTC agreements are the "principal-to-principal" master agreements published by the Forward Freight Agreement Brokers Association (FFABA) and ISDA. For transactions using the ISDA master agreement, the **2005 ISDA Commodity Derivatives Definitions** also provide confirmation templates and additional provisions in **Sub-Annex I. Sub-Annex A** [which is **Section 7(1)(c)** of the 2005 ISDA Commodity Derivatives Definitions] sets out certain reference prices relating to freight derivatives. In addition to a master agreement structure, the FFABA 2005 standard contract also incorporates provisions the 1992 ISDA master agreement (without the schedule).

Freight Derivatives Clearing Services

Clearing services are provided by IMAREX/NOS, NYMEX/LCH. Clearnet, Clarkson Investor Services and the Singapore Stock Exchange.

FFABA Freight Derivatives Brokers

- Clarkson's Securities
- Simpson, Spence & Young
- Ifchor
- Freight Investor Services
- BGC Partners
- GFI
- BRS Futures
- Tradition-Platou
- ICAPHYDE

Freight Derivatives Exchanges and Associations

- International Maritime Exchange IMAREX
- Norwegian Futures and Options Clearinghouse NOS
- New York Mercantile Exchange NYMEX
- LCH.Clearnet
- Forward Freight Agreement Brokers Association (FFABA)
- International Association of Independent Tanker Owners (Intertanko) and its useful collection of presentations.

CHAPTER 3

Trading of Oil in International Market

3.1 INTRODUCTION TO OIL TRADING

Crude oil and petroleum products are traded globally 24 hours a day, every business day in both the physical and paper markets. With the daily physical consumption of oil of almost 84 million barrels and annual trade valued at over $1 trillion, the growth in paper energy trading seems assured, with new financial products evolving to meet the needs of producers, refiners, marketers, and consumers. Paper trading for oil has grown on established futures exchanges to almost 400 million bbl/d with much greater growth on the *Over the Counter (OTC)* markets. Moreover, because major oil companies must now buy and trade on the spot markets to meet more of their supply needs that were previously met by their own production, their active involvement in paper trading has increased over time. Greater oil price volatility shows that energy price risk management must be increasingly managed by a wide variety of existing and emerging financial instruments for the longer term.

Because of the great price volatility endured by the energy business during the past decade, the industry adopted short-term hedging tools, such as forward and futures contracts, as well as the use of longer-dated instruments including price swaps and OTC options to manage price risk. (Swaps and OTC options are also used to manage short-term price risk). Changes in management thinking have made it unacceptable in today's volatile price environment to remain unhedged and acceptable to use the financial tools available to manage price risk. This sea change is becoming paramount in Asia today. These financial tools can not only be used as means to hedge price risk but are beginning to be employed to manage corporate assets more effectively for the longer term.

Thus, energy companies can utilize their corporate assets and extract their present value rather than simply wait for earnings through future oil production and retail sales. Because risk management tools are now more widely accepted and available as a means to reduce financial risk in the commodities markets, their application in the oil markets has accelerated over the past 20 years. They provide a degree of certainty for oil producers, refiners, airlines, shipping companies, and manufacturers who use them. The off-exchange OTC instruments allow custom tailoring of agreements to meet the needs of the user while the financial intermediary (oil companies, banks or insurance companies) assumes the risk. Moreover, these agreements arrange fixed price deals for the energy producer or consumer for periods ranging from one month to 10 years forward. These advantages over the shorter-term futures contracts demonstrate that price swaps and OTC options are not only complementary to futures in meeting different needs for their users, but also have become competitive and superior to exchange-traded contracts because of their flexibility and customization attributes.

In effect, in many oil markets, the OTC markets set prices not futures contracts. The application of technologies of the financial markets to the energy markets in a wide variety of strategies has made it necessary to think of oil as money: in other words, a fungible commodity. Oil has been stripped of its physical delivery and has become the province of financial specialists, oil traders, and oil companies who can now assist the industry to tap into their corporate assets more efficiently. The selective use of financial instruments ranging 'from forwards, futures, and options to price swaps and oil linked debentures can not only aid in managing price risk, but also protect cash flow, meet operational needs, finance prospective projects, and service debt obligations for producers, refiners, marketers, and end users. These parties now face economic, political, and environmental uncertainties that impact on how they conduct business under present market conditions and in the future. The use of currency trading tools and techniques indicate that oil has become a substitute for money due to its fungibility as a global commodity.

OTC oil trading is not conducted on regulated, established financial exchanges such as New York Mercantile Exchange (NYMEX), International Petroleum Exchange (IPE), or Tokyo Commodities Exchange (Tocom); yet it is a market that is monitored but not regulated by government agencies. OTC markets can also be cleared on existing energy futures exchanges. As the energy industry continues to reintegrate vertically once again with new supply arrangements between producing and consuming countries, the uses of financial instruments of the off-exchange markets and regulated exchanges provide both financial price risk management and the access to the tremendous capital needs of the energy industry. Conservatively estimated $2 to $3 trillion notionally for all price deals, OTC energy swaps and options markets are dwarfed by the $120 trillion global interest rate and currency swaps markets. The maturation of the energy swaps and OTC options market has brought to the fore the fundamental concerns of credit risk and contractual performance.

Credit risk concerns the ability of swaps counterparties to perform for the length of the transaction. It has become a leading market issue due to the Enron financial debacle and is continuing to resonate in energy markets. Credit risk has always been a significant factor in all swaps markets, as the pressure to preserve capital and reduce risk are preeminent factors among many market makers. It is now of rising importance due to the fact that the energy industry companies have less creditworthiness. The proprietary and somewhat secretive nature of the price swaps and OTC options off-exchange markets makes it very difficult for public disclosure of individual transactions. Some of these financial transactions are an "off balance sheet" item (again highlighted by Enron), and thus there is no ability to investigate the outstanding commitments of the financial intermediaries. Today, we are seeing the proliferation of energy risk management services, primarily offered by banks, serving as principals and the growth of swaps brokerage services for crude oil, gas, petroleum products, electric power, weather and emissions. After transactions are undertaken, the matching transaction of the counterparty is often unknown by other market participants. In fact, banks should be the clear beneficiaries of the Enron disaster since they have stronger balance sheet than many energy companies.

The OTC oil markets exist in Asia primarily because financial futures for energy have failed on the Singapore Exchange and are quite illiquid on the Tocom, and customized OTC financial instruments can be developed more rapidly to meet changing market conditions. However, both markets count on the financial performance of their participants to honor their commitments.

OTC Markets and Trading Disasters

OTC energy derivatives create both new opportunities and challenges for companies with price exposure. OTC trading practices have therefore been a matter for study by boards of directors and senior management of energy companies. If defined at the highest level of policy and decision making within an energy company, the policies set forth can prevent trading disasters which still occur noticeably in energy markets. Some very basic guidelines in the establishment of a risk management portfolio came from the Group of Thirty report on "Derivatives, Practices and Principles" released in July 1993. Their recommendations for the OTC markets included the following: that derivatives should be valued at the market (marked to market), market risks should be quantified, credit risks assessed, and management information systems that are sophisticated enough to measure, manage and report risks should be established.

All derivatives trading disasters have been due to the breakdown of controls, inadequate systems, management failure, and human error. Rogue traders exist because of the lack of internal controls to ensure that the risks of OTC transactions are appropriately managed.

3.2 ASIAN OIL MARKETS AND TRADING: A PERSPECTIVE

Net Importer

The Asia Pacific markets are net crude oil and petroleum product importer, primarily from the Middle East. The region also sources crude and products from West Africa, Alaska, and the North Sea. The increased arbitrage opportunities have significantly increased the need to use risk management tools to hedge growing price exposures over different time horizons. Thus, the market drivers of increased physical oil trading activity, shifting trade patterns, and fuel specification changes due to environmental imperatives are all pushing the region's oil producers and consumers to manage price risk more proactively.

Lack of Price Benchmarks

One glaring problem in Asian oil markets has been the lack of liquid price benchmarks, particularly for crude oil. The role of a price benchmark is useful as it provides price transparency, reference, discovery, and facilitates risk management through price certainty. Successful and liquid benchmarks provide market efficiency by lowering costs and increasing price information. They also allow risk shifting (not elimination) with more players participating in the markets.

Lack of Depth

Benchmarks provide a mechanism to separate price from supply. In more mature markets, arbitrage opportunities are more complex and price spreads narrow over time. Asian oil markets, however, remain relatively immature paper markets at the present time due to the lack of trading volume and a lack of active participation of Asian oil producers and consumers; however, this is starting to change.

Paper Trading and Reference Indices

Regarding crude oil trading, Asia has never had a liquid marker crude. The two relatively liquid Asian crude oil swaps markers for the past decade have been Malaysian Tapis and the United Arab Emirates' Dubai crude oils. Tapis and Dubai production is significantly less than Brent blend and is declining. Nevertheless, they have been the most visible Asian crude oil price indicators, but are prone to periodic supply problems or pricing patterns that are sometimes detached from other price markers. Paper Tapis trades in 50,000 to 100,000 barrel lots, and enables traders to lock in price spreads typically in Tapis/Brent swaps trading. The paper players in Tapis include Morgan Stanley, Vitol, Koch, and Elf Trading.

The Tapis swaps are priced at Tapis APPI (Asian Petroleum Price Index) related prices and book out against the APPI average. The Malaysian Malacca refinery has hampered liquidity in Tapis as it uses the crude for the refinery. The APPI has been used for many years as the standard for crude oil pricing in Asia. This index was launched in January 1986. It is a biweekly panel assessment by industry players, and is used intensively in Asian crude term contracts by oil producers. Twenty-one Asian crudes including Tapis, Australian, Chinese, and Indonesian crudes are assessed in this index. The APPI is administered by SeaPac Services Ltd in Hong Kong.

However, the APPI index is not used for pricing OTC paper deals. Dubai is traded in 50,000-barrel increments which are called the partial Dubai contract. It has been a major Middle East pricing indicator since most Persian Gulf oil producers do not want to disclose their spot prices and prefer to price their crude at their official selling price. Unfortunately, Dubai production is continuing to decline so its viability as price market remains in doubt. Dubai production has declined from 420,000 bbl/day in 1991 to 170,000 bbl/d in 1999. Japanese refiners have been active in locking in Dubai swaps for their crude acquisition costs. Dubai is being replaced by Omani crudes. Incidentally, Dubai has already been tried as an Asian futures contract on the IPE and SIMEX when they launched a contract, unfortunately during the Gulf War. The contract failed to gain any liquidity.

3.3 PLATTS ASSESSMENT METHODOLOGIES

About Platts

Platts is planning to reflect market values at the close of the market for its US Gasoline and Distillates assessments effective November 15th, 2006. The new methodology is commonly referred to as a market-on-close assessment.

Platts, a division of The McGraw-Hill Companies (NYSE: MHP), is the world's leading provider of energy information. With over 10,000 customers across 150 countries, 600 employees in 16 offices around the world, and nearly a century of business experience, Platts is the most recognized and respected information company in the energy industry. Our real-time news, pricing, analytical services and conferences are essential to enabling markets to operate with transparency and efficiency. Traders, Risk Managers, Analysts, and Executive Management depend on Platts to help them make better, faster, more effective decisions.

Clearly defined and transparent methodologies are the backbone of Platts' price assessment processes. Platt's methodologies provide detailed descriptions and explanations of how we assess the market value of commodities so that customers and other interested parties understand those processes. In this section, we post documentation by commodity and region for all published price assessments and indices.

Price Assessment

Platts, US dominate oil market pricing and it quotes daily price assessments for a number of products. If the quality or grades assessed change, Platts informs subscribers in advance through the various wires, faxes, printed products, and electronic systems, Platt's price specialists focus on price discovery and follow prices in all main trading markets. In markets where there is continuous trading activity, Platt's price specialists cover those markets continuously as well, with the editorial responsibilities transferred across the globe.

In Asia, Platt's energy reporters in Singapore, Tokyo, Sydney, and Hong Kong track down prices starting at 9:00 am Singapore time. The Platts methodology reflects the activity of the market players and takes into account industry practices in each reporting region. FOB Singapore assessments reflect trading activity up to 5:30 pm Singapore time.

Industry practices include quality specifications, typical volumes, loading timing, and location. The Platts assessment reflects the standard grade and therefore market participants can determine more efficiently what the quality differentials should be versus the benchmark grades. Platts also assesses paper values for two months forward. The typical trading volumes also vary greatly and the prices for the parcels vary depending on the volume.

For the gas oil assessments, Platts focuses on volumes for a minimum of 150,000 barrels, for jet fuel 100,000 barrels, for fuel oil *20,000* metric tones, gasoline 50,000 barrels and naphtha C & F Japan 25,000 metric tones. The price assessments for Singapore reflect pricing for products loading 15-30 days from the date of publication. For instance, on July 16, Platt's assessments reflect the product market 15-30 days forward, namely July 31-August 15.

The timing specification is important due to variability in pricing in the various windows, particularly when a market is steeply in contango or backwardation. It is not uncommon in times of tight supply for the backwardation to be over $2.00/month or roughly 6 cents per day. This means that cargoes loading one week apart can vary in price by close to 30 cents/bbl, or even more, depending on the steepness of the forward price curve. Thus it is extremely important to follow pricing windows very methodically. In a contango market, the excess of prompt barrels causes the front end to be significantly cheaper than forward barrels, while in backwardation the tightness causes the front barrels to be much higher.

Liquidity Profile

Singapore product price assessments take into account the prevailing trading practices in the Asian region. Most companies tend to cover their requirements far in advance with sellers also selling forward. As a result, most of the liquidity is centered on this time period for physical transactions (only up to two months forward). By not taking the first two weeks into consideration, transactions reflecting distressed prices are excluded.

Location Profile

The locations used on the Asian assessment are Singapore, the Arab Gulf, South Korea, and Japan on a FOB basis, China, Japan, and Australia on a C + F basis. Spot transactions in the region are typically done at a differential versus the Mean of Platts Singapore (MOPS). The main benchmarks are FOB

Singapore and bids/offers FOB from any terminal or refinery are accepted as a basis for pricing. All Asia-Pacific/Arab Gulf product cargo assessments are assessed up until 17:30 (09:30 GMT) Singapore standard time. Transactions and bids/offers after 17:30 Singapore time deadline will not be taken into account. All Singapore assessments are FOB and are assessed 15-30 days from date of publication. Assessments reflect merchantable grades.

Contract Denominator

All product prices are quoted in US dollars. All FOB Singapore product assessments, with the exception of fuel oil and MTBE are quoted on a dollars per barrel basis. FOB Singapore fuel oil and MTBE are quoted on a dollars per metric tonne basis. The FOB Singapore premium/discount assessment takes into account physical cargo activities 15-30 days from date of publication. Both the Singapore gas oil paper market and the jet paper assessment reflect a minimum of 50,000 barrels paper transactions. Assessments for paper are for one and two months forward and are quoted on a full calendar basis. As an example, during the month of February, Platts would quote paper for March and April. Both paper gas oil and jet kerosene are non-deliverable contracts.

In conclusion, Platt's price assessments will reflect the closing levels, and the three main factors in price determination are: fixed price, premiums, and paper/swaps.

Fixed Price

The ultimate question in the mind of an end user, a producer, refiner, trader, or broker is price, which in turn determines expense, refining margin, profit, or loss. The spot market trades actively on a FOB Singapore basis on a fixed price basis. Activity has been commoditized and most players now can focus fully on prices as the qualities, volumes, and loading procedures have become more standardized each year.

Premiums

Many transactions are carried out in relation to a standard commodity. In this case a differential, also known as premium in the Asian markets, is generated. Premiums will arise if the quality, volumes, or loading times differ from the benchmark. In addition, the floating transactions (variable prices) are done in relation to the assessments. Premiums will usually rise in those times when the market is backwardated and the steeper the curve the greater the premium. In a contango situation, premiums will have a tendency to turn into discounts.

Paper/Swaps

Paper/swaps are another major determinant in price. Swaps trade freely in an OTC market and can trade at any time. Paper markets are very reactive and provide players with an instant feedback of market conditions. Swaps react to arbitrage conditions or movements in overseas market as well as local conditions. Paper hedges allow users to lock in price because swaps players can transform floating prices to fixed or fixed to floating.

3.4 ENERGY FUTURE EXCHANGE AND OTC TRADING

Energy trading began after the end of the Official Selling Price (OSP) programs by the major oil companies and OPEC nations after the 1973 Oil Embargo and coincided with the development of a spot market for crude oil and petroleum products. In 1978, the changing structure of the physical spot market for oil brought about the development and launch of a successful oil futures contract for home heating oil on the 132-year-old NYMEX which was tied to physical delivery of the oil in New York Harbor.

Successive oil futures contracts and the development of an active and viable OTC market for forward oil trading in the 1980s brought significant structural changes to the global oil industry. Oil had become a commodity. In effect, price transparency accelerated both the physical and financial trading of crude oil and petroleum products on an international scale.

Asia, however, has been a nascent and emerging oil futures and OTC market. Its exchanges have been slow to react to the changes underway in the physical oil world because its energy risk management infrastructure barely exists as compared to New York and London. Following are the main Global Energy Futures Exchanges:

1. NYMEX (New York Mercantile Exchange), New York http://www.nymex.com
2. ICE (Intercontinental Exchange), London http://www.theice.com
3. TOCOM, (Tokyo Commodity Exchange),Tokyo http://www.tocom.or.jp
4. SGX (Singapore Exchange), Singapore http://www.sgx.com
5. SFE (Sydney Futures Exchange), Sydney

Following are the main Indian Energy Futures Exchanges:

1. MCX (Multi Commodity Exchange), Mumbai
2. NCDEX (National Commodity Exchange), Mumbai

Presented below a detailed perspective of major future exchanges of the world :

NEW YORK MERCANTILE EXCHANGE (NYMEX)

NYMEX is the world's largest physical commodity futures exchange and the preeminent trading forum for energy and precious metals. Transactions executed on NYMEX avoid the risk of counterparty default because the Exchange clearinghouse acts as the counterparty to every trade. Trading is conducted through two divisions, the NYMEX Division, home to the energy, platinum, and palladium markets; and the COMEX Division, on which all other metals trade. NYMEX pioneered the development of energy futures and options contracts 26 years ago as a means of bringing price transparency and risk management to this vital market.

Energy Future Contracts

Energy futures contracts, or simply energy futures, are exchange traded energy derivatives. A futures contract gives the holder the obligation to buy or sell. Both parties of a "futures contract" must fulfill the contract on the settlement date. The seller delivers the energy commodity to the buyer, or, if it is a cash-settled future, then cash is transferred from the futures trader who sustained a loss to the one who made a profit. To exit the commitment prior to the settlement date, the holder of a futures position has to offset his position by either selling a long position or buying back a short position, effectively closing out the futures position and its contract obligations.

Following are the major specifications of a Futures Contract

(1) Trading Period/Trading Session: It signifies the days and hours during which the trading at a particular exchange takes place.
(2) Unit of Trading: It signifies the lot size.
(3) Grade and Quality Specifications: It signifies the quality specifications of the commodity that is deliverable under a particular contract.
(4) Quotation: It tells us the base value of commodity in which it is traded on the exchange.
(5) Tick Size: It is the Minimum Price Fluctuation that is allowed in a contract. It gives the minimum amount by which the price of the contract is allowed to increase or decrease.
(6) Maximum Daily Price Fluctuation: It is the maximum amount by which the price of a contract is allowed to increase or decrease in a day.
(7) Initial Margin: Parties have to deposit margins with the exchange's clearing house in order to initiate positions in Futures.
(8) Special Margin: Parties may have to deposit an special margin, in an amount deemed fit the exchange's clearing house, in case of additional volatility in the price of futures contract.
(9) Position Limits: It is the maximum number of contracts of a commodity, a client or a clearing member can hold at a time.
(10) Settlement/Delivery Terms and Procedure: It spells out the terms and conditions on the basis of which a futures contract is settled (either by cash or physical delivery) on expiry.

NYMEX trades energy futures contracts including crude oil, natural gas, electric power, coal, propane, home heating oil, and gasoline. Crude oil is the world's most actively traded commodity, and the NYMEX Division's light, sweet crude oil futures contract is the world's most liquid forum for crude oil trading, as well as the world's largest-volume futures contract trading on a physical commodity. Because of its excellent liquidity and price transparency, the contract is used as a principal international pricing benchmark.

Additional risk management and trading opportunities are offered through options on the futures contract, calendar spread options, crack spread options on the pricing differential of heating oil futures, and crude oil futures and gasoline futures and crude oil futures, and average price options. The contract trades in units of 1000 bbls, and the delivery point is Cushing, Oklahoma, which is also accessible to the international spot markets via pipelines. The contract provides for delivery of several grades of domestic and internationally traded foreign crudes, and serves the diverse needs of the physical market.

Natural Gas Contract

NYMEX launched its highly successful natural gas futures contact in April 1990. Natural gas accounts for almost a quarter of US energy consumption, and the NYMEX Division's natural gas futures contract is widely used as a national benchmark price. The futures contract trades in units of 10,000 million British thermal units (mBtu). The price is based on delivery at the Henry Hub in Louisiana, the nexus of 16 intra- and interstate natural gas pipeline systems that draw supplies from the regions prolific gas deposits. The pipelines serve markets throughout the US East Coast, the Gulf Coast, the Midwest, and up to the Canadian border.

An options contract and calendar spread options contracts provide additional risk management opportunities. The spread between natural gas futures and electricity futures - the spark spread - can be used to manage price risk in the power markets. Because of the volatility of natural gas prices, a vigorous basis market has developed in the pricing relationships between Henry Hub and other important natural gas market centers in the continental United States and Canada. NYMEX makes available for trading a series of basis swap futures contracts that are quoted as price differentials between approximately 30 natural gas pricing points and Henry Hub. The basis contracts trade in units of 2500 mBtu on the NYMEX ClearPortSm trading platform. Transactions can also be consummated off-Exchange and submitted to the Exchange for clearing via the NYMEX ClearPort clearing website as an exchange of futures for physicals or exchange of futures for swaps transaction.

Heating Oil Contract

Heating oil, also known as No.2 fuel oil, accounts for about 25% of the yield of a barrel of crude, the second largest "cut" after gasoline. The heating oil futures contract trades in units of 42,000 gallons (1000 bbls) and is based on delivery in New York harbor, the principal cash market trading center. Options on futures, calendar spread options contracts, crack spread options contracts, and average price options contracts give market participants even greater flexibility in managing price risk. The heating oil futures contract is also used to hedge diesel fuel and jet fuel, both of which trade in the cash market at an often stable premium to New York harbor heating oil futures.

Gasoline Contract

Gasoline is the largest single volume refined product sold in the United States and accounts for over 40% of national oil consumption. It is a highly diverse market, with hundreds of wholesale distributors and thousands of retail outlets, often making it subject to intense competition and price volatility. The New York harbor gasoline futures contract trades in units of 42,000 gallons (1000 bbls). It is based on delivery at petroleum products terminals in the harbor, the major East Coast trading center for imports and domestic shipments from refineries in the New York harbor area or from the Gulf Coast refining centers.

Clearing House

The NYMEX clearinghouse provides a system of guarantees that mitigates counterparty credit risk. NYMEX provides a financially settled (not physically delivered) monthly futures contract for on-peak electricity transactions based on the daily floating price for each peak day of the month at the PJM western hub. The PJM western hub consists of 111 delivery points, primarily on the Pennsylvania Electric Co. and the Potomac Electric Co. utility transmission systems. Additional risk management and trading opportunities are offered through options on the PJM monthly futures contract.

OTC Clearing and ClearPort

NYMEX also clears OTC trades for market participants who wish to avoid counterparty credit risk by using standardized contracts for natural gas, crude oil, refined products, and electricity. By clearing OTC energy contracts through ClearPort, NYMEX provides guaranteed financial performance. In effect, these contracts become quasi-futures contract since they are cleared on the exchange.

e-miNY

The Exchange also lists e-miNY energy futures"", fractional light, sweet crude oil, and natural gas futures contracts that offer smaller investors and traders the opportunity for an efficient means of participating in energy markets. The contracts trade via the Chicago Mercantile Exchange's GLOBEX@ electronic trading system and clear through the NYMEX clearinghouse. The close of the open outcry session on the trading floor does not postpone the potential for risk in the energy and metals markets until the next morning.

NYMEX ACCESS

NYMEX ACCESS@ allows participants to protect themselves from that exposure as the energy futures markets are available for trading for another 17 hours through the internet-based NYMEX ACCESS electronic trading system. With NYMEX ACCESS, the exchange is open virtually around the clock. The system operates via the Internet, so traders can log on from any internet-enabled computer almost anywhere in the world. NYMEX ACCESS trading opens at 3:15 pm New York Time, Mondays through Thursdays, and runs as late as 9:30 am the next morning, depending on the contract. Sunday evening sessions open at 7 pm NYMEX ACCESS trades are executed on the basis of price and time with total anonymity and absolute fairness. The first order in at the best price is the first executed. NYMEX ACCESS offers enhanced views of the market at any given time, and gives traders unparalleled flexibility in executing trading strategies.

Today, NYMEX is actively interested in growing its energy futures business in Asia for oil and gas trading. It is developing relationships with both the Singapore Exchange and Tocom to clear oil and gas products in Asia.

INTERNATIONAL PETROLEUM EXCHANGE (IPE) / INTERCONTINENTAL EXCHANGE (ICE)

Background

The London-based International Petroleum Exchange (IPE) was acquired by the Intercontinental Exchange (ICE) in June 2001. The IPE was founded by group of energy and futures companies in 1980. The IPE trades crude oil, gas oil, and natural gas futures contracts.

The first futures contract, for gas oil, was launched the following year. In June 1988, the IPE successfully launched Brent crude oil futures. Volumes rapidly took off, and the exchange has experienced incremental growth, year-on-year for almost its entire history. New trading instruments such as swaps, futures, and options have grown up. Since its inception crude oil futures and latterly options have been traded in pits on the market floor using the open outcry system. As business volumes have grown the IPE has moved location several times to accommodate new pits and increasing number of traders.

Natural Gas Contracts

In 1997 the IPE began trading its first non-oil contract with the launch of the IPE natural gas futures. These contracts were launched on the IPE's Electronic Trading Platform (IPE ETS), This platform has been further enhanced since its first launch and now also provides the medium for parallel trading of oil contracts. WE gas oil options equip IPE market participants with the tools to carry out a range of sophisticated and highly efficient price risk management strategies, with the security and transparency of a regulated trading exchange. On the back of its success with futures contracts the IPE launched the gas oil options contracts in July 1987. The contract is based on the underlying futures and if exercised will result in a corresponding futures position and is for American style exercise, allowing the buyer to exercise an option any time up to, and including the expiry day. The IPE launched a natural gas futures contract in January 1997. It is used primarily to manage price risk in the underlying physical natural gas market and to effect physical delivery into the UK gas grid. The contract also allows users to separate / reference 'the price element within a long-term physical gas contract. The natural gas contract results in physical delivery only if position(s) are not closed out prior to expiry. Physical delivery takes place within the UK natural gas grid at the National Balancing Point (NBP). Delivery must be made equally throughout the delivery period and equivalent to the number of lots open at the time of expiry.

Trading Platform

IPE crude oil and gas oil futures and options contracts and IPE natural gas futures contracts are traded electronically on the ICE platform using either WebICE the ICE's proprietary Internet based trading front end, or Independent Software Vendors (ISVs). WebICE is a fully functional browser based front-end to the ICE electronic trading platform, it provides instant access to the IPE markets through a web browser using industry standard data security and encryption technology.

Future of IPE

IPE has recently become more active in Asia once again. It is known that electronic trading growth on the ICE has come from Asia. It will be interesting to see how both exchanges gain momentum in the vast Asia Pacific energy region. Since the energy commodities are the most volatile of all commodities and Asia is the growth region for oil and gas consumption, it seems likely that Internet trading will grow exponentially both on- and off-exchange in the region.

DUBAI MERCANTILE EXCHANGE (DME)

The Dubai Mercantile Exchange (DME) is a 50/50 joint venture between the New York Mercantile Exchange (NYMEX) and a subsidiary of Dubai Holding. The DME will be the region's first energy futures exchange. The DME will be designed based on the needs of the industry and with the input of key participants in the sector. The DME will trade Middle East based energy futures contracts, with physical or financial settlement.

The DME is expected to begin trading during the 4th quarter of 2006, subject to receiving the appropriate regulatory approvals. The next steps include speaking directly with market participants in order to determine their specific needs for various futures contracts. Initially, the DME plans on listing a Middle-Eastern based sour crude oil futures contract. In the future, the DME may trade futures and options on other energy commodities. All contracts chosen for trading on the DME will be determined based on market research. The DME will be driven by the market; it will respond to market needs and will not try and create demand where there is none.

Trading requirements are currently being developed for the DME. Intention is to enable anyone, who meets the minimum membership requirements set by the DME and is backed by an approved clearing member, to trade on the DME.

The DME will be located within the Dubai International Financial Centre (DIFC), a financial free zone situated in Dubai, United Arab Emirates and will be regulated by the Dubai Financial Services Authority (DFSA).The DFSA is a regulatory body established within the DIFC and charged with the creation of a world-class regulatory regime which promotes integrity, transparency and efficiency. The DFSA has been modeled after the Financial Services Authority from the United Kingdom and is the integrated regulator of all financial and ancillary services undertaken in or from the DIFC. The DME will have to be accepted as an Authorized Market Institution (AMI) before it may operate the exchange.

SINGAPORE EXCHANGE LIMITED (SGX)

Singapore Exchange Limited (SGX) is Asia-Pacific's first demutualised and integrated securities and derivatives exchange. SGX was inaugurated on 1 December 1999, following the merger of two established and well-respected financial institutions - the Stock Exchange of Singapore (SES) and the Singapore International Monetary Exchange (SIMEX).

On 23 November 2000, SGX became the first exchange in Asia-Pacific to be listed via a public offer and a private placement. Listed on our own bourse, the SGX stock is a component of benchmark indices such as the MSCI Singapore Free Index and the Straits Times Index.

Home to Singapore's leading listed companies, SGX is also at the forefront of exchanges globally in attracting international issuers and is rapidly emerging as Asia's offshore risk management centre for international derivatives.

SHANGHAI FUTURES EXCHANGE

History

The last time a futures contract operated in China was when The Shanghai futures Exchange (SHFE) ran a fuel oil futures contract from 1993 until 995, when it was finally delisted as part of a drive by the Chinese government against extreme financial market speculation. However, the last time an internationally utilized liquid fuel oil futures contract was traded in Asia was the SIMEX Fuel Oil Futures contract which actively traded from 1989 to around 1992. This contract was eventually delisted. There seemed little support from the energy industry to the futures contract that competed with an increasing and dynamic Singapore Platts pricing related OTC Swaps market that was until August 25, 2004, when the SHFE was launched.

Market Development

China's fledging fuel oil futures market appears to have taken off and looks like it will offer traders selling fuel oil in one of the fastest growing economies in the world, an opportunity to hedge. It may be fair to expect in 2005/2006 an OTC market to develop indexed against this futures contract. Adding further liquidity to the futures contract as OTC market makers re-hedge their positions in to the Shanghai futures market. One criticism of the Shanghai futures contracts is its small contract size (in terms of tonnage per futures contract), and this may spark the development of an OTC market which meets the contract specification and currency needs of the international trading community. It merits close attention.

A combination of soaring crude oil prices and complex regulations governing specifications and physical delivery arrangements have proved a barrier to most traders involved in the fuel oil physical market.

As the contracts quality standards are stricter than the active OTC; Platts-related fuel oil swaps, in terms of density, carbon residue, and other specifications, the amount of fuel oil available for physical delivery is limited and this puts a strain on the strength of the underlying contract.

Exchange officials were unavailable to give formal comment on the question of specifications, delivery, or volumes. Sources close to the exchange argue that the regulations are necessary and correct. They say the top priority for the exchange is not the trading volume, but the steady and healthy development of the futures market. China companies want to wait to see if there is any problem in physical delivery before plunging into the market. In the meantime, majority of international and Chinese traders are continuing to use the Singapore paper market as their major hedging tool.

Fuel Oil Contract

When fuel oil futures were launched, the fuel oil front month contract was five-months forward, compared with one-month for other futures contracts, as the Shanghai Exchange hoped to give the market plenty of time to get familiar with the new tool. This is probably a key reason why physical traders had been reluctant to trade the January 2005 contract, which was the front month contact as of August 2004. The Shanghai fuel oil futures market still needs more liquidity, which in turn should increase its price correlation and reliability with real local physical market fundamentals. It is too early to promote it as a key hedging instrument, but also too early to write it off as, mother failed oil futures contract in Asia.

TOKYO COMMODITY EXCHANGE (TOCOM)

Background

TOCOM is a non-profit membership organization as defined under the Commodity Exchange Law (1950), which regulates all commodities futures and options trading in Japan. The Tokyo Commodity Exchange (TOCOM) was formed on November 1, 1984, through the integration of the Tokyo Textile Exchange, the Tokyo Rubber Exchange, and the Tokyo Gold Exchange, in order to establish a comprehensive commodity exchange in Japan. Thanks to the continued support and cooperation of the exchange's membership and others concerned, and with careful guidance from the competent authorities.

Performance

TOCOM has achieved significant progress over the past two decades. In 2004 it celebrated its twentieth anniversary and is considered the largest futures exchange in Japan, and world's second largest commodity futures exchange. TOCOM lists 10 commodities: crude oil, gasoline, kerosene, gas oil, gold, silver, platinum, palladium, aluminum, and rubber futures contracts. The 2003 annual trading volume exceeded 87 million contracts, setting a new record for the sixth consecutive year. The change and development of the past two decades have made a significant mark on the history of TOCOM, and of commodity futures industry in Japan. In the course of these 20 years, TOCOM has succeeded in listing gasoline, kerosene, and crude oil, among other contracts, on the back of deregulation in the oil sector. The oil market today occupies a key position among TOCOM's mainstay markets, alongside precious metals.

Clearing House

In June 2003, TOCOM also became the first commodity exchange in Japan to establish an in-house clearinghouse, which further enhanced its market reliability and attractiveness to professional foreign investors and traders looking for hedging instruments. On November 4, 2003, The IPE, Europe's leading energy futures and options exchange, and The TOCOM, Asia's leading commodity futures market announced that they have entered into a Memorandum of Understanding (MOU).

Under the terms of the MOU, both parties will cooperate with a view to helping the other party to make available trading in financial instruments traded on each exchange via computer terminals in the country of the other party. Also to secure the appropriate regulatory approvals from the UK and Japanese authorities. Both parties will explore mechanisms which will enhance the liquidity, efficiency, and integrity of each market and discuss areas of cooperation, including new business opportunities or developments when they arise.

TOCOM and NYMEX ACCESS

Then on July 13, 2004, The NYMEX, Inc., and the TOCOM announced that energy and metals futures contracts would become available for NYMEX ACCESS, internet-based trading in Japan on July 20, 2004. When the oil contracts such as Gasoline were first launched in Tokyo, trading houses such as Mitsui, Marubeni, and Sumitomo etc. were nowhere to be seen, trading volume was concentrated in the hands of small retail investors/traders. However times have changed and TOCOM's futures contracts have proven that oil futures can exist in Asia, (at least where a dynamic existing OTC market does not exist) since the demise of the Singapore Fuel Oil Futures contract c. 1993. An established market that clearly serves the needs of the market will be difficult to replace with another new market place/mechanism that just serves nice to have ideals of some market participants. This has been the downfall of failed futures contracts in energy across Asia for the past decade.

INDIAN FUTURE EXCHANGES

In India there are 25 recognized future exchanges, of which there are three national level multi-commodity exchanges. After a gap of almost three decades, Government of India has allowed forward transactions in commodities through Online Commodity Exchanges, a modification of traditional business known as Adhat and Vayda Vyapar to facilitate better risk coverage and delivery of commodities. The three exchanges are:

- National Commodity & Derivatives Exchange Limited (NCDEX)
- Multi Commodity Exchange of India Limited (MCX)
- National Multi-Commodity Exchange of India Limited (NMCEIL)

All the exchanges have been set up under overall control of Forward Market Commission (FMC) of Government of India.

NATIONAL COMMODITY & DERIVATIVES EXCHANGE LIMITED (NCDEX)

National Commodity & Derivatives Exchange Limited (NCDEX) located in Mumbai is a public limited company incorporated on April 23, 2003 under the Companies Act, 1956 and had commenced its operations on December 15, 2003.This is the only commodity exchange in the country promoted by national level institutions. It is promoted by ICICI Bank Limited, Life Insurance Corporation of India (LIC), National Bank for Agriculture and Rural Development (NABARD) and National Stock Exchange of India Limited (NSE). It is a professionally managed online multi commodity exchange. NCDEX is regulated by Forward Market Commission and is subjected to various laws of the land like the Companies Act, Stamp Act, Contracts Act, Forward Commission (Regulation) Act and various other legislations.

MULTI COMMODITY EXCHANGE OF INDIA LIMITED (MCX)

Headquartered in Mumbai Multi Commodity Exchange of India Limited (MCX), is an independent and de-mutulised exchange with a permanent recognition from Government of India. Key shareholders of MCX are Financial Technologies (India) Ltd., State Bank of India, Union Bank of India, Corporation Bank, Bank of India and Canara Bank. MCX facilitates online trading, clearing and settlement operations for commodity futures markets across the country.

Multi Commodity Exchange of India Ltd., established in 2003 and already the largest commodity futures exchange in India, provides the premier forum for managing the price risk associated with oil market. MCX is a dominant center for gold and silver futures trading. MCX is the first exchange to launch Crude Oil futures in India. MCX's liquidity, price transparency and financial integrity will make it a benchmark for crude oil markets in India

MCX has built strategic alliances with Bombay Bullion Association, Bombay

Metal Exchange, Solvent Extractors' Association of India, Pulses Importers Association and Shetkari Sanghatana.

NATIONAL MULTI-COMMODITY EXCHANGE OF INDIA LIMITED (NMCEIL)

National Multi Commodity Exchange of India Limited (NMCEIL) is the first de-mutualized, Electronic Multi-Commodity Exchange in India. On 25th July, 2001, it was granted approval by the Government to organise trading in the edible oil complex. It has operationalised from November 26, 2002. It is being supported by Central Warehousing Corporation Ltd., Gujarat State Agricultural Marketing Board and Neptune Overseas Limited. It got its recognition in October 2002.

Commodity exchange in India plays an important role where the prices of any commodity are not fixed, in an organized way. Earlier only the buyer of produce and its seller in the market judged upon the prices. Others never had a say. Today, commodity exchanges are purely speculative in nature. Before discovering the price, they reach to the producers, end-users, and even the retail investors, at a grassroots level. It brings a price transparency and risk management in the vital market.

A big difference between a typical auction, where a single auctioneer announces the bids and the Exchange is that people are not only competing to buy but also to sell. By Exchange rules and by law, no one can bid under a higher bid, and no one can offer to sell higher than someone else's lower offer. That keeps the market as efficient as possible, and keeps the traders on their toes to make sure no one gets the purchase or sale before they do.

Finally, In a global perspective the oil and gas trading takes place in two distinct markets; the exchanged based organized market and the Unorganized OTC market. While the latter is customized, the former is standardized and regulated. Some of the international exchange based markets are NYMEX, IPE, TOCOM, SGX and SHFE, while MC, NCDEX and NMCE are Indian exchange market. Asian pacific reason is potential market for all market participants but one glaring problem in Asian oil market has been the lack of price benchmark. Platts, the worlds leading emerging information provider enables market to operating the transparency and efficiency by providing the real time used, pricing assessment, analytical services and conferences. Plats methodology reflects the activity of market players and takes into account industry practices in each region. Platt's methodology involves price assessment, liquidity profile, location profile, contract denomination, fixed price premium and papers (swaps). Major contacts on NYMEX are natural gas, heating oil, crude oil, electricity, clearing, propane and coal contracts.

Challenges before India's Energy Futures Market

(1) There is no structured spot market in India. Thus, lack of transparent spot price availability makes it difficult to calculate Due Date Rate of futures contracts.

(2) MCX has taken the initiative to open National Spot Market to facilitate spot trading in the country as well as benefit the futures market but it is mostly for agricultural products.

(3) Currently cash settlement of outstanding contracts at maturity is not allowed by the Forward Contracts (Regulation) Act (FCRA). This means that at maturity a party can ask for delivery. To avoid this, traders square-off their positions before maturity, which is practically equivalent to cash settlement. Options' trading is also not allowed in commodities on Indian futures exchanges. The parliamentary committee's report recommending options trading in commodities is lying with the government but it has not taken any step to implement it so far.

(4) With the growing refining business in the country, it is the demand of time to have crack spread contracts to be introduced in country's futures exchanges.

(5) The PSUs are still reluctant to trade on MCX and NCDEX because of the following reasons:
   - Absence of refined product futures like Aviation Turbine Fuel, Naphtha etc. These products can be used for crack spread hedging. Refiners find it very useful to lock their margins.
   - MCX and NCDEX have limited trading hours. They have trading hours from 10:00 a.m. to 11:30 p.m. which is not the ideal way to go when one is trading in commodities that are truly global.
   - Moreover, the contract size at these exchanges is also too small to be good enough for physical delivery.

(6) As the PSUs are not interested in trading on Indian futures exchanges, there is practically no physical delivery in futures contracts of energy commodities on these exchanges.

(7) Lack of an appropriate marker for Asia:
Pricing of crude oil is done through the use of marker crudes. Some of the major marker crudes are:
   - West Texas Intermediate (USA)
   - Brent (Europe and Africa)
   - Dubai and Oman (Asia)
   - Tapis (Australia)

The main criteria for a marker crude is for it to be sold in sufficient volumes to provide liquidity (many buyers and sellers) in the physical market as well as having similar physical qualities of alternative crudes. In addition, the marker crude should provide pricing information. WTI does this through its use on the New York Metals Exchange (NYMEX) as the basis of a futures contract. Brent offers pricing information based more on the physical trading of oil through spot trading.

In Asia there is no futures exchange where crude oil is traded and which would provide pricing information to the same extent as WTI and Brent. Because of this the prices of Light Sweet crude for the Asian market have been higher than those for U.S./European markets. The prices for supplies to Europe are linked to the average spot price of North Sea Brent crude and that for U.S. supplies to WTI crude. Prices for supplies to Asia are linked to a 30-day spot average of Oman and Dubai crude during the month of delivery.

Major reason for this is the absence of strong trading markets in Asia. Also, today the production of Dubai crude has declined substantially to less than four cargoes per month (less than 100,000 bpd). It is difficult to determine the price of Dubai crude on spot trading at present because of the decline in its production.

India should join other Asian countries to use their bargaining power for a better price marker for Asia as the markers have been decided by OPEC on the basis of consumption and buying power. The fact that Asian nations led by India and China have led the global oil demand growth in recent years makes the case even more strong.

(8) Despite the size and importance of Asian market, price of oil in the continent is dependent on less effective crudes like, Tapis, Minas and Dubai crude. India should work with other Asian nations to develop stronger markets in the region. This will help in real price discovery.

(9) For the success of trading in India, it is a must that government deregulates the markets. In India, we still have government controlling the prices of important petroleum products like, petrol, diesel and LPG.

3.5 CRUDE OIL FUTURES AND CONTRACTS

Crude oil traded futures and options have several economic benefits like managing the price risk of market positions. These are open markets where large number of buyers and sellers compete for the best prices. Companies discover and establish competitive prices on futures markets. A large amount of risk capital is involved in futures markets and they are among the most liquid of all global financial markets as they have low transaction costs and ease of entry and exit. Today, futures markets have clearing house guarantees and exchange self-regulation along with other institutional arrangements.

Facts and Features of Futures Markets

1. A futures contract is a legally binding agreement to buy or sell a specific quantity and grade of oil, gas or any other specified commodity or financial instrument sometime in the future at a price agreed upon at the time of the trade.
2. The expiry of the contract provides either for delivery of the underlying physical energy product or a final cash payment (cash settlement).
3. The contracts are traded on regulated exchanges, mostly electronically via a computerized network. However, in some specified trading areas called pits or rings they are still traded by open outcry.
4. In a futures contract, all open positions are marked-to-market daily at the settlement price and the resulting daily gains and losses are passed to the gaining or losing futures accounts held by brokers for their customers.

5. In case of delivery, the tenderable goods must meet the contract specifications. In addition, these goods must be certified for quantity and quality by an approved surveyor.
6. Because futures contracts are standardized, sellers and buys can exchange one contract for another and "offset" their obligation to take or give delivery of a commodity or instrument underlying the futures contract. Offset in the futures market means taking another futures position opposite or equal to one's initial futures transaction. For e.g. if a trader bought one December MCX Brent Crude Oil contract, he must sell one December MCX Brent Crude Oil contract before the contracts call for delivery. This will offset his position.
7. Majority of futures contracts are terminated by offset or a final cash payment rather than by physical delivery (only about 2%).

Mechanics of a Futures Trading

Buyers and sellers deposit initial margin (a sort of good faith deposit) with a brokerage firm who would be a clearing member of the exchange where the futures contract is being traded. This initial margin amount is around 5-10 percent of the total notional contract value.

Table 1: Margins in Futures Trading

Position	Price of Underlying	Profit/Loss	Amount
Buy a futures contract	Price goes up	Profit	Price Increase × Contract Size
Buy a futures contract	Price goes down	Loss	Price Decrease × Contract Size
Sell a futures contract	Price goes up	Loss	Price Increase × Contract Size
Sell a futures contract	Price goes down	Profit	Price Decrease × Contract Size

These profits and losses are paid daily via the variation futures margin which a clearing broker must deposit with the clearing house every day on behalf of its customers. The broker either finances this for its customer or calls its customer for collateral against unrealized losses.

Commercial Applications of the Exchange's Energy Futures and Options Contracts

The Exchange provides buyers and sellers with price insurance and arbitrage opportunities that can be integrated into cash market operations. Trading Exchange contracts can improve the credit worthiness and add to the borrowing capacity of natural resource companies, thus augmenting the companies' financial management and performance capabilities.

Cash vs. Futures Price Relationships

Cash prices are the prices for which the commodity is sold at the various market locations. The futures price represents the current market opinion of what the commodity will be worth at some time in the future. Under normal circumstances of adequate supply, the price of the physical commodity for future delivery will be approximately equal to the present cash price, plus the amount it costs to carry or store the commodity from the present to the month of delivery. These costs, known as carrying charges, determine the normal premium of futures over cash.

As a result, one would ordinarily expect to see an upward trend to the prices of distant contract months. Such a market condition is known as *contango* and is typical of many futures markets. In most physical markets, the crucial determinant of the price differential between two contract months is the cost of storing the commodity over that particular length of time. As a result, markets which compensate an individual fully for carry charges interest rates, insurance, and storage - are known as full contango markets, or full carrying charge markets.

Under normal market conditions, when supplies are adequate, the price of a commodity for future delivery should be equal to the present spot price plus carrying charges. The contango structure of the futures market is kept intact by the ability of dealers and financial institutions to bring carrying charges back into line through arbitrage.

Futures markets are typically contango markets, although seasonal factors in energy markets play an important role in market relationships. For example, during the summer, heating oil futures are often in contango as the industry begins to build inventory for the approaching cold weather. On any given day, prices in the forward contract months are progressively higher through the fall, reflecting the costs of storage, interest rates, and the assumption of increased demand.

The opposite of contango is *backwardation*, a market condition where the nearby month trades at a higher price relative to the outer months. Such a price relationship usually indicates a tightness of supply; a market can also be in backwardation when seasonal factors predominate.

Convergence of Futures and Cash Markets

As a futures contract approaches its last day of trading, there is little difference between it and the cash price. The futures and cash prices will get closer and closer, a process known as convergence, as any premium the futures have had has disappeared over time. A futures contract nearing expiration becomes, in effect, a spot contract.

Hedging in Futures Markets

We use various hedging strategies with the purpose of avoiding the risk of losses due to adverse market moves. However, there is no 'Perfect Hedge' because the physical cash markets and the futures markets do not always have a perfect price correlation. Hence, almost always there is a profit or loss in hedging.

In futures markets, we take a futures position opposite to that of a cash market position for the purpose of hedging. That is, an energy producer or consumer might want to buy or sell energy futures against their price risk exposure in anticipation of a market price increase/decrease.

Basis

As noted earlier, futures contracts are standardized instruments that stipulate the quantity, quality, and delivery points for a wide cross section of the underlying industry. Basis is the differential that exists between the cash price of a given commodity and the price of the nearest futures contract for the same, or a related commodity.

The predictability and size of the basis can involve three price relationships:

1. The difference between the futures contract and the spot price of the underlying commodity.
2. The difference between the price at the futures contract delivery point and the price at a different location.
3. The price at the futures contract delivery point and the price of a similar, but not identical, quality commodity at the same location.

The cash/futures basis can be effectively minimized if delivery is made or accepted at the same time that the trading of a futures contract nears expiration. Because of price convergence, a futures contract nearing expiration becomes, in effect, a spot contract.

Locational basis is a consideration for firms that desire to hedge but do not make deliveries at the futures contract location. In theory, the price relationship between two different markets will be based on the cost of transportation between them. Sudden local shifts in supply or demand, however, can distort this price relationship.

The extent to which these changes in relative market conditions are predictable will determine the hedged firms' exposure to locational basis risk. The product basis concerns those firms that desire to hedge the purchase or sale of a specific commodity not offered as a liquid futures contract. Firms base their hedge on the historical relationship of the commodity underlying the contract to the commodity to be hedged.

For example, the NYMEX Division heating oil futures contract is widely used as a hedging proxy for jet fuel, as the products are chemically similar and often trade within a narrow price range in relation to each other. The NYMEX Division light, sweet crude oil futures contract, likewise, is used to hedge crudes not specified for delivery by the contract, but which nevertheless trade at a high correlation to the futures contract. For the futures trader, the critical period for assessing basis is between the time that the futures transaction is entered into and the time it is closed, or liquidated.

If a short hedger (a seller of futures) experiences a widening in the basis during the time the hedge is held, a basis loss may result. This narrowing implies means that the cash prices have declined relative to futures prices. That is, they have fallen more or risen less than futures prices. The short hedger's cash loss in a declining market would exceed the gain on the short futures transaction. Or, conversely, the cash gain in a rising market would be exceeded by the loss on the futures transaction. On the other hand, a basis gain would occur in the case of a long hedge and a widening basis because the futures price would rise relative to the cash price. Thus, a narrowing basis yields gains for the short hedger and losses for the long hedger.

Table 2: Potential Basis Changes

Cash/Futures Position	Rising Market		Falling Market	
	Cash Less than Futures	Cash Rises More than Futures	Cash Falls Less than Futures	Cash Falls More than Futures
Bought the cash /Sold the futures	Loss	Gain	Gain	Loss
Sold the cash /bought the futures	Gain	Loss	Loss	Gain

Exchange of Futures for Physicals (EFP) and Deliveries via Futures Markets

It is estimated that more than 95% of all energy futures contracts do not directly result in physical delivery. Nevertheless, significant volumes of crude oil, heating gasoline, natural gas, coal, and propane are delivered, and companies using energy futures for delivery have several options. They can choose standard delivery as per the specifications of the futures contracts; attempt to arrange an Alternative Delivery Procedure (ADP) or effect an Exchange of Futures for Physicals (EFPs). Alternate Delivery Procedure means two parties that are stuck with a position decide to cash settle by negotiating with each other. ADP also includes a case when a party wishes to deliver some energy commodity which is not as per the specifications of the futures contract. Such an energy commodity that does not meet futures contract specifications cannot be delivered through the exchange or clearing house. In ADP transactions, the exchange and the clearing brokers do not have any liability related to the delivery negotiated between parties. Traders and hedgers can also execute an Exchange of Futures for Physicals (EFP).

In order to engage in an EFP, it is not necessary for both sides to be in the futures market when the EFP is initiated except during EFP-only sessions. A futures market long, for example, might effect an EFP with a cash market long. The net result of the transaction would be delivery of a commodity to the futures hedger and the assumption by the physical market participant of the hedger's futures market contracts (next Example)

Why companies go for EFP?

1. Companies that use energy futures contracts for hedging purposes are mostly not interested in making or taking delivery at the specified locations.
2. They also do not prefer being matched to a trading partner by the futures exchange.

3. Mostly, it is more economical for a hedger using futures to make or take delivery of physical energy at a location as per his need and logistics rather than taking delivery at a location specified in the futures contract.
4. An EFP provides the mechanism for such transactions and is usually the preferred method of delivery because it allows flexibility in choosing the delivery location, the trading counterpart, time of delivery and the grade of the commodity.
5. EFP mechanism also allows the two parties to negotiate price for their physical energy market transaction.

Note: The major concern of the futures exchanges is to ensure that the quantity of the cash commodity involved in an EFP must be approximately equal to the quantity specified by the number of futures contracts involved in the EFP.

Practical Mechanics of an EFP Transaction

To understand the mechanics of an EFP transaction lets consider the following example:

A party having long position in crude oil in the futures market would take delivery of the crude oil from a party which has a short position in crude oil in the futures market. In this transaction both party's' hedge is liquidated and actual transfer of crude oil takes place between the parties but not through the exchange.

Companies looking for some counterpart to an EFP transaction usually contact the futures brokers first, as futures brokers are the ones among whom EFP market is normally quoted.

Exchange of Futures for Swaps (EFS)

Exchange of Futures for Swaps (EFS) is also an extremely useful trading instrument. The mechanics of the transaction are similar to that of the EFP. EFS permits the substitution of a futures position for an over-the counter cash-settled position, letting market participants avoid having to liquidate a physical delivery contract. It also gives market participants the ability to liquidate a swaps position in a market which may have limited liquidity. Two parties are allowed to privately negotiate the execution of an integrated over-the-counter swaps and related futures transaction on pricing terms agreed upon by the involved parties. The transaction must involve approximately equal but opposite side-of-market quantities of futures and swap exposures in the same or related commodities and are be permitted until two hours after trading terminates in the underlying futures contract. EFS transactions are permitted to liquidate, initiate, and transfer futures market positions between the two parties involved in the transaction.

Future Contracts

Oil trading occurs through several contract arrangements and spot transactions in the international oil markets. Oil futures contracts also play a significant role in determining the international oil prices.

How are Oil Prices Determined?

The demand and supply cycle drives oil prices in the international economy. Since there are so many varieties of oil in the market, it is essential to consider a benchmark quality, called the marker crude. In the London-based oil market, Brent Crude is used as the benchmark quality against which the price of other oil varieties are determined. Brent Crude determines the prices of almost two-third of the world's oil trade. Gulf countries refer to Dubai Crude as marker quality. Dubai Crude is available in spot transactions rather than long-term supply contracts. The US refers to West Texas Intermediate or 'sweet crude' to determine oil prices. Owing to its high quality, West Texas Intermediate is the highest priced crude.

All the oil trading markets consider the US dollar as the trading currency. Thus, the role of the US gets highlighted in regulating oil commerce.

The Organization of Petroleum-Exporting Countries (OPEC) has developed the OPEC Basket Price index. This is an average of the oil prices in OPEC nations, including Algeria, Angola, Ecuador, Indonesia, Iran, Libya and Saudi Arabia. Member nations tend to regulate oil production to keep the oil prices within the desired range. The OPEC Basket Price is the lowest priced crude due to high sulfur content, which is removed while refining.

Oil Trading: Futures Contracts

The two major trading exchanges for the oil sector are the New York Mercantile Exchange in the US and the Intercontinental Exchange in London, UK. In 2008, the Iranian Oil Bourse emerged as another international oil exchange center. These international exchanges act as the hub to buy oil futures and options. Changes in the trading prices at these centers affect global oil prices and the international economy.

Futures contracts also play a dominant role in international oil commerce. Oil futures involve an agreement wherein the buyer promises to buy a particular quantity of oil and the seller offers to deliver the oil at a fixed price on a specific future date. Like other commodity futures, oil futures contracts provide the opportunity to mitigate risk factors associated with the global oil economy. It acts a protective tool for producers and distributors against unpredictable losses in the oil industry. Following are the standard oil futures contracts used in the oil trading:

(a) NYMEX Light, Sweet Crude Oil Futures Contract
(b) Heating Oil Futures Trading
(c) Gasoline Futures Trading
(d) ICE Brent crude Oil Future Contract
(e) Standard Contract of Fuel Oil of SHFE
(f) Brent Crude Oil Contract Specification at NCDEX
(h) Furnace Oil Contract Specification at NCDEX
(i) Dubai Mercantile Exchange (DME)
(j) Brent Crude Oil MCX-India

NYMEX Light, Sweet Crude Oil Futures Contract

Trading Unit: 1,000 U.S. barrels (42,000 gallons).

Price Quotation: U.S. dollars and cents per barrel.

Trading Hours *(All times are New York time)* Open outcry trading is conducted from 10:00 AM until 2:30 PM.

Electronic trading is conducted from 6:00 PM until 5:15 PM via the CME Globex® trading platform, Sunday through Friday. There is a 45-minute break each day between 5:15PM (current trade date) and 6:00 PM (next trade date).

Trading Months: The current year and the next five years. A new calendar year will be added following the termination of trading in the December contract of the current year. Additionally, trading can be executed at an average differential to the previous day's settlement prices for periods of two to 30 consecutive months in a single transaction. These calendar strips are executed during open outcry trading hours.

Trading at Settlement (TAS): Trading at settlement is available for the front two months except on the last trading day and is subject to the existing TAS rules. Trading in all TAS products will cease daily at 2:30 PM Eastern Time. The TAS products will trade off of a "Base Price" of 100 to create a differential (plus or minus) in points off settlement in the underlying cleared product on a 1 to 1 basis. A trade done at the Base Price of 100 will correspond to a "traditional" TAS trade which will clear exactly at the final settlement price of the day.

Minimum Price Fluctuation: $0.01 (1¢) per barrel ($10.00 per contract).

Maximum Daily Price Fluctuation: $10.00 per barrel ($10,000 per contract) for all months. If any contract is traded, bid, or offered at the limit for five minutes, trading is halted for five minutes. When trading resumes, the limit is expanded by $10.00 per barrel in either direction. If another halt were triggered, the market would continue to be expanded by $10.00 per barrel in either direction after each successive five-minute trading halt. There will be no maximum price fluctuation limits during any one trading session.

Last Trading Day: Trading terminates at the close of business on the third business day prior to the 25th calendar day of the month preceding the delivery month. If the 25th calendar day of the month is a non-business day, trading shall cease on the third business day prior to the business day preceding the 25th calendar day.

Settlement Type: Physical.

Delivery: F.O.B. seller's facility, Cushing, Oklahoma, at any pipeline or storage facility with pipeline access to TEPPCO, Cushing storage, or Equilon Pipeline Co., by in-tank transfer, in-line transfer, book-out, or inter-facility transfer (pumpover).

Delivery Period: All deliveries are ratable over the course of the month and must be initiated on or after the first calendar day and completed by the last calendar day of the delivery month.

Alternate Delivery Procedure (ADP): An alternate delivery procedure is available to buyers and sellers who have been matched by the Exchange subsequent to the termination of trading in the spot month contract. If buyer and seller agree to consummate delivery under terms different from those prescribed in the contract specifications, they may proceed on that basis after submitting a notice of their intention to the Exchange.

Exchange of Futures for Physicals (EFP): The commercial buyer or seller may exchange a futures position for a physical position of equal quantity by submitting a notice to the Exchange. EFPs may be used to either initiate or liquidate a futures position.

Deliverable Grades: Specific domestic crudes with 0.42% sulfur by weight or less, not less than 37° API gravity nor more than 42° API gravity. The following domestic crude streams are deliverable: West Texas Intermediate, Low Sweet Mix, New Mexican Sweet, North Texas Sweet, Oklahoma Sweet, South Texas Sweet. Specific foreign crudes of not less than 34° API nor more than 42° API. The following foreign streams are deliverable: U.K. Brent and Forties, for which the seller shall receive a 30 cent per barrel discount below the final settlement price; Norwegian Oseberg Blend is delivered at a 55¢–per-barrel discount; Nigerian Bonny Light, Qua Iboe, and Colombian Cusiana are delivered at 15¢ premiums.

Inspection: Inspection shall be conducted in accordance with pipeline practices. A buyer or seller may appoint an inspector to inspect the quality of oil delivered. However, the buyer or seller who requests the inspection will bear its costs and will notify the other party of the transaction that the inspection will occur.

Position Accountability Levels and Limits: Any one month/all months: 20,000 net futures, but not to exceed 2,000 contracts in the last three days of trading in the spot month.

Margin Requirements: Margins are required for open futures positions.

Trading Symbol: CL CLT (TAS Code)

HEATING OIL FUTURES TRADING

Unit: 42,000 U.S. gallons (1,000 barrels).

Price Quotation: U.S. dollars and cents per gallon.

Trading Hours (All times are New York Time): Open outcry trading is conducted from 10:05 AM until 2:30 PM.

Electronic trading is conducted from 6:00 PM until 5:15 PM via the CME Globex® trading platform, Sunday through Friday. There is a 45-minute break each day between 5:15PM (current trade date) and 6:00 PM (next trade date).

Trading Months: 18 consecutive months.

Trading at Settlement (TAS): Trading at settlement is available for the front two months except on the last trading day and is subject to the existing TAS rules. Trading in all TAS products will cease daily at 2:30 PM Eastern Time. The TAS products will trade off of a "Base Price" of 100 to create a differential (plus or minus) in points off settlement in the underlying cleared product on a 1 to 1 basis. A trade done at the Base Price of 100 will correspond to a "traditional" TAS trade which will clear exactly at the final settlement price of the day.

Minimum Price Fluctuation: $0.0001 (0.01¢) per gallon ($4.20 per contract).

Maximum Daily Price Fluctuation: $0.25 per gallon ($10,500 per contract) for all months. If any contract is traded, bid, or offered at the limit for five minutes, trading is halted for five minutes. When trading resumes, the limit is expanded by $0.25 per gallon in either direction. If another halt were triggered, the market would continue to be expanded by $0.25 per gallon in either direction after each successive five-minute trading halt. There will be no maximum price fluctuation limits during any one trading session.

Last Trading Day: Trading terminates at the close of business on the last business day of the month preceding the delivery month.

Settlement Type: Physical.

Delivery: F.O.B. seller's facility in New York harbor, ex-shore. All duties, entitlements, taxes, fees, and other charges paid. Requirements for seller's shore facility: capability to deliver into barges. Buyer may request delivery by truck, if available at the seller's facility, and pays a surcharge for truck delivery. Delivery may also be completed by pipeline, tanker, book transfer, or inter- or intra-facility transfer. Delivery must be made in accordance with applicable federal, state, and local licensing and tax laws. Delivery Period Deliveries may only be initiated the day after the fifth business day and must be completed before the last business day of the delivery month.

Alternate Delivery Procedure (ADP): An alternate delivery procedure is available to buyers and sellers who have been matched by the Exchange subsequent to the termination of trading in the spot month contract. If buyer and seller agree to consummate delivery under terms different from those prescribed in the contract specifications, they may proceed on that basis after submitting a notice of their intention to the Exchange.

Exchange of Futures for Physicals (EFP): The commercial buyer or seller may exchange a futures position for a physical position of equal quantity by submitting a notice to the Exchange. EFPs may be used to either initiate or liquidate a futures position.

Grade and Quality Specifications: Generally conforms to industry standards for fungible No. 2 heating oil.

Inspection The buyer may request an inspection for grade and quality or quantity for all deliveries, but shall require a quantity inspection for a barge, tanker, or inter-facility transfer. If the buyer does not request a quantity inspection, the seller may request such inspection. The cost of the quantity inspection is shared equally by the buyer and seller. If the product meets grade and quality specifications, the cost of the quality inspection is shared jointly by the buyer and seller. If the product fails inspection, the cost is borne by the seller.

Position Accountability Levels and Limits: 7,000 contracts for all months combined, but not to exceed 1,000 in the last three days of trading in the spot month.

Margin Requirements: Margins are required for open futures positions.

Trading Symbol:
HO
HOT (TAS Code)

GASOLINE FUTURES TRADING

Unit: 42,000 U.S. gallons (1,000 barrels).

Price Quotation: U.S. dollars and cents per gallon.

Trading Hours (All times are New York time): Open outcry trading is conducted from 10:05 AM until 2:30 PM.

Electronic trading is conducted from 6:00 PM until 5:15 PM via the CME Globex® trading platform, Sunday through Friday. There is a 45-minute break each day between 5:15PM (current trade date) and 6:00 PM (next trade date).

Trading Months: 12 consecutive months.

Trading at Settlement (TAS): Trading at settlement is available for the front two months except on the last trading day and is subject to the existing TAS rules. Trading in all TAS products will cease daily at 2:30 PM Eastern Time. The TAS products will trade off of a "Base Price" of 100 to create a differential (plus or minus) in points off settlement in the underlying cleared product on a 1 to 1 basis. A trade done at the Base Price of 100 will correspond to a "traditional" TAS trade which will clear exactly at the final settlement price of the day.

Minimum Price Fluctuation: $0.0001 (0.01¢) per gallon ($4.20 per contract).

Maximum Daily Price Fluctuation: $0.25 per gallon ($10,500 per contract) for all months. If any contract is traded, bid, or offered at the limit for five minutes, trading is halted for five minutes. When trading resumes, the limit is expanded by $0.25 per gallon in either direction. If another halt were triggered, the market would continue to be expanded by $0.25 per gallon in either direction after each successive five-minute trading halt. There will be no maximum price fluctuation limits during any one trading session.

Last Trading Day: Trading terminates at the close of business on the last business day of the month preceding the delivery month.

Settlement Type: Physical.

Delivery: F.O.B. seller's facility in New York harbor ex-shore. All duties, entitlements, taxes, fees, and other charges paid. Requirements for seller's shore facility: capability to deliver into barges. Delivery may also be completed by pipeline, tanker, book transfer, or inter- or intra-facility transfer. Delivery must be made in accordance with applicable federal, state, and local licensing and tax laws. Delivery shall comply with all state laws related to oxygen content.

Delivery Period: Deliveries may only be initiated the day after the fifth business day and must be completed before the last business day of the delivery month.

Alternate Delivery Procedure (ADP): An alternate delivery procedure is available to buyers and sellers who have been matched by the Exchange subsequent to the termination of trading in the spot month contract. If buyer and seller agree to consummate delivery under terms different from those prescribed in the contract specifications, they may proceed on that basis after submitting a notice of their intention to the Exchange.

Exchange of Futures for Physicals (EFP): The commercial buyer or seller may exchange a futures position for a physical position of equal quantity by submitting a notice to the Exchange. EFPs may be used to either initiate or liquidate a futures position.

Grade and Quality Specifications: Generally conforms to industry standards for Phase II Complex Model Reformulated Gasoline in accordance with Colonial Pipeline Co. specifications for fungible A grade, 87 octane index gasoline.

Month	Reid Vapor Pressure (RVP) pounds per square inch	Volatile Organic Compounds (VOC) Percentage Reduction
January	15.0 max	N/A
February	15.0 max	N/A
March	13.5 max	N/A
April	N/A	23.4 min
May	N/A	23.4 min
June	N/A	23.4 min
July	N/A	23.4 min
August	N/A	23.4 min
September 1-15	N/A	23.4 min
September 16-30	13.5 max	N/A
October	13.5 max	N/A
November	15.0 max	N/A
December	15.0 max	N/A

Margin Requirements: Margins are required for open futures positions.

Trading Symbol :
HU
HUT (TAS Code)

ICE BRENT CRUDE OIL FUTURE CONTRACT

Price Units: USD / bbl

Quantity Units: bbl

Price Increment: 0.01

Min Lot Increment: 1

Granularity: monthly

Contract Description: The IPE Brent Crude futures contract is a deliverable contract based on EFP delivery with an option to cash settle. Trading Period/ Strip: Thirty consecutive months then half yearly out to a maximum of seven years.

Expiration Date: The close of business on the business day immediately preceding the 15th day prior to the first day of the delivery month, if such 15th day is a banking day in London. If the 15th day is a non-banking day in London (including Saturday), trading shall cease on the business day immediately preceding the first business day prior to the 15th day.

Contract Security: LCH Clearnet Ltd. (LCH) acts as central counterparty to all trades thereby guaranteeing the financial performance of IPE contracts registered in the name of its Members up to and including delivery, exercise and/or settlement

Trading Hours: 01:00 – 22:00, London Local Time (LLT), Monday-Friday

Contract Size: 1,000 barrels

Minimum Price Flux: One tick

Maximum Price Flux: There are no limits

Settlement Prices: The weighted average price of trades during a three minute settlement period from 19:27:00, London time.

Daily Margin: All open contracts are marked-to-market daily.

Margin Rates: Margin rates are set by LCH.Clearnet Ltd.

Position Limits: There are no position limits.

Trading Methods: Electronic futures, Exchange of futures for physical (EFP), Exchange of futures for swap (EFS) and Block Trades are available for this contract.

Delivery/Settlement Basis: The IPE Brent Crude futures contract is a deliverable contract based on EFP delivery with an option to cash settle, i.e. the IPE Brent Index price for the day following the last trading day of the futures contract.

Standard Contract of Fuel Oil of SHFE

Underlined Product	Fuel Oil
Trading Unit	10 tons/lot
Quotation Unit	Yuan (RBM)/ton
Tick Size	1 Yuan/ton
Daily Price Limit	±5% of the settlement price of the previous business day
Contract Months	Jan. to Dec. (Except the Spring Festival)
Trading Hours	9:00-11:30 a.m. to 1:30-3:00 p.m.
Last Trading Day	The last trading day of the month before the spot month
Delivery Period	Five consecutive business days after the last trading day
Delivery Grades	Fuel oil, 180 CST or higher standard (refer to the Attachment for detailed quality requirement)
Delivery Sites	Designated by the Exchange
Minimum Margins	8% of the contract value
Transaction Fee	No more than 0.20% of the trading value (including reserve for risks)
Delivery Method	Physical delivery
Symbol	FU
Listing Exchange	Shanghai Futures Exchange

The Attachment of SHFE Standard Contracts of Fuel Oil

Delivery Unit: The delivery unit of Fuel Oil prescribed in the standard contract is 10 lots (100 tons). And the delivery quantity shall be integral times of delivery unit.

Quality Standard of Fuel Oil of SHFE

Physical and Chemical Properties	Limit	Test Method
Density (15°C, kg/1)	Maximum 0.985	ASTM D1298
Kinematic Viscosity (50°C, CST)	Maximum 180	ASTM D445
Ash on Weight Basis (m/m, %)	Maximum 0.10	ASTM D482
Couradson Carbon Residue (m/m, %)	Maximum 14	ASTM D189
Pour Point (°C)	Maximum 24	ASTM D97
Water by Distillation Volume (V/V,%)	Maximum 0.5	ASTM D95
Flash Point (°C)	Minimum 66	ASTM D93
Sulfur (m/m, %)	Maximum 3.5	ASTM D4294/D1552
Total Existent Sediment (m/m, %)	Maximum 0.10	ASTM D4870
Vanadium (PPM)	Maximum 150	ICP

The Designated Oil Depots: Designated and announced by the Exchange separately.

BRENT CRUDE OIL CONTRACT SPECIFICATION AT NCDEX

Futures Contract Specifications. Updated as on September 15, 2005 (Please refer to notes, if any, listed below this contract specifications)

Type of contract: Futures Contract Specifications
Name of commodity: Brent Crude Oil
Ticker symbol: BRENTCRUDE
Trading system: NCDEX Trading System
Unit of trading: 100 barrels
Delivery unit: 50,000 barrels
Quotation/base value: Rs per barrel* [*1 Barrel =42 US gallons=158.98 liters]
Basis: Sullom Voe, Shetland Islands, United Kingdom exclusive of all levies and taxes.
Tick size: Re 0.50

Quality specification:	*Crude Type*	*API Gravity*	*Sulphur Content*
	Brent	38.5 degrees	0.36%

Also Deliverable:	*Crude Type*	*API Gravity*	*Sulphur Content*
	Forties	$41.5\text{-}42.5^{\circ}$	0.25-0.3%
	Osberg	35.9 degrees	0.32%

Quantity variation: +/- 1% by volume

Delivery center: Mumbai Port/Jawaharlal Nehru Port Trust (JNPT)

The Buyer will be responsible for the freight cost, insurance, import duty and all other taxes & levies on actual basis. Freight and insurance will be paid on actual basis on production of satisfactory documentary evidence from the seller.

Trading hours: As per directions of the Forward Markets Commission from time to time, currently (with effect from May 16, 2005)

Mondays through Fridays: 10:00 a.m. to 11:30 p.m.

Saturdays: 10:00 a.m. to 2:00 p.m.

The Exchange may vary the above timing with due notice.

Delivery specification: The buyers and seller shall give intentions of delivery/receipt through the delivery request window during last three trading days including the day of expiry of the contract. This will be matched by the Exchange for physical delivery.

Opening of contracts: The first contract would be launched on September 15, 2005 as per the approval of the Forward Markets Commission. New contracts would open as per schedule given in Exhibit 1.

Closing of Contract: All open positions for which delivery intentions have not been received or for which delivery intentions have been rendered but remain unmatched for want of counterparty to settle delivery, will be cash settled at Final settlement Price on the expiry of the contract.

No. of Active Contracts: Three consecutive months running concurrently

Price Limit: The daily price limit will be 6% and will be raised to 9% after a 15-minute cooling period if the price limit of 6% is reached. If the price limit reaches 9%, trading will continue within the 9% limit.

Position Limits: Member: 6,00,000 barrels or 20% of open interest, whichever is higher.

Client: 1,50,000 barrels.

Special Margin: Special margin of 5% of the value of the contract will be applied whenever the rise or fall in price from the first day's closing price is 20%. This is payable by buyer or seller depending on whether prices rise or fall respectively. The margins shall stay in force so long as price stays beyond the 20% limit and will be withdrawn as soon as the price is within the 20% band.

Quality Allowance (for Delivery): No variation allowed

* 1 Barrel =42 US gallons=158.98 liters

FURNACE OIL CONTRACT SPECIFICATION AT NCDEX

Type of Contract: Futures Contract Specifications

Name of Commodity: Furnace Oil

Ticker Symbol: FURNACEOIL

Trading System: NCDEX Trading System

Unit of Trading: 10 Metric Tons (MT)

Delivery Unit: 10 Metric Tons (and multiples thereof)

Quotation/Base Value: Rs per Metric Ton

Basis: Ex-Mumbai, exclusive of all levies and taxes

Tick Size: Re 1.00

Quality Specification: Refer to Exhibit 1 for detailed quality specifications

Quantity Variation: +/- 1% by weight or 10 Metric Tons, whichever is lower

Delivery Center: NCDEX accredited storages in Mumbai and such other delivery centers as announced by exchange from time to time

Trading Hours: As per directions of the Forward Markets Commission from time to time, currently :

Mondays through Fridays: 10:00 a.m. to 11:30 p.m.

Saturdays: 10:00 a.m. to 2:00 p.m.

The Exchange may vary the above timing with due notice.

Delivery Specification: All open positions at the expiry of the contract would have to be physically delivered.

No. of Active Contracts: Three consecutive months and one six month contract expiring in March/September of every calendar year

Opening of Contracts: The first set of contracts will be launched immediately upon approval of the Forward Markets Commission. New contracts would open as per schedule updated every month by circular.

Initial Margin: 5%

Due date/Expiry date: As per schedule announced by the exchange at the time of launching the first set of contracts.

Daily Price Fluctuation Limit: Limit of 6%

Position Limits (Maximum): Client: 50,000 MT; Member: 400,000 MT or 15% of market open position whichever is higher

Special Margin: Special margin of 3% of the value of the contract will be applied whenever the rise or fall in price from the first day's closing price is 20%. This is payable by buyer or seller depending on whether prices rise or fall respectively. The margins shall stay in force so long as price stays beyond the 20% limit and will be withdrawn as soon as the price is within the 20% band.

Quality Allowance (for Delivery): No variation allowed.

Exhibit 1: Quality Specification for Furnace Oil

Sl. No.	Characteristics	Test Method Ref to [P:] of IS : 1448	Specifications
1	Acidity, inorganic	P-2	NIL
2	Ash, % wt. max.	P-4	0.1
3	Density at 15°C, g/ml, Max	P-32	0.99
4	Flash point, (PMCC) °C, Min.	P-21	66
5	Kinematic viscosity in centistokes at 50°C, max	P-25	180
6	Sediment, % wt. max.	P-30	0.25
7	Sulphur, total, % by wt., max.	P-33 or P-35	3.5
8	Water Content, % by vol., max.	P-40	1.0
9	Pour Point, °C, max	P-10	30
10	Condarson Carbon residue/micro carbon residue, % by mass	P-8	20

DUBAI MERCANTILE EXCHANGE (DME)

Commodity: Brent-Oman Financial Spread

Settlement: Financial

Duration of Contract: 12 months

Trading Unit: 1,000 bbl

Price Quotation: US dollars and cents per barrel

Minimum Price Movement: One cent ($0.01) per barrel

Maximum Daily Price Fluctuation: There shall be no maximum price fluctuation limit.

Working: Contract is based on the spread between Brent Crude Oil and Oman Crude Oil. Performance under the contract shall be by cash settlement. Final settlement shall take place following termination of trading for a Contract Month and will be based on the Floating Price.

Final Settlement Price: The Exchange shall publish the Final Settlement Price on the last Trading Day in that Contract Month.

The Exchange shall determine a Settlement Price for each Trading Day of a Contract Month of the Brent-Oman Financial Spread Contract. The Settlement Price for each Trading Day of a Contract Month (including the final Trading Day of a Contract Month) shall be determined as at 14.30 (New York time) on a Trading Day using:

1. the DME Daily Post Close Price determined in respect of:
   - (i) the same Trading Day; and
   - (ii) the corresponding Contract Month in the DME Oman Crude Oil Futures Contract (the **Relevant DME Oman Price**); and
2. the NYMEX Brent Financial Crude Oil Daily Settlement Price determined in respect of:
   - (i) the same day; and
   - (ii) the corresponding contract month on NYMEX, Inc. in the Brent Financial Crude Oil Contract (the **Relevant NYMEX Brent Price**).

The daily Settlement Price for each Trading Day of a Contract Month (including the final Trading Day of a Contract Month) shall be the Relevant NYMEX Brent Price minus the Relevant DME Oman Price.

Delivery Center: No Physical Delivery.

Note: **Corresponding NYMEX Brent Financial Crude Oil Daily Settlement Price** means, in relation to the DME Daily Post Close Price published on the last Trading Day of a Contract Month, the NYMEX Brent Financial Contract (Code BB) first nearby month daily settlement price published in respect of trading up to the same time on the same day as such DME Daily Post Close Price is published in respect thereto.

BRENT CRUDE OIL MCX-INDIA

Commodity: Brent Crude Oil

Settlement: Both cash as well as physical delivery

Duration of Contract: 3 months

Trading Unit: 100 bbl

Price Quotation: Rs. per bbl

Minimum Price Movement: Rs. 1 per bbl

Maximum Daily Price Fluctuation: 4%

Maximum Allowable Open Position:

(i) For individual clients: 150,000 barrels

(ii) For a member collectively for all clients: 600,000 bbl or 20% of open interest, whichever is higher

Margins: Margins are required to initiate a position on the exchange.

1. *Initial Margin:* Buyers and Sellers have to submit a 5% initial margin to initiate a position on the Exchange.
2. *Special Margin:* In case of additional volatility, a special margin at such percentage, as deemed fit, will be imposed immediately on both buy and sale side in respect of all outstanding position, which will remain in force for next 2 days, after which the special margin will be relaxed.
3. *Delivery Period Margin:* A Delivery Period Margin of 25% will be imposed during tender and delivery period on both buyers and sellers on matched quantity.

Working

1. On the contract expiry day by 6.00 p.m. sellers and buyers will submit Notices of Intention to Deliver and Notices of Intention to Accept, respectively with the exchange.
2. Sellers will also submit copies of relevant documents as a proof of holding stock at the time of giving Notice of Intention to Deliver.
3. On the basis of Notices of Intentions to Deliver and Accept the Exchange will match the total quantity offered by the buyers and sellers and with respect to the matched quantity, the allocation of delivery between the buyers and sellers will be done.
4. *Penalty:* After receipt of intentions from the buyer and seller to take or

give delivery, if any of the party fails to honor his obligations, a penalty of 1% of the DDR will be imposed on him, out of which 90% will be passed on to the other party and 10% will be retained by the Exchange.

5. *Cash Settlement:*

   (i) All positions that are offset before the expiry of the contract are cash settled.

   (ii) All outstanding positions on the expiry of contract where Notice of Intention to Deliver or Accept has not been received and such positions where Notices of Intention have been received but have not found the counter party for honoring the intentions, shall be closed out at due date rate and respective pay-in and pay-out of funds of such close out positions shall be take place on the following day of contract expiry day by 11.00 a.m.

6. *Physical Delivery:* Physical Delivery takes place for the open positions after the expiry date where the Exchange is able to match the Notices of Intention to Deliver and Accept.

Note: This contract is mainly used for hedging/speculation purposes by players in the market. Also, trading participants are still not doing physical trading on exchanges in India. They go for OTC market for physical trade as there they know and can choose the counter party and can negotiate the terms of contract. Hence, all the contracts are cash settled even though there is a provision for physical settlement. Another reason for absence of physical delivery is that there is no supply of Brent Crude Oil in India. Moreover, the contract size is also too small to attract traders to go for physical delivery.

Due Date Rate: Due date rate of Brent Crude Oil will be calculated by taking Bloomberg fixings i.e., international spot price of Brent Crude Oil as reported by Bloomberg. The Bloomberg price will pertain to the last trading day (expiry date) of the MCX Brent Crude Oil contract and it would be multiplied by Rupee - US Dollar rate as notified by the Reserve Bank of India on that particular day.

Delivery Center: Port installation at Mumbai/JNPT port

Delivery Grades: Brent Blend confirming to the following quality specification is deliverable:

(i) API Gravity: Between 38 degree – 39 degree

(ii) Sulfur: 0.4% by weight or less

The buyer has no option to select a particular grade and the delivery offered by the seller and allocation by the Exchange shall be binding on buyer.

Delivery & Payment: The seller has to make the delivery by 11:00 a.m. on the 3rd day from the day of expiry of contract. The buyer submits the payment with the clearing house by 11:00 a.m. on the 2nd working day from the day of expiry of the contract. The clearing house will release this amount to the seller by 11:00 a.m. on the 3rd working day from the day of expiry of contract.

Extension of Delivery Period: As per the Exchange decision due to a force majeure or otherwise.

Endorsement of Delivery Order: The buyer member can endorse delivery order to a client or any third party with full disclosure given to MCX. Responsibility for contractual liability would be with the original assignee.

CHAPTER 4

Risk Management in the Oil Industry

4.1 INTRODUCTION TO RISK AND ITS SOURCES

Risk is the potential impact (positive or negative) to an asset or some characteristic of value that may arise from some present process or from some future event. In everyday usage, "risk" is often used synonymously with "probability" and restricted to negative risk or threat. In professional risk assessments, risk combines the probability of an event occurring with the impact that event would be. Financial risk is often defined as the unexpected variability or volatility of returns, and thus includes both potential worse than expected as well as better than expected returns.

In most financial markets there are a fairly small number of fundamental price drivers which can be easily translated into pricing and risk management models. In currency markets, for example, the commodity that has to be delivered is cash, a piece of paper which is easily stored, transferred and not sensitive to weather conditions. But energy markets are concerned with bulky, dangerous commodities that have to be transported over vast distances through some of the most politically unstable regions of the world. This means that there are a large number of factors that can affect energy prices. A fairly short list might include: the weather, the balance of supply and demand, political tensions, comments from country leaders, decisions taken by OPEC, analysts' reports, shipping problems, and changes to tax and legal systems. All these contribute to the high levels of volatility in energy markets which often experience sudden price movements from one day to the next or even from one minute to the next.

Various types of risks in energy sector are :

(1) Financial Risks

(2) Basis Risk

(3) Legal risk
(4) Tax Risk
(5) Operational Risk

(1) Financial Risks

Price risk

This is the risk of losing money as a result of price movements in the energy markets and is sometimes referred to as 'market risk'. Typically, producers will lose money when prices fall, while users will find themselves out of pocket when prices increase.

Credit risk

Credit risk is the risk of financial losses due to the counterpart to a contract defaulting. It is often said that a hedge contract is only as reliable as the credit standing of the counterpart and credit risk management has moved to the top of the priority list for the energy industry. The credit crunch felt in the USA energy sector in the aftermath of the Enron disaster has prompted energy traders to review credit policies and also review effective methods to control and reduce credit risk wherever possible.

Liquidity risk

In the context of this book, this is the risk of losses caused by a derivatives market becoming illiquid. This happened during the Gulf War when there was so much volatility in the markets that many banks and oil traders would not give a bid or offer price. Companies who were exposed to those markets at the time were sometimes unable to close out their positions or could only do so at great cost to themselves.

Cash flow risk

This is the risk that an organization will not be able to produce the cash to meet its derivatives obligations. In the late 1990s, Korean Airlines found itself in this kind of situation and suffered heavy losses as a result. The company had been hedging against movements in the jet fuel price by using derivatives which were denominated in dollars. When the Korean won suddenly fell in value against the dollar, the company found that the cost of the dollars needed to service its derivatives contracts had soared. The company lost out because it had not hedged against the risk of a negative movement in the currency differential between the won and the US dollar.

(2) Basis Risks

Basis risk is the risk of loss due to an adverse move or the breakdown of expected differentials between two prices (usually different products). In the context of price risk management, basis risk describes the risk that the value of a hedge (using a derivative contract or structure) may not move up or down in sync with the value of the price exposure that is being managed.

In the energy market, these market movements may be triggered by factors such as poor weather conditions, political developments, physical events or changes in regulation. These can lead to basis risk occurring in circumstances such as the following:

- Physical material in one location cannot be delivered to relieve a shortage in another location.
- A different quality of product cannot be substituted for an energy product in severe shortage. This often happens in the pipeline gas and power markets if there are any problems with transmission networks.
- There is not enough time to transport or produce an energy product to alleviate a shortage in the market.

When conducting price risk management, the ideal derivatives contract is one that has a zero risk or the lowest basis risk with the energy price that protection is needed from. The larger the basis risk, the less useful the derivative is for risk management purposes.

The attraction of over-the-counter (OTC) swaps and options is that basis risk can at times be zero, as OTC contracts can often price against the same price reference as the physical oil. However futures contracts (sometimes referred to as 'On-exchange' derivatives) traded on exchanges like the International Petroleum Exchange, the New York Mercantile Exchange and the Tokyo Commodity Exchange all have their pricing references and terms fixed in the exchange's regulations. This means that if their pricing reference does not match the underlying physical exposure, the basis risk must either be accepted or an OTC alternative needs to be sought.

Components of Basis Risk

(a) Locational basis

- You utilize a derivatives contract which prices against exactly the same specification *of* energy you are hedging price risk against. However, the derivatives contract is pricing against the same energy contract but in a different geographic region.
- You have locational basis risk. Localized supply/demand factors, political tension, grid problems or, in the case *of* hydrocarbons/gas, pipeline problems, in either the location used for pricing the derivatives contract or the location where your physical supply is located, could make your derivatives contract a liability rather than a risk-reducing benefit.

(b) Time basis

This is a common exposure in many markets. In energy markets a time basis exposure can be very dangerous, particularly when there is a sudden shift in demand or transportation problems occur.

(c) Mixed basis risk

Mixed basis risk occurs when an underlying position is hedged with more than one type of mismatch between the energy that is the subject of the price risk management and the pricing index reference of the derivatives instrument that is being used. For example, if a January Gasoil (heating oil) Cargo is hedged with a March Jet Kerosene swap, it would leave both time and product basis exposures.

Example: BRENT CRUDE FUTURES AND THE CUSHING CUSHION

The success of the Brent Crude Oil Futures contract is an interesting example of the importance of basis risk in the energy markets. This contract was first traded on London's International Petroleum Exchange (IPE) in 1983, two years after the West Texas Intermediate (WTI) crude futures contract had been launched on the NYMEX in New York. On the surface, both contracts do similar jobs, for hedging purposes, at least. *So*, over the years, why have international companies chosen to hedge with the IPE Brent futures contract rather than its better established and more liquid American rival?

The answer is, a particular kind of basis risk, known in the industry as the 'Cushing Cushion' (after the Cushing refinery in Oklahoma, the destination of several of the south-east USA's major oil pipelines). The 'Cushing Cushion' means that WTI crude price in the USA can act totally independently from international market prices. This can be because pipeline bottlenecks at the Gulf coast are preventing additional foreign crude from reaching the mid-continent refineries or it can be because bad weather has closed the Louisiana Offshore Offloading Point (LOOP), halting the offloading of foreign crude from carriers into the pipeline system.

In situations like these, the first reaction of speculators and refineries which depend on oil in the pipeline system is to buy WTI NYMEX Futures. Sometimes WTI Premiums of US$3.00 a barrel over the IPE Brent price have been seen due to LOOP problems, pipeline problems or both.

So for anyone hedging international crudes such as West African, Brent, Middle East crude oils, Dubai or Tapis, the WTI NYMEX contract carries a significant basis risk. The IPE Brent future, on the other hand, is exempt from this basis risk, which is almost certainly one of the keys to its success.

(3) Legal Risk

This is the risk that derivatives contracts may be not be enforceable in certain circumstances. The most common concerns in this area surround clauses on netting of settlements, netting of trade, bankruptcy and the concern that the liquidation of contracts may be unenforceable. Opinions on many jurisdictions around the world can be obtained from the International Swaps Dealers Asscciation (ISDA).

(4) Operational Risk

The risk that may occur through errors or omissions in the processing and settlement of derivatives is known as operational risk. Internal controls alongside an appropriate back office system (whether manual or computerized) should be employed to reduce this risk.

(5) Tax Risk

Tax risk can occur when there are changes to taxation regulations that affect either the derivatives market directly or the physical underlying energy market in some way. This can create additional costs to the trade.

For derivatives contracts the issue of imposed withholding taxes on any settlement payments is normally an issue covered by ISDA contracts.

OTC Energy Derivative Markets

In the energy industry, derivatives can be bought and sold in two main ways: on-exchange and over-the-counter (OTC). On-exchange refers to the futures markets which are found on regulated financial exchanges such the New York Mercantile Exchange (NYMEX) and/London's International Petroleum Exchange (IPE). The OTC market is specific to the non-standard swaps and OTC options. These are usually traded directly between two companies (principals, players) in the energy markets. Over-the-counter (OTC) Derivatives and Exchange-traded Derivatives. They are different in the way they are traded in the market.

Over-the-counter (OTC) Derivatives are contracts that are traded (and privately negotiated) directly between two parties, without going through an exchange or other intermediary. The counterparts negotiate the key terms of the contracts like pricing reference, the payment terms and the volume. It is beneficial for the organizations with unique price risk to use OTC derivatives for hedging. Today, there is a fair amount of standardized OTC contracts in oil, gas and electricity markets. Standardized here means they have a standard floating price reference and the minimum traded volume.

OTC Derivatives market has now become a well established and liquid market. It is the main reason why the futures exchanges are not launching new energy futures contracts as the needs of the market are increasingly being met by the OTC Derivatives market. This is evident from the example of Asian market. Asia is now second only to US in energy consumption but still Asia does not have a well established and liquid energy futures exchange. The reason behind this is that Singapore, the hub of a well established Asian OTC market, serves the needs of energy derivatives contracts in Asia.

Tokyo Commodity Exchange (TOCOM) with its Middle East crude futures contract and product futures is the only major energy futures exchange in Asia but it is still far from achieving the kind of liquidity and range of market participants that is there on New York Mercantile Exchange (NYMEX) or Intercontinental Exchange (ICE).

In an attempt to strengthen the core functions of futures markets – to hedge risk and achieve price discovery, Singapore Exchange Limited (SGX) and the Tokyo Commodity Exchange (TOCOM) signed an agreement in 2002, to launch the Middle Eastern Crude Oil (MECO) futures on SGX. The MECO futures contract on SGX follows similar contract specifications as the very successful TOCOM MECO futures contract, except that it is denominated in USD/barrel instead of JPY/kilolitre. The contract's final settlement is based on the Monthly Average of the Daily Dubai and Oman crude oil prices assessed by Platts, a unit of McGraw-Hill Companies, Inc. The aim was to bring together the liquidity in TOCOM's oil derivatives contracts and the participation of international oil refiners and traders based in Singapore and the Asia-Pacific region. However, the contract was not a big success on SGX as it was on TOCOM. Traders on SGX had to deal with the already present oil price volatility in the market and the additional risk of volatility in the exchange rates of Japanese Yen (JPY) and United States Dollar (USD).

Pricing of OTC Oil Derivatives Market

Platt's, which is a division of McGraw-Hill, publishes a daily price assessment of any given crude oil or oil product in any given location, according to Platt's own specifications. It also publishes an assessment of the forward curve. This is an industry recognized publication and is used for pricing the majority of physical transactions and OTC swaps. These daily value assessments are based on the aggregated bids and offers from many brokers and dealers around the world during a specified time window for each geographic region (Usually towards the end of each business day in each major time zone: Asia (Singapore), Europe (London), USA (New York, and then the West Coast).

For pricing of swaps the monthly average of these Platt's assessments is used. The difference between the traded fixed price and the calculated average floating price from the Platt's assessments is found and multiplied by the monthly quantity of the contract. The amount hence found is the monthly financial payment that is paid and there is no exchange of physical energy.

Following are the main Global Oil Trading/Pricing hubs:

1. Asia – Singapore is the main pricing hub.
2. Europe – Mediterranean and Arab Gulf Prices, North West Europe and Amsterdam-Rotterdam-Antwerp (ARA) are the main pricing hubs.
3. USA – New York Harbor, US Gulf Coast, US West Coast are the main pricing hubs.

Although the futures markets are important to the energy industry, it relies much more heavily on OTC derivatives. This is because OTC derivatives are customized transactions, whereas their on-exchange counterpart, the 'futures' contract, is a standard contract. In theory, each deal on the OTC market is unique, so it is important to be alert to contract terms, pricing mechanisms and price reference when using OTC derivatives.

Some companies find that the measurement and control of risks can be more difficult with an OTC contract because of the lack of price and liquidity transparency in the OTC market (unlike regulated futures exchanges, which publish public real-time price data) and this can create the possibility of an unexpected loss. There are also sometimes additional legal, credit and operational risks with OTC derivatives compared to on exchange futures contracts. However, the OTC market remains a popular option for price risk management purposes. Many companies find that there are benefits in the flexibility of an OTC derivative because it can be valued against the same price reference as the energy which is being produced or consumed.

As the pie chart in Figure shows, energy OTC derivatives markets are far less liquid than most other financial derivatives markets accounting for less than one per cent of the value outstanding on derivatives markets worldwide. This means that those who take part in energy markets whether as market makers, traders or end-users (usually companies with underlying price risk in the energy being hedged either as a producer or consumer), need to have clear policies for derivatives usage, including strong management controls and organizational reporting structures effective before derivatives are employed. They should also provide shareholders with information that will put to rest any unjustified fears associated with their company's use of derivatives. Indeed, as a result of the concerns of regulators and public shareholders around the world, more and more information is now required by international accounting standards.

OTC Derivative Contracts (Total Amount Outstanding) December 2005

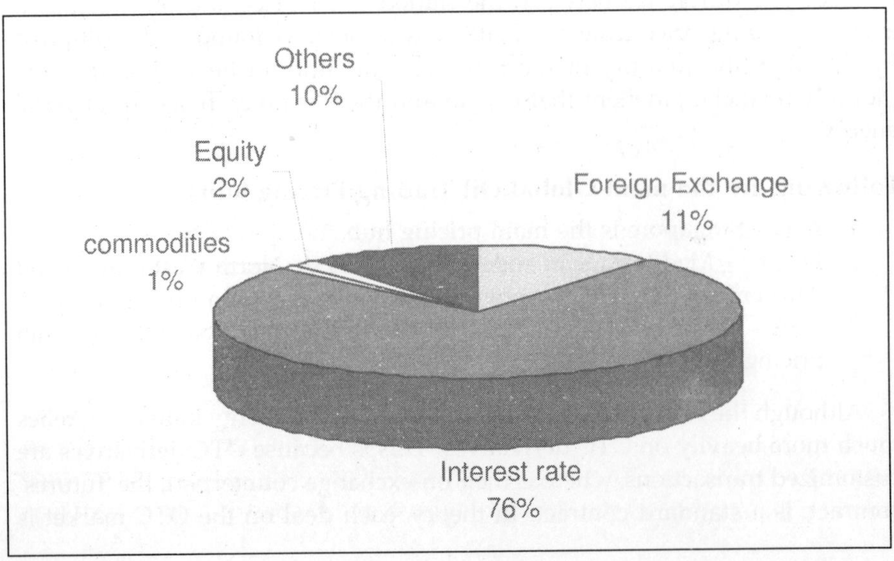

FUTURE MARKETS

A Brief History of the Futures Markets

Oil futures contracts have been traded on financial exchanges since the 1970s, although *ad hoc* negotiated physical supply contracts have been around since oil was drilled in the USA in the 1850s. The first formalized regulated futures exchange for oil was the New York Mercantile Exchange (NYMEX) which started contracts on heating oil in 1977 (relaunched as the current contract in 1979) which was followed by West Texas Intermediate contract (WTI crude). On the other side of the Atlantic, the International Petroleum Exchange (IPE) of London was launched in 1981 and now boasts, in the Brent Crude Oil Futures contract, the leading international benchmark for the pricing of physical crude markets around the world; approximately 70% of the world's crude oil markets price in some way against Brent Crude Oil. Both NYMEX and IPE also operate futures markets for Natural Gas and Electricity/Power.

In recent times, there has been concern that the liquidity in the physical Brent crude oil market has been getting smaller and as a result some oil 111ajors have taken the initiative to participate in the development of Brent-Forties-Osenberg (BFO)-related trading rather than just Brent related trading. Pricing information services such as Platts have already modified the crudes they include in their Brent price reporting and even the International Petroleum Exchange is examining (September 2002) the use of BFO prices in its price index for settling the Brent Futures contract.

In the Far East, SIMEX (now merged into SGX in Singapore) ran a popular Fuel Oil Futures contacts in Singapore until the early 1990s, when it was overtaken in popularity by the OTC (off-exchange over-the-counter derivatives market) and Asia is now totally dependent on OTC derivatives for risk management purposes in energy markets. However, a Tokyo Commodity Exchange (TOCOM) contract for Middle East crude oil has been attracting both interest and trading volume, and this could become a useful on-exchange futures contract tool for Middle East crude hedging, particularly since Asia is heavily dependent on Middle East crude imports for oil refinery operation.

Futures versus OTC

At one time it was easy to distinguish the futures market from the OTC market and also to establish the pros and cons of using one or the other. As Figure 4.2 shows, when risk managers or traders used futures contracts they knew that the contract would be traded on an exchange, that they would have an account with their futures broker and that they were operating in a highly regulated market. They could also see the price of the contract on a screen and they could be sure that the security of the contract and its performance would be guaranteed by the clearing house of the exchange. This in turn was guaranteed by 'margins' (good faith payments by everyone with a futures position on that particular exchange), plus, the funding the exchange raised itself and the funds contributed by its clearing broker members.

Margins on a futures exchange can be split into two types: 'initial margins' and 'variation margins'. Initial margins are the good faith deposit that is placed with the clearing house or that a broker finances (at a cost) when a trade is opened, A variation margin is the daily revaluation of a portfolio with the clearing house. If the valuation is negative, you or your broker (if you have a credit line) will have to place a margin to cover that negative variation margin, if the next day the portfolio has a positive variation margin (i.e. it is showing an unrealized profit), because the position has not been traded or closed out yet, some of that margin will be returned.

Following are the features of a Futures Contract:

1. Contract is traded on an exchange.
2. Highly regulated market.
3. Price of the contract can be seen on a screen.
4. Clearing house of the exchange guarantees the security and performance of the contract.
5. Margins are deposited by everyone with a futures position on that particular exchange. Margins are of 2 types – **Initial Margins** are the deposits that are placed with the clearing house when a trade is opened. The clearing house does a daily revaluation of the portfolio. A **Variation Margin** is the payment made by clearing members to their respective clearing houses based upon adverse price movements of the futures contracts that these members hold. It is paid by clearing members on a daily or intraday basis in order to reduce the exposure created by carrying highly risky positions. By demanding variation margin from its members, clearing organizations are able to maintain a suitable level of risk and cushions against significant devaluations.

Following are the features of the OTC market:

1. No guarantor is involved.
2. There is credit risk of the counterpart to the contract.
3. Liquidity risk is there.
4. Lack of price transparency as there is no screen to display the real-time price.

However when OTC contracts are used there is always the credit risk of the other company in the transaction, as well as a liquidity risk and a lack of price transparency because there is no screen to display a real-time price.

The Convergence of OTC and Futures

The OTC market and the Futures Market are now stating to converge despite of the distinction between them. This has been made possible because more and more clearing houses are accepting to act as guarantor in the OTC

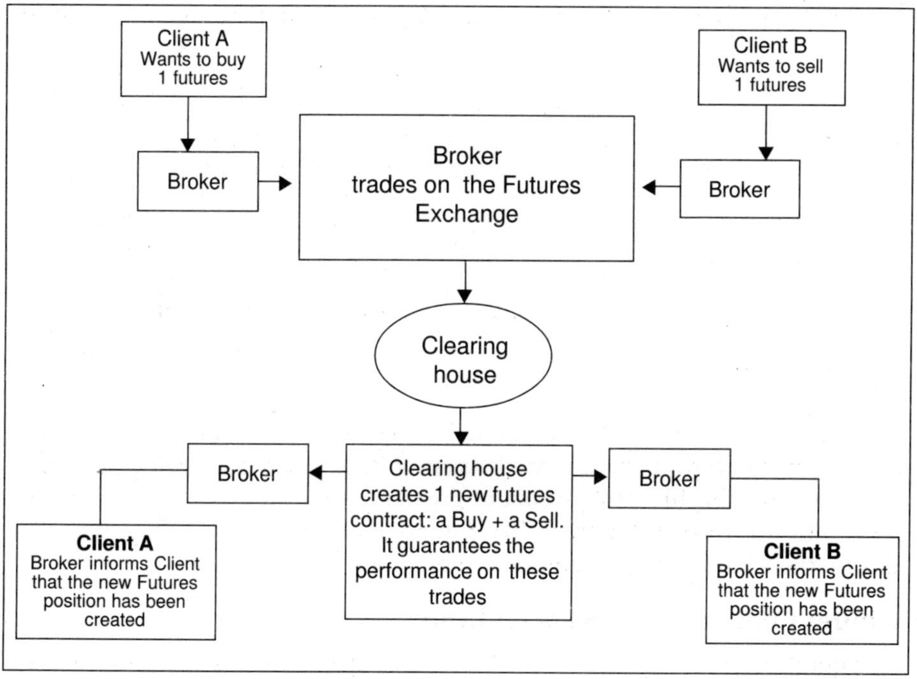

Figure 4.2: Basic Future Trade Transaction Flow

trades. A clearing house decouples the risk involved in a trade for counterparty A, from the risk involved in the trade for counterparty B. As the whole function of credit risk and collateral is centralized, it makes a more level playing field for all parties. The clearing house is responsible for defining the rules, and setting the margin rates, based on an assessment of risks.

The clear distinction between the OTC energy market and the futures markets is now disappearing as the two markets converge. Clearing houses around the world have started to accept OTC trades into their guarantee umbrella. This means that after executing bilateral OTC trades with one another, both counterparts can agree to 'give-in' their OTC deal to a clearing house. This process basically makes the clearing house the counterpart to the OTC deal, so that the two OTC counterparts can benefit from the higher credit quality of the clearing house as well as getting other benefits such as more netting opportunities on settlement and offsetting of positions.

The usual market approach for two OTC counterparts to trade an OTC derivative contract with one another directly and to take on one another's credit risk. In the new convergence environment we now sometimes have a situation like this:

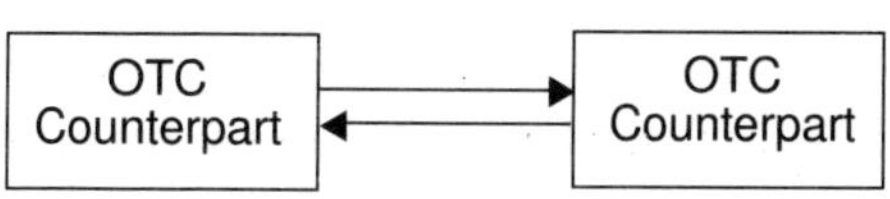

In the new convergence environment we now sometimes have a situation like this:

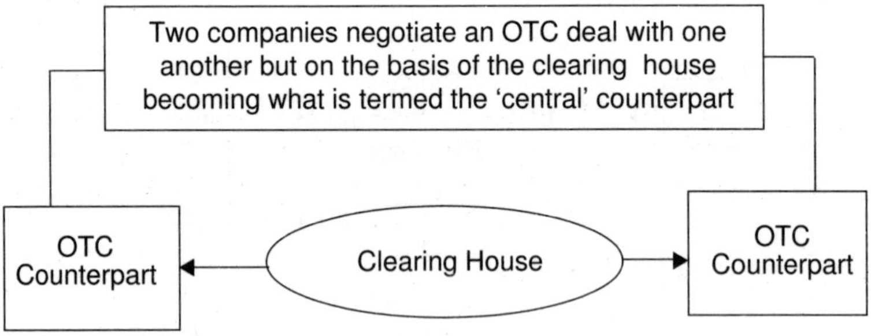

Although market share penetration has been slow in the oil sector, we have seen the newer power and gas markets embracing electronic trading platforms in a big way. This has brought about greater price transparency as users can view and trade prices onscreen like futures markets. As a result, power and gas markets have been the quickest to embrace OTC clearing.

Futures Contracts Settlement on Expiry

Energy futures contracts all entail physical and cash delivery on expiry (apart from IPE Brent Crude Futures in London). So if a seller (someone who is short in the market) holds the futures contract to expiry he will have to deliver the underlying physical energy (oil, gas, power) and if a buyer (someone who is long in the market) holds the contract to expiry, he will have to take delivery of the underlying physical energy. However, actual delivery via futures markets like the NYMEX or IPE is very small, normally less than 2% of the total open interest (the total amount of outstanding contracts in the market). Only about 2% of the total outstanding contracts on a Futures exchange are settled by physical delivery

Both parties of a "futures contract" must fulfill the contract on the settlement date. The seller delivers the commodity to the buyer, or, if it is a cash-settled future, then cash is transferred from the futures trader who sustained a loss to the one who made a profit. To exit the commitment prior to the settlement date, the holder of a futures position has to offset his position by either selling a long position or buying back a short position, effectively closing out the futures position and its contract obligations.. The majority of trades on these markets are for hedging and or speculative purposes, with consumers or producers of energy preferring to make delivery via the normal physical markets rather than through the futures markets.

Contracts Study

Commodity Markets

Commodity markets are markets where raw or primary products are exchanged. These raw commodities are traded on regulated commodities exchanges, in which they are bought and sold in standardized Contracts.

Commodity Exchanges

A commodities exchange is an exchange where various commodities and derivatives products are traded. There are many commodity markets across the world that trade energy products and contracts based on them. These contracts can include spot prices, forwards, futures and options on futures.

Commodities exchanges, usually trade futures contracts on commodities. Such as trading contracts to receive something, say gasoline, in a certain month. A refiner can sell a future contract on his gasoline, which will not be produced for several months, and guarantee the price he will be paid when he delivers; a large gasoline consuming company buys the contract now and guarantees the price will not go up when it is delivered. This protects the refiner (seller) from price drops and the consuming company (buyer) from price rises.

SWAPS AND OPTIONS

Swaps contracts settlement on expiry

Swaps are contracts which, unlike futures, never go to physical delivery. They are by their very legal structure purely financially based contracts, which allow companies to benefit from the price/value movement of the underlying asset that the swaps price is derived from. It is called a swap because the two counterparts to the deal, the buyer and the seller (the long and the short) exchange an agreed fixed price today for the unknown floating price In the future. When traders are negotiating an OTC deal they focus on:

- The fixed price
- The floating price reference
- Pricing period (e.g. one month, quarterly, calendar year)
- Start date or effective date
- End date or termination date
- Payment due date.

For example, for a swap priced against an American or European floating price reference, payment due date is normally the fifth business day after the last pricing day of each pricing period. In energy and generally commodity markets, OTC derivatives will price out monthly, so even if a quarterly contract is traded, after each month during the pricing period, one third of the volume will price out and a settlement will become due by or a payment received by the organization. For contracts pricing against an Asian-based floating price reference, payment for settlement is generally due 10 business days (sometimes up to 14 business days) after each pricing period.

Option Contracts on Expiry

What happens to an option contract on expiry and when or whether it is exercised (transfers into its underlying) depends very much on the type of option it is and also whether it is a futures option (traded on a futures exchange, referred to as traded options) or whether it is an OTC option.

When a traded options position is held on a futures exchange, if the option is 'in the money' on expiry, the clearing house will prompt clearing brokers to notify their customers that their option is in the money and request whether they wish to exercise the option as it is profitable to do so. An option is in the money when it has intrinsic value; i.e. exercising the option into its underlying futures contract (in respect of traded options) and then trading out (closing out) that futures contract would bring a profit. In some instances, if the traded option is heavily in the money, the clearing house of the futures exchange may even exercise it automatically, which acts as a safety net for users of the market.

However, there are no safety nets in the OTC world of derivatives. If you have a profitable swaption that could exercise into a profitable swaps position for you and you forget to tell your counterpart that you wish to exercise it by the cut-off time (written in the original option contract), you will be left to negotiate with that other counterpart. It will be up to your counterpart whether they will still let you exercise the swaption and if they do, it will most probably come at a price.

Types of Swaps in Energy Markets

(I) Plain vanilla – terms used to describe a simple averaging swap

A plain vanilla swap (Figure 4.3) is a monthly averaging swap with the following features.

- Fixed prices versus floating prices in the future are exchanged; i.e. they are swapped.
- They are used extensively in Oil, LPG and LNG related hedging and trading.
- When executing the deal counterparts discuss the fixed price agreed today and which floating price reference they will use to calculate the settlement.

The following is a cash flow example of a plain vanilla deal:

- Counterpart A buys fixed price 15.00 (buys fixed, sells floating)
- Counterpart B sells fixed price 15.00 (sells fixed, buys floating)
- Floating price reference is chosen, e.g. Platts average during the price period (say) March 2003 = 16.00

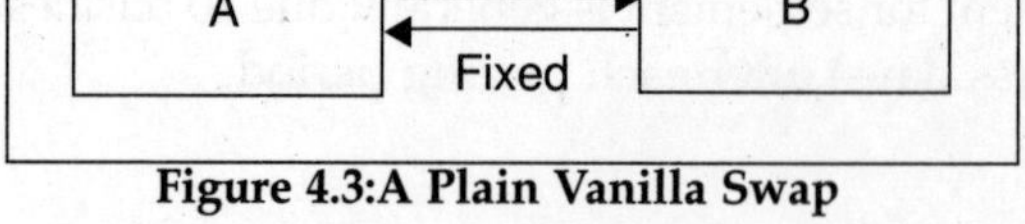

Figure 4.3:A Plain Vanilla Swap

- Net result:
- Counterpart A = +$1.00 (difference between fixed/floating)
- Counterpart B = -$1.00
- Counterpart B pays Counterpart A US$1.00. Only the difference is exchanged, *not the* principal notional amount.

(II) Differential swap

A differential swap is like a plain vanilla swap except that instead of one fixed price versus a floating price, it is based on the difference between a fixed prices in two products. In the oil sector, the most popular differential swap is the Jet Kero versus Gasoil (Figure 4.4), commonly termed the 'regrade' swap.

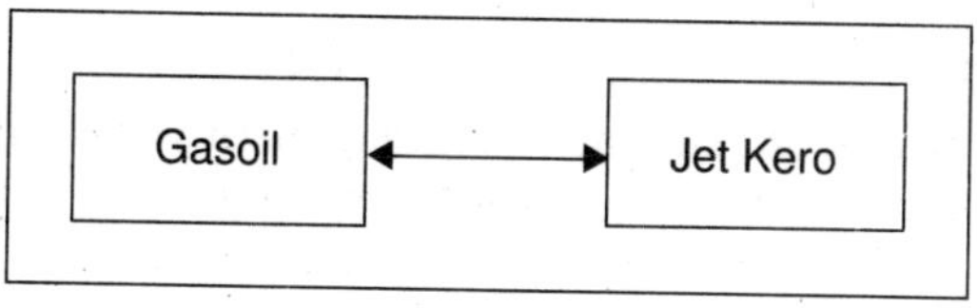

Figure 4.4: A Differential Swap

The following is a cash flow example of a differential swap:

- Counterpart A buys fixed price Kero and sells fixed price Gasoil at a difference of US$0.50 per barrel Kero premium.
- Counterpart B sells fixed price Kero and buys fixed price Gasoil at a difference of US$0.50 per barrel Kero Premium.
- Floating price reference is chosen, e.g. Platts Kero and. Gasoil average difference during the price period (say) March 2003 = 0.60 Kero premium.
- Net result:
  - Counterpart A = +$.10 (difference between fixed differential and the floating differential)
  - Counterpart B = -$0.10 cents per barrel
  - Counterpart B pays Counterpart A US$0.10. Only the difference is exchanged, *not* the principal notional amount.

Differential swaps are used across the whole energy spectrum. In the power and gas markets we see' spark spreads', where hedgers and traders use derivatives pricing against the difference (referred to as the 'dif') between power and gas markets. This is based on the amount to be made by burning gas and selling power in a perfect world using a standard percentage efficiency of the conversion of energy. The normal efficiency used is 49.13% and the spark spread is quoted in megawatt hours (MWh).

In coal versus power there is the so-called' dark spread', which works on the same principle as the spark spread with two fixed prices and two floating prices documented in swaps confirmation. However, the net exposure is only on the differential between the two products/instruments. In the UK, dark spreads use an energy conversion efficiency of 5,000 MT of coal producing 55 MW of electricity at an efficiency of 38%.

Spark Spread

Month	Gas price		Power price £/MWh	Spark spread £/MWh
	Pence per therm	£/MWh		
September	12.15	4.15	12.7	4.26
October-December	20.55	7.01	16.35	2.08
July-September	16.5	5.63	14.55	3.09

(III) Participation Swaps

Participation swaps are similar to regular plain vanilla fixed for floating swaps as the fixed price buyer can be 100% protected when prices rise above the agreed fixed price or the fixed price seller can be 100% protected when prices move down below the agreed fixed price.

However, unlike an ordinary swap, the client 'participates' in the down-side by only an agreed percentage. The percentage of participation affects the starting fixed price of the swap. A fixed price buyer who only wants to participate in a percentage of any price move lower may find that the fixed price quoted for a participation swap would be higher than a normal swap. On the other hand, if you were a seller of fixed price who wanted to participate in only a percentage of any move higher that would incur a loss 0,", the short swap position, you might find that the fixed price quoted for a Participation Swap would be lower than a normal swap.

(IV) Double up Swaps

By using the double up swap, swap users can achieve a swap price which is better that the actual market price, but the swap provider will retain the option to double the swap volume before the pricing period starts. If a company has price exposure to energy prices going higher, but the current plain vanilla swap is not being quoted around its budgeted level, it may find that a double up swap will let it hedge some of its required volume closer to its hedging budget level. The risk is that the market price could move against the derivatives position and the swap could price out against twice the original executed volume. Double up swaps are not commonly used in the market for price risk management purposes (hedgers tend to use options more if the current swap price is not interesting for them). However, double up swaps could offer an interesting opportunity for speculators who have a strong price direction view on their particular energy focus and want to get a head start by buying at a better price level than the current plain vanilla swap, or selling at a higher level than the plain vanilla swap quotes.

(V) Margin Swaps

This is where an organization can take its overall price risks from several energy inputs and outputs of the business process and get a complete swap structure that guarantees its profit margin. Organizations could construct complex hedges themselves to protect their energy inputs/ outputs. This has a cost in terms of managing many individual positions with perhaps several counterparts. It can therefore be more cost efficient and easier to enter into a margin swap with one counterpart who is willing to provide a contract that covers all the price risks (Figure 4.5).

However, it should be remembered that price risk management is never free and there are always costs attached to any control function in a company internally or externally. Administration and human resources available in an organization will have to be reviewed and adjusted if necessary depending on the level of activity the organization expects to have in derivatives.

Knock-ins and Knock-outs – Integration with Swaps and Options

Knock-in and knock-out options are two types of barrier option which are activated if the underlying moves through a trigger price level, in the case of a knock-in, and is cancelled or deactivated in the case of a knock-out. A market maker or trader will normally offer a more attractive price on this kind of option because the buyer is giving the seller of the option the opportunity to cancel it before its original expiry/termination date. This adds another dimension of opportunity or potentially reduced risk to the seller of the option;

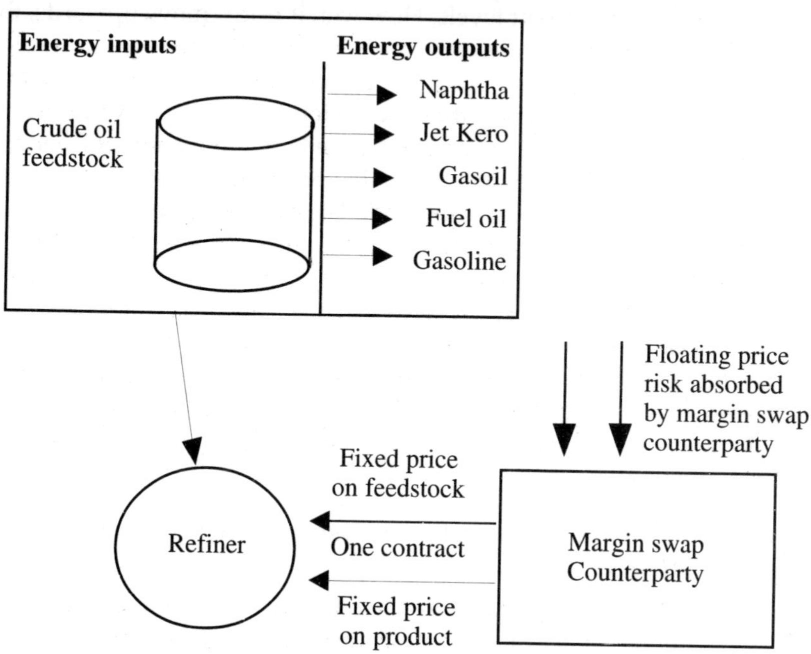

Figure 4.5: Example of a Margin Swap for an Oil Refiner

hence the lower price than an option of same strike price, tenure and underlying price reference without such a barrier option structure. Knock-in and knock-out triggers can be integrated with both swaps and options.

Figure 4.6 shows how the barrier option either comes to life (is knocked in) or is extinguished (knocked out) under certain conditions. In practice, the event which activates or kills the options is defined in terms of a price level (the barrier).

Price	Type of option: *Out*	Type of option: *In*
Up	Up-and-out	Up-and-in
Down	Down-and-out	Down-and-in

Figure 4.6: Barrier Options – caps/floors

A common example is the up-and-out floor (put) which is typically purchased by an energy producer to hedge their natural long position in the energy markets.

Up-and-out floor (put) may be an attractive alternative to the normal floor or put option, as it is less expensive and provides the same price protection if prices move down from current levels. However, if prices move upwards, the increase in the underlying commodity's price reduces the need for downside risk protection at the original strike price. If the price moves up sufficiently to cross the selected 'barrier' price, then the option is cancelled/extinguished.

The owner may consider re-entering a hedge by buying another floor at a higher strike price, which gives more valuable protection than the floor with the lower strike price which was cancelled. The barrier option may also be combined with a rebate. For a knock-out option, the rebate is paid when the option is cancelled prior to its normal expiry as compensation to the holder.

The up-and-out barrier is less expensive than a standard Asian, European or American option because the underlying price may fall below the strike price after initially rising, hitting the barrier and canceling the option. However, there may be liquidity issues with this strategy, as there are a limited number of traders in the market who may be able to quote you this more complicated option strategy.

The common and more liquid option markets in energy – calls and puts, caps and floors

On futures exchanges, traded options are referred to as calls and puts, while in the OTC market the same sort of contracts are referred to as caps and floors. If an organization buys a call or cap, it gives the buyer of the option price protection against the market moving above the agreed price, the 'strike price', in return for the payment of a premium or fee. The strike price is the level at which the players can participate in the market via the option contract.

If an organization buys a put or floor, it gives the buyer of the option protection against the market moving below the strike price, again in return for the payment of a premium or fee. Figure 4.7 illustrates the various possibilities.

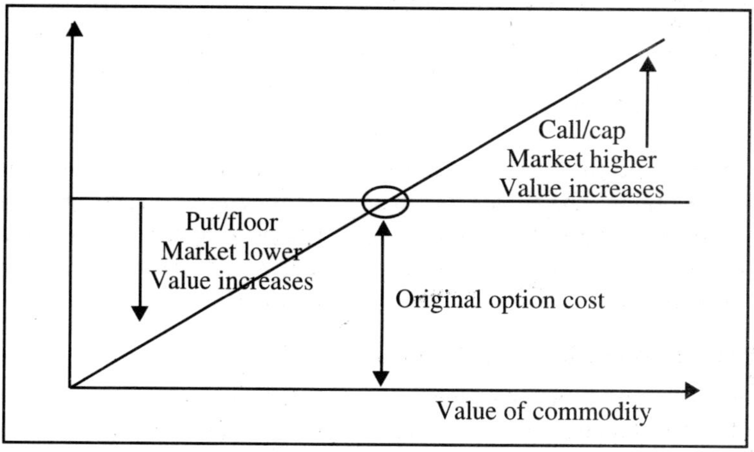

Figure 4.7: Calls and Puts, Caps and Floors

Options strategies can be very flexible and can help companies achieve exactly the risk reduction or risk exposure profile they want to have. When a buyer purchases an option, the cost of the contract is the premium paid, and the buyer will not be required to pay any more than whatever the market price demands.

If an option is sold on its own, it is called a naked option. This means that the seller does not own the underlying physical commodity or does not have another future or swaps position against the option that has been sold. In this case, there is unlimited risk if the market price moves in an adverse way (Figure 2.8).

If the option is sold as a naked option:	
Cap/call	Potential unlimited loss exposure on the option if the underlying market price moves above the strike price of the option
Floor/put	Potential unlimited loss exposure on the option if the underlying market price moves below the strike price of theoption

Figure 4.8: The Risks of Naked Options

Main Option Styles

(1) *American style* – An American style option is one that may be exercised into its underlying instrument (i.e. a futures contract) on any business day until expiry. All the IPE and NYMEX traded options on energy futures contracts are American style. These options are more expensive than European options because they give so much flexibility to the buyer as to when the option can be exercised.

(2) *European style* – These are not very common in the energy markets, as they only permit the buyer to exercise the option on expiry. European options are cheaper than American options, but generally more expensive than Asian-style.

(3) *OTC Asian style* – This is the most common option style in the OTC market and they are sometimes called 'retrospective' or 'path-dependent' options. The reason for this is that they are average price options, with their profit being dependent o_ the price history of the underlying energy market that is being used as the price reference, either overall or sometimes at a specific stage in the life of the option.

The Cost of an Option – Its Premium

There are many types of option models available and each one has its own particular use depending on the type of option that is being used. It is not necessary to have an in-depth understanding of the mathematics of these models. However, users should understand what needs to be put into the model in order to obtain the right answers and they should be able to interpret the output results. It is fair to say, though, that the core factors that play an important role in determining the value of any option are generally those shown in Figure 4.9.

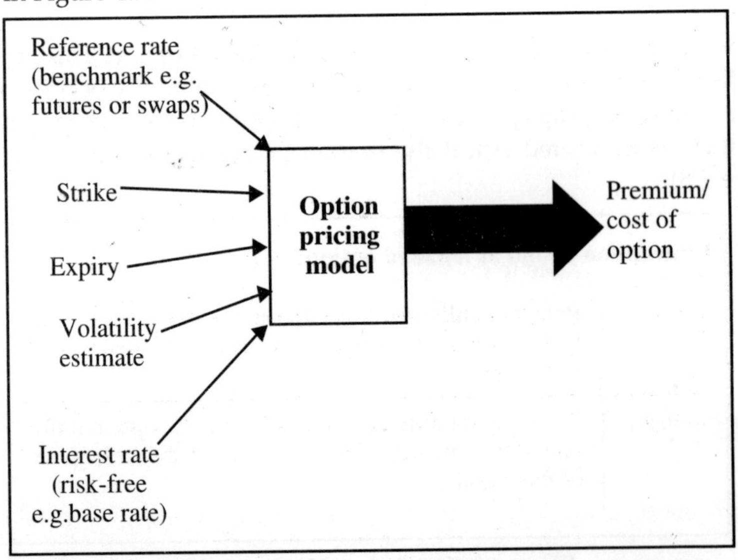

Figure 4.9: Option Premium Calculation Principles – Premium

Margin Options

These are options that can price against a complex structure of differentials instead of pricing against a single floating price reference. Earlier in the chapter, we looked at the illustration of a margin swap for an oil refiner. That oil refiner could have bought an option on its refining margin instead of using the swap strategy. The option strategy might at first appear less attractive, as it contains an up-front premium cost, whereas the swap strategy would not require any such cost. However, the flexibility offered by the option strategy becomes apparent if the margin gets better. If the refiner had used the swap strategy its profitability would be fixed, although if the margin improved, any loss on the swap would be offset by better prices on the resale of its physical assets: the petroleum products. In this case, it would just have the opportunity cost. But if it paid a premium for an option strategy (e.g. a margin option) then if the margin improved more than the cost of its option strategy, it would still be able to benefit from that margin improvement. This is most useful when dealing with a present day or even forward negative margin - a situation which has been experienced by oil refiners in some of the fuel oil markets for a long time. In this case, the refiners might have good margins which they wish to lock in by using swaps on the middle distillates (e.g. Naphtha, Gasoil, Jet) but they are still faced with the need to halt any further exposure in the Fuel Oil margin becoming more negative (and at the same time, they do not want to lock in a negative margin). In this instance, the refiner could look at a Crack Option (Crude (feedstock) versus Fuel Oil) and still have the potential to benefit and profit from any improvement in the margin on this product in the future.

Option Premium Cash Flow

When a traded option is purchased on a futures exchange, it is normal to put up 'margin', in the form of a good faith deposit (approximately 10% of the notional value, subject to market volatility at the time of the trade). After that, the position will be marked to market on a daily basis and there will be an obligation to finance any negative 'variation margin'. In the case of an OTC option, the buyer normally pays the premium up-front to the seller. This generation of cash premium is where OTC options can offer interesting opportunities for linkage to commodity- or energy-linked projects that require financing. It is possible to create structures that offer a price risk hedge at the same time as generating prompt cash flow which can be reinvested in the project or in other business activities of the organization (these activities are usually associated with the structured finance departments of banks).

For traders who are trying to make money by speculating in the very risk of the energy price moving or not (as the case may be), options offer the ability to do the following:

- Create trading strategies that profit from price direction moves
- Create trading strategies that profit from the price moving in a particular price band

- Create trading strategies that profit from the price staying the same by using volatility trades; money can be made not from the market price moving up or down, but on volatility increasing or decreasing.

4.2 PRICE RISK AND DERIVATIVES

Diversification and insurance are the major tools for managing exploration risks and protecting firms from property loss and liability. Firms manage volume risk - not having adequate supplies - by maintaining inventories or acquiring productive assets. Derivatives are particularly appropriate for managing the price risk that arises as a result of highly volatile prices in the petroleum and natural gas industries.

Table: Petroleum and Natural Gas Price Risks and Risk Management Strategies

Participants	Price Risks	Risk Management Strategies and Derivative Instruments Employed
Oil Producers	Low crude oil price	Sell crude oil future or buy put option
Petroleum Refiners	High crude oil price	Buy crude oil future or call option
	Low product price	Sell product future or swap contract, buy put option
	Thin profit margin	Buy crack spread
Storage Operators	High purchase price or low sale price	Buy or sell calendar spread
Large Consumers		
Local Distribution Companies (Natural Gas)	Unstable prices, wholesale prices higher than retail	Buy future or call option, buy basis contract
Power Plants (Natural Gas)	Thin profit margin	Buy spark spread
Airlines and Shippers	High fuel price	Buy swap contract

Source: Energy Information Administration.

The typical price risks faced by market participants and the standard derivative contracts used to manage those risks are shown in the table. Price risk in the petroleum and natural gas industries is naturally associated with each participant's stage of production. Some companies integrate their operations from exploration through final sales to eliminate the price risks that arise at the intermediate stages of processing. For example, for an integrated producer, an increase in the cost of crude oil purchased at its refinery will be offset by revenue gains from its sales of crude oil. Other, smaller companies usually do not have integrated operations. Independent producers want protection from low crude oil prices, and they sell to refiners who want protection from high prices. Refiners want protection from low product prices, and they sell to storage facilities and customers who are concerned about high prices. At each stage, suppliers and purchasers can split the risk in order to allay their concerns. They typically supplement exchange-traded futures and options with over-the-counter (OTC) products to manage their price risks.

Risk managers in the petroleum and natural gas industries commonly use derivatives to achieve certainty about the prices they pay or receive. Depending on their circumstance, they may be concerned with the price paid *per se*, with price spreads (differences between prices), with ceilings and floors, and/or with price changes over time. In addition, volumetric production payment contracts - a variant of a standard swap - may be used to reduce uncertainty about cash flows and credit. Some of the instruments particular to the oil and gas industries are described below.

The principal difficulty in using exchange-traded products is they often do not exactly correspond to what the trader is attempting to hedge or to speculate in. For examples, price movements in premium gasoline are not identical to those in unleaded gasoline. Similarly, the price of natural gas at Henry Hub is not identical to that at Chicago. The distinction between what exchange products can hedge and what the user wants to hedge is the source of *basis risk*. Basis risk is the risk that the price difference between the exchange contract and the commodity being hedged will widen (or narrow) unexpectedly. To a large extent, the OTC market exists to bridge the gap between exchange-traded products and the needs of individual traders, so that the two markets in effect have a symbiotic relationship.

Basis Contracts

Price certainty in a unified market can be bought with forward sales, futures contracts, or swaps (contracts for differences). When one or both parties face a spot market price that differs from the price in reference market, however, other derivative contract instruments may be needed to manage the resulting basis risk. For example, a local distribution company (LDC) in Tennessee could enter into a swap contract with a natural gas producer, using the Henry Hub price as the reference price; however, the LDC would lose price certainty if the local spot market price differed from the Henry Hub price. In this example, when the Henry Hub price is higher than the Tennessee price by more than it was at the initiation of the swap contract, the LDC gains, because its payment from the producer will exceed the amount it pays to buy gas in its local market. Effectively, the LDC will pay less per thousand cubic feet than the fixed amount the LDC pays the producer. Conversely, if the Tennessee price is lower, the producer's payment will not cover the LDC's gas bill in its local market.

A variety of *basis contracts* are available in OTC markets to hedge locational, product, and even temporal differences between exchange-traded standard contracts and the particular circumstances of contract users. The simplest is a *basis swap*. In the example above, the OTC trader would pay the LDC the difference between the Tennessee price and the Henry Hub price (for the nominal amount of gas) in exchange for a fixed payment. The variety of contractual provisions is unlimited. For example, the flexible payment could be defined as a daily or monthly average (weighted or unweighted) price difference; it could be capped; or it could require the LDC to share the costs when the contract's ceiling price is exceeded. What this OTC contract does is to close the gap between the Henry Hub price and the price on the LDC's local spot market, allowing the LDC to achieve price certainty.

The traders supplying basis contracts can survive only if the basis difference they pay - averaged over time and adjusted for both financing charges and the time value of money - is less than the fixed payment from the LDC. Competition among OTC traders can only reduce the premium for supplying basis protection. Reducing the underlying causes of volatile price differences would require more pipeline capacity, more storage capacity, cost-based transmission pricing, and other physical and economic changes to the delivery system itself.

Crack Spread Contracts

In the petroleum industry, refinery managers are more concerned about the difference between their input and output prices than about the level of prices. Refiners' profits are tied directly to the spread, or difference, between the price of crude oil and the prices of refined products. Because refiners can reliably predict their costs other than crude oil, the spread is their major uncertainty. One way in which a refiner could ensure a given spread would be to buy crude oil futures and sell product futures. Another would be to buy crude oil call options and sell product put options. Both of those strategies are complex, however, and they require the hedger to tie up funds in margin accounts.

To ease this burden, NYMEX in 1994 launched the *crack spread contract*. NYMEX treats crack spread purchases or sales of multiple futures as a single trade for the purposes of establishing margin requirements. The crack spread contract helps refiners to lock in a crude oil price and heating oil and unleaded gasoline prices simultaneously in order to establish a fixed refining margin. One type of crack spread contract bundles the purchase of three crude oil futures (30,000 barrels) with the sale a month later of two unleaded gasoline futures (20,000 barrels) and one heating oil future (10,000 barrels). The 3-2-1 ratio approximates the real-world ratio of refinery output - 2 barrels of unleaded gasoline and 1 barrel of heating oil from 3 barrels of crude oil. Buyers and sellers concern themselves only with the margin requirements for the crack spread contract. They do not deal with individual margins for the underlying trades.

An average 3-2-1 ratio based on sweet crude is not appropriate for all refiners, however, and the OTC market provides contracts that better reflect the situation of individual refineries. Some refineries specialize in heavy crude oils, while others specialize in gasoline. One thing OTC traders can attempt is to aggregate individual refineries so that the trader's portfolio is close to the exchange ratios. Traders can also devise swaps that are based on the differences between their clients' situations and the exchange standards.

Crack Spread Options

Some industry participants may be comfortable with price variation so long as prices do not get too high or too low. An LDC that cannot readily pass along natural gas price increases to its residential customers may want to ensure that wholesale prices do not exceed what regulators allow it to charge. Holders of heating oil inventories may want to protect against price declines, but without giving up the opportunity to profit from price increases. A call option, which allows the holder to buy the commodity at a fixed strike price, sets a price ceiling. A put option, which allows the holder to sell at a fixed price, sets a floor.

NYMEX *crack spread options* are unusual because they protect against the growth or shrinkage in the *difference* between prices. A refiner, fearing that a currently profitable spread will disappear, can buy a crack spread put option. A large user of refined products, fearing that the spread will grow while the price of crude oil is stable, can buy a crack spread call option to compensate for potentially large increases in petroleum product prices when refinery margins grow.

Refiners who use crack spread options pay in advance for the price protection they desire. Options can be expensive when the terms are more favorable to the buyer, and the longer their lifespan, the more they cost. An alternative strategy is for a refiner to simultaneously buy a put and sell a call, so that the cost of the put is offset by the premium earned on the call. In essence, such a *collar* pays for the desired downside protection by selling off the opportunity for a windfall when the crack spread increases.

Calendar Spread Options

Storage facilities play an important role in the crude oil and refining supply chain. Facilities near producing fields allow the producers to store crude oil temporarily until it is transported to market. Facilities at or near refining sites allow refiners to store crude oil and refined products. Heating oil dealers build inventories during the summer and fall for winter delivery. Natural gas storage facilities allow producers to inject excess supply during "shoulder months" for withdrawal during peak demand months and provide producers with the convenience of a shortened injection and withdrawal cycle (a day or a few days), giving the producers and traders the ability to capitalize on the differential between forward prices and spot prices.

For most non-energy commodities, the cost of storage is one of the key determinants of the differential between current and future prices. Although storage plays a smaller role in price determination in some energy markets (most notably, for electricity), it can be important for heating oil and natural gas. For example, natural gas prices in the winter months could be established by the prices in the preceding shoulder months plus storage expenses and an uncertainty premium to account for the possibility of a colder than normal winter. If the price differential between winter months and shoulder months substantially exceeds storage expenses, traders can buy and store gas and sell gas futures. Such arbitrage tends to narrow the price differential.

The owners of storage facilities can use excess capacity both to manage the price risk that often exists between months and to make additional income. Assuming the market is in *contango* – i.e., when near-term prices (for "prompt months" are lower than prices for the months further in the future – owners of underground natural gas storage facilities with excess capacity that can be used to store natural gas for less than the difference between the prices can purchase futures contracts for the prompt months and sell futures contracts for the further future months. The storage facility can then take delivery of the natural gas on the nearby contract and deliver it against the distant contract, earning an arbitrage profit equal to the difference between the sale and purchase of the futures contracts less the facility's cost of storage.

Such arbitrage can also be accomplished by using a *calendar spread call option*. NYMEX offers calendar spread options on crude oil, heating oil, and unleaded gasoline. Buying a call on the calendar spread options contract will represent a long position (purchase) in the prompt months of the futures contract and a short position (sale) in the further months of the contract. Thus, the storage facility can buy a call on a calendar spread that will allow it to lock in a storage profit or to arbitrage a spread that is larger than its cost of storage.

If the market is in *backwardation* – i.e., when the prices for prompt months are higher than the prices for further months – storage facilities with excess capacity cannot arbitrage the calendar spread. In this case, storage facilities can sell put options on calendar spreads to earn income from the option premium. The buyer of a calendar spread put option, when the option is exercised, will receive a short position (sale) in the prompt months of the futures contract and a long position (purchase) in the further months of the contract. Thus, if the storage facility that sold (wrote) the put option is forced to accept delivery because the buyer has exercised the option, it will receive a long position in the prompt futures and a short position in the further futures. If the facility has excess storage capacity, however, it can take delivery on the prompt contract and then deliver on the later dated contract. If the put option is not exercised, the facility can keep the option premium without any further obligation. In summary, storage facilities can use futures contracts and calendar spread options to optimize utilization by arbitraging the difference in the prices specified for different months of a futures contract.

Volumetric Production Payment Contracts

A *volumetric production payment contract* (VPP) is both a prepaid swap and a synthetic loan. Unlike a normal swap, where the differences between the fixed and variable payments are periodically settled in cash, the buyer (usually a producer) is paid the present value of the fixed payments in advance. In exchange, the seller receives an agreed-upon amount of natural gas or other product over time. These deals typically last for 3 to 5 years. VPPs have been purchased by natural gas producers in the past, and in some cases they appear to have been used in project finance. In function, VPPs are identical to loans paid off with product.

The obvious problem with VPPs is that the seller, usually an energy trader, invests a large amount in advance, risking both buyer default and adverse price movements. In addition, VPPs can be used in place of loans to hide debt. What Enron and others often did was to find users of the product who were willing to pay up front in exchange for a price guarantee, use part of those payments to make the advance payment on the VPP, and then hedge their price risks by securing guarantees in the event of default.

Risk is the potential impact to and asset that may arise from some present / future events. Some of the important risk involves in oil trading are financial risk, basis, legal, tax & operational risk. Oil futures contracts have been traded on financial exchanges since 1970's. The first formalized regulated future exchange for oil was the NYMEX. Diversification and insurance are major tools for managing exploration risk and protecting firms from property loss and liability. Derivatives are particularly appropriate for managing the price risks that arises as a result of highly volatile price. Various strategies are being used for risk management. Crack spread contract, basis contract, crack spread option, and calendar spread option are among of them.

4.3 ADVANCE TECHNIQUE-SHORT HEDGES

Futures contracts have been used to manage cash market price risk for more than a century in the United States. Hedging allows a market participant to lock in prices and margins in advance and reduces the potential for unanticipated loss. Hedging reduces exposure to price risk by shifting that risk to those with opposite risk profiles or to investors who are willing to accept the risk in exchange for profit opportunity. Hedging with futures eliminates the risk of fluctuating prices, but also means limiting the opportunity for future profits should prices move favorably.

A hedge involves establishing a position in the futures or options market that is equal and opposite of a position at risk in the physical market. For instance, a crude oil producer who holds (is "long") 1,000 barrels of crude can hedge by selling (going "short") one crude oil futures contract. The principle behind establishing equal and opposite positions in the cash and futures or options markets is that a loss in one market should be offset by a gain in the other market.

Hedges work because cash prices and futures prices tend to move in tandem, converging as each delivery month contract reaches expiration. Even though the difference between the cash and futures prices may widen or narrow as cash and futures prices fluctuate independently, the risk of an adverse change in this relationship (known as basis risk) is generally much less than the risk of going unhedged and the larger a group of participants in the market, the greater the likelihood that the futures price will reflect widely held industry consensus on the value of the commodity. Because futures are traded on exchanges that are anonymous public auctions with prices displayed for all to see, the markets perform the important function of price discovery. The prices displayed on the trading floor of the Exchange, and disseminated to information vendors and news services worldwide, reflect the marketplace's collective valuation of how much buyers are willing to pay and how much sellers are willing to accept.

The purpose of a hedge is to avoid the risk of adverse market moves resulting in major losses. Because the cash and futures markets do not have a perfect relationship, there is no such thing as a perfect hedge so there will almost always be some profit or loss. However, an imperfect hedge can be a much better alternative than no hedge at all in a potentially volatile market. One of the most common commercial applications of futures is the short hedge, or seller's hedge, which is used for the protection of inventory value. Once title to a shipment of a commodity is taken anywhere along the supply chain, from wellhead, barge, or refinery to consumer, its value is subject to price risk until it is sold or used. Because the value of commodity in storage or transit is known, a short hedge can be used to essentially lock in the inventory value. A general decline in prices generates profits in the futures market, which are offset by decline in the value of the physical inventory. the opposite applies when prices rise.

Example 1 – Crude Oil Producer's Short Hedge

A crude oil producer agrees to sell 30,000 barrels a month for each of six months at the posted prices prevailing at delivery. When he agrees to the deal, posted prices are $20.50 barrel, but as market conditions appear to be weakening, he wants to protect his revenues against a decline, and executes a short hedge. The example shows how the producer's revenue is protected from the full brunt of a declining market. In this example, the oil producer establishes hedges for the second, third, fourth, fifth, sixth, and seventh contract months against his production during the first, second, third, fourth, fifth, and sixth months ahead. Near-month futures positions are liquidated after a price posting is established (normally on the first day of the calendar month). In a surplus crude market, spot prices generally fall faster than postings, regardless of whether prices decline more slowly, as in Case 1, or more rapidly, as in Case 2.

Date	Cash Market	Futures Market	Futures Results $/bbl.	Net Price Received $/bbl
Dec. 1	Commits to sell 30,000 barrels in each month for January, February, March, April, May, June crude at the posted price	Sells 30 crude contracts in each month for February, $20.00; March, $19.75; April, $19.50; May, $19.50; June, $19.25; July, $19.00		
Case 1: Slowly Declining Prices				
Jan. 1	Posted price for January crude: $21.00/bbl.	Buys back February contracts at $20.50	($0.50)	$20.50
Feb. 1	Posted price for February crude: $20.50/bbl.	Buys back March contracts at $19.75	0	$20.50
Mar. 1	Posted price for March crude: $20.00/bbl.	Buys back April contracts at $19.00	$0.50	$20.50
Apr. 1	Posted price for April crude: $19.50/bbl.	Buys back May contracts at $18.50	$1.00	$20.50
1-May	Posted price for May crude: $19.50/bbl.	Buys back June contracts at $18.75	$0.50	$20.00
Jun. 1	Posted price for June crude: $20.00/bbl.	Buys back July contracts at $19.50	($0.50)	$19.50
Case 2: Rapidly Declining Prices				
Jan. 1	Posted price for January crude: $20.00/bbl.	Buys back February contracts at $19.50	$0.50	$20.50
Feb. 1	Posted price for February crude: $19.50	Buys back March contracts at $18.75	$1.00	$20.50
Mar. 1	Posted price for March crude: $19.00	Buys back April contracts at $18.00	$1.50	$20.50
Apr. 1	Posted price for April crude: $18.50	Buys back May contracts at $17.50	$2.00	$20.50
1-May	Posted price for May crude: $18.50	Buys back June contracts at $17.75	$1.50	$20.00
Jun. 1	Posted price for June crude:$19.00/bbl.	Buys back July contracts at $18.50	$0.50	$19.50

Selling Prices ($/bbl.)

		Unhedged	
Month	**Hedged**	**Case 1**	**Case 2**
January	$20.50	$21.00	$20.00
February	$20.50	$20.50	$19.50
March	$20.50	$20.00	$19.00
April	$20.50	$19.50	$18.50
May	$20.00	$19.50	$18.50
June	$19.50	$20.00	$19.00
Average	**$20.25**	**$20.08**	**$19.08**

Increased Cash Flow
Case 1: $20.25 – $20.08 = $0.17 × 180,000 barrels = $30,600
Case 2: $20.25 – $19.08 = $1.17 × 180,000 barrels = $210,600

If the producer could not lock in revenue, he could be faced with shutting in all or part of the production.

The example shows two possible outcomes. Case 1, with relatively high posted and futures prices, and Case 2, with relatively low posted and futures prices. Short hedges for February, March, April, May, June, and July (against January, February, March, April, May, and June production) are initially established on December 1.

Assuming the futures hedge is placed on December 1, the near-month contract is January and the second month out is February. Because the January crude futures contract expires three business days prior to December 24, and the posted prices for January are not finally established until January 2, the example attempts to have the liquidation of the futures coincide with the setting of the posted price.

In summary, the nearby contract is used to hedge current production. For example, the February futures contract is utilized to hedge January production because the timing is better matched.

Example 2 – Petroleum Marketer's Long Hedge, Rising and Falling Markets

On September 7, the New York Harbor price for heating oil is 55 cents and the cash market price at the fuel dealer's location is 54 cents a gallon, a 1 cent differential, or basis, between New York Harbor and the retailer's location.

The dealer agrees to deliver 168,000 gallons to a commercial customer in December at 70 cents per gallon. On September 7, he buys four December heating oil contracts (42,000 gallons each) at 57 cents, the price quoted that day on the Exchange's NYMEX Division. Total cost: $95,760 (42,000 × 4 × $0.57).

Case 1 – Rising Prices

On November 25, the fuel dealer buys 168,000 gallons in the cash market at the prevailing price of 59 cents a gallon, a 1 cent differential to the New York Harbor cash quotation of 60 cents, Cost: $99,120.

He sells his four December futures contracts (initially purchased for 57 cents) at 60 cents a gallon, the current price on the Exchange, realizing $100,800 on the sale, for a futures market profit of $5,040 (3 cents a gallon).His cash margin is 11 cents (the difference between his agreed-upon sales price of 70 cents and his cash market acquisition cost of 59 cents for a total of $18,480 ($0.11 per gallon × 168,000 gallons)

	Cash Market	**Futures Market**
7-Sep		Buys 4 December futures contracts for 57¢ per gallon
Nov. 25	Buys 168,000 gallons at 59¢ per gallon	Sells 4 December heating oil futures for 60¢ per gallon
A cash margin of:	$18,480 or 11¢/gallon plus	
A futures profit of:	$5,040 or 3¢/gallon equals	
A total margin of:	$23,520 or 14¢/gallon	

Case 2 – Falling Prices

On November 25, the dealer buys 168,000 gallons at his local truck loading rack for 49 cents a gallon, the prevailing price on that day, based on the New York Harbor cash quotation of 50 cents a gallon.

He sells his four December futures contracts for 50 cents a gallon, the futures price that day, realizing $84,000 on the sale, and experiencing a futures loss of $11,760 (7 cents a gallon).

	Cash Market	**Futures Market**
7-Sep		Buy 4 December heating oil futures at 57¢ per gallon
Nov. 25	Buys 168,000 gallons	Sells 4 December heating oil futures for 49¢ per gallon for 50¢ per gallon
A cash margin of:	$35,280 minus	
A futures loss of:	($11,760) (7¢/gallon) equals	
A total margin of:	$23,520 or 14¢/gallon	

In summary, the fuel retailer guarantees himself a margin of 14 cents a gallon regardless of price moves upwards or down in the market.

With the differential between cash and futures stable, as in Cases 1 and 2, spot-price changes in either direction are the same for both New York and the marketer's location. As a result, a decline in the futures price, which causes a loss in the futures market, is offset cent-for-cent by the increase in the cash margin.

Example 3 – Utility Protects Acquisition Costs of Future Wholesale Purchases without a Price Commitment From a Seller

On February 1, a gas utility decides that the prices reflected in the futures market for the second quarter are less expensive than the current cash market prices. The utility does not want to wait until the second quarter to buy gas because it fears that its acquisition costs may move significantly higher than its sales prices. The utility buys natural gas futures contracts for April, May, & June.

To fix its acquisition costs, the utility might take a long hedge against its forward sales. If wholesale buying prices, or production costs, increase, profits on the futures market will offset the rising costs in the cash market, keeping the retail margin constant.

Similarly, if wholesale costs decrease, the lower acquisition costs will be offset by a loss on the futures.

Case 1 – Rising Prices

On February 1, the gas utility buys 10 Henry Hub natural gas futures contracts in each of three months, April, May, and June for $4.75, $5.00 and $5.25 per mmBtu, respectively. The cost of the futures purchases are $475,000 for the April contract, $500,000 for May, and $525,000 for June, for a total cost of $1,500,000.

On March 27, the utility buys 100,000 mmBtu of natural gas in the cash market for the then prevailing price of $5.25 per mmBtu, and pays $525,000. That is $50,000 more than it budgeted when it anticipated a price of $4.75 per mmBtu. The utility also sells back its 10 April natural gas futures contracts at the then-current price of $5.25 so it doesn't have to take delivery through the Exchange. The contracts originally purchased for $4.75 ($475,000), are now worth $5.25 ($525,000) yielding a gain in the futures market of $50,000.

Cash market purchase of:	$525,000 plus
A futures gain of:	$50,000 equals
A net amount of	$475,000 or $4.75 per mmBtu, the budgeted sum for April.

As cash prices continue to be strong during the second quarter, his hedge looks like this:

Cash Market Futures Market

	Cash Market	Futures Market
1-Feb		Buys 10 natural gas futures contracts in each of April, May, June for $4.75, $5.00, $5.25, respectively
27-Mar	Buys 100,000 mmBtu at $5.25	Sells back 10 April contracts, $5.25
26-Apr	Buys 100,000 mmBtu at $5.50	Sells back 10 May contracts, $5.50
26-May	Buys 100,000 mmBtu at $6.00	Sells back 10 June at $6.00

Financial Result	April	May	June	Quarter
Expected Cost	$475,000	$500,000	$525,000	$1,500,000
Cash Market Expense	$525,000	$550,000	$600,000	$1,675,000
Futures Mkt Gain (Loss)	$50,000	$50,000	$75,000	$175,000
Actual Cost	$475,000	$500,000	$525,000	$1,500,000
				$5.00 per mmBtu

Case 2 – Falling Prices

What happens to the utility's hedge if prices fall instead of rise? In that case, assume the market falls to $4.50 in April, $4.75 in May, and $5.00 in June.

	Cash Market	Futures Market
1-Feb		Buys 10 natural gas futures contracts in each of April, May, June for $4.75, $5.00, $5.25, respectively.
27-Mar	Buys 100,000 mmBtu at $4.50	Sells back 10 April contracts, $4.50
26-Apr	Buys 100,000 mmBtu at $4.75	Sells back 10 May contracts, $4.75
26-May	Buys 100,000 mmBtu at $4.90	Sells back 10 June at $4.90

Financial Result	April	May	June	Quarter
Expected Cost	$475,000	$500,000	$525,000	$1,500,000
Cash Market Expense	$450,000	$490,000	$525,000	$1,415,000
Futures Mkt Gain (Loss)	($25,000)	($25,000)	($35,000)	($85,000)
Actual Cost	$475,000	$500,000	$475,000	$1,500,000

The average cost of $5.00 per mmBtu represents an opportunity cost of 28.3¢ per mmBtu because cash market prices averaged $4.717 during the period of the hedge. The utility is comfortable with this because it is within the tolerance for risk that the company's risk management committee had set at the time the positions were opened. Managing a hedge strategy is an evolving process. While hedges serve to stabilize prices, risk management targets can be reevaluated in future periods as market and financial circumstances change.

4.4 HEDGING STRATEGIES INVOLVING MULTIPLE CONTRACTS

Strip Trades

Strip trading is a flexible strategy that energy futures market participants use when hedging positions for several consecutive months forward. A market participant can lock in an average price for several months at a time by simultaneously opening a futures position in each of the months to be hedged through a single Exchange transaction. The average of the futures contracts over the period is the price level of the hedge. A six-month strip, for example, consists of an equal number of futures contracts for each of six consecutive contract months.

Strip trades in the futures contracts for crude oil; heating oil; gasoline and natural gas are executed as a single transaction during the open outcry trading session, after being bid and offered at an agreed-upon differential to the previous day's settlement price. The strip and the differential are calculated based on the average value of those months currently versus the average of the previous day's settlement prices for those months. The ability to obtain an average price for multiple months enables a hedger to average his cash flow over a period of time. Positions can be hedged for as little as two consecutive months, or can go forward for up to 12 months in gasoline; 18 months in heating oil, 30 months in light, sweet crude oil; and 72 months in Henry Hub natural gas.

The futures positions assumed in a strip trade are like any other futures position. Any single month's position can be liquidated by an offsetting futures trade, an exchange of futures for physicals (EFP), or, if desired, physical delivery through the Exchange clearing-house. Strips let a hedger retain the flexibility to change a strategy by buying or selling additional futures contracts in any month, or liquidating the position of any month of the strip, something that cannot be done easily with over-the-counter instruments.

Regular margin requirements apply to strip trades. The participant will be required to post and maintain margin levels for each month in the strip as if it were a separate position.

Example 4 – Petroleum Refiner's Use of a Strip Trade

A refiner anticipates the purchase of 60,000 barrels of crude oil, the volume distributed evenly over a six-month period beginning in October. The use of the strip allows the company to hedge its expenditures for crude oil evenly throughout the period.

The refiner's risk management committee is comfortable with locking in this strip's price level over the period of their purchases, and, considering their fundamental view of the crude oil market, is even willing to pay a higher price, if necessary.

The risk manager determines that those months are currently trading at an average that is 10 cents over the average of the previous day's settlement price, or $18.35. Assuming he wishes to hedge 100% of his physical requirements, he buys 10 contracts for each of the six months, or 60 contracts representing 60,000 barrels since each futures contract is for 1,000 barrels.

The refiner's hedge, which has locked in a price of $18.35 for a total of 60,000 barrels over the period, looks like this:

Month	Transaction Price	Market Position
October	$19	Long 10 contracts
November	$18	Long 10 contracts
December	$18	Long 10 contracts
January	$18	Long 10 contracts
February	$18.26	Long 10 contracts
March	$18.24	Long 10 contracts
Average	$18.35	

Assuming the refiner takes delivery from its traditional suppliers, it will liquidate the futures positions in the relevant months, offsetting its physical market transactions. If the refiner chooses, however, it can take delivery through the Exchange for any, or all, of the months involved in the strategy.

As with any other hedge, there may be a loss in the futures market for a particular month that is compensated for by a gain in the cash market, conversely a gain in the futures market will offset a loss in the cash market.

Regular margin requirements apply to strip trades. The participant will be required to post and maintain margin levels for each month in the strip as if it were a separate position.

Example 5 – Petroleum Marketer's Long Hedge; Guaranteeing Retail Prices by Purchasing a "Strip" of Futures

By allowing market participants to lock in an average price over a period of time, strip trading strategies can be used by vendors of refined products to offer their customers seasonal price stability through fixed-price programs. These programs have become increasingly popular with heating oil retailers who are faced with increasingly competitive market conditions, and their customers, who desire stable pricing, thus avoiding the price spikes that can wreak havoc with household budgets.

Retailers are able to offer seasonal fixed pricing with minimal risk to their operating margins by locking-in ahead of time during the off-season the price required to service those customers during the winter. This can be accomplished with strips of NYMEX Division heating oil futures contracts.

A retailer who wants to offer a fixed price program would in the spring and summer solicit as many customers as possible whose consumption averages 1,000 gallons each over the period from October to the beginning of April, with the bulk of consumption occurring during December, January, and February.

A large dealer typically sells about 10 million gallons in a season. Assume 20% of his customers choose to take advantage of the fixed price program, representing 2 million gallons, during the summer, he would buy a portfolio of futures contracts:

Customer Delivery Month	Futures Contract Month	Heating Oil Futures Price ¢/gallon
October	November	50.02¢
November	December	50.77¢
December	January	51.17¢
January	February	50.92¢
February	March	49.87¢
March	April	48.92¢

Weighted average price: 50.28¢ per gallon.

During the period, he will need 50 futures contracts (representing 2.1 million gallons). However, he has to weight the volume toward the contract months of December, January, and Februaıy, which represent, say, 60% of the volume.

He would buy five contracts per month for October, November, March, and April, and 10 per month for December, January, and February.

As the distributor sells his product to his customers, he liquidates his futures contracts, one contract for each 42,000 gallons of product purchased.

May 25

Commitment to offer fixed price for heating oil in October through March

	Contracted Price
October	50.02¢
November	50.77¢
December	51.17¢
January	50.92¢
February	49.87¢
March	48.92¢
Avg Price	50.27¢

	Physical (Cash)		Futures		Profit /(loss)
Sept. 15	Buy Oct.	48.00¢	Sell Oct.	48.00¢	2.02¢
Oct. 15	Buy Nov.	48.50¢	Sell Nov.	48.50¢	2.27¢
Nov. 15	Buy Dec.	50.00¢	Sell Dec.	50.00¢	1.17¢
Dec. 15	Buy Jan.	50.00¢	Sell Jan.	50.00¢	0.92¢
15-Jan	Buy Feb.	50.08¢	Sell Feb.	50.08¢	(0.21)¢
15-Feb	Buy Mar.	48.76¢	Sell Mar.	48.76¢	0.16¢
Avg price		49.22¢	Avg profit		1.05¢

Physical P&L Net Price 50.27¢.

Several other factors must be taken into account in constructing a guaranteed price offering: transaction costs, the relationship between rack and futures prices, volume risk, and the dealer's operating costs and profit broken down to a cents-per-gallon figure.

The retailers will hedge, that is, purchase their physical supply from their traditional suppliers while liquidating their futures positions. The difference between the rack price and the futures price is the basis. (In the New York area, for example, rack prices might be 2¢ per gallon over the futures price).

The distributor must compare his own historical rack prices to futures prices. There are two important pieces to this analysis: The average difference, which represents the basis, and the standard deviation, a statistical measure of the variation of the basis around the average, which represents basis risk.

Components of Guaranteed Price	
Weighted Average Futures Price	$0.50
plus Operating Costs (insurance, permits, mortgage/rent, salaries, taxes, profit)	$0.45
plus Basis (rack over futures)	$0.02
plus Transaction Costs	$0.002*
Minimum guaranteed price offer	$0.9748 per gallon

Spread Trades

Spread positions offer another way of using futures. There are many types of spreads, but they all have two things in common. First, a spread always involves at least two futures positions, which are maintained simultaneously. For example, a trader may be long 10 June coal contracts and short 10 September coal contracts. Second, the price changes in the two or more legs of the position are expected to have a reasonably predictable relationship, and the potential profitability of the spread lies in that relationship or expected changes to that relationship. For example, the trader who is long 10 June contracts (the near-term contract) and short 10 September contracts (the distant contract) will benefit if market forces cause the near-term contract to make a larger advance than the more distant contract – or if market forces cause the distant contract to drop more sharply than the near-term contract.

Crack Spreads

A petroleum refiner, like most manufacturers, is caught between two markets: the raw materials he has to purchase and the finished products he offers for sale. It is the nature of these markets for prices to be independently subject to variables of supply, demand, transportation, and other factors. This can put refiners at enormous risk when crude oil prices rise while refined product prices stay static or even decline, thus narrowing the spread.

The Exchange facilitates crack spread trading in its futures trading rings by treating them as a single transaction for the purpose of determining a market participant's margin requirement.

To calculate the theoretical refining margin, first calculate the combined value of gasoline and heating oil, and then compare the combined value to the price of crude. Since crude oil is quoted in dollars per barrel and the products are quoted in cents per gallon, heating oil and gasoline prices must be converted to dollars per barrel by multiplying the cents per gallon price by 42 (there are 42 gallons in a barrel). If the combined value of the products is higher than the price of the crude, the gross cracking margin is positive. Conversely, if the combined value of the products is less than that of crude, then the cracking margin is negative.

Using a ratio of two crude to one gasoline plus one heating oil, the gross cracking margin is calculated as follows:

(Assume heating oil is $0.5450 per gallon, gasoline is $0.5750 per gallon and crude is $18.50 per barrel.)

$0.5450 per gallon × 42 = $22.89 per barrel of heating oil

$0.5750 per gallon × 42 = $24.15 per barrel of gasoline

The sum of the products is: $47.04

Two barrels of crude ($18.50 × 2) = $37.00

Therefore, $47.04 – $37.00= $10.04

$10.04/2 = $5.02 (margin)

A refiner expects crude prices to hold steady, or rise somewhat, while products will fall. In this case, the refiner would "sell the crack"; that is, he would buy crude futures and sell gasoline and heating oil futures.

Conversely, buying the crack means buying gasoline and heating oil and selling crude.

Whether a hedger is selling the crack or buying the crack reflects what is done on the product side of the spread.

Once the hedge is in place, the refiner need not worry about movements in absolute futures prices. He need be concerned only with how the combined value of products moves in relation to the price of crude.

The following example shows a refiner locking in a margin between crude oil and heating oil.

Example 6 – Fixing Refiner Margins through Crack Spreads

In January, a refiner reviews his crude oil acquisition strategy and his potential distillate margins for the spring.

In January, he sees that distillate prices are strong, and plans a two-month crude-to-distillate spread strategy that will allow him to lock in his refinery margins.

On January 22, the spread between April crude ($18 per barrel) and May heating oil ($0.4925 per gallon or $20.69 per barrel) presents what he believes to be a favorable $2.69 per barrel.

The refiner sells the April/May crude-to-heating oil spread, thereby locking in the $2.69 margin.

In March, he purchases the crude for refining into products. Crude oil futures are $19 per barrel in March, $1 higher than the original crude futures position. Heating oil is trading at $0.4950 per gallon ($20.79 per barrel), which equals a margin of $1.79.

Had the refiner been unhedged, his margin would have totaled only the $1.79. Instead, the net margin from the combination of the futures position and the cash position is the $2.69 he originally sought.

Date	Cash	Financial Effect Futures	Financial Effect Cash	Futures ($/bbl)
Jan.	–	Sell crack spread:		
		Buy crude	–	($18.00)
		Sell heating oil at $0.4925/gal.		$20.69
		Net		$2.69
Mar.	Buy crude at $19		($19.00)	
	Sell heating oil at $0.4950/gal.		$20.79	
	Net		$1.79	
		Buy crack spread: Sell crude		$19.00
		Buy heating oil at $0.4925/gal		($20.79)
		Net		($1.79)
		Futures gain (loss)		$0.09
	Cash refining margin (loss) without hedge		$1.79	
	Final net margin with hedge			$2.69

Example 7 – Refiner with a Diversified Slate, 3:2:1 Crack Spread

An independent refiner who is exposed to the risk of increasing crude oil costs and falling refined product prices runs the risk that his refining margin will be less than anticipated.

The refiner initiates a long hedge in crude and short hedges in heating oil and gasoline to fix a substantial portion of his refining margin.

On September 15, the refiner incurs an obligation to buy 6,000 barrels of crude oil on May 16 at prevailing cash prices. He is also obliged to sell 84,000 gallons (2,000 barrels) of heating oil and 168,000 gallons (4,000 barrels) of gasoline on November 28 at prevailing spot prices.

The crack spread has ensured that refining crude oil will be at least as profitable in November as it was in September, regardless of whether the actual cash margin narrows or widens. A decline in the cash margin is offset by a gain in the futures market; conversely, any gain in the cash market is offset by a loss in the futures market. The example assumes a crack spread of three crude, two gasoline, and one heating oil contract.

Date	Prices	Action	Futures Market
Sept 15.	Sweet crude: Cushing - $18.90	Agrees to buy at prevailing prices: 6,000 bbl. sweet crude on Oct 16	Buys 6 Nov sweet crude contracts at $18.45/ bbl.
	Heating Oil: Gulf Coast, $0.4875/gal, $20.47/bbl N.Y. Harbor, $0.5125/gal, $21.52/bbl	Commits to sell at prevailing prices: 84,000 gal. heating oil on 28-Nov	Sells 2 Dec. heating oil contracts, $0.5255/ gal, $22.07/bbl
	Gasoline:	Commits to sell at prevailing prices: 168,000 gal. gasoline on Nov. 28	Sells 4 Dec. New York Harbor gasoline contracts $0.5275/gal, $22.15/ bbl
	Gulf Coast, $0.5450/gal, $22.89/bbl		
	N.Y. Harbor, $0.5850/gal, $24.57/bbl		
	Gulf Coast cash margin: (6 × $18.90)-[(2 × $20.47)+ (4 × $22.89)]/6 = $3.18/bbl.		
	Cushing/NY Harbor cash margin: {[(2 × $21.52)+(4 × $24.57)]- (6 × $18.90)}/6 = $4.65/bbl.		
	Crack spread: {[(2 × $22.07)+(4 × $22.15)]-(6 × $18.45)}/6 = $3.67		
Cash basis	- $1.47/bbl ($3.18-$4.65)		Futures Crack Spread:
Expected margin	$2.20/bbl ($3.67-$1.47)		$3.67/bbl.

Case A: Rising Crude, Falling Product Prices, Stable Basis

16-Oct	**Crude:**	**Buys 6,000/bbl crude at $19**	**Sells 6 Nov sweet crude contracts at $19/bbl.**
	NYMEX Div. crude oil futures,$19/bbl.		

28-Nov	**Heating Oil:** Gulf Coast, $0.46/gal, $19.32/bbl. N.Y. Harbor, $0.4850/gal, $20.37/bbl	Sells 84,000 heating oil $0.4850/gal.	Buys 2 Dec heating oil contracts at $0.4942/gal
	Gasoline: Gulf Coast, $0.4625/gal, $19.42/bbl N.Y. Harbor, $0.4875/gal, $20.47/bbl.	Sells 168,000 gasoline at $0.4875/gal.	Buys 4 Dec N.Y. Harbor gasoline contracts at $0.4936/gal.
Results Cash margin	$1.44	{[(2 × $20.37) + (4 × $20.47)] – (6 × $19)} / 6	
Futures profit	$1.94	($3.68 – $1.74)	Crack spread:
Realized margin	$3.38		$1.74/bbl.

Case B: Falling Crude, Rising Product Prices, Stable Basis

16-Oct	Crude: NYMEX Div. crude oil futures, $17.50/bbl.	Buys 6,000/bbl crude at $17.50	Sells 6 Nov sweet crude contracts at $17.50
28-Nov	**Heating Oil:** Gulf Coast, 0.50/gal, $21/bbl. N.Y. Harbor, $0.5250/gal, $22.05/bbl.	Sells 84,000/gal. Heating oil at $0.5250/gal.	Buys 2 Dec heating oil contracts at $0.52/gal.
	Gasoline: Gulf Coast, $0.56/gal, $23.52/bbl N.Y. Harbor, $0.60/gal, $25.20	Sells 168,000/gal. Gasoline at $0.60/gal.	Buys 4 Dec N.Y. Harbor gasoline contracts at $0.5950/gal
Results Cash margin	$6.65	{[(2 × $22.05) + (4 × $25.20)] – (6 × $17.50)}/6	
Futures profit	($2.66)	($3.68 – $6.34)	Crack spread:
Realized margin	$3.99/bbl.		$6.34/bbl.

Timing risk and basis risk can be quantified and are usually less than the absolute price risk to which the refiner is subjected. The example assumes fixed points in time of obligation to buy and sell in the cash market. In practice, these may not be entirely known or fixed.

Purchasing a Crack Spread

The purchase of a crack spread is the opposite of the crack spread hedge. It entails a short hedge in crude oil and long hedges in products. Refiners are naturally long the crack spread as they buy crude and sell products. At times, however, refiners do the opposite, they buy products and sell crude and thus find purchasing a crack spread a useful strategy.

When refiners are forced to shut down for repairs, they often have to enter the crude oil and product markets to honor existing purchase and supply contracts. Unable to produce enough products to meet term supply obligations, the refiner must buy products at spot prices for resale to his term customers. Furthermore, lacking adequate storage space for incoming supplies of crude oil, the refiner must sell the excess on the spot market.

If the refiner's supply and sales commitments are substantial and if he is forced to make an unplanned entry into the spot market, it is possible that prices might move against him. To protect himself from increasing product prices and decreasing crude oil prices, the refiner uses a short hedge against crude and a long hedge against products.

Frac Spreads

Natural gas and propane are also likely candidates for a market-related spread since natural gas processing is a major source of propane production. The natural gas processor can partially hedge the price risk of processing natural gas and extracting propane through the use of a fractionation spread, much the same way refiners use crack spreads to establish margins when refining crude oil into heating oil and gasoline.

Balancing the Frac Spread by Equating Heating Value

The frac spread is quoted in heating value terms, dollars per million British thermal units (MMBtu), to equate propane to natural gas. The natural gas futures contract is comprised 10,000 MMBtu, and is quoted in dollars and cents per MMBtu.

Propane is quoted in cents per gallon. One gallon in gaseous form contains approximately 91,500 Btu. Dividing the price of propane by 0.0915 gives the equivalent price per MMBtu. If propane were trading at 35¢ per gallon, the cost would be $3.825 per MMBtu.

One propane futures contract, 42,000 gallons, represents about 38% of the heating value of one natural gas contract of 10,000 MMBtu. The two most popular ratios used to create a balance heating value position are a 3:1 or 5:2 propane to natural gas spread.

Once the price of propane, quoted in dollars per gallon, has been converted into the price per MMBtu, the frac spread can be calculated by subtracting the price of natural gas from the calculated value of propane to yield the gross manufacturing margin. At this point, the fractionator has only paid for the value of natural gas consumed, or reduced, in processing. There are many additional costs including processing, transportation, fractionation, and marketing that must be paid out of the gross manufacturing margin.

Example 8 – Gas Processors Frac Spread with a 5:2 Ratio

Assume propane futures are trading at $0.47 per gallon ($5.136 per MMBtu). If natural gas futures were trading at $3.086 per MMBtu, then the frac spread would have a positive margin of $2.05. From this margin, other operating costs could be covered.

The most frequent application of frac spreads is found among gas processors. In the example below, a processor wishes to lock in the $2.05 per MMBtu spread between the two products for the March contract. He enters into the market on December 3, and liquidates his position on February 10, several days before the March contract ceases trading.

Gas Processor's Frac Spread with a 5:2 Ratio

	3-Dec Sold PN	**6-Jan Brought NG**	**10-Feb Brought PN**	**10-Feb Sold NG**	**Gain (Loss) PN**	**Gain (Loss) NG**
Contract Month	March	March	March	March	March	March
$ per Contract Unit	$0.47	$3.09	$0.36	$2.17	$0.12	($0.92)
# of Units per Contract	42,000	10,000	42,000	10,000	42,000	10,000
$ per Mmbtu NG=PN/.0915	$5.136*	$3.09	$3.879*	$2.17	$1.256*	($0.92)
Heat Value per Contract	3,843	10,000	3,843	10,000	3,843	10,000
Spread Ratio	5	2	5	2	5	2
Ratio × Heat Value	19,215	20,000	19,215	20,000	19,215	20,000
$ Value per Contract	$19,740	$30,860	$14,910	$21,670	$4,830	($9,190)
Ratio × $ Contract Value	$98,700	$61,720	$74,550	$43,340	$24,150	($18,380)
Spread PN-NG	$2.05		$1.71			$0.34
Net Futures						$5,770

* Propane price divided by 0.0915.

A profit of 34¢ per MMBtu on 19,215 MMBtu is realized in the futures market. In March, the gas processor can no longer realize the high margin he received in December for fractionating propane and "dry" natural gas out of wet gas. The 34¢ per MMBtu futures gain is used to supplement, or offset, the reduced processing margin received in the cash market. The net effect on the processor's balance sheet is a maintained margin.

If the frac spread had increased beyond $2.05 instead of decreasing, the processor would have lost an opportunity for further gain but, as a hedger, he was content in locking in the $2.05 per MMBtu margin. He was willing to forego a potentially higher margin in exchange for eliminating the chance of a lower margin.

Finally, Hedging is one of the important tools to avoid the risk of adverse market moves resulting in major losses. Short hedge, also known as sellers hedge which is used for the protection of inventory value. A general decline in prices generates profit in the futures market, which are offset by decline in the value of the physical inventory. The opposite applies when prices arise. Hedging strategy involves mainly strip trades and spread trade. Frac spread is also used in a risk management and quoted in hitting value terms, dollar per million British Thermal Units. The most frequent application of frac spread is found among gas processors.

4.5 CRUDE OIL PRICE HEDGING

Crude oil prices have become more and more volatile since the 1973 oil crisis. Particularly long since the recent Gulf crisis (crude oil prices rose and fell sharply between August 1990 and March 1991), producers, refiners, and consumers have been interested in acquiring more assurances about the prices they would pay or receive over future periods. Increasingly they have used such risk management instruments as futures, options, and swaps to protect themselves against adverse oil price movements. The main benefit of risk management is reduced uncertainty about the oil prices consumers and the state oil-importing company will pay rather than lower average crude oil import prices. There are two scenarios the short-term hedge, in which the state oil-importing company locks in a price for its imports for one month ahead, and the long term hedge, in which it locks in the price for six months ahead. In one of the study over the simulation, it has been found that the short-term hedge reduces oil price volatility a potential 72 percent to 85percent and the long-term hedge reduces volatility of potential 65 percent to 81 percent. For these reductions to be realized, the prices of the crude oils hedged must move together with the futures prices. Simple regression tests were carried out to see if this is so. Apparently oil-importing developing countries could gain considerably from using financial risk management instruments. But several constraints – particularly negative publicity and legal obstacles – can impede a state oil-importing company's use of risk management instruments. Educating government policymakers and state enterprise officials about the proper use, limits, and benefits of risk management instruments will make them more acceptable. Oil supply disruption due to resources depletion, Middle East conflict and the political decisions are some major factors which have turned oil most volatile commodity. Oil producers, importers and refiners plan their budget on the basis of prevalent oil price and estimate their cost and profit, but frequent fluctuations in the price have made them to bear huge losses in the oil market. Therefore, they need protection from such loss. Hedging facilitates the process, which provides derivative instruments to protect them from such price risks.

'Hedging crude oil' prices have great importance for the oil importing country like India, China. It has been seen that due to oil-price fluctuations oil-importing country faces critical situations. It raises import bill and inflation, which ultimately leads to adverse 'balance of payments' situations. Till now, there is no appropriate substitute for the crude oil, so it is compulsion for each country to bear such adverse price movements.

Here author has described about the problems faced by oil importing country due to oil price fluctuations and approached its solution based on simple regression methodology. Further I have discussed about the constraints over the use of hedging (futures, option contracts).To mitigate the impact of these constraints, I have suggested some concrete ideas.

Risk Management Problems

A typical SOC (State- owned Oil Company) is involved in purchasing crude and petroleum products for domestic energy needs. While SOCS are exposed to international oil price volatility, domestic prices are not always immediately adjusted in line with international price movements. The SOC often has to absorb the consequences of this imperfect adjustment and has therefore an exposure to international oil price risk.

On the basis of a typical SOC's operating procedures, two types of risk management problems can be distinguished: transaction risk and long-term risk. Transaction risk refers to the price uncertainty between the time of ordering crude oil and the time of delivery. Long-term risk refers to the price uncertainty over a longer planning horizon, say six months or a year. For each of the problems a different risk management strategy is applicable. However, the two risk management strategies follow a logical sequence: the SOC would first engage in transaction risk management (i.e., short-term risk management) and the in long-term risk management.

A. Transaction Risk

The typical SOC commits to a purchase of crude oil considerable period ahead of actual loading, while the price of crude is determined at the time of loading. The time difference between the purchasing commitment and the date of loading varies, but usually is from 20 to 50 days. To analyze more concretely the transaction risk, we describe the case of the SOC of Costa Rica (RECOPE) which is representative of many other SOCs. Indian oil companies are allowed to do hedging under the stipulated guidelines of G.O.I.

The following time-axis illustrates the transaction risk management problem for RECOPE.

Time:	(50-20)	(20-15)	0	+30	(38-40) days
	A	B	C	D	E

Date Activity

A: RECOPE decides on its production program and contracts the carrier
B: RECOPE pays in colones (local currency) to the Central Bank
C: RECOPE date of loading (bill-of-lading) and setting of price
D: Payment to the seller (by the Central Bank through a foreign bank)
E: Return of excess payment (in colones) by the Central Bank to RECOPE.

RECOPE decides on its program of production (whether to run the refinery and which product mix to produce) on the 10th of every month, based on expected demand, prices and market sentiment, and makes an assessment of the future prices of the different crudes and refined products. It then decides which crudes and refined products to import and contracts with its suppliers. Five days later the suppliers confirm availability, the contract is finalized and RECOPE agrees with the supplier of the crude (or refined product) to take delivery.

RECOPE then Contracts with independent shipping companies for loading of the difference crudes at various ports. There is a minimum of 15 days between the time of contracting the shipper (A) and actual loading. The actual loading (C) therefore does not happen until 20 to 50 days after RECOPE decides its production program (A). The actual price paid for the crude is set at the date of loading at the then prevailing international spot price. Approximately 20 to 15 days before the bill-of-lading date, RECOPE transfers the expected sum due on the bill-of-lading date to the Central Bank, adding 10% for possible variation in oil price and exchange rate over the 20 to15 days period (B). The Central Bank then arranges a letter-of-credit with a foreign bank to be used for payment at the date of loading.

The letter of credits cashed about 30 days after the bill-of-lading date when payments are made to the seller (D). About 8 to 10 days after the letter of credit is cashed, any excess payments 'the difference between the actual payments and the expected payments plus the 10%) are returned by the Central Bank to RECOPE in colones (E). RECOPE is exposed to two types of risks over the period from contracting the shipper (A) and receiving the excess payments (E): dollar-colone exchange risks and oil price risks. The exchange rate risk is as follows. Between points in time A and D RECOPE faces the risk that the colone exchange rate depreciates, in which case its foreign currency payment at time D will be more in colones terms than expected and it will receive less back of the 10% variation margin. In addition, RECOPE faces exchange rate risks between points D and E on any excess payments. RECOPE faces oil price risks since it has to decide where to obtain its supplies50 to 20 days (A) before the actual price to be paid is determined (C). If oil prices increase in this period, RECOPE will have to incur the extra costs since it cannot change or cancel its order (or only at very high costs).

B. Long-Term Risk

Long-term risk concerns the transmission of changes in international oil prices from the SOC to the domestic economy over some planning period. The allocation of the long-term risk between the SOC and the rest of the economy depends on the internal pricing system for energy products. At one extreme, the internal pricing system may transmit all international price volatility directly to the final consumer. In this case, the SOC does not face any risk but the economy does. At the other extreme, the internal pricing system insulates the domestic consumer from world price fluctuations by maintaining a fixed domestic price. In this case, all price risk lies with the SOC. In reality, maintenance of fixed domestic prices over a long period is impossible without undermining the viability of the SOC, imposing budget deficits, and distorting consumption and investment decisions. More commonly, domestic prices are fixed for some period of time, say six months, in which case the risk within that period lies entirely with the SOC. Whether the risk lies with the SOC or with the domestic consumer, the SOC is in a position to manage it for its own benefit as well as for the benefit of the economy as a whole. The SOC is the logical place to manage such risks since it has a physical exposure through purchasing oil) and can thus "match' physical with futures or options trading. It also has a good knowledge of the oil markets.

C. Hedging Transaction Risks

The transaction exposure to oil price risks could be managed using futures on oil or options on futures on oil.

Hedging with Futures

The SOC has a short position in oil because it has a commitment to satisfy domestic demand and domestic prices cannot be adjusted instantaneously. To hedge the resulting price risk, we seek an asset whose return is positively correlated with oil prices. Then as oil prices go up, the return on the hedging asset increases to offset the additional cost of higher import prices for the SOC. The purchase of futures contracts is a good example of such an asset. The basic concept of futures contracts is that the buyer of a futures contract purchases a specific asset at a specified price on a specific date. Contract terms (amounts, grades, delivery dates, etc.) are all standardized in futures contracts, and futures are traded only on organized exchanges through clearinghouse systems. To ensure good performance on these contracts profits and losses on futures are settled daily, and futures contracts require a small "margin" as Collateral (see further Annex 1). If spot oil prices rise over the planning horizon, so will futures prices. The profit from selling futures contracts back at the higher price will offset the extra cost of importing oil. Conversely, if prices fall lower oil import costs offset the loss on the futures.

The transaction risk of the SOC could thus be managed using futures as follows. On the date of arranging the purchase of oil, the SOC could buy an oil futures contract. The maturity date of the contract should be as close as possible to the expected date of loading, but not before it. The futures contract would be for the same volume of oil as the oil contracted for purchase. The contract would be terminated through a reverse transaction in the futures market (i.e., selling the futures contract) on the date of loading (note that even though futures contracts specify physical delivery, by reversing contracts this does not need to happen). In this way the effective price paid for the oil will be locked in on the date of entering the agreement with the shipping company.

For example, assume that in March 1991 the SOC plans to buy crude oil two months ahead i.e., in May 1991, in the amount of 10,000 barrels, where the price to be paid would be the spot Mexico (Istmo) price two months from now. The SOC would then buy 10 futures contracts of 1,000 barrels each, maturing in two months. The price of the May 1991 futures contract is quoted as say $18.50/barrel. The SOC would have to put up about $12,000 in margins. Two months later, a gain or loss in the import cost of the crude arising from fluctuations in the Istmo oil price is offset by the loss or gain in the futures contracts. The offset may not be exact because of basis risks. Between initiating the futures contracts and closing them out, transfers of cash to and from a broker will be required almost every day because the gains and losses in the futures contracts are settled daily. *Basis Risk is the risk that the cash price and the futures price will differ by more or less than same expected amount.*

Hedging with Options

The main disadvantage of hedging with futures is that the potential benefit of an import price decline is eliminated along with the potential loss of an import price increase. Yet falling import prices are clearly desirable because this reduces the SOC's cost. Hedging with option provides protection against upward price movements while still allowing the SOC to gain from a price fall. But this asymmetry has a cost. A number of technical terms are involved in options. We explain the most important:

(a) If the option gives the right to buy, the option is a "call' option; if it gives the right to sell, it is a "put" option;

(b) The asset on which the option is written is referred to as the "underlying" asset;

(c) The price at which a buyer of the option can buy or sell the underlying asset is called the "strike" or "exercise" price; (d) If the right to buy or sell is exercised by the buyer, the option is "exercised";

(e) The date on (or before) which the buyer can buy or sell the underlying asset is called the "maturity" or" expiration" date;

(f) An option which can be exercised only on the expiration date is called a "European" option; an option which is exercisable on or before the expiration date is an "American" option; and

(g) The price of the option is called a "premium." The buyer pays the premium to the seller at the time of contracting. Oil options are written on futures contracts on oil with a maturity equal to the maturity of the option and are American-style options. Oil call options are the most relevant for an oil-importing SOC.

An oil call option gives the SOC the right, but not the obligation, to buy a futures contract for oil at a predetermined price called the strike price. If the futures price at maturity is higher than the strike price then the call option is valuable and will be exercised (because the SOC can buy at the lower strike price and sell at the higher market price). But if the futures price at maturity is lower than the strike price then the option will expire without value. Purchasing call options allows the SOC to lock in maximum prices for imported oil over the planning horizon (subject to basis risk), at the cost of the premiums on the options.

Compared to futures contracts, options have the following three interesting characteristics. First, futures "lock-in" a price, but options contract limit the maximum loss (equal to the premium paid up-front) and leave an opportunity to take advantage of favorable price movements. Options thus have flexibility in payoff profiles. Second, the buyer has to pay the premium up-front; this often requires a significant amount of cash. Third, while the buyer of options faces the credit risk or default risk of the counter-party, the seller does not. Options are not subject to margin calls. It is the seller who is liable, not the buyer. The SOC would buy call options at the same time as making purchase arrangements. The maturity-date of the option would be as close as possible to the expected date the purchase is settled but not before it. For example, if the SOC contracted in January 1991 to purchase one month ahead, paying the then current oil price, the SOC could have bought a March call option with an exercise price of say $22 dollar/barrel. On January 21, 1991, it would have paid a premium for this option of $1.10. If oil price move above $22/barrel, the gain on the call option would have been offset by the increase in import costs and the net price paid would have been $22/barrel, plus the cost of the premium, or$23.10/barrel. If the oil price fell, the net price paid would be the actual oil price, plus the premium.

Options are available with several exercise prices: the lower the exercise price, the more insurance the SOC obtains, but the higher the premium for the call option. As an example, Figure 1 plots the premium as a function of the exercise price for January 21, 1991 options. As can be observed, option premiums are decreasing in the exercise price (and, in general, increasing in the maturity of the contract.) Which exercise price of the call option will suit the needs of the SOC best will depend on, among others, the following two factors? First, the impact of higher oil prices on the SOC's cash flows and profitability. Through sensitivity derived scenarios the SOC can determine a profile of cash flows and profitability under different oil prices. This will provide the SOC with an Indication of which oil price levels are acceptable in terms of profitability (for which no insurance needs to be obtained) and those price levels which are not acceptable (for which price insurance needs to be obtained).

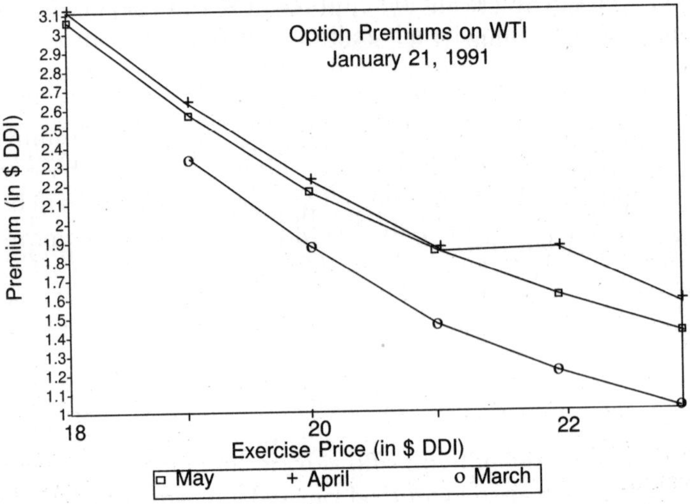

Source of Data: Wall Street Journal

Price level insurance could then be obtained by buying call options with exercise prices equal to the price level at which prices become unacceptable in terms of cash flow and profitability. For example, the SOC could determine that prices above $25/barrel lead to negative cash flows and that call options with exercise prices of $25/barrel need to be bought. The second main factor determining the choice of options will be the SOC's expectations about future oil prices. On this basis SOC may decide that a particular option is attractively priced. It should be realized, however, that prices of options already reflect market sentiment regarding future oil price movements and options are efficiently priced, i.e., provide no opportunity for profitable arbitrage and no consistent rewards for speculators. It would thus be difficult to make profits by buying certain options at certain times and, in general, the SOC's own price forecasts should play a minor role in deciding whether to buy an option or which option to buy since it cannot be expected that the SOC is able to consistently beat the market.

Hedging Long-Term Price Risks

Management of the transaction risks will only reduce oil price risk over a short horizon (one to three months), will mainly protect the SOC, and will not hedge the risks for the entire economy of developing country. Strategy for managing the long-term oil price risks would be to use longer-dated futures contracts. This strategy would involve buying a series of futures contracts with maturities varying from say three to 12 months. The amounts to be bought would reflect the anticipated imports of crude oils in each month. The hedges would be put in place, for example, at the start of each fiscal year, and not be changed during the year, or could be rolled forward during the year and thus cover overlapping fiscal years. *Hedge ratio higher than one means that in excess of one contract of WTI is needed to hedge 1,000 barrels of imports.*

If this strategy is pursued, the short-term, transaction hedges would not be necessary (or, at least, they would have to be modified). We simulate here the following long-term hedging strategies. We assume that in July and January of each year the SOC wants to lock in import prices over a six-month planning horizon. Since trading in the nearby futures contract is not very liquid, we assume that the SOC buys contracts two to seven months ahead. For example, in July of each year the SOC buys contracts for each month in equal amounts, starting from September and up to February (of the next year). As the contracts approach their maturity month, the SOC reverses the contract by selling in the month before expiration an identical contract, e.g., in August of each year the SOC sells the September contract of that year. We assume also that the SOC buys and sells all contracts gradually throughout the month so that it pays or obtains the average price for the contracts in each month. This is to assure comparability to the SOC's cash transactions, which, since we use average cash price data, are also assumed to take place gradually throughout the month. We have 4-$^1/_2$ years of observations on futures prices (starting July 1986 through December 1990) and can therefore simulate eight six-monthly hedges, starting the first in July 1986 and going through January 1990. We use a hedge ratio of one, which was shown above to be a conservative, yet efficient hedge ratio for short-term hedging. Table 2, in the analysis chapter reports the important results from the six-month hedging strategy applied to each of the crude oils. The last row (Futures) reports some statistics on the N.Y. futures contract.

Futures and options on oil are available on several exchanges, including the New York Mercantile Exchange (NYMEX) where they are written on West Texas Intermediate (WTI) (contract terms are listed in Annexure1, certain restrictions apply. However, most SOCs in developing countries do not import WTI crude oil. Using futures or options based on the WTI to hedge other crude oil price risks could expose the SOC to so-called "basis risks" *The possibility that the WTI futures price moves differently than the hedged crude oil price.*

The hedge ratio, the amount of futures or options contracts to buy as a fraction of the amount of physical crude oil bought, could be adjusted to minimize the degree of basis risk. A hedge ratio of one would imply that for each 1,000 barrels of imports one WTI futures contract (equivalent to 1,000 barrels) would be bought. The optimal hedge ratios are calculated in accordance to the following simple regression methodology:

Simple Regression

$\Delta S = S_T - S_t$	change in spot prices over the life of the contract
$\Delta F = F_T - F_t$	change in futures price over the life of the contract
$\sigma_S =$	standard deviation on ΔS
$\sigma_F =$	standard deviation on ΔF

A hedger holds the portfolio P, consisting of quantities Q_s and Q_F of spot and futures with the respective variances of the prices of spot and futures are σ_S^2 and σ_F^2 and covariance in price is σ_{SF}.

Then, $Var(p) = \sigma_S^2 Q_S^2 + \sigma_F^2 Q_F^2 + 2\sigma_{SF} Q_S Q_F$

Let, h be the proportion of the total holdings of the spot commodities that is hedged, so that

$$h = \frac{Q_F}{Q_S}$$

Then, $var\ (p) = Q_S^2 (\sigma_S^2 + h\sigma_F^2 - 2h\sigma_{SF})$

The risk minimizing hedge- ratio will be, if

$$\frac{\delta\ var\ (p)}{\delta h} = 0$$

And, $$\frac{\delta^2\ var\ (p)}{\delta h^2} > 0 \quad \text{(for minimization)}$$

This can be deduce to

$$h = \frac{cov\ (S,F)}{\sigma F^2} = \rho x \left(\frac{\sigma S}{\sigma F}\right)$$

The approach of this methodology is risk-reduction, however the approach should have been risk-return trade-offs.

In this methodology the effectiveness of hedging is defined as 1 minus the ratio of the variance of basis (basis risk) denoted by σ_B^2 to the variance of spot prices denoted by σ_S^2. The value of co-efficient of determination is denoted by ρ^2. It varies between 0 and 1.

Thus,

$$\rho^2 = 1 - \frac{\sigma_B^2}{\sigma_B^2}$$ Is the mathematical representation of hedging efficiency.

The slope coefficient h from this equation is the optimal hedge ratio. Results from estimating hedge ratios for the 15 crudes are given in Table 1. The relevant futures are the two nearby New York crude oil futures contracts based on WTI. Monthly averages of daily futures price data for the period August 1986to December 1990 are used as independent variables. For all crudes, the monthly averages of the cash prices are used as the dependent variables.

Tab1e 1: Hedge Ratios for the Different Crudes

Crude's Marker	H	h_i=0 t- Stat	h_i=1,2 t- Stat	R^2
Lagotreco (Ven)	0.85	10.73	-1.89	0.69
Lagotreco	0.92	9.63	-0.84	0.78
Oriente (Ecu)	1.01	12.24	0.13	0.77
Cano Limon(Col)	1.05	15.03	0.73	0.85
Bachaquero (Ven)	0.64	8.95	4.94	0.60
WTI (USA)	1.03	30.75	1.03	0.97
Dubai (Abu Dubai)	1.00	17.79	0.00	0.86
Brent (U.K.)	1.07	19.13	1.28	0.88
Alaska North Slope (USA)	1.06	18.11	1.03	0.86
Es Sider (Libya)	1.09	14.39	1.18	0.84
Arab Light (Saudi Arabia)	1.01	14.40	0.21	0.80
Attaka (Indo)	1.03	14.60	0.36	0.80
Bonny Light *(Nig)*	1.08	16.10	1.17	0.83
Sahara (Algeria)	1.12	14.48	1.55	0.84
Khafji (Neutral Zone)	0.97	13.22	-0.41	0.74

Signifies the Hedge Ratio

(i) This column shows the t-stat for this coefficient h_i in the equation s = $h_{i.}$ $F_{i\,+}$ $U_{i.}$ The hedge ratio is equal to 1. The column shows the t-statistics from testing the hypothesis h_i = 0. A high t-statistics implies that the hypothesis h_i = k (k = 0, 1) may be rejected, a low statistics indicates that this hypothesis may be accepted.

(ii) Regression results for the sub-period August 1998 to December 1990. Table 1 indicates that, with the exception Bachaquero, all 15 crudes can be hedged effectively using the WTI futures contract. The high R^2 from the regressions indicates that basis risk is small (between 3% and 23%). Excluding (which is expected to have extremely low basis risk since the N.Y. contract is written on it) most of the remaining crudes have basis risk between 15% and 20%. The closest relationships with the N.Y. futures contract are exhibited by Dubai, Brent, Alaska North Slope, Istmo, Cano Limon, Es Sider, Bonny Light and Saharan (basis risk between 12% and 17%). The worst performers are Bachaquero, Khafji, and Lagotreco (for the period January 1986 to December 1990). However, in the recent period (after July 1988), the Lagotreco price has followed the N.Y.futures contract price more closely, and basis risk has declined from 31% to 22%.Bachaquero and Khafji are so-called heavy crudes, with smaller price variability, since they are relatively more price-inelastic. The WTI contract cannot be expected to hedge the price risk of these two heavy crudes well. Khafji performs better in terms of basis risk, probably due to the fact that it is not as heavy as Bachaquero (API gravity of 250 versus 17).

We examined also what happens to basis risks if instead of buying the contract with an expiration date nearest to delivery, the SOC buys a contract with a later expiration (second to fourth contract from delivery). The results indicate that this would lead to only a small increase in basis risk [an average of 1.1%, with 4% being the highest (Lagotreco)]. Again the hedge ratio is not significantly different from one. The hypothesis that the hedge ratio is equal to one is accepted for all crudes, except Bachaquero, as indicated by the figures in the third column of Table 1, Using a hedge ratio of one, rather than the estimated ratio, basis risk increased at most by 2% and there was no increase in basis risk for Dubai, Oriente and Arab Light. For most crude, basis risks increased by only 1 to 1.5%. The hedge ratios derived so far refer to hedging the physical (barrels of oil). When hedging dollar expenditures, hedge ratios can be obtained by estimating the h based on spot and future prices, expressed as percentage differences rather than as absolute dollar amounts. This was done for all 15 crudes and the results are very similar to those obtained for hedging the physical. The hedge ratios are found to be not significantly different from one but there was an increase in basis risk compared to hedging the physical.

In summary, therefore, the results indicate that the SOC can reduce its transaction exposure by between 74% and 86% for all light and Khafji crudes and that a hedge ratio of one can be used for WTI options as well as for futures contracts. Only in the case of Bachaquero crude did we find a weak relationship with the WTI futures price and thus it would be inadvisable to use WTI for hedging Bachaquero crude oil imports.

Table 2: Long- term Hedging Strategies

($/bbl)

PRICES	CASH	STD CASH	NET	STD NET	% REDUCT
AVERAGE	-0.357	1.92	0.743	0.60	68.75
DUBAI	-0.361	1.91	0.792	0.46	75.92
UK BRENT	-0.372	1.83	0.708	0.60	67.21
ALASKA NS	-0.365	2.04	0.698	0.83	59.31
TIA JUANA	-0.858	2.62	0.022	0.95	63.74
SAHARAN	-0.149	1.71	0.914	0.35	79.53
KHAFJI	-0.255	2.06	0.769	0.57	72.33
BONNY	-0.238	1.79	0.853	0.33	81.56
ATTAKA	-0.304	2.04	0.749	0.70	65.69
ARAB	-0.246	2.15	0.674	0.71	66.98
ES SIDER	-0.256	1.80	0.825	0.35	80.56
ISTNO	-0.032	2.34	0.717	0.72	69.23
LAGOTREC	0.167	2.32	0.916	1.50	35.34
BACHAQ	-0.034	2.39	0.714	1.09	54.39
ORIENTE	0.074	2.58	0.822	0.89	65.50
CANO LIM	0.043	2.53	00.792	0.74	70.75
WTI	-0.050	2.31	00.699	00.71	69.26
FUTURES			0.884	1.99	

In Table 2, the column 'Cash' reports the average loss or gain (all figures in dollar per barrel) the SOC incurred over the eight planning periods as a result of spot oil price movements during the six-monthly period. For example, the entry -0.032 for the Istmo strategy reflects an average loss per barrel of3.2 cents from buying Istmo at higher prices during the eight six-monthly periods. It shows that on average the SOC did not pay much higher prices over the planning period when compared to the price in the month when the plan was started. The column "STD CASH" reports the standard deviation of the unexpected higher or lower cash payments and is thus a reflection of the uncertainty in the spot market. For Istmo, for example, the standard deviation of spot prices was $2.33/bbl. This indicates that the SOC experienced substantial uncertainty over the planning periods. The column "NET" represent the net results of hedging. It is thus equal to the gain or loss on the cash positions minus the gain or loss on the futures positions. In the case of Istmo, the hedging strategy led to an effective price paid for oil which was 71.7 cents per barrel lower than average spot prices for the planning periods. The column "STD NET" indicates the riskiness of the net result: the standard deviation of the net gain or loss. The column "% REDUCT" indicates the percentage risk reduction achieved through using the futures. In the case of Istmo again, the risk reduction as a result of hedging is 69%; in other words, the SOC is able to lay off 69% of the risk of unexpected price movements during the six-month planning horizons using the long-term futures hedges. Finally, the column "MONTH" indicates the standard deviation of the monthly net gains and losses from unwinding the futures contracts and buying the crudes imports. This standard deviation is different from the standard deviation "STD NET" since this is a standard deviation of (48) monthly figures and the "STD NET" is a standard deviation of only eight averages. Table 2 indicates that, on average, the six-month hedging strategy achieved a risk reduction of about 69% in the case of Istmo, while also leading to a lower purchase price for oil by about US$0.72 per barrel. The gains in terms of lower import prices could, of course, have been losses if a different period was chosen, but for any period the six-month hedging strategy would have led to a substantial decline in oil price volatility. For most crude oils, the risk reduction was between 65% and 80%. The best performers were the same as in the short-term (transactions) hedge, which is to be expected given the fact that these crudes follow the WTI, contract closer by than the others. The results indicate that long-term hedging of Lagotreco and Bachaquero (based on the N.Y. futures contract) could be problematical. However, as reported above, Lagotreco has followed the N.Y. futures contract more closely since August 1988 and thus hedging Lagotreco in this way may be possible now. Other risk management techniques exist to hedge long-term risk. The most Applicable to a SOC is a commodity swap that allows effective long-term price insurance. A commodity swap is a series of commodity forward contracts however; it does not involve deliveries of physical commodities since transactions are made as pure financial transactions. Assume that a SOC wants to 'lock-in" the price of oil imports for the next five years, and that it expects to import 2.5 million barrels a year. The SOC could arrange with a commercial bank the following commodity swap agreement.

Amount: The U.S. dollar equivalent of 2.5 million barrels of oil a year.

Fixed Price Payer: SOC.

Floating Price Payer: Commercial bank.

Maturity: 5 years, with annual, semi-annual or monthly payments.

Fixed Price: US$ 20.00 per barrel (indicative).

Floating Price: The average daily closing spot price of WTI oil over the year, semester, or month preceding each payment date.

Settlement: Netting-out.

"Netting out" means that a settlement amount is calculated by multiplying the contracted amount by the difference between the fixed and floating price. The floating price payer pays the calculated amount if the floating price is higher than the fixed price, and vice-versa. For details see Masuoka (1990)

In this example, the SOC would buy oil from any third party from time to time at the spot price. However, the SOC's expenses for the next 5 years would effectively be fixed by means of the oil swap at a price of oil of US$20.00/barrel since the floating rates payments from the commercial bank will offset the variable costs of import payments.

For creditworthiness reasons, the SOC may have to put up some "collateral" to assure lenders of performance under the terms of the swap agreement. One way to do this is by using a "marked-to-market" swap which would thus operate similarly to a futures contract. Other methods are also possible such as escrowing some foreign exchange at a foreign bank. An important point is that the oil swap reduces the SOC's probability of financial distress.

Even if the price of oil increases to, say, $40.00/barrel, expenses will stay the same. This improved credit risk may lower the cost of financing SOC's working capital or may give the SOC access to new lenders.

Swaps do not necessarily refer to long-term coverage but can also cover shorter periods such as three or six months. A SOC may prefer swaps for shorter term hedging over futures/option if liquidity for margin calls or option premiums becomes a problem. Also, futures and options can be very labor intensive, requiring frequent monitoring of the market. Swaps require little monitoring once they are in place. Swaps for Shorter periods are now commonly available in the over-the-counter market.

Finally, we have examine the applicability of risk management techniques to a typical SOC importing crude oil and to provide estimates of the gains if such techniques were implemented. We examined the relationship between the different crude oils, which are imported by developing countries, and the

N.Y futures contract. We found a significant relationship between all the crude studied and the N.Y. futures contract, with the exception of Bachaquero crude. The basis risk in using the N.Y. futures contract to hedge this crude was found to be in the range of 3-26%, and the hedge ratio was found to be not significantly different from one. The results from hedging dollar oil expenditures instead of barrels (physical hedge) are quite similar. In summary, we recommend the use of N.Y. futures contracts to hedge crude oil imports (with the exception of Bachaquero crude), particularly the lighter crude, with a hedge ratio of one for all crude. This recommendation holds for either future or options based on the N.Y. futures contract

We also identified the risk exposures that a typical SOC faces. These are transactions risk, and long-term risk. Transactions risk refers to the risk from price fluctuations between ordering and shipping. Long-term risk refers to the exposure of the SOC and the national economy to longer-term trends in oil prices. Here longer term was taken to be a six-month planning horizon. Within such a period the SOC and the economy should be able to adjust smoothly to any shift in international oil prices. We recommend the use of futures and options to hedge both transactions and long-term risks. Table 4 summarizes our recommendations and major findings. If a six-month (approximate) hedge is undertaken, there is no need for additional hedging activity to cover the transactions risk.

Table 3

Length of potential	Instrument	Potential	Risk Reduction
Transaction risks	Approx. one month	Futures	72% - 85%
Long-term risk	Six months	Futures/options	64% - 81%
Long-term risk	More than one	Commodity swaps	n.a.

In studying hedging of transactions risk we simulated a scenario where one-month exposures were hedged using the second nearby N.Y. futures contract. The potential gains from pursuing such a strategy are between 73% and 86% risk reduction for all crude except Bachaquero. The hedging potential for Bachaquero is more limited; only about 60% risk reduction can be achieved. For long-term risk we simulated a scenario in which the SOC buys at six monthly intervals the six nearby futures contracts, which it sells gradually throughout the following six-month period. We choose a six-month horizon since these futures are actively traded. The potential risk reduction gains from pursuing such a strategy are on the average about 70%. However, much less risk reduction can be achieved hedging Lagotreco and Bachaquero crude imports in this way (only about 35% to 54%). For hedging over longer periods we recommend oil swap as the best risk management tool. We have also identified four constraints to the use of risk management techniques by SOCs. These are:

(i) *Cash Flow*: the need for foreign funds to purchase futures, margin calls and option premiums;

(ii) *Negative Publicity*: the possibility of criticism which could arise from the use futures. To counter this, options can be used, but at a price. Furthermore, it is advisable to distribute the responsibility for hedging across the government and spread knowledge about hedging and hedging instruments through education;

(iii) *Flexibility of Action*: where applicable, there is need for a close relationship between the SOC and the Central Bank to assure foreign exchange availability to carry-out the risk management program; and (iv) Legal System: where applicable, the legal restrictions that prevent SOCs from using futures and options and other financial instruments should be removed.

CRUDE OIL: OPTIONS CONTRACT

Trading Unit: One NYMEX crude oil futures contract.

Tradng Hours: 9:45 a.m. – 3:10 p.m. (New York Time)

Trading Months: Fifteen consecutive months.

Minimum Price; Fluctuation: $.01 (1 cent) per barrel ($10 per option contract).

Maximum Daily Limit: None

Last Trading Day: (Expiration Day): Second Friday of the month prior to the delivery month of the underlying crude oil futures contract, provided there are at least five days remaining to trade in the underlying futures contract.

Exercise: By a Clearing Member to the NYMEX Clearing House not later than 6:00 p.m., or 45minutes after the price of the option being exercised or the underlying futures settlement price is posted. Whichever is later, on any day up to and including the option's expiration.

Strike Prices: Strike prices are in increments of $1.00 per barrel. At all times at least eleven strike prices are available for puts and calls on the underlying futures contracts. The middle strike price is closest to the previous day's close of the underlying futures contract. Strike price boundaries are adjusted according to the future price movements.

CHAPTER 5

Upstrems Oil Contracts

5.1 PRODUCTION SHARING CONTRACTS

There are various mechanisms and systems in the field of upstream oil business of the world on which host government award right to explore and exploit petroleum resources. For exploration and exploitation of natural resources, India has adopted a mixture of concessionary and production sharing regimes which has been adopted under a Model Production Sharing Contract..

Development of Production Sharing Concept

In 1960, Law No. 44 was passed in Indonesia providing that Oil companies will be known as "contractors" rather then "concessionaires". The Western Oil Companies were very upset and much worried because of its possible impact on their Middle East concessions. The companies working in Indonesia refused to accept this Law. Negotiations failed and Government of Indonesia declared that in case agreements are not reached, the companies will be nationalized. In this backdrop and after the intervention of USA, an agreement was reached through which these companies were allowed to carry on the operations under a "Contract of Work" arrangement. The sharing of production and control under a production sharing arrangement was not agreed to by major oil companies operating in Indonesia but when some Japanese and American Independents entered in to such contracts, others followed suit. So, after 1970, the things were concretized and a new era of joint ventures of oil companies with the host government started. With the passage of time these Contract of Work graduated in to Production Sharing Contracts.

In India, PSC incorporates the charging of Petroleum Exploration License fees and royalty as well as sharing of profit oil left after cost recovery. Thus, it a hybrid of both the concessionary and production sharing regime. The

Indian PSC makes following declarations as to its legal position in the Recital clause, which put PSC in proper legal perspective:

For offshore areas Recital in the Contract provides as follows:

By virtue of article 297 of the Constitution of India, Petroleum in its natural state in the Territorial Waters and the Continental Shelf of India is vested in the Union of India. The Oil Fields (Regulation and Development) Act, 1948 (53 of 1948) and the Petroleum and Natural Gas Rules, 1959, made there under (hereinafter referred to as "the Rules") make provisions, inter alia, for the regulation of Petroleum Operations and grant of licenses and leases for exploration, development and production of Petroleum in India.

The Territorial Waters, Continental Shelf, Exclusive Economic Zone and other Maritime Zones Act, 1976 (No. 80 of 1976) provides for the grant of a license or letter of authority by the Government to explore and exploit the resources of the continental shelf and exclusive economic zone and any Petroleum Operation under this Contract shall be carried out under a license or letter of authority granted by the Central Government;

The above Acts and Rules provide for the grant of exploration license/ letter of authority and mining lease in respect of any land or mineral underlying the ocean, within the territorial waters, the continental shelf and exclusive economic zone of India by the Central Government. Rule 5 of the Rules provides for an agreement between the Government and the Licensee or Lessee containing additional terms and conditions with respect to the license or lease;

Clause 12 of the Oilfield Regulation and Development Act provides that Government in certain cases can award a license on terms different from those given in Petroleum and Natural Gas Rules. The above statement has been incorporated to reflect this aspect of the matter.

For Onshore areas following has been provided in the Recital of PSC

The Oil Fields (Regulation and Development) Act, 1948 (53 of 1948) (hereinafter referred to as "the Act") and the Petroleum and Natural Gas Rules, 1959, made hereunder (hereinafter referred to as "the Rules") make provision, inter alia, for the regulation of Petroleum Operations and grant of licenses and leases for exploration, development and production of Petroleum in India;

The Rules provide for the grant of exploration licenses and mining leases in respect of land vested in a State Government by that State Government with the prior approval of the Central Government. Rule 5 of the Rules provides for an agreement between the Central Government and the Licensee or Lessee containing additional terms and conditions with respect to the license or lease. The provisions of PSC are such additional terms.

Initially it started with a concession arrangement but changed world scenario and big oil discoveries after World War II, forced a rapid change in the policies and practices followed and joint venture type of relationship started taking shape wherein oil companies jointly operate petroleum discovery and develop it at its risk and cost with the right of pay back from the production. This was the initial form and starting point of production sharing arrangement.

The clauses of PSC deal with technical, administrative, financial and legal aspects. There are also clauses, which manage uncertainties, as duration is 20 years, like force- majuere, assignment, guarantees, termination etc.

Management Committee

Article 6 of the PSC deals with the issue of Management Committee as follows:

There shall be constituted a committee to be called the Management Committee. Government shall nominate two (2) members representing Government in the Management Committee, whereas each Company constituting Contractor shall nominate one member each to represent Company in the Management Committee. Each Party may nominate alternate members with full authority to act in the absence and on behalf of the members nominated under Article 6.2 and may, at any time, nominate another member or alternate member to replace any member nominated earlier by notice to other members of the Management Committee.

One representative of the Government is designated as the Chairman of the Management Committee and the second representative of the Government shall be designated as the Deputy Chairman. The member of the Operator in the Management Committee shall be designated as the Secretary of the Committee.

Operator on behalf of Contractor with the approval of Operating Committee shall submit important matters to the Management Committee. But during exploration operations Management Committee has review and advisory functions only. During Development and production operations, it has approving functions as well.

The Management Committee is required to meet at least once every six (6) months during the Exploration Period and thereafter if exploration results in commercial discovery and development and production operations are carried on then, at least once every three (3) months or more frequently at the request of any member.

The Chairman or Deputy Chairman, as may be the case, shall preside over the meetings of the Management Committee and, in their absence, any other member representing Government and present shall preside over the meetings. The Chairman shall appoint the member nominated by the Operator as Secretary to the Management Committee with responsibility, inter alia, for preparation of the minutes of every meeting in the English language and provision to every member of the Management Committee with two copies of the minutes approved by the Chairman not later than fourteen (14) days after the date of the meeting.

Unanimous Vote

The important matters with few exceptions, requiring the approval of the

Management Committee, are generally approved by a unanimous vote of the members of the Management Committee present as well as the views of the members received by some other mode of communication.

The Management Committee, if it considers necessary, may appoint legal, financial or technical subcommittees.

General Rights and Obligations of the Parties

A Contractor has right to carry out Petroleum Operations to recover costs and expenses as provided in this Contract. The right shall exclude exploitation of coal/lignite bed methane (CBM) by the Contractor in the Contract Area; the right to use, free of charge, such quantities of Petroleum produced as are reasonably required for conducting Petroleum Operations in the Contract Area in accordance with generally accepted Good International Petroleum Industry Practices; the right to lay pipelines, build roads, construct bridges, ferries, aerodromes, landing fields, radio telephones and related communication and infrastructure facilities and exercise other ancillary rights as may be reasonably necessary for the conduct of Petroleum Operations subject to such approvals as may be required and the applicable laws in force from time to time for the regulation and control thereof;

The Contractor will use all available technical data, seismic and well information, maps, samples etc. of the Contract Area as on the Effective Date, free of charge, subject to nominal copying/reproduction costs for further Petroleum Operations. The Contractor shall submit the list of data required by them to Directorate General of Hydrocarbons (DGH) and the same, if available and reproducible, shall be made available to the Contractor within 120 days of the Effective Date in the office of DGH. In addition to these Contractor shall have such other rights as are specified in the Contract.

Position of Operator

Operator carries out all the operations on behalf of the Contractor. In the preparation and implementation of Work Program and in the conduct of Petroleum Operations, Good International Petroleum Industry Practices have to be followed with such degree of diligence and prudence reasonably and ordinarily exercised by experienced parties engaged in a similar activity under similar circumstances and conditions and he shall always be mindful of the rights and interests of India in the conduct of Petroleum Operations etc.

The Contractor is to conduct all Petroleum Operations at its sole risk, cost and expense and is required to provide all funds necessary for the conduct of Petroleum Operations including funds for the purchase or lease of equipment, materials or supplies required for Petroleum Operations as well as for making payments to employees, agents and Subcontractors. He is to conduct all Petroleum Operations are to be conducted diligently, expeditiously, efficiently and in a safe and workmanlike manner pursuant to the Work Program formulated in accordance with Contract.

Petroleum Mining Lease

For offshore area

On submission of a plan for development of a Commercial Discovery, the Contractor submits an application to Government for a Lease in respect of the proposed Development Area. Where a part of a Reservoir in respect of which a Commercial Discovery has been declared extends beyond the Contract Area, such area may be included in the proposed Development Area, in relation to which application for a Lease is made, on terms and conditions as decided by the Government; provided that such area is:

(a) not subject to a license or lease granted to any other person;
(b) not the subject of negotiations/bidding for a license or lease;
(c) available for licensing (i.e. is not an area over which Petroleum Operations are excluded).

Where a Development Plan has been approved and the Contractor has complied with the terms and conditions of the License and the Contract and is not in breach, or violated provisions of any law, the Government grants to the Contractor a Lease over the Development Area, to enable the Contractor to carry out Petroleum Operations in the Development Area in accordance with the Development Plan.

The Lease is granted for an initial period of twenty (20) years from the date of grant. For onshore area some procedure is followed but application for lease is made to the concerned State Government

Protection of the Environment

In the performance of the Contract, the Contractor is to conduct its Petroleum Operations with due regard to concerns with respect to protection of the environment and conservation of natural resources. (This aspect has been dealt within greater detail in the chapter on Environment)

Insurance and Indemnification

As per Article 24 of the Contract, the Contractor shall, during the term of the Contract, maintain and obtain insurance coverage for and in relation to Petroleum Operations for such amounts and against such risks as are customarily or prudently insured in the international petroleum industry in accordance with Good International Petroleum Industry Practices, and shall furnish to the Government certificates evidencing that such coverage is in effect. Such insurance policies shall include the Government as additional insured and shall waive subrogation against the Government. A list of these insurance's has been given in the PSC.

Assignment of Interest (Art. 28)

Certain conditions have been prescribed for assignment of interest by a party to another party to the Contract or to a third parties.

Any Party comprising the Contractor may assign, or transfer, a part or all of its Participating Interest, with the prior written consent of the Government, which consent shall not be unreasonably withheld, provided that the Government is satisfied that:

(a) the prospective assignee or transferee is of good standing, has the capacity and ability to meet its obligations hereunder, and is willing to provide an unconditional undertaking to the Government to assume its Participating Interest share of obligations and to provide guarantees in respect thereof as provided in the Contract;
(b) the prospective assignee or transferee is not a company incorporated in a country with which the Government, for policy reasons, has restricted trade or business;
(c) the prospective assignor or transferor and assignee or transferee respectively are willing to comply with any reasonable conditions of the Government as may be necessary in the circumstances with a view to ensuring performance under the Contract; and
(d) the assignment or transfer will not adversely affect the performance or obligations under this Contract or be contrary to the interests of India.

In case of any material change in the status of a party or its shareholding or the relationship with any guarantor, such party is to seek the consent of the Government for assigning the participating interest under the changed circumstances.

An assignment or transfer shall not be made where the Participating Interest to be retained by the proposed assignor or the percentage interest of assignee shall be less than ten per cent (10%) of the total Participating Interest of all the constituents of the Contractor, except where the Government, on the recommendations of the Management Committee may, in special circumstances, so permit.

Creation of Charge for Raising Finance

A Party comprising Contractor can mortgage, pledge, charge or otherwise encumber at its own risk and cost all or part of its Participating Interest for the purposes of security relating to finance to the extent required for performing its obligation under the Contract.

The financing arrangement are subject to the rights of Government under the Contract and the pre-emptive rights of the Parties as contained in Operating Agreement. Any Party that wishes to exercise the said pre-emptive rights will explicitly assume the obligation on the same terms and conditions as the Borrower.

Guarantees (Art. 29)

Each of the Companies constituting the Contractor shall procure and deliver to the Government on the Effective Date of the Contract:

(a) An irrevocable, unconditional bank guarantee from a reputable bank of good standing in India, acceptable to the Government, in favor of the Government, for the amount specified hereinafter

(b) Performance guarantee, in favor of the Government from a parent company of good financial standing acceptable to the Government, or, where there is no such parent company from an insurance company or financial institution of good financial standing acceptable to the Government.

(c) A legal opinion from its legal advisors, in a form satisfactory to the Government, to the effect that the aforesaid guarantees have been duly signed and delivered on behalf of the guarantors with due authority and is legally valid and enforceable and binding upon them.

(d) Public sector enterprises, as defined in the Companies Act and the Companies having a net worth of US$ one (1) billion or more (as per the latest audited account) shall not be required to furnish bank guarantee towards its minimum work obligations

The amount of the bank guarantee shall be an amount equal to a Company's Participating Interest share and shall be thirty five percent (35%) of the total estimated expenditure in respect of the Work Program to be undertaken by the Contractor in the Contract Area during each Contract Year of a Phase.

The bank guarantee shall provide that at the end of each Contract Year it shall be automatically renewed for an amount equal to a Company's Participating Interest share of 35% per cent of the total estimated expenditure in respect of the Work Program to be undertaken for the following Contract Year of an Exploration Phase, unless the Contractor has terminated the Contract in accordance with the terms thereof. The guarantee shall be renewed at the end of each Contract Year positively thirty (30) days before the expiry of the guarantee period.

Term and Termination of Contract (Art. 30)

The term of Contract is equal or co-terminus to the period of the License and any Lease granted there under, unless the Contract is terminated earlier in accordance with its terms, and shall be deemed to have been terminated, if for any reason, the Contractor ceases to hold such License or Lease.

Subject to the provision of Contract, the Contractor and in certain circumstances Government has the right to terminate the Contract.

Force-Majeure (Art. 31)

Any non-performance or delay in performance by any Party of any of its obligations under this Contract, or in fulfilling any condition of any License or Lease granted to such Party, or in meeting any requirement of the Act, the Rules or any License or Lease, shall, except for the payment of monies due under this Contract or under the Act and the Rules or any law, be excused if, and to the extent that, such non-performance or delay in performance under this Contract is caused by Force-majeure as defined in this Article.

The term Force-majeure has been defined as meaning any cause or event, other than the unavailability of funds, whether similar to or different from those enumerated herein, lying beyond the reasonable control of, and unanticipated or unforeseeable by, and not brought about at the instance of, the Party claiming to be affected by such event, or which, if anticipated or foreseeable, could not be avoided or provided for, and which has caused the non-performance or delay in performance. Without limitation to the generality of the foregoing, the term Force majcurc shall include natural phenomena or calamities, earthquakes, typhoons, fires, wars declared or undeclared, hostilities, invasions, blockades, riots, strikes, insurrection and civil disturbances but shall not include the unavailability of funds.

If an event of Force-majeure occurs and is likely to continue for a period in excess of thirty days, the Parties shall meet to discuss the consequences of the Force-majeure and the course of action to be taken to mitigate the effects thereof or to be adopted in the circumstances.

This clause has definition of force-majeure, notice, burden of proof, extension of time and lastly resolution of course of action if force-majeure continues for a specific period.

Applicable Law and Language of the Contract (Art. 32)

The Contract is governed and interpreted in accordance with the laws of India and nothing in this Contract shall entitle the Contractor to exercise the rights, privileges and powers conferred upon it by this Contract in a manner which will contravene the laws of India. The English language shall be the language of this Contract and shall be used in arbitral proceedings. All communications, hearing or visual materials or documents relating to this Contract shall be written or prepared in English. The laws will also include amendments, revisions, modifications etc.

Sole expert, Conciliation and Arbitration (Art. 33)

The Parties shall use their best efforts to settle amicably all disputes, differences or claims arising out of or in connection with any of the terms and conditions of this Contract or concerning the interpretation or performance thereof.

The Parties may refer any controversy difference, disagreement or claim for damages, compensation or otherwise (hereinafter in this Clause referred to as a "dispute") arising between the Parties, which cannot be settled amicably within ninety (90) days after the dispute arises, to sole expert or to conciliation or submit for decision by an arbitral tribunal.

The arbitral tribunal shall consist of three arbitrators. Each Party to the dispute shall appoint one arbitrator and the Party or Parties shall so advise the other Parties. The two arbitrators appointed by the Parties shall appoint the third arbitrator.

Any Party may, after appointing an arbitrator, request the other Party (ies) in writing to appoint the second arbitrator. If such other Party fails to appoint an arbitrator within thirty (30) days of receipt of the written request to do so, such arbitrator may, at the request of the first Party, be appointed by the Chief

Justice of India or by a person authorized by him or within thirty (30) days of the date of receipt of such request, from amongst persons who are not nationals of the country of any of the Parties to the arbitration proceedings.

If the two arbitrators appointed by on behalf of the Parties fail to agree on the appointment of the third arbitrator within thirty (30) days of the appointment of the second arbitrator and if the Parties do not otherwise agree, at the request of either Party, the third arbitrator shall be appointed in accordance with Conciliation and Arbitration Act, 1996.

If any of the arbitrators fails or is unable to act, his successor shall be appointed by the Party or person who originally appointed such in the manner set out in this Article as if he was the first appointment.

The decision of the arbitral tribunal, and, in the case of difference among the arbitrators, the decision of the majority, shall be final and binding upon the Parties.

The arbitration agreement shall be governed by the Arbitration and Conciliation Act, 1996 (Arbitration Act). Arbitration proceedings shall be conducted in accordance with the rules for arbitration provided in Arbitration Act and the United Nations Commission on International Trade Law (UNCITRAL) rules may apply to the extent where corresponding rules are not provided in the Act. In case of any conflict between the Arbitration Act and the provisions of this Article, the provisions of the Arbitration Act shall prevail.

The right to arbitrate disputes under this Contract shall survive expiry or the termination of this Contract.

The venue of the sole expert, conciliation or arbitration proceedings has been kept in India unless otherwise agreed between the parties. Insofar as practicable, the Parties shall continue to implement the terms of the Contract notwithstanding the initiation of arbitral proceedings before a sole expert, conciliator or arbitral tribunal and any pending claim or dispute.

Change of Status or Share holding of Companies

The Parties comprising the Contractor shall have to notify the Government of any material change in their status, shareholding or relationship of that of any guarantor of the Companies, in particular, where such change would impact on performance of obligations under this Contract.

This clause has been included so that in case of change in the parent company Government is kept informed.

Certificates

A Company shall furnish, prior to execution of this Contract, a duly authorized copy of a resolution properly and legally passed by the Board of Directors of the Company authorizing its President or any Vice-President or any other representative to execute this Contract along with a certificate duly signed by the Secretary or an Assistant Secretary of the Company under its seal in this regard and to the effect that the Company has the power and authority to enter into this Contract and to perform its obligations thereunder and has taken all necessary action to authorize the execution, delivery and performance of the Contract

Accounting Procedure to Production Sharing Contract

Generally, the purpose of the Accounting Procedure is to set out principles and procedures of accounting which will enable the Government of India to monitor effectively the Contractor's costs, expenditures, production and income so that the Government's entitlement to Profit Petroleum can be accurately determined pursuant to the terms of the Contract. More specifically, the purpose of the Accounting Procedure is to classify costs and expenditures and to define which costs and expenditures shall be allowable for cost recovery and profit sharing and participation purposes:

- specify the manner in which the Contractor's accounts shall be prepared and approved; and
- address numerous other related accounting matters.

The Accounting Procedure is intended to apply to the provisions of the Contract and is without prejudice to the computation of income tax under applicable provisions of the Income-Tax Act, 1961, as amended.

Procedure for Acquisition of Goods and Services

The objectives of the procedures are to ensure the goods and services acquired by the Operator for the carrying out of the Petroleum Operations are acquired at the optimum cost on competitive rates taking into consideration all relevant factors including price, quality, delivery times and the reliability of potential suppliers. It also ensures that goods and services are delivered in a timely manner taking into consideration the consequences of delays in the acquisition of these goods and services on the project as a whole.

NELP-VI
MODEL PRODUCTION SHARING CONTRACT
SIXTH OFFER OF BLOCKS
MINISTRY OF PETROLEUM & NATURAL GAS
GOVERNMENT OF INDIA
2006
MODEL PRODUCTION SHARING CONTRACT
FOR OFFSHORE AREAS

This Contract made this _____ day of _____ between:

1. The President of India, acting through the ____________, Ministry of Petroleum and Natural Gas (hereinafter referred to as "the Government") of the FIRST PART;

AND

XYZ Company, a Company incorporated under the laws of ____ (hereinafter referred to

as "XYZ") having its registered office at ____________________ _________which expression shall include its successors and such assigns as are permitted under Article 28 hereof, of the SECOND PART;

WITNESSETH:

WHEREAS

(1) By virtue of article 297 of the Constitution of India, Petroleum in its natural state in the territorial waters and the continental shelf of India is vested in the Union of India;

(2) The Oilfields (Regulation and Development) Act, 1948 (53 of 1948) (hereinafter referred to as "the Act") and the Petroleum and Natural Gas Rules, 1959, made there under (hereinafter referred to as "the Rules") make provisions, inter alia, for the regulation of Petroleum Operations and grant of Licenses and Leases for exploration, development and production of Petroleum in India;

(3) The Territorial Waters, Continental Shelf, Exclusive Economic Zone and other Maritime Zones Act, 1976 (80 of 1976) provides for the grant of a license by the Government to explore and exploit the resources of the continental shelf and exclusive economic zone and any Petroleum Operation under this Contract shall be carried out under a license granted by the Central Government;

(4) The above Acts and Rules provide for the grant of License and Lease in respect of any land or mineral underlying the ocean, within the territorial waters, the continental shelf and exclusive economic zone of India by the Central Government;

(5) Rule 5 of the Rules provides for an agreement between the Government and the Licensee or Lessee containing additional terms and conditions with respect to the License or Lease;

(6) The Government desires that the Petroleum resources which may exist in the territorial waters, the continental shelf, and the exclusive economic zone of India be discovered and exploited with the utmost expedition in the overall interest of India and in accordance with modern oilfield and petroleum industry practices;

(7) XYZ Company(ies) has/have committed that it has/they have, or will acquire and make available, the necessary financial and technical resources and the technical and industrial competence and experience necessary for proper discharge and/or performance of all obligations required to be performed under this Contract in accordance with modern oilfield and petroleum industry practices and will provide guarantees as required in Article 29 for the due performance of its obligations hereunder; and

(8) As a result of discussions between representatives of the Government and XYZ Company(ies) on the proposal of XYZ Company(ies), the Government has agreed to enter into this Contract with XYZ Company(ies) with respect to the Contract Area identified as block __________ and detailed in Appendix A and Appendix B, on the terms and conditions herein set forth.

NOW, THEREFORE, in consideration of the premises and covenants and conditions herein contained, IT IS HEREBY AGREED between the Parties as follows:

FOR ONLAND AREAS

This Contract made this _____ day of ______between:

1. The President of India, acting through the __________, Ministry of Petroleum and Natural Gas (hereinafter referred to as "the Government") of the FIRST PART;

AND

XYZ Company, a Company incorporated under the laws of _____ (hereinafter referred to as "XYZ") having its registered office at ______________ ________________ which expression shall include its successors and such assigns as are permitted under Article 28 hereof, of the SECOND PART;

WITNESSETH:

WHEREAS

(1) The Oilfields (Regulation and Development) Act, 1948 (53 of 1948) (hereinafter referred to as "the Act") and the Petroleum and Natural Gas Rules, 1959, made there under (hereinafter referred to as "the Rules") make provision,

inter alia, for the regulation of Petroleum Operations and grant of Licenses and Leases for exploration, development and production of Petroleum in India;

(2) The Rules provide for the grant of Licenses and Leases in respect of land vested in a State Government by that State Government with the prior approval of the Central Government;

(3) Rule 5 of the Rules provides for an agreement between the Central Government and the Licensee or Lessee containing additional terms and conditions with respect to the License or Lease;

(4) The Government desires that the Petroleum resources which may exist in India be discovered and exploited with the utmost expedition in the overall interest of India in accordance with modern oilfield and petroleum industry practices;

(5) XYZ Company(ies) has/have committed that it has/they have, or will acquire and make available, the necessary financial and technical resources and the technical and industrial competence and experience necessary for proper discharge and/or performance of all obligations required to be performed under this Contract in accordance with modern oilfield and petroleum industry practices and will provide guarantees as required in Article 29 for the due performance of its obligations hereunder; and

(6) As a result of discussions between representatives of the Government and XYZ Company(ies) on the proposal of XYZ Company(ies), the Government has agreed to enter into this Contract with XYZ Company(ies) with respect to the Contract Area identified as block __________ and detailed in Appendix A and Appendix B, on the terms and conditions herein set forth.

NOW, THEREFORE, in consideration of the premises and covenants and conditions herein contained, IT IS HEREBY AGREED between the Parties as follows:

ARTICLE 1
DEFINITIONS

In this Contract, unless the context requires otherwise, the following terms shall have the meaning ascribed to them hereunder:

1.1 "Accounting Procedure" means the principles and procedures of accounting set out in Appendix C.

1.2 "Act" means Oilfields (Regulation and Development) Act, 1948 as amended from time to time.

1.3 "Affiliate" means a company or a body;

(a) which directly or indirectly controls or is controlled by a Company which is a Party to this Contract; or

(b) which directly or indirectly controls or is controlled by a company which directly or indirectly controls or is controlled by a Company which is a Party to this Contract.

For the purpose of this definition it is understood that "control" means:

(i) ownership by one company of more than fifty percent(50%) of the voting securities of the other company; or

(ii) the power to direct, administer and dictate policies of the other company even where the voting securities held by such company exercising such effective control in that other company is less than fifty percent(50%) and the term "controlled" shall have a corresponding meaning.

1.4 "Appendix" means an Appendix attached to this Contract and made a part thereof.

1.5 "Appraisal Programme" means a programme, carried out following a Discovery in the Contract Area for the purpose of appraising Discovery and delineating the Petroleum Reservoirs to which the Discovery relates in terms of thickness and lateral extent and determining the characteristics thereof and the quantity of recoverable Petroleum therein.

1.6 "Appraisal Well" means a Well drilled pursuant to an Appraisal Programme.

1.7 "Approved Work Programme" and "Approved Budget" means a Work Programme or a Budget that has been approved by the Management Committee pursuant to the provisions of this Contract.

1.8 "Arms Length Sales" means sales made freely in the open market, in freely convertible currencies, between willing and unrelated sellers and buyers and in which such buyers and sellers have no contractual or other relationship, directly or indirectly, or any common or joint interest as is reasonably likely to influence selling prices and shall, inter alia, exclude sales (whether direct or indirect, through brokers or otherwise) involving Affiliates, sales between Companies which are Parties to this Contract, sales between governments and government-owned entities, counter trades, restricted or distress sales, sales involving barter arrangements and generally any transactions motivated in whole or in part by considerations other than normal commercial practices.

1.9 "Article" means an article of this Contract and the term "Articles" means more than one Article.

1.10 "Associated Natural Gas" or "ANG" means Natural Gas produced in association with Crude Oil either as free gas or in solution, if such Crude Oil can by itself be commercially produced.

1.11 "Barrel" means a quantity or unit equal to 158.9074 litres (forty two (42) United States gallons) liquid measure, at a temperature of sixty (60) degrees Fahrenheit (15.56 degrees Celsius) and under one atmosphere pressure (14.70 psia).

1.12 "Basement" means any igneous or metamorphic rock, or rocks or any stratum of such nature, in and below which the geological structure or physical characteristics of the rock sequence do not have the properties necessary for the accumulation of Petroleum in commercial quantities and which reflects the maximum depth at which any such accumulation can be reasonably expected in accordance with the knowledge generally accepted in the international petroleum industry.

1.13 "Budget" means a budget formulated in relation to a Work Programme.

1.14 "Business Day" means any of the Calendar Day excluding holidays.

1.15 "Calendar Day" means any of the seven (7) days of a week.

1.16 "Calendar Month" means any of the twelve (12) months of the Calendar Year.

1.17 "Calendar Quarter" or "Quarter" means a period of three (3) consecutive Calendar Months commencing on the first day of January, April, July and October of each Calendar Year.

1.18 "Calendar Year" means a period of twelve (12) consecutive Months according to the Gregorian calendar, commencing with the first (1st) day of January and ending with the thirty-first (31st) day of December.

1.19 "Commercial Discovery" means a Discovery of Petroleum reserves which has been declared as a Commercial Discovery in accordance with the provisions of Article 10 and/or Article 21.

1.20 "Commercial Production" means production of Crude Oil or Condensate or Natural Gas or any combination of these from the Contract Area (excluding production for testing purposes) and delivery of the same at the relevant Delivery Point under a programme of regular production and sale.

1.21 "Company" for the purpose of this Contract means a company which is a Party to this Contract and, where more than one Company is Party to the Contract, the term "Companies" shall mean all such Companies collectively, including their respective successors and permitted assigns under Article 28.

1.22 "Condensate" means those low vapour pressure hydrocarbons obtained from Natural Gas through condensation or extraction and refers solely to those hydrocarbons that are liquid at normal surface temperature and pressure conditions provided that in the event Condensate is produced from a Development Area and is segregated at the Delivery Point or transported, then the provisions of this Contract shall apply to such Condensate as if it were Crude Oil.

1.23 "Contract" means this agreement and the Appendices mentioned herein and attached hereto and made an integral part hereof and any amendments made thereto pursuant to the terms hereof.

1.24 "Contract Area" means, on the Effective Date, the area described in Appendix-A and delineated on the map attached as Appendix B or any portion of the said area remaining after relinquishment or surrender from time to time pursuant to the terms of this Contract (including any additional area as provided under Article 11.3).

1.25 "Contract Costs" means Exploration Costs, Development Costs and Production Costs as provided in Section 2 of the Accounting Procedure and allowed to be cost recoverable in terms of Section 3 of the Accounting Procedure.

1.26 "Contract Year" means a period of twelve (12) consecutive months counted from the Effective Date or from the anniversary of the Effective Date.

1.27 "Contractor" means the Company(ies).

1.28 "Cost Petroleum" means, the portion of the total value of Petroleum Produced and Saved from the Contract Area which the Contractor is entitled to take in a particular period, for the recovery of Contract Costs as provided in Article 15.

1.29 "Crude Oil" or "Oil" or "Crude" means all kinds of hydrocarbons and bitumen, both in solid and in liquid form, in their natural state or obtained from Natural Gas by condensation or extraction, including distillate and Condensate when commingled with the heavier hydrocarbons and delivered as a blend at the Delivery Point but excluding Natural Gas.

1.30 "Deepwater Area" (for deepwater blocks/areas) means area falling beyond four hundred (400) metre isobath, provided, however, that for the purposes of this Contract, the Contract Area as on Effective Date, as described in the Appendix-A and Appendix-B shall be deemed to be Deepwater Area falling beyond four hundred (400) metre isobath.

1.31 "Delivery Point" means, except as otherwise herein provided or as may be otherwise agreed between the Parties having regard to international practice, the point at which Petroleum reaches the outlet flange of the delivery facility, either offshore or onshore and different Delivery Point(s) may be established for purposes of sales. Delivery Point(s) shall be approved by the Management Committee.

1.32 "Development Area" means part of the Contract Area which encompasses one or more Commercial Discovery(ies) and any additional area that may be required for proper development of such Commercial Discovery(ies) and established as such in accordance with the provisions of the Contract.

1.33 "Development Costs" means those costs and expenditures incurred in carrying out Development Operations, as classified and defined in Section 2 of the Accounting Procedure and allowed to be recovered in terms of Section 3 thereof.

1.34 "Development Operations" means operations conducted in accordance with the Development Plan and shall include, but not be limited to the purchase, shipment or storage of equipment and materials used in developing Petroleum accumulations, the drilling, completion and testing of Development Wells, the drilling and completion of Wells for Gas or water injection, the laying of gathering lines, the installation of offshore platforms and installations, the installation of separators, tankages, pumps, artificial lift and other producing and injection facilities required to produce, process and transport Petroleum into main Oil storage or Gas processing facilities, either onshore or offshore, including the laying of pipelines within or outside the Contract Area, storage at Delivery Point(s), the installation of said storage or Gas processing facilities, the installation of export and loading facilities and other facilities required for the development and production of the said Petroleum accumulations and for the delivery of Crude Oil and/or Gas at the Delivery Point and also including incidental operations not specifically referred to herein but required for the most efficient and economic development and production of the said Petroleum accumulations in accordance with modern oilfield and petroleum industry practices.

1.35 "Development Plan" means a plan submitted by the Contractor for the development of a Commercial Discovery, which has been approved by the Management Committee or the Government pursuant to Article 10 or Article 21.

1.36 "Development Well" means a Well drilled, deepened or completed after the date of approval of the Development Plan pursuant to Development Operations or Production Operations for the purposes of producing Petroleum, increasing production, sustaining production or accelerating extraction of Petroleum including production Wells, injection Wells and dry Wells.

1.37 "Directorate General of Hydrocarbons or DGH" means an organisation, including its successors under the Ministry of Petroleum and Natural Gas.

1.38 "Discovery" means the finding, during Petroleum Operations, of a deposit of Petroleum not previously known to have existed, which can be recovered at the surface in a flow measurable by conventional petroleum industry testing methods.

1.39 "Discovery Area" means that part of the Contract Area about which, based upon Discovery and the results obtained from a Well or Wells drilled in such part, the Contractor is of the opinion that Petroleum exists and is likely to be produced in commercial quantities.

1.40 "Effective Date" means the later of the date on which this Contract is executed by the Parties or the date of issue of License or date from which License has been made effective by the Central Government or State Government(s) as the case may be.

1.41 "Environmental Damage" means soil erosion, removal of vegetation, destruction of wildlife, pollution of groundwater or surface water, land contamination, air pollution, noise pollution, bush fire, disruption to water supplies to natural drainage or natural flow of rivers or streams, damage to archaeological, palaeontological and cultural sites and shall include any damage or injury to, or destruction of, soil or water in their physical aspects together with vegetation associated therewith, aquatic or terrestrial mammals, fish, avi-fauna or any plant or animal life whether in the sea or in any other water or on, in or under land.

1.42 "Exploration Costs" means those costs and expenditures incurred in carrying out Exploration Operations, as classified and defined in Section 2 of the Accounting Procedure and allowed to be recovered in terms of Section 3 thereof.

1.43 "Exploration Operations" means operations conducted in the Contract Area pursuant to this Contract in searching for Petroleum and in the course of an Appraisal Programme and shall include but not be limited to aerial, geological, geophysical, geochemical, palaeontological, palynological, topographical and seismic surveys, analysis, studies and their interpretation, investigations relating to the subsurface geology including structural test drilling, stratigraphic test drilling, drilling of Exploration Wells and Appraisal Wells and other related activities such as surveying, drill site preparation and all work necessarily connected therewith that is conducted in connection with Petroleum exploration.

1.44 "Exploration Period" means the period mentioned in Article 3 during which Exploration Operations may be carried out by the Contractor as provided in Article 3 hereof.

1.45 "Exploration Phase" or "Phase" means any of the periods specified in Article 3 in which the Contractor is required to complete the Minimum Work Programme specified therein.

1.46 "Exploration Well" means a Well drilled for the purpose of searching for undiscovered Petroleum accumulations on any geological entity (be it of structural, stratigraphic, facies or pressure nature) to at least a depth or stratigraphic level specified in the Work Programme.

1.47 "Field" means an Oil Field or a Gas Field or combination of both as the case may be.

1.48 "Financial Year" means the period from the first (1st) day of April to the thirty-first (31st) day of March of the following Calendar Year.

1.49 "Foreign Company" means a Company within the meaning of Section 591 of the Companies Act, 1956.

1.50 "Frontier Area" means any area identified, demarcated and so notified by the Government or its authorised agency(ies) for the purpose of exploration and exploitation of Oil and Gas, which is logistically and technically difficult and lacks infrastructural and/or marketing facilities, etc.

1.51 "Gas" means Natural Gas.

1.52 "Gas Field" means, within the Contract Area, a Natural Gas Reservoir or a group of Natural Gas Reservoirs within a common geological structure or feature.

1.53 "Government" or "Central Government" means Government of India unless otherwise stated.

1.54 "Investment" shall have the meaning ascribed to that expression in paragraph 3 of Appendix D.

1.55 "Investment Multiple" means, the ratio of accumulated Net Cash Income to accumulated Investment by the Contractor, as determined in accordance with Appendix D.

1.56 "Lease" means a petroleum mining lease referred to in the Rules and shall, unless otherwise stated therein, exclude right for exploration and exploitation of coal/ lignite bed methane (CBM).

1.57 "Lessee" means the Contractor to whom a Lease is issued under the Rules for the purpose of carrying out Petroleum Operations in a Development Area or Contract Area.

1.58 "LIBOR" means the London Inter-Bank Offer Rate for six-month maturates of United States Dollars as quoted by the International Swaps and Derivative Association or such other bank being a BBA LIBOR contributor panel bank as the Parties may agree.

1.59 "License" means a petroleum exploration license referred to in the Rules.

1.60 "Licensee" means the Contractor to whom a License is issued under the Rules for the purpose of carrying out Petroleum Operations in the Contract Area.

1.61 "Minimum Work Programme" means with respect to each Exploration Phase, the Work Programme specified in Article 5 with respect to such Phase.

1.62 "Management Committee" means the committee constituted pursuant to Article 6 hereof.

1.63 "Month" means Calendar Month.

1.64 "Natural Gas" means wet gas, dry gas, all other gaseous hydrocarbons, and all substances contained therein, including sulphur, carbondioxide and nitrogen but excluding extraction of helium, which are produced from Oil or Gas Wells, excluding those condensed or extracted liquid hydrocarbons that are liquid at normal temperature and pressure conditions, and including the residue gas remaining after the condensation or extraction of liquid hydrocarbons from gas.

1.65 "Net Cash Income" shall have the meaning assigned in paragraph 2 of Appendix- D.

1.66 "Non Associated Natural Gas" or "NANG" means Natural Gas which is produced either without association of Crude Oil or in association with such quantities of Crude Oil which by itself cannot be commercially produced.

1.67 "Oil Field" means, within the Contract Area, an Oil Reservoir or a group of Oil Reservoirs within a common geological structure or feature.

1.68 "Operator" means one of the Parties comprising the Contractor, appointed as the Operator pursuant to Article 7.

1.69 "Operating Agreement" means the joint operating agreement entered by the constituents of the Contractor in accordance with Article 7, with respect to conduct of Petroleum Operations.

1.70 "Operating Committee" means the Committee established by that name in the Operating Agreement pursuant to Article 7.

1.71 "Participating Interest" means, in respect of each Party constituting the Contractor, the undivided share expressed as a percentage of such Party's participation in the rights and obligations under this Contract.

1.72 "Parties" means the parties signatory to this Contract including their successors and permitted assigns under this Contract and the term "Party" means any of the Parties.

1.73 "Petroleum" means Crude Oil and/or Condensate and/or Natural Gas existing in their natural condition but excluding helium occurring in association with Petroleum or shale.

1.74 "Petroleum Operations" means, as the context may require, Exploration Operations, Development Operations or Production Operations or any combination of two or more of such operations, including construction, operation and maintenance of all necessary facilities, plugging and abandonment of Wells, safety, environmental protection, transportation, storage, sale or disposition of Petroleum to the Delivery Point, Site Restoration and any or all other incidental operations or activities as may be necessary.

1.75 "Petroleum Produced and Saved" means gross Petroleum produced minus impurities such as water or solids produced along with Petroleum, Petroleum recycled to the reservoir, Petroleum used in Petroleum Operations or flared or otherwise unavoidably lost under the provisions of the Contract.

1.76 "Production Costs" means those costs and expenditures incurred in carrying out Production Operations as classified and defined in Section 2 of the Accounting Procedure and allowed to be recovered in terms of Section 3 thereof.

1.77 "Production Operations" means all operations conducted for the purpose of producing Petroleum from the Development Area after the commencement of production from the Development Area including the operation and maintenance of all necessary facilities therefor.

1.78 "Profit Petroleum" means, the total value of Petroleum Produced and Saved from the Contract Area in a particular period, as reduced by Cost Petroleum and calculated as provided in Article 16.

1.79 "Recompletion" means an operation whereby a completion in one zone is abandoned in order to attempt a completion in a different zone within an existing Well bore.

1.80 "Reservoir" means a naturally occurring discrete accumulation of Petroleum.

1.81 "Rules" means the Petroleum and Natural Gas Rules, 1959 and any amendments made thereto from time to time.

1.82 "Section" means a section of the Accounting Procedure.

1.83 "Self-sufficiency" means, in relation to any Year, that the volume of Crude Oil and Crude Oil equivalent of Petroleum products exported from India during that Year either equals or exceeds the volume of Crude Oil and Crude Oil equivalent of Petroleum products imported into India during the same Year, as determined by Government.

1.84 "Site Restoration" shall mean all activities required to return a site to its state as of the Effective Date pursuant to the Contractor's environmental impact study and approved by the Government or to render a site compatible with its intended after-use (to the extent reasonable) after cessation of Petroleum Operations in relation thereto and shall include, where appropriate, proper abandonment of Wells or other facilities, removal of equipment, structures and debris, establishment of compatible contours and drainage, replacement of top soil, re-vegetation, slope stabilisation, in-filling of excavations or any other appropriate actions in the circumstances.

1.85 "Statement" or "Statements" refers to the statements required to be furnished in accordance with Appendix-C of this Contract.

1.86 "State Government" means any government of a state of the Union of

India, which has control over the Contract Area for the purpose of grant of Licenses/Leases. In case the Contract Area covers more than one state, the State Government shall include all such governments of those states.

1.87 "Subcontractor " means any company or person contracted by the Contractor or Operator to provide goods or services with respect to Petroleum Operations.

1.88 "US $" or "USD" or "US Dollar" or "United States Dollar" means the currency of the United States of America.

1.89 "Well" means a borehole, made by drilling in the course of Petroleum Operations, but does not include a seismic shot hole.

1.90 "Work Programme" means a work programme formulated for the purpose of carrying out Petroleum Operations.

1.91 "Year" means a Financial Year.

ARTICLE 2
PARTICIPATING INTERESTS

2.1 The initial Participating Interest of the Parties comprising the Contractor shall be as follows:

X Company__________ : (%)
Y Company__________ : (%)
Z Company__________ : (%)

2.2 Except as provided in this Article or elsewhere in this Contract, the rights and obligations of the Parties comprising the Contractor shall include but not be limited to:

(a) the right to take Cost Petroleum in accordance with the provisions of Article 15;

(b) the right to take its Participating Interest share of Profit Petroleum in accordance with the provisions of Article 16;

(c) the right to receive its Participating Interest share of any incidental income and receipts arising from Petroleum Operations; and

(d) the obligation to contribute its Participating Interest share of costs and expenses including Contract Costs.

ARTICLE 3
LICENSE AND EXPLORATION PERIOD

3.1 The Exploration Period shall begin on the Effective Date and shall consist of two Exploration Phases. First Exploration Phase shall be for a period not exceeding four (4) Contract Years and second Exploration Phase shall be for a period not exceeding three (3) Contract Years provided that in case of deepwater and Frontier Area blocks, first Exploration Phase will have additional one (1) Contract Year i.e. a total first Exploration Phase not exceeding five (5) Contract Years. In case of onland and shallow water blocks, the Exploration Period shall not exceed seven (7) consecutive Contract Years and eight (8) consecutive Contract Years in case of Deepwater Areas and Frontier Areas unless extended pursuant to the terms of this Contract. (**Explanatory Note : The blocks offered as deepwater blocks and frontier area have been specified in Notice Inviting Offers (NIO). Bidders are advised to carefully refer the NIO)**

3.2 Except as otherwise provided in this Contract, the term of the first Exploration Phase shall not exceed ________ (to be taken from the accepted bid and subject to Article 3.1) consecutive Contract Years (hereinafter referred to as the first Exploration Phase).

3.3 Except as otherwise provided in this Contract, the term of the second Exploration Phase shall not exceed ________ (to be taken from the accepted bid and subject to Article 3.1) consecutive Contract Years from the end of the first Exploration Phase (hereinafter referred to as the second Exploration Phase).

3.4 At the expiry of any Exploration Phase of the Exploration Period, provided that the Contractor has completed the Minimum Work Programme for that Exploration Phase, the Contractor shall have the option, exercisable by giving a written notice to the Government at least thirty (30) days prior to the expiry of the relevant Phase, either:

(a) to proceed to the next Exploration Phase on presentation of the requisite guarantees as provided for in Article 29; or

(b) to relinquish the entire Contract Area except for any Discovery Area and any Development Area and to conduct Development Operations and Production Operations in relation to any Commercial Discovery in accordance with the terms of this Contract, and the Contractor shall have no further obligation in respect of the Minimum Work Programme under Article 5 for any subsequent Exploration Phases of the Exploration Period. If neither of the options provided for in paragraphs (a) and (b) hereof is exercised by the Contractor, this Contract shall terminate at the end of the then current Exploration Phase and the License shall be automatically cancelled.

3.5 If at the end of an Exploration Phase the Minimum Work Programme for that phase is not completed, the time for completion of the said Minimum Work Programme shall be extended for a period necessary to enable completion thereof but not exceeding six (6) months, provided that the Contractor submits his request by giving a written notice to the Government at least thirty (30) days prior to the expiry of the relevant Phase and can show technical or other good reasons for non-completion of the Minimum Work Programme and the Management Committee gives its consent to the said extension and provided further that the period of such extension shall be subtracted from the next succeeding Exploration Phase, if any. In case the Minimum Work Programme of any particular Exploration Phase is completed before stipulated time as provided in the Article 3.2, the time so saved will be added to the next Exploration Phase, if so requested by the Contractor giving a notice in writing to the Government thirty (30) days prior to such early completion of the Phase and in that event the provision of the Article 3.4 (a) shall apply immediately after such early completion of the Phase.

3.6 If, at the end of an Exploration Phase, execution of any Work Programme is in progress and which is in addition to the Minimum Work Programme, such Exploration Phase shall be extended for a period not exceeding six (6) months to enable completion thereof provided that the Minimum Work Programme for such Phase has been completed and the Management Committee gives its consent to the said extension as provided in the Article 3.5. In the event of an extension as provided for herein, the notice referred to in Article 3.4 shall be given at least thirty (30) days prior to the expiry of the relevant extension.

3.7 Where sufficient time is not available prior to the expiry of the Exploration Period to complete an Appraisal Programme, at the request of the Contractor, the Government shall extend the Exploration Period for such period, not exceeding eighteen (18) months for onland and shallow water blocks and thirty (30) months for deepwater blocks, as may be mutually agreed between the Parties for the Appraisal Programme to be carried out and for the Contractor and the Management Committee, to comply with the provisions of Article 10 and Article 21.

3.8 If no Commercial Discovery has been made in the Contract Area by the end of the Exploration Period, the Contract shall terminate.

3.9 If this Contract is terminated in accordance with its terms, the License shall be automatically cancelled.

3.10 If at the expiry of the Exploration Period a development plan for development of a Commercial Discovery and an application for Lease is under consideration by the Management Committee or Government, as the case may be, pursuant to Articles 10, 11 and 21 respectively, the License shall continue in force with respect to that part of the Contract Area to which the application for the Lease relates, pending a decision on the proposed development plan and the application for the Lease, but shall cease to be in force and effect with respect to the remainder of the Contract Area.

ARTICLE 4
RELINQUISHMENT

4.1 If at the end of the first Exploration Phase, the Contractor elects, pursuant to Article 3.4, to continue Exploration Operations in the Contract Area in the second Exploration Phase, the Contractor shall have the option to relinquish a part of area in simple geometrical shape. In case, the Contractor exercises its option to relinquish a part of the Contract Area at the end of first Exploration Phase, then such area to be relinquished shall not be less than twenty five percent (25%) of the original Contract Area and the Contractor shall be entitled to retain the balance area including any Development Area and Discovery Area in not more than three (3) areas of simple geometrical shapes. Notwithstanding the provision of this Article 4.1, the Contractor shall be permitted to retain the Development Areas and Discovery Areas in accordance with Article 3.4.

4.2 At the end of the second Exploration Phase, the Contractor shall retain only Development Areas and Discovery Areas.

4.3 If the Contractor exercises the option provided for in paragraph (b) of Article 3.4, the Contractor shall, after any Discovery Areas or Development Areas have been designated, relinquish all of the Contract Area not included within the said Discovery Areas or Development Areas.

4.4 As and when the Contract is terminated under the provisions of Article 3 or in accordance with any other provisions of this Contract, the entire Contract Area remaining with the Contractor shall be deemed to have been relinquished by the Contractor as on the date on which the Contract is terminated.

4.5 Relinquishment of all or part of the Contract Area or termination of the Contract shall not be construed as absolving the Contractor of any liability undertaken or incurred by the Contractor in respect of the Contract Area during the period between the Effective Date and the date of such relinquishment or termination.

4.6 Subject to Article 14.9, the liability of the Contractor shall be limited to any liability undertaken or incurred in respect of, relating to or connected with the Contract, and/or any claim arising out of or in relation to the act of negligence, misconduct, commission or omission in carrying out Petroleum Operations during the period between the Effective Date and the date of relinquishment of the Contract Area or termination or expiry of the Contract, as the case may be.

ARTICLE 5
WORK PROGRAMME

5.1 The Contractor shall commence Petroleum Operations not later than six (6) months from the Effective Date.

5.2 In addition to the Bid Work Programme Commitment in first Exploration Phase specified below in Article 5.2.1 and 5.3, the Contractor Shall be required to undertake and complete the 2D seismic – API, in grid size of ______ LKM × ________LKM covering the entire Contract Area (herein after referred to as " Mandatory Work Programme") during First Exploration Phase. In case due to any reason intrinsic to the Contract Area reason(s), the Contractor is not able to cover any part of the Contract Area by 2D seismic survey of grid size specified in this Article, the Contractor shall submit a proposal for substitution of the short fall in the Mandatory Work Programme to the Management Committee for approval. The Management Committee shall consider and take a reasoned unanimously decision on the proposal of the Contractor in a timely manner. The Management Committee may ask the Contractor by giving a reasonable time, any relevant information/ details/data to enable it to take a decision on the proposal. The Management Committee shall ensure that the substituted work programme shall be atleast equal to the shortfall in the Mandatory Work Programme when evaluated in terms of efforts and expenditure. In case Management Committee is not able to decide unanimously, the matter may be referred for approval of the Government. In case no proposal is received from the Contractor for the substitution and fails to complete Mandatory Work Programme the provision of Article 5.6 shall apply.

5.2.1 During the currency of the first Exploration Phase, as per Article 3.2, the Contractor shall complete the following Work Programme:

(a) a seismic programme consisting of the acquisition, processing and interpretation of __________line kilometres of 2D and/ or ________ sq. kms. of 3D seismic data in relation to the exploration objectives; and

(b) ______________Exploration Wells shall be drilled to at least one of the following depths:

i) __________________metres and ________ (geological objective);

ii) to Basement; and

iii) that point below which further drilling becomes impracticable due to geological conditions encountered and drilling would be abandoned by a reasonable prudent operator in the same or similar circumstances. Abandonment of drilling under this provision by the Contractor, would require unanimous approval of the Management Committee.

5.3 During the currency of the second Exploration Phase, as per Article 3.3, the Contractor shall complete the following Work Programme:

(a) a seismic programme consisting of the acquisition, processing and interpretation of __________line kilometres of 2D and/or ________ sq. kms. of 3D seismic data in relation to the exploration objectives; and

(b) ______________Exploration Wells shall be drilled to at least one of the following depths:

(i) _________metres and ______________(geological objective);

(ii) to Basement; and

(iii) that point below which further drilling becomes impracticable due to geological conditions encountered and drilling would be abandoned by a reasonable prudent operator in the same or similar circumstances. Abandonment of drilling under this provision by the Contractor, would require unanimous approval of the Management Committee.

Note : The details of the work programme to be filled in the Articles

5.2, 5.2.1 & 5.3 shall be as per the NIO and the bid of Company(ies) and accepted by the Government.

5.4 The actual depth objective for each of the Wells shall be determined by the Contractor in the light of the advice of the Management Committee before the commencement of the drilling. Each Well which reaches the geological objective for which the depth objective was determined shall be deemed to have been drilled to the depth objective or to actual total depth, whichever is greater. The Contractor shall ensure that all relevant subsurface, geological, geochemical and geophysical information necessary for the attainment of the exploration objectives in accordance with modern oilfield and petroleum industry practices is obtained during exploratory drilling.

5.5 If the depth/geological objective of the Well is not achieved for any reason, a substitute Well shall be drilled of the same specifications as stipulated in and subject to Articles 5.2 & 5.3, as the case may be.

5.6 Subject to Article 31, the Contractor undertakes to complete the Mandatory Work Programme and Minimum Work Programme in accordance with Articles 5.2, 5.2.1, 5.3 and 5.5, as the case may be. In the event that the Contractor fails to fulfill the said Mandatory Work Programme or Minimum Work Programme or both by the end of the relevant Exploration Phase or early termination of the Contract by the Government for any reason whatsoever, each Company constituting the Contractor shall pay to the Government, within sixty (60) days following the end of the relevant Exploration Phase or early termination of the Contract, as may be the case, its Participating Interest share for an amount which, when evaluated in terms of the Mandatory Work programme or Minimum Work Programme specified for the relevant Phase, is equal to the amount which would be required to complete the said Mandatory Work Programme or Minimum Work Programme or both, as may be the case. For determination of this amount, available relevant information including the Budget and modern oilfield and petroleum industry practices may be taken into account.

5.7 If the Minimum Work Programme for the second Exploration Phase has been completed earlier than eighteen months from the end of the Phase, the Contractor shall meet with the Government to discuss the possibility of early relinquishment, unless the Contractor undertakes further work with the approval of the Management Committee.

5.8 In the event that the Contractor has carried out work in excess of the Minimum Work Programme in any Exploration Phase, the excess exploration work done shall be set off against the Minimum Work Programme for the following Exploration Phase.

5.9 As soon as possible after the Effective Date and thereafter within ninety (90) days before commencement of each following Year, the Contractor shall submit to the Management Committee the Work Programmes and the Budgets relating to Petroleum Operations to be carried out during the relevant Year. Work Programme and Budgets for the Exploration

Period shall include work sufficient to meet the relevant Mandatory Work Programme and Minimum Work Programme with respect to each Exploration Phase specified in this Article 5.

5.10 The Contractor may propose modifications or revisions to the details of a reviewed or an approved Work Programme and Budget, as the case may be, in the light of the then existing circumstances and shall submit to the Management Committee modifications or revisions to the Work Programme and Budget referred to in Article 5.9.

5.11 Work Programmes and Budgets and any modifications or revisions thereto relating to Exploration Operations shall be submitted to the Management Committee for review and advice as provided in Article 6.5. Work Programmes and Budgets related to Development Operations and Production Operations and any modifications or revisions thereto shall be submitted to the Management Committee for approval as provided in Article 10 and Article 21.

ARTICLE 6
MANAGEMENT COMMITTEE

6.1 There shall be constituted a committee to be called the Management Committee with functions as stated herein below.

6.2 Government shall nominate two (2) members representing Government in the Management Committee, whereas each Company constituting the Contractor shall nominate one (1) member each to represent Company in the Management Committee provided that in case the Contractor constitutes only one Company, that Company shall have two (2) members. The Parties shall nominate the members to the Management Committee within thirty (30) days of the Effective Date.

6.3 Each Party may nominate alternate members with full authority to act in the absence and on behalf of the members nominated under Article 6.2 and may, at any time, nominate another member or alternate member to replace any member nominated earlier by notice to other members of the Management Committee.

6.4 One representative of the Government shall be designated as the Chairman of the Management Committee and the second representative of the Government shall be designated as the Deputy Chairman. The member of the Operator, or the member designated by the Operator where Operator has two (2) members in the Management Committee shall be designated as the Secretary of the Committee.

6.5 Operator on behalf of the Contractor with the approval of Operating Committee, if constituted under the Article 7.4, or in case of a single Party constituting the Contractor, then that Party shall submit following matters to the Management Committee for review and it shall have advisory functions:

(a) the annual Work Programmes and Budgets in respect of Exploration Operations and any revisions or modifications thereto;

(b) annual work progress and costs incurred thereon;
(c) proposals for surrender or relinquishment of any part of the Contract Area by the Contractor;
(d) proposals for an Appraisal Programme or revisions or additions thereto and the declaration of a Discovery as a Commercial Discovery;
(e) any other matter required by the terms of this Contract to be submitted to it for review or advice; and
(f) any other matter which the Contractor decides to submit for review or advice including matters concerning inter-Party relationships.

6.6 The following matters shall be submitted by Operator on behalf of the Contractor with the approval of Operating Committee, if constituted under the Article 7.4, or in case of single Party constituting the Contractor, then by that Party to the Management Committee for approval:
(a) Annual Work Programmes and Budgets in respect of Development Operations and Production Operations and any modifications or revisions thereto;
(b) proposals for the approval of development plans as may be required under this Contract, or modifications or revisions to a Development Plan;
(c) determination of a Development Area;
(d) appointment of auditors along with scope of audit, approval and adoption of audited report submitted under Article 25.4.3;
(e) collaboration with licensees or contractors of other areas;
(f) claims or settlement of claims for or on behalf of or against the Contractor in excess of limits fixed by the Management Committee from time to time;
(g) proposal about abandonment plan/Site Restoration as required to be submitted under Article 14.10;
(h) any other matter required by the terms of this Contract to be submitted for the approval of the Management Committee;
(i) any other matter which the Contractor decides to submit to it; and
(j) any matter, which Government refers to the Management Committee for its consideration and reasoned opinion.

6.7 Unless agreed otherwise by all the members of the Management Committee, the Management Committee shall meet at least once every six (6) months during the Exploration Period and thereafter at least once every three (3) months or more frequently at the request of any member. The Secretary, with the approval of the Chairman, shall convene each meeting by notifying the members twenty eight (28) days prior to such a meeting (or a shorter period of notice if the members unanimously so agree) of the time and place of such meeting and the purpose thereof and shall include in such notice a provisional agenda for such meeting. The Chairman shall be responsible for processing the final agenda for such meeting and the agenda shall include all items of business requested by the members to be included, provided such requests are received by the Secretary at least ten (10) days prior to the date fixed for the meeting. The Secretary shall forward the agenda to the members at least seven (7) Business Days prior to the date fixed for the meeting. Matters not included in the agenda may be taken up at the meeting by any member with the unanimous consent of all the members whether present or not present at the meeting.

6.8 The Chairman or the Deputy Chairman, as may be the case, shall preside over the meetings of the Management Committee and, in their absence, any other member representing Government and present shall preside over the meetings.

6.9 Secretary to the Management Committee shall be responsible, inter alia, for preparation of the minutes of every meeting in the English language and provision to every member of the Management Committee with two (2) copies of the minutes approved by the Chairman within three (3) Business Days of the meeting. Unless agreed otherwise by all the members of the Management Committee, the minutes of a meeting shall be finalized by the Management Committee within three (3) Business Days thereafter. Members shall notify the Chairman and the other members of their approval of the minutes by putting their signatures on one copy of the minutes and returning the same to the Chairman. Members may suggest any modification to the minutes while returning the signed copy. Members may also communicate with the Chairman through telex, cable, or facsimile or any other effective mode of communication agreed by all the members of the Management Committee. If the Chairman or any other member does not agree with the modification to the minutes suggested by any member, the matter shall be brought to the attention of the other members and resubmitted to the Management Committee at the next meeting and the minutes shall stand approved as to all other matters. If a member fails to respond within the aforesaid three (3) Business Day period, unless agreed otherwise by the Management Committee as herein provided, the minutes shall be deemed to be approved by such member.

6.10 Any member shall be entitled, if either he/she or his/her alternate is unable to attend a meeting, to cast his vote by telex, cable, facsimile transmission or any other effective mode of communication agreed by all the members of the Management Committee and received by the Chairman prior to the date on which the vote is taken in the course of the meeting or by giving a prior written notice to all other members, appoint a member, with his/her prior consent, representing another Party in the Management Committee as its proxy to attend a meeting and to exercise the appointing member's right to vote at the meeting whether as directed by the appointing member or otherwise. A member appointed as a proxy and attending a meeting shall be present in two separate capacities and vote accordingly. All such votes shall have the same effect as if that member had been present and so voted at the meeting.

6.11 In case of urgency, where Operating Committee has made a recommendation together with reasons to the Chairman requiring consideration of a matter by the Management Committee without delay, Chairman, after being satisfied may waive the requirements of notice period for the meeting and circulation of agenda to such extent as would be consistent with the urgency and consideration of the matter by the Management Committee. Alternatively, Chairman may approve submission of notice and agenda to members by telex or facsimile transmission or any other effective mode of communication agreed by all the members of the Management Committee, receipt of which shall be confirmed by telephone by the Chairman requiring the members to confirm their decision by these modes of communication not later than three (3) Business Days from confirmation of receipt of notice and agenda by the member. Any member failing to convey the decision within the time limits of three (3) Business Days shall be deemed to have voted in favour of the proposal. The result of any such vote shall be notified by the Chairman to all the members.

6.12 The meetings of the Management Committee shall be held in India, unless otherwise mutually agreed by the members of the Management Committee. All expenses of the members of the Management Committee attending meetings shall be borne by the respective Party and shall in no event be cost recoverable.

6.13 All matters requiring the approval of the Management Committee shall be generally approved by a unanimous vote of the members of the Management Committee present as well as the views of the members received by some other mode of communication. In case, unanimity is not achieved in decision making process within a reasonable period as may be required under the circumstances, the decision of the Management Committee shall be approved by the majority Participating Interest of seventy percent (70%) or more with Government representative having a positive vote in favour of the decision.

6.14 There shall be a quorum of the Management Committee for holding a meeting and making decisions with each Party to the Contract represented by at least one of its nominated members in the Management Committee either present in person or represented as per Article 6.10. If there is no quorum in a meeting, the meeting shall stand postponed to the same day and time in the next week and if quorum is not present or represented even in the next meeting and subject to a Government member being present, the members present and represented will constitute the quorum and take decisions and decisions taken by such quorum shall be final and binding to all the absenting Parties or Parties not represented, notwithstanding the provisions of Article 6.13.

6.15 The Management Committee, if it considers necessary, may appoint legal, financial or technical subcommittees comprised of such representatives as may be agreed by the Management Committee to consider any matter requiring approval or decision of the Management Committee. Such sub-committee expenses shall form part of Contract Cost with relevant cost classification as decided by the Management Committee pursuant to the Section 2 of the Accounting Procedure and will be cost recoverable.

6.16 In the event a Party to the Contract is not entitled to vote in the Operating Committee meetings being in default under the Operating Agreement, and Operator notifies Chairman of the default by the Party, then the issue of exercising voting right by such defaulting Party in the Management Committee meetings shall be discussed by the Management Committee. The Management Committee excluding the defaulting Party, after duly hearing the views of the defaulting Party on the matter of their default under Operating Agreement, shall take unanimous decision on exclusion or otherwise of the defaulting Party from voting in the Management Committee meetings. For avoidance of any doubt, it is clearly understood that unanimous decision by the Management Committee referred to in this Article 6.16 excludes defaulting Party from such decision. Accordingly, if the Management Committee decides to exclude the defaulting Party from voting in the Management Committee, then the said Party shall not be entitled to vote in the meetings of the Management Committee under Contract. In that event, notwithstanding the provisions of Article 6.13, decisions of the Management Committee shall be made by vote of the members of the Management Committee excluding the member appointed by the said Party in default and any vote or purported vote by such member in the Management Committee shall be ignored. The said Party in default shall be bound by all decisions of the Management Committee. The non-defaulting Parties under the Operating Agreement shall indemnify Government against any claims of whatsoever nature which may arise due to exclusion of defaulting Party from voting in the Management Committee.

ARTICLE 7
OPERATORSHIP, JOINT OPERATING AGREEMENT AND OPERATING COMMITTEE

7.1 XYZ shall be the Operator for the purpose of carrying out Petroleum Operations pursuant to this Contract during the term of the Contract.

7.2 No change in the operatorship shall be effected without the consent of the Government and such consent shall not be unreasonably withheld.

7.3 The functions required of the Contractor under this Contract shall be performed by the Operator on behalf of all constituent(s) of the Contractor subject to, and in accordance with, the terms and provisions of this Contract and generally accepted modern oilfield and petroleum industry practices, provided, however, that this provision shall not be construed as relieving the constituent(s) of the Contractor from any of its obligations or liability under the Contract.

7.4 Within forty five (45) days of the Effective Date or such longer period as may be agreed to by Government, the Companies constituting the Contractor shall execute a Joint Operating Agreement. The said agreement shall be consistent with the provisions of this Contract and shall provide for, among other things:

(a) the appointment, resignation, removal and responsibilities of the Operator;

(b) the establishment of an Operating Committee comprising of an agreed number of representatives of the Companies chaired by a representative of the Operator;

(c) functions of the said Operating Committee taking into account the provisions of the Contract, procedures for decision making, frequency and place of meetings; and

(d) contribution to costs, default, sole risk, disposal of Petroleum and assignment as between the Parties to the Joint Operating Agreement.

7.4.1 Operator shall provide to the Government a copy of the duly executed Joint Operating Agreement within thirty (30) days of its execution date or such longer period as may be agreed to by the Government.

7.4.2 In case a single Company constitutes the Contractor, the provisions of Article 7.4 and 7.4.1 shall not be applicable. However, in case of increase in the number of constituents of the Contractor, the provisions of Article 7.4 and 7.4.1 shall apply from the date of such increase in the number of the constituents.

ARTICLE 8
GENERAL RIGHTS AND OBLIGATIONS OF THE PARTIES

8.1 Subject to the provisions of this Contract, the Contractor shall have the following rights:

(a) subject to the provisions of Article 12, the exclusive right to carry out Petroleum perations to recover costs and expenses as provided in this Contract. The right hall exclude exploitation of coal/lignite bed methane (CBM) by the Contractor in the Contract Area;

(b) the right to use, free of charge, such quantities of Petroleum produced as are reasonably required for conducting Petroleum Operations in the Contract Area in accordance with generally accepted modern oilfield and petroleum industry practices;

(c) the right to lay pipelines, build roads, construct bridges, ferries, aerodromes, landing fields, radio telephones and related communication and infrastructure facilities and exercise other ancillary rights as may be reasonably necessary for the conduct of Petroleum Operations subject to such approvals as may be required and the applicable laws in force from time to time for the regulation and control thereof;

(d) the right to use all available technical data, seismic and well information, maps, samples etc. of the Contract Area as on the Effective Date, free of charge, subject to nominal copying/reproduction costs for further Petroleum Operations. The Contractor shall submit the list of all data required by them to Directorate General of Hydrocarbons (DGH) based on the list of data provided in the information docket for the block pertaining to the Contract Area as soon as possible but not later than one hundred and eighty (180) days from the execution of the Contract and the same, if available and reproducible, shall be made available to the Contractor in the office of DGH within ninety (90) days from the submission of such request for data by the Contractor, provided the Effective Date of the Contract has commenced and the Contractor has furnished relevant guarantees under Article 29 of the Contract.

(e) such other rights as are specified in this Contract.

8.2 The Government reserves the right to itself, or to grant to others the right, to prospect for and mine minerals or substances other than Petroleum within the Contract Area; provided, however, that if after the Effective Date, others are issued rights, or the Government proceeds directly to prospect for and mine in the Contract Area any minerals or substances other than Petroleum, the Contractor shall use its best efforts to avoid obstruction to or interference with such operations within the Contract Area and the third parties and/or the Government, as the case may be, shall use best efforts to ensure that operations carried out do not obstruct or unduly interfere with Petroleum Operations in the Contract Area.

8.3 The Contractor shall having due regard to modern oilfield and petroleum industry practices :

(a) except as otherwise expressly provided in this Contract, conduct all Petroleum Operations at its sole risk, cost and expense and provide all funds necessary for the conduct of Petroleum Operations including funds for the purchase or lease of equipment, materials or supplies required for Petroleum Operations as well as for making payments to employees, agents and Subcontractors;

(b) conduct all Petroleum Operations within the Contract Area diligently, expeditiously, efficiently and in a safe and workmanlike manner pursuant to the Work Programme formulated in accordance with Contract;

(c) ensure provision of all information, data, samples etc. which may be required to be furnished under the applicable laws or under this Contract;

(d) ensure that all equipment, materials, supplies, plant and installations used by the Contractor, the Operator, and Subcontractors comply with generally accepted standards and are of proper construction and kept in safe and good working order;

(e) in the preparation and implementation of Work Programmes and in the conduct of Petroleum Operations, follow modern oilfield and petroleum industry practices with such degree of diligence and prudence reasonably and ordinarily exercised by experienced parties engaged in a similar activity under similar circumstances and conditions;

(f) the procedure for acquisition of goods and services, as of the Effective Date, shall be as per the Appendix-F of this Contract. Based on economic considerations and generally accepted practices in the international petroleum industry with the objective of ensuring cost and operational efficiency in the conduct of Petroleum Operations, the Appendix-F to this Contract may be modified or changed with the prior approval of the Management Committee when circumstances so justify;

(g) after the designation of a Development Area, pursuant to this Contract, forthwith proceed to take all necessary action for prompt and orderly development of the Development Area and for the production of Petroleum in accordance with the terms of this Contract;

(h) appoint a technically competent and sufficiently experienced representative, and, in his absence, a suitably qualified replacement therefor, who shall be resident in India and who shall have full authority to take such steps as may be necessary to implement this Contract and whose name(s) shall, on appointment within ninety (90) days after commencement of the first Contract Year, be made known to the Government;

(i) provide acceptable working conditions, living accommodation and access to medical attention and nursing care for all personnel employed in Petroleum Operations;

(j) carry out such other obligations as are specified in this Contract, in particular those specified in Article 14; and

(k) be always mindful of the rights and interests of India in the conduct of Petroleum Operations.

ARTICLE 9
GOVERNMENT ASSISTANCE

9.1 Upon application in the prescribed manner, and subject to compliance with applicable laws and relevant procedures, the Government or its nominee will:

(a) use their good offices to provide the right of ingress and egress from the Contract Area and any facilities used in Petroleum Operations, wherever located, and which may be within their control;

(b) use their good offices, when necessary, to assist the Contractor in procurement or commissioning of facilities required for execution of Work Programmes including necessary approvals, permits, consents, authorisations, visas, work permits, Licenses including Licenses and Leases, rights of way, easement, surface rights and security protection at the Contractor's cost, required pursuant to this Contract and which may be available from resources within its control; and

(c) in the event that onshore facilities are required outside the Contract Area for Petroleum Operations including, but not limited to, storage, loading and processing facilities, pipelines and offices, use their good offices in assisting the Contractor to obtain from the authorities of the state in which such facilities are required, such licenses, permits, authorizations, consents, security protection at the Contractor's cost, surface rights and easements as are required for the construction and operation of the said facilities by the Contractor.

ARTICLE 10
DISCOVERY, DEVELOPMENT AND PRODUCTION

10.1 If and when a Discovery is made within the Contract Area, the Contractor shall:

(a) forthwith inform the Management Committee and Government of the Discovery;

(b) promptly thereafter, but in no event later than a period of thirty (30) days from the date of the Discovery, furnish to the Management Committee and Government particulars, in writing, of the Discovery; and

(c) promptly run tests to determine whether the Discovery is of potential commercial interest and, within a period of sixty (60) days after completion of such tests, submit a report to the Management Committee containing data obtained from such tests and its analysis and interpretation thereof, together with a written notification of whether, in the Contractor's opinion, such Discovery is of potential commercial interest and merits appraisal.

10.2 If the Contractor determines to conduct a drill stem or production test, in open hole or through perforated casing, with regard to any Exploration Well, it shall notify the Government of the time of such test at least forty eight (48) hours prior to the proposed test, and the Government shall have the right to have a representative present during such test.

10.3 If, pursuant to Article 10.1 (c), the Contractor notifies the Management Committee that the Discovery is of potential commercial interest, the Contractor shall prepare and submit to the Management Committee within one hundred and twenty (120) days of such notification, a proposed Appraisal Programme with a Work Programme and Budget to carry out an adequate and effective appraisal of such Discovery designed to achieve both the following objectives:

(i) determine without delay, and, in any event, within the period specified in Article 10.5, whether such Discovery is a Commercial Discovery and (ii) determine, with reasonable precision, the boundaries of the area to be delineated as the Development Area.

10.4 The proposed Appraisal Programme shall be reviewed by the Management Committee within thirty (30) days after submission thereof pursuant to Article 10.3. The said Appraisal Programme, together with the Work Programme and Budget submitted by the Contractor, which may be revised or modified or amended by the Contractor in light of the Management Committee review, shall be adopted as the Appraisal Programme and the Contractor shall promptly commence implementation thereof; and the annual Budget for the Exploration Period, adopted pursuant to Article 5, shall be revised accordingly.

10.5 The Contractor shall in respect of a Discovery of Crude Oil advise the Management Committee by notice in writing within a period of eighteen (18) months for onland and shallow water blocks and thirty (30) months for deepwater blocks from the date on which the notice provided for in Article 10.1 (c) was delivered, whether such Discovery should be declared a Commercial Discovery or not. Such notice shall be accompanied by a report on the Discovery setting forth all relevant technical and economic data including estimated recoverable reserves, sustainable production levels, estimated development and production expenditures, prevailing and forecasted prices, and other pertinent technical and economic factors according to modern oilfield and petroleum industry practices as well as all evaluations, interpretations and analyses of such data and feasibility studies relating to the Discovery prepared by or for the Contractor, with respect to the Discovery and any other relevant information. If the Contractor is of the opinion that Crude Oil has been discovered in commercial quantities, it shall submit the proposal to the Management Committee for review that the Discovery be declared a Commercial Discovery. In the case of a Discovery of Gas, the provisions of Article 21 shall apply.

10.6 The Management Committee shall, within forty (40) days of the date of the notice referred to in Article 10.5, review the proposal of the Contractor and request any other additional information it may reasonably require so as to complete the review of the proposal made by the Contractor. The Contractor shall furnish the additional information within thirty (30) days from the date of the request. The review by the Management Committee shall be made and conveyed to the Contractor within the later of (a) ninety (90) days from the date of notice referred to in Article 10.5 or (b) forty (40) days of receipt of such other information as may be required under this Article.

10.7 If the Contractor declares the Discovery a Commercial Discovery after taking into account the advice of the Management Committee as referred in the Article 10.6, within two hundred (200) days of the declaration of the Discovery as a Commercial Discovery, the Contractor shall submit to the Management Committee a comprehensive development plan of the Commercial Discovery which shall:

(a) relate to the Discovery Area and contain a Reservoir or part thereof and the boundaries of the proposed Development Area;
(b) be designed to ensure the most efficient, beneficial and timely use of the Petroleum resources discovered; and
(c) be prepared in accordance with sound engineering, economic, safety and environmental principles recognised in the generally accepted modern oilfield and petroleum industry practices.

Such plan shall contain detailed proposals by the Contractor for the construction, establishment and operation of all facilities and services for and incidental to the recovery, storage and transportation of the Petroleum from the proposed Development Area to the Delivery Point together with all data and supporting information including but not limited to:

(i) description of the nature and characteristic of the Reservoir, data, statistics, interpretations and conclusions on all aspects of the geology, Reservoir evaluation, Petroleum engineering factors, Reservoir models, estimates of reserve in place, possible production magnitude, nature and ratio of Petroleum fluids and analysis of producible Petroleum;
(ii) outlines of the development project and/or alternative development projects, if any, describing the production facilities to be installed and the number of Wells to be drilled under such development project and/or alternative development projects, if any;
(iii) estimate of the rate of production to be established and projection of the possible sustained rate of production in accordance with modern oilfield and petroleum industry practices under such development project and/or alternative development projects, if any, which will ensure that the area does not suffer an excessive rate of decline of production or an excessive loss of Reservoir pressure;
(iv) estimates of Development Costs and Production Costs under such development project and/or alternative development projects, if any;
(v) Contractor's recommendations as to the particular project that it would prefer;
(vi) Work Programme and Budget for development proposals relating to the proposed Development Area;
(vii) anticipated adverse impact on the environment and measures to be taken for prevention or minimisation thereof and for general protection of the environment in conduct of operations;

(viii) measures to be taken for the health and safety of persons employed in Petroleum Operations;

(ix) the information required in Article 21.

10.8 A proposed development plan submitted by the Contractor pursuant to Article 10.7 may be approved by the Management Committee within one hundred and ten (110) days of submission thereof or eighty (80) days of receipt of any additional information requested by the Management Committee. In case the Management Committee requires any reasonable additional information, the same shall be requested by it within eighty (80) days from the submission of the development plan. The Contractor shall provide such additional information within thirty (30) days from the request by the Management Committee. If, within a period of one hundred and ten (110) days after submission of a proposed development plan or eighty (80) days from the receipt of any additional information, where asked by the Management Committee, the Management Committee fails to convey a decision to the Contractor, the Contractor shall have option to submit the proposal to the Government. Also, where, the Management Committee rejects the development plan of the Contractor, the Contractor can submit the development plan for the approval of the Government. The Government shall respond on the proposed development plan submitted by the Contractor within one hundred and ten (110) days. In case Government refuses to approve the proposed development plan, it shall convey the reasons for such refusal and the Contractor shall be given opportunity to make appropriate modifications to meet concerns of Government and the provisions of the foregoing Article and re-submit the plan within ninety (90) days from the date of receipt of refusal from the Government.

10.9 A Development Plan approved by the Management Committee or Government, as may be the case, from time to time shall commit the Contractor to the obligations stipulated in Articles 10.10 to 10.12.

10.10 Work Programmes and Budgets for Development and Production Operations shall be submitted to the Management Committee as soon as possible after the approval of a Development Plan under Article 10.8 and thereafter not later than 31st December each Year in respect of the Year immediately following.

10.11 The Management Committee, when considering any Work Programme and Budget, may require the Contractor to prepare an estimate of potential production to be achieved through the implementation of the said Work Programme and Budget for each of the three (3) Years following the Year to which the Work Programme and Budget relate. If major changes in yearly estimates of potential production are required, these shall be based on evidence necessitating such changes.

10.12 Not later than the fifteenth (15th) of January each Year, in respect of the Year immediately following commencement of Commercial Production, the Contractor shall determine the "Programme Quantity" with the approval of the Management Committee. The Programme Quantity for any Year shall be the maximum quantity of Petroleum based on Contractor's estimates, as approved by the Management Committee, which can be produced from a Development Area consistent with modern oilfield and petroleum industry practices and minimising unit production cost, taking into account the capacity of the producing Wells, gathering lines, separators, storage capacity and other production facilities available for use during the relevant Year, as well as the transportation facilities up to the Delivery Point.

10.13 Proposed revisions to the details of a Development Plan or an annual Work Programme or Budget in respect of Development and Production Operations shall, for good cause and if the circumstances so justify, be submitted for approval to the Management Committee.

10.14 In the event the area encompassing the Commercial Discovery extends beyond the Development Area designated in the Development Plan, either within the original Contract Area but subsequently relinquished or, outside the original Contract Area, the Management Committee may make recommendations to the Government concerning enlargement of the Development Area, provided the same was not awarded to any other company by the Government or is not held by any other party or not on offer by the Government and no application for a License or Lease is pending with the Government. However, in case the area is held by any other party or on offer by the Government or application for License or Lease is pending with the Government, the Management Committee shall notify the same to the Government for further action on the matter. Government may consider such request for extension at its sole discretion and on terms and conditions, which it may consider fit.

ARTICLE 11
PETROLEUM EXPLORATION LICENSE AND MINING LEASE FOR OFFSHORE AREA

11.1 The Contractor shall submit an application for grant of License in respect of the Contract Area, as early as possible, but not later than fifteen (15) Business Days from the date of execution of this Contract.

11.2 On submission of a development plan of a Commercial Discovery pursuant to Article 10.7, the Contractor shall submit an application for a Lease in respect of the proposed Development Area.

11.3 Where a part of a Reservoir in respect of which a Commercial Discovery has been declared extends beyond the Contract Area, subject to Article 10.14 such area may be included in the proposed Development Area, in relation to which application for a Lease is made, on terms and conditions as decided by the Government; provided that such area is:

(a) not subject to a license or lease granted to any other person;
(b) not the subject of negotiations/bidding for a license or lease; and
(c) available for licensing (i.e. is not an area over which Petroleum Operations are excluded).

11.4 Where a Development Plan has been approved pursuant to Article 10 and the Contractor has complied with the terms and conditions of the License and this Contract and is not in breach of any of the terms thereof, or the provisions of any law and subject to normal Government clearances/approvals being obtained by the Contractor as applicable before grant/issue of the Lease, the Government shall grant to the Contractor a Lease over the Development Area as agreed, subject to Article 11.5 to enable the Contractor to carry out Petroleum Operations in the Development Area in accordance with the Development Plan.

11.5 The Lease shall be granted for an initial period of twenty (20) years from the date of grant thereof subject to:

(a) cancellation in accordance with its terms or for termination of this Contract in accordance with its terms;

(b) extension by mutual agreement between the Parties for five (5) years or such period as may be agreed after taking into account the balance recoverable reserve and balance economic life of the Field/ Development Area from the expiry of the initial period provided that in the event of a Commercial production of Non Associated Natural Gas the extension may be for a period of ten (10) years or such period as may be mutually agreed between the Parties after taking into account the balance recoverable reserves and balance economic life of the Field/ Development Area from the date of expiry of the initial term; and

(c) the terms of this Contract and other terms and conditions as set forth in such Lease be consistent with this Contract and the relevant legislation.

ARTICLE 12
UNIT DEVELOPMENT

12.1 If a Reservoir in a Discovery Area is situated partly within the Contract Area and partly in an area in India over which other parties have a contract to conduct petroleum operations and both parts of the Reservoir can be more efficiently developed together on a commercial basis, on receiving information in writing from any party to these contracts or any information on this from any bonafide source, the Government may, for securing the more effective recovery of Petroleum from such Reservoir, by notice in writing to the Contractor, require that the Contractor:

(a) collaborate and agree with such other parties on the joint development of the Reservoir;

(b) submit such agreement between the Contractor and such other parties to the Government for approval within one hundred and eighty (180) days; and

(c) prepare a plan for such joint development of the said Reservoir, within one hundred and eighty (180) days of the approval of the agreement referred to in (b) above.

12.2 If no plan is submitted within the period specified in Article 12.1 (c) or such longer period as the Government and the Contractor and the other parties referred to in Article 12.1 may agree, or, if such plan as submitted is not acceptable to the Government and the Parties cannot agree on amendments to the proposed joint development plan, the Government may cause to be prepared, at the expense of the Contractor and such other parties a plan for such joint development consistent with generally accepted modern oilfield and petroleum industry practices which shall take into consideration any plans and presentations made by the Contractor and the aforementioned other parties.

12.3 If the parties are unable to agree on the proposed plan for joint development, the Government may call for a joint development plan from an independent agency, which agency, may make such a proposal after taking into account the position of the parties in this regard. Such a development plan, if approved by Government, shall be binding on the parties, notwithstanding their disagreement with the plan. However, the Contractor may in case of any disagreement on the issue of joint development or the proposed joint development plan, prepared in accordance with Article 12.2 or within forty five (45) Business Days of the plan approval as aforesaid in this Article, notify the Government that it elects to surrender its rights in the Reservoir/Discovery in lieu of participation in a joint development.

12.4 If a proposed joint development plan is agreed and adopted by the parties, or adopted following determination by the Government, the plan as finally adopted shall be the approved joint development plan and the Contractor shall comply with the terms of the said development plan as if the Commercial Discovery is established.

12.5 The provisions of Articles 12.1, 12.2 and Article 12.3 shall apply mutatis mutandis to a Discovery of a Reservoir located partly within the Contract Area, which, although not equivalent to a Commercial Discovery if developed alone, would be a Commercial Discovery if developed together with that part of the Reservoir which extends outside the Contract Area to the areas subject to contract for petroleum operations by other parties.

ARTICLE 13
MEASUREMENT OF PETROLEUM

13.1 Petroleum used for internal consumption for Petroleum Operations, flared, saved and sold from the Contract Area shall be measured by methods and appliances generally accepted and customarily used in modern oilfield and petroleum industry practices and approved by the Management Committee and the Government.

13.2 The Government may, at all reasonable times, inspect and test the appliances used for measuring the volume and determining the quality of Petroleum, provided that any such inspection or testing shall be carried out in such a manner so as not to unduly interfere with Petroleum Operations.

13.3 Before commencement of production from the Contract Area, the Parties shall mutually agree on:

(a) methods to be employed for measurement of volumes of Petroleum production;

(b) the point or points at which Petroleum shall be measured and the respective shares allocated to the Parties in accordance with the terms of this Contract;

(c) the frequency of inspections and testing of measurement appliances and relevant procedures relating thereto; and

(d) the consequences of a determination of an error in measurement.

13.4 The Contractor shall undertake to measure the volume and quality of the Petroleum Produced and Saved from the Contract Area at the agreed measurement point consistent with generally accepted modern oilfield and petroleum industry practices with the frequency and according to procedures agreed pursuant to Article 13.3. The Contractor shall not make any alteration in the agreed method or procedures for measurement or to any of the approved appliances used for that purpose without the written consent of the Management Committee and the Government.

13.5 The Contractor shall give the Government timely notice of its intention to conduct measuring operations or any agreed alteration for such operations and the Government shall have the right to be present at and supervise, either directly or through authorized representatives, such operations.

13.6 The Contractor shall keep all the records of analysis and measurement of hydrocarbons calibrations and proving of measurement system and make available to Government or its authorized agency such records on request.

ARTICLE 14
PROTECTION OF THE ENVIRONMENT

14.1 The Government and the Contractor recognise that Petroleum Operations will cause some impact on the environment in the Contract Area. Accordingly, in performance of the Contract, the Contractor shall conduct its Petroleum Operations with due regard to concerns with respect to protection of the environment and conservation of natural resources and shall in particular;

(a) employ modern oilfield and petroleum industry practices and standards including advanced techniques, practices and methods of operation for the prevention of Environmental Damage in conducting its Petroleum Operations;

(b) take necessary and adequate steps to:

(i) prevent Environmental Damage and, where some adverse

impact on the environment is unavoidable, to minimise such damage and the consequential effects thereof on property and people;

(ii) ensure adequate compensation for injury to persons or damage to property caused by the effect of Petroleum Operations; and

(c) comply with the requirements of applicable laws and the reasonable requirements of the Government from time to time.

14.2 If the Contractor fails to comply with the provisions of paragraph (b)(i) of Article 14.1 or contravenes any relevant law, and such failure or contravention results in any Environmental Damage, the Contractor shall forthwith take all necessary and reasonable measures to remedy the failure and the effects thereof.

14.3 If the Government in accordance with the laws has good reason to believe that any works or installations erected by the Contractor or any operations conducted by the Contractor are endangering or may endanger persons or any property of any person, or are causing or may cause pollution, or are harming or may harm fauna or flora or the environment to a degree which the Government deems unacceptable, the Government may require the Contractor to take remedial measures within such reasonable period as may be determined by the Government and to repair any such damage. If the Government deems it necessary, it may also require the Contractor to discontinue Petroleum Operations in whole or in part until the Contractor has taken such remedial measures or has repaired any damage caused.

14.4 The measures and methods to be used by the Contractor for the purpose of complying with the terms of paragraph (b)(i) of Article 14.1 shall be determined in timely consultation with the Government upon the commencement of Petroleum Operations or whenever there is a significant change in the scope or method of conducting Petroleum Operations and shall take into account the international standards applicable in similar circumstances and the relevant environmental impact study carried out in accordance with Article 14.5 below. The Contractor shall notify the Government, in writing, of the measures and methods finally determined by the Contractor and shall cause such measures and methods to be reviewed from time to time in the light of prevailing circumstances.

14.5 The Contractor shall cause a person or persons with special knowledge on environmental matters, to carry out two environmental impact studies in order:

(a) to determine at the time of the studies the prevailing situation relating to the environment, human beings and local communities, the flora and fauna in the Contract Area and in the adjoining or neighbouring areas; and

(b) to establish the likely effect on the environment, human beings and local communities, the flora and fauna in the Contract Area and in the adjoining or neighbouring areas in consequence of the relevant phase of Petroleum Operations to be conducted under this Contract, and to submit, for consideration by the Parties, methods and measures contemplated in Article 14.4 for minimizing Environmental Damage and carrying out Site Restoration activities.

14.5.1 The first of the aforementioned studies shall be carried out in two parts, namely, a preliminary part which must be concluded before commencement of any field work relating to a seismographic or other survey, and a final part relating to drilling in the Exploration Period. The part of the study relating to drilling operations in the Exploration Period shall be approved by Government before the commencement of such drilling operations, it being understood that such approval shall not be unreasonably withheld.

14.5.2 The second of the aforementioned studies shall be completed before commencement of Development Operations and shall be submitted by the Contractor as part of the Development Plan, with specific approval of Government being obtained before commencement of Development Operations, it being understood that such approval shall not be unreasonably withheld.

14.5.3 The studies mentioned in Article 14.5 above shall contain proposed environmental guidelines to be followed in order to minimize Environmental Damage and shall include, but not be limited to, the following, to the extent appropriate to the respective study taking into account the phase of operations to which the study relates:

(a) proposed access cutting;
(b) clearing and timber salvage;
(c) wildlife and habitat protection;
(d) fuel storage and handling;
(e) use of explosives;
(f) camps and staging;
(g) liquid and solid waste disposal;
(h) cultural and archaeological sites;
(i) selection of drilling sites;
(j) terrain stabilization;
(k) protection of freshwater horizons;
(l) blowout prevention plan;
(m) flaring during completion and testing of Gas and Oil Wells;
(n) abandonment of Wells;
(o) rig dismantling and site completion;
(p) reclamation for abandonment;
(q) noise control;
(r) debris disposal; and
(s) protection of natural drainage and water flow.

14.5.4 Subject to the provision of all applicable laws and notifications on protection of environment, any new project or expansion or modernization projects for petroleum operations for which a proposal is submitted by the Contractor, the Government shall complete the assessment of the project within a period of ninety (90) days from the receipt of the requisite documents and data from the project authorities and completion of public hearing. The decision of the Government on the proposal of the Contractor for environmental clearance shall be conveyed within thirty (30) days thereafter. However, wherever forest land is involved; the Contractor/user agency shall have to obtain approval of the Central Government through the State Government concerned, under the Forest (Conservation) Act, 1980. The decision of the Central Government on such proposal of the Contractor/user agency shall be taken within sixty (60) days from the date of receipt of proposal of the Contractor/user agency from the State Government. The State Government is required to convey its recommendations on the proposal complete in all respect to the Central Government or otherwise within ninety (90) days from the date of receipt of the same from the applicant.

14.6 The Contractor shall ensure that:

(a) Petroleum Operations are conducted in an environmentally acceptable and safe manner consistent with modern oilfield and petroleum industry practices and that such Petroleum Operations are properly monitored;

(b) the pertinent completed environmental impact studies are made available to its employees and to its contractors and Subcontractors to develop adequate and proper awareness of the measures and methods of environmental protection to be used in carrying out the Petroleum Operations; and

(c) the contracts entered into between the Contractor and its contractors and Subcontractors relating to its Petroleum Operations shall include the provisions stipulated herein and any established measures and methods for the implementation of the Contractor's obligations in relation to the environment under this Contract.

14.7 The Contractor shall, prior to conducting any drilling activities, prepare and submit for review by the Government contingency plans for dealing with Oil spills, fires, accidents and emergencies, designed to achieve rapid and effective emergency response. The plans referred to above shall be discussed with the Government and concerns expressed shall be taken into account.

14.7.1 In the event of an emergency, accident, Oil spill or fire arising from Petroleum Operations affecting the environment, the Contractor shall forthwith notify the Government and shall promptly implement the relevant contingency plan and perform such Site

Restoration as may be necessary in accordance with modern oilfield and petroleum industry practices.

14.7.2 In the event of any other emergency or accident arising from the Petroleum Operations affecting the environment, the Contractor shall take such action as may be prudent and necessary in accordance with modern oilfield and petroleum industry practices in such circumstances.

14.8 In the event that the Contractor fails to comply with any of the terms contained in Article 14.7 within a period specified by the Government, the Government, after giving the Contractor reasonable notice in the circumstances, may take any action which may be necessary to ensure compliance with such terms and to recover from the Contractor, immediately after having taken such action, all costs and expenditures incurred in connection with such action together with such interest as may be determined in accordance with Section 1.7 of Appendix C of this Contract.

14.9 On expiry or termination of this Contract or relinquishment of part of the Contract Area, the Contractor shall:

(a) subject to Article 27, remove all equipment and installations from the relinquished area or former Contract Area in a manner agreed with the Government pursuant to an abandonment plan; and

(b) perform all necessary Site Restoration in accordance with modern oilfield and petroleum industry practices and take all other action necessary to prevent hazards to human life or to the property of others or the environment.

14.10 The Contractor shall prepare a proposal for the restoration of site including abandonment plan and requirement of funds for this and the annual contribution. This will be submitted along with the annual Budget for the consideration and approval of the Management Committee. The annual contribution shall be deposited by the Contractor in the Site Restoration fund which will be established, in accordance with the scheme notified by the Government.

14.11 Subject to Section 3.2 of Accounting Procedure, any Site Restoration fund scheme formulated by Government and subject to provisions of this Contract, any and all costs incurred by the Contractor pursuant to this Article shall be cost recoverable including but not limited to sinking funds established for abandonment and restoration of the Contract Area.

14.12 In this Article, a reference to Government includes the State Government.

14.13 Where the Contract Area is partly located in areas forming part of certain national parks, sanctuaries, mangroves, wetlands of national importance, biosphere reserves and other biologically sensitive areas passage through these areas shall generally not be permitted. However, if there is no passage, other than through these areas to reach a particular point beyond these areas, permission of the appropriate authorities shall be obtained.

14.14 The obligations and liability of the Contractor for the environment hereunder shall be limited to damage to the environment which:

(a) occurs after the Effective Date; and

(b) results from an act or omission of the Contractor.

ARTICLE 15
RECOVERY OF COST PETROLEUM

15.1 The Contractor shall be entitled to recover Contract Costs out of a percentage of the total value of Petroleum Produced and Saved from the Contract Area in the Year in accordance with the provisions of this Article.

15.2 Exploration Costs incurred by the Contractor in the Contract Area upto the date of first Commercial Production shall be aggregated, and the Contractor shall be entitled to recover the aggregate of such Exploration Costs out of the Cost Petroleum at the rate of one hundred percent (100%) per annum of such Exploration Costs beginning from the date of such Commercial Production.

15.3 The Contractor shall be entitled to recover out of the Cost Petroleum from the Contract Area the Exploration Costs which it has incurred in any Year after the date of Commercial Production at the rate of **one hundred percent (100%)** per annum of such Exploration Costs beginning from the date such Exploration Costs are incurred.

15.4 Development Costs incurred by the Contractor in the Contract Area upto the date of first Commercial Production shall be aggregated, and the Contractor shall be entitled to recover out of the Cost Petroleum the aggregate of such Development Costs at the rate of one hundred percent (100%) per annum of such Development Costs beginning from the date of such Commercial Production.

15.5 The Contractor shall be entitled to recover out of the Cost Petroleum from the Contract Area the Development Costs which it has incurred after the date of first Commercial Production at the rate of one hundred percent (100%) per annum of such Development Costs beginning from the date such Development Costs are incurred.

15.6 The Contractor shall be entitled to recover in full during any Year the Production Costs incurred in that Year out of the Cost Petroleum.

15.7 The Contractor shall be entitled to recover in full during any Year the royalty payments to the Government/State Government(s) in that Year out of the Cost Petroleum.

15.8 If during any Year the Cost Petroleum is not sufficient to enable the Contractor to recover in full the Contract Costs due for recovery in that Year in accordance with the provisions of Articles 15.1 to 15.7 then, subject to the provisions of Article 15.12:

(a) recovery shall first be made of royalty payments; and

(b) recovery shall next be made of the Production Costs; and

(c) recovery shall next be made of the Exploration Costs; and

(d) recovery shall then be made of the Development Costs.

The unrecovered portions of Contract Costs shall be carried forward to the following Year and the Contractor shall be entitled to recover such Contract Costs in such Year or the subsequent Years as if such Contract Costs were due for recovery in that Year, or the succeeding Years, until the unrecovered Contract Costs have been fully recovered out of Cost Petroleum from the Contract Area.

15.9 The maximum amount of Cost Petroleum to which the Contractor shall be entitled, in accordance with the provisions of this Article, shall be X% (to be taken from the accepted bid) of the total value of the Petroleum Produced and Saved from the Contract Area.

15.10 For the purposes of this Article, as well as Article 16, costs, receipts and income shall be converted into production unit equivalents, and vice versa viz both in physical and monitory terms, using the relevant prices established pursuant to Article 19 for Crude Oil and Article 21 for Natural Gas.

15.11 Pending completion of the calculations required to establish definitively the Contractor's entitlement to Cost Petroleum from the Contract Area in any Year, the Contractor shall take delivery, provisionally, of volumes of Crude Oil or Natural Gas representing its estimated Cost Petroleum entitlement calculated with reference to estimated production quantities, costs and prices as established by the Contractor and approved by the Management Committee. Such provisional determination of Cost Petroleum shall be made every Quarter on an accumulative basis. Within ninety (90) days of the end of each Year, a final calculation of the Contractor's entitlement to Cost Petroleum, based on actual production quantities, costs and prices for the entire Year as reflected in audited accounts under Article 25.4.3, shall be undertaken and any necessary adjustments to the Cost Petroleum entitlement shall be agreed upon between the Government and the Contractor within thirty (30) days and made within thirty (30) days thereafter.

15.12 Where more than one Party constitutes the Contractor, the percentage of the total Cost Petroleum from the Contract Area which shall be available to each such Party in any Year for recovery of its share of Contract Costs shall be determined on the basis of the respective Participating Interest of each such Party.

ARTICLE 16
PRODUCTION SHARING OF PETROLEUM

16.1 The Parties to this Contract shall share in the Profit Petroleum in each Year in accordance with the provisions of this Article. A Party's share of Profit Petroleum in any Year, shall be calculated on the basis of the Investment Multiple actually achieved by the Contractor at the end of the preceding Year for the Contract Area as provided in Appendix-D.

16.2.1 When the Investment Multiple of the Contractor at the end of any

Year is less than one and one half (1.5), the Government shall be entitled to take and receive [___] percent (__%) and the Contractor shall be entitled to take and receive [___] percent (__%) of the total Profit Petroleum from the Contract Area with effect from the start of the succeeding Year.

16.2.2 When the Investment Multiple of the Contractor at the end of any Year is equal to or more than one and one half (1.5) but is less than two (2.0) the Government shall be entitled to take and receive [___] percent (__%) and the Contractor shall be entitled to take and receive [___] percent (__%) of the total Profit Petroleum from the Contract Area with effect from the start of the succeeding Year.

16.2.3 When the Investment Multiple of the Contractor at the end of any Year is equal to or more than two (2.0) but is less than two and one half (2.5), the Government shall be entitled to take and receive [___] percent (__%) and the Contractor shall be entitled to take and receive [___] percent (__%) of the total Profit Petroleum from the Contract Area with effect from the start of the succeeding Year.

16.2.4 When the Investment Multiple of the Contractor at the end of any Year is equal to or more than two and one half (2.5) but is less than three (3.0), the Government shall be entitled to take and receive [___] percent (__%) and the Contractor shall be entitled to take and receive [___] percent (__%) of the total Profit Petroleum from the Contract Area with effect from the start of the succeeding Year.

16.2.5 When the Investment Multiple of the Contractor at the end of any Year is equal to or more than three (3.0) but is less than three and one half (3.5), the Government shall be entitled to take and receive [___] percent (__%) and the Contractor shall be entitled to take and receive [___] percent (__%) of the total Profit Petroleum from the Contract Area with effect from the start of the succeeding Year.

16.2.6 When the Investment Multiple of the Contractor at the end of any Year is equal to or more than three and one half (3.5), the Government shall be entitled to take and receive [___] percent (__%) and the Contractor shall be entitled to take and receive [___] percent (__%) of the total Profit Petroleum from the Contract Area with effect from the start of the succeeding Year.

Note : These figures shall be as per the bid of the Contractor and accepted by Government.

16.3 Any balance left to the credit of the Parties in any Site Restoration account, opened pursuant to the provision of Article 14.10, after Site Restoration has been done by the Contractor in accordance with the provisions of this Contract and the laws in this regard, shall be shared between the Government and the Contractor as per the Investment Multiple reached at the time of ceasing of production from the Contract Area.

16.4.1 The Government shall have the option to take its entitlement to

Profit Petroleum other than "ANG" or "NANG" either in cash or in kind in any Year. In case of "ANG" or "NANG", as the case may be, the Government shall have the option to take its entitlement to Profit Petroleum in cash or in kind and such option shall be exercised at interval of every five (5) Years from the commencement of first Commercial Production from the Contract Area.

16.4.2 In accordance with the Article 16.4.1, The Government shall exercise such option by giving a written notice to the Contractor not later than thirtieth (30th) June in the preceding Year in which the entitlement is due. Once the Government has exercised its option, the same shall continue unless the Government informs the Contractor otherwise.

16.4.3 Where the Government has informed the Contractor of its intention to take its share in kind, the Parties shall mutually agree on a procedure for delivery of the Government's share of Profit Petroleum and, where relevant, the composition of the Petroleum which is to be delivered.

16.5 The value of the Contractor's Investment Multiple at the end of any Year in respect of the Contract Area shall be calculated in the manner provided for, and on the basis of the net cash flows specified in Appendix-D to this Contract. However, the amount of Profit Petroleum to be shared between the Government and the Contractor shall be determined for each Quarter on an accumulative basis. Pending finalisation of accounts, Profit Petroleum shall be shared between the Government and the Contractor on the basis of provisional estimated figures of Contract Costs, production, prices, receipts, income and any other income or allowable deductions and on the basis of the value of the Investment Multiple achieved at the end of the preceding Year. All such provisional estimates shall be approved by the Management Committee. When it is necessary to convert monetary units into physical units of production equivalents or vice versa, the price or prices determined pursuant to Articles 19 and 21 for Crude Oil, Condensate and Natural Gas respectively shall be used. Within ninety (90) days of the end of each Year, a final calculation of Profit Petroleum based on actual costs, quantities, prices and income for the entire Year shall be completed and any necessary adjustments to the sharing of Petroleum shall be agreed upon between the Government and the Contractor within thirty (30) days and made within thirty (30) days thereafter. (Explanation : The Profit Petroleum due to the Government shall be deposited with "Pay & Accounts officer or its successor, Ministry of Petroleum & Natural Gas, Government of India, Shastri Bhavan, New Delhi by 10th of the Month following each Quarter)

16.6 The Profit Petroleum due to the Contractor in any Year from the Contract Area shall be divided amongst the Parties constituting the Contractor, in proportion to their respective Participating Interest.

ARTICLE 17
TAXES, ROYALTIES, RENTALS, DUTIES ETC.

17.1 Companies, their employees, persons providing any materials, supplies, services or facilities or supplying any ship, aircraft, machinery, equipment or plant (whether by way of sale or hire) to the Companies for Petroleum Operations or for any other purpose and the employees of such persons shall be subject to all fiscal legislation in India except where, pursuant to any authority granted under any applicable law, they are exempted wholly or partly from the application of the provisions of a particular law or as otherwise provided herein.

17.2 Pursuant to the provisions of section 42 of the Income-tax Act, 1961, the allowances specified herein shall apply in computing income tax payable by a Company on its profits and gains from the business of Petroleum Operations in lieu of (and not in addition to) corresponding allowances provided for under the heading "Profits and Gains of Business or Profession" in the Income-tax Act, 1961. Any other allowance, which are not specified herein, shall be treated in accordance with the provisions of Income-tax Act, 1961.

17.2.1 Subject to the provisions herein below, deductions at the rate of one hundred percent (100%) per annum shall be allowed for all expenditures, both capital and revenue expenditures, incurred in respect of Exploration Operations and drilling operations. The expenditure incurred in respect of Development Operations, other than drilling operations, and Production Operations will be allowable as per the provisions of the Income-tax Act, 1961. The expenses so incurred are subject to the following:

(a) where any expenditure is not solely incurred on Petroleum Operations or is incurred as part of or in conjunction with any other business, only that proportion of the total expenditure which can be proved to the assessing officer to represent a fair proportionate part thereof, having regard to all relevant facts and circumstances, shall be allowed;

(b) sections 40A and 44C of the Income-tax Act, 1961, shall apply.

17.2.2 A Company shall be entitled, for income tax purposes only, to deduct all its unsuccessful Exploration Costs in contract areas covered by other contracts from the aggregate value of Petroleum allocable to the Company from any Field(s) in the Contract Area in the manner as follows:

(a) unsuccessful Exploration Costs incurred in contract areas other than the Contract Area where a Commercial Discovery has been made up to the date of commencement of Commercial Production shall be aggregated and the Company shall be entitled to deduct such costs at the rate of one hundred per cent (100%) per annum;

(b) unsuccessful Exploration Costs incurred in contract areas other than the Contract Area where a Commercial Discovery has been made, after the commencement of Commercial Production, shall be deductible at the rate of one hundred per cent (100%) per annum of such costs beginning from the Year such costs are incurred.

17.2.3 All allowable expenditure incurred prior to the Year in which Commercial Production commences shall be aggregated and the assessed loss for that Year as well as the assessed loss, if any, incurred in the assessment year relevant to the Year in which Commercial Production commences, or in any subsequent assessment year, shall be carried forward to succeeding assessment years and set off as provided in the Income-tax Act, 1961.

17.2.4 For any or all accumulated expenditures incurred in respect of Exploration Operations and drilling operations prior to the date of commercial production, Company(ies) shall have option to amortize such expenditures over a period of ten (10) years from the date of first commercial production.

17.2.5 The profits and gains of the business of the Parties comprising the Contractor consisting of Petroleum Operations shall, for the purpose of levy of income tax under the Incometax Act, 1961, be computed on the basis of the value, determined in accordance with Article 19, of its Participating Interest share of Crude Oil produced and saved and sold, or otherwise disposed of, from the Contract Area and from any revenue realised on the sale of Associated or Non Associated Natural Gas referred to in Article 21 as well as any other gains or receipts from Petroleum Operations as reduced by the deductions as specified herein, and, except as herein provided, all the provisions of the Income-tax Act, 1961, shall apply.

17.2.6 Company(ies) shall be eligible for benefits available under section 80 IB of the Incometax Act, 1961 as applicable from time to time.

17.3 For the purposes of Article 17.2 and section 42 of the Income-tax Act, 1961:

17.3.1 The following terms used in section 42 of the Income-tax Act, 1961, shall have the meanings corresponding to the terms used in this Contract and defined in Article 1 as follows:

(a) "agreement" means this Contract as defined in Article 1;

(b) "commercial production" shall have the meaning assigned in Article 1.

17.3.2 The terms "assessing officer", "assessed loss", and "assessment year" shall have the meaning as defined in the Income-tax Act, 1961.

17.3.3 The other terms used herein and defined in Article 1 shall have the meaning therein ascribed.

17.4 Companies (Lessee) shall be required to pay royalty to the Government (Lessor) for offshore areas at the rate of ten percent (10%) of the well-head value of Crude Oil and Natural Gas. In case of an onshore area, Companies shall be required to pay to the State Government(s) (Lessor) at the rate of twelve point five zero percent (12.5%) of the well-head value of Crude Oil and ten percent (10%) of the well-head value of Natural Gas. In case of an offshore area falling beyond four hundred (400) metre isobath, the rate of royalty payable by Companies (Lessee) to the Government (Lessor) shall be at the rate of five percent (5%) of the well-head value of Crude Oil and Natural Gas for the first seven years from the date of commencement of Commercial Production in the Field. The valuation of Crude Oil and Natural Gas shall be as per the Article 19 and Article 21 respectively. The royalty amount due to Government/State Government(s) shall be payable latest by the end of the succeeding Month.

17.5 Machinery, plant, equipment, materials and supplies imported by the Contractor and its Subcontractors solely and exclusively for use in Petroleum Operations under this Contract or similar contracts with the Government where customs duty has been exempted by the Government shall be exempt from customs duties and export duties or other charges on re-exportation of the said items in accordance with applicable legislation.

17.6 The Government shall have the right to inspect the records and documents of the physical item or items for which an exemption has been provided pursuant to Article 17.5 to determine that such item or items are being or have been imported solely and exclusively for the purpose for which the exemption was granted. The Government shall also be entitled to inspect such physical items wherever located to ensure that such items are being used for the purpose herein specified and any item not being so used shall immediately become liable to payment of the applicable customs duties.

17.7 Subject to Article 27, the Contractor and its Subcontractors may sell or otherwise transfer in India all imported items which are no longer required for Petroleum Operations, subject to applicable laws including rules, regulations, procedures, notifications etc. governing customs duties and sale or disposal of such items.

17.8 Any sales tax or tax of similar nature payable on the sale(s) of Petroleum under this Contract shall be borne/reimbursed by the buyer(s).

17.9 Subject to the provisions herein above provided, the Contractor shall be liable for payment of:

(a) annual license charges and rental fees and other charges under the Rules;

(b) charges payable by specified industries or in connection with Petroleum Operations under applicable legislation;

(c) payments for purchase, lease or rental of land or land rights in connection with Petroleum Operations;

(d) taxes, fees or charges for specific services rendered on request or to the public generally;

(e) customs duties, except for those items subject to exemption as provided in Article 17, applicable at the rates specified from time to time; and

(f) stamp duties, registration fees, license fees, taxes such as taxes on property or assets (not calculated by reference to income or otherwise exempted) or other levies, fees or charges of a non-discriminatory nature and generally applicable in India or in the State where Petroleum Operations are being conducted.

17.10 If any change in or to any Indian law, rule or regulation dealing with income tax or other corporate tax, export/import tax, excise, customs duty or any other levies, duties or taxes imposed on Petroleum or dependent upon the value of Petroleum results in a material change to the expected economic benefits accruing to any of the Parties after the date of execution of the Contract, the Parties shall consult promptly in good faith to make necessary revisions and adjustments to the Contract in order to maintain such expected economic benefits to each of the Parties, provided, however, that the expected economic benefits to the Parties shall not be reduced as a result of the operation of this Article.

ARTICLE 18
DOMESTIC SUPPLY, SALE, DISPOSAL AND EXPORT OF CRUDE OIL AND CONDENSATE

18.1 Until such time as the total availability to the Government of Crude Oil and Condensate from all Petroleum production activities in India meets the total national demand, each Company comprising the Contractor, shall be required to sell in the domestic market in India all of the Company's entitlement to Crude Oil and Condensate from the Contract Area in order to assist in satisfying the national demand.

18.2 If, during any Year, India attains Self-sufficiency, the Government shall promptly thereafter, but in no event later than the end of the first Quarter of the following Year, so advise the Company(ies) by written notice. In such event, as from the end of the second Quarter of the following Year, or such earlier date as the Parties may mutually agree, domestic sale obligation shall be suspended and the Company shall have the right to lift and export its Participating Interest share of Crude Oil and Condensate until such time, if any, as Self-sufficiency shall have ceased to exist. If Self-sufficiency ceases to exist during a Year, the Government shall recover its position to ask Company(ies) under Article 18.1 in respect of the following Year by giving ninety (90) days notice thereof to the Company(ies) to sell Crude Oil and Condensate in the Indian domestic market.

18.3 Upon India achieving Self-sufficiency, the Company(ies) shall be entitled to freely lift and export any Crude Oil and Condensate pursuant to this

Article 18, subject to Government's generally applicable destination restrictions to countries with which the Government, for policy reasons, has severed or restricted trade.

18.4 No later than sixty (60) days prior to the commencement of production in a Field, and thereafter no less than sixty (60) days before the commencement of each Year, the Contractor shall cause to be prepared and submitted to the Parties a production forecast setting out the total quantity of Crude Oil that it estimates can be produced from a Field during the succeeding Year, based on a maximum efficient rate of recovery of Crude Oil from that Field in accordance with modern oilfield and petroleum industry practices. No later than thirty (30) days prior to the commencement of each Quarter, the Contractor shall inform its estimate of production for the succeeding Quarter and shall endeavour to produce the forecast quantity for each Quarter.

18.5 Each Company comprising the Contractor shall, throughout the term of this Contract, have the right to separately take in kind and dispose of all its share of Cost Petroleum and Profit Petroleum and shall have the obligation to lift the said Petroleum on a current basis and in such quantities so as not to cause a restriction of production or inconvenience to the other Company(ies).

18.6 The Government shall, throughout the term of this Contract, have the right to separately take in kind and dispose of its share of Crude Oil and shall have the obligation to lift the said Oil on a current basis and in such quantities so as not to cause a restriction of production or inconvenience to the Contractor.

18.7 For the purpose of implementing the provisions of this Article, a Crude lifting procedure and Crude sales agreement based on generally acceptable international terms shall be agreed upon by the Contractor with buyer(s) no later than six (6) months or such shorter period as may be mutually agreed between the Contractor and buyer(s) with the consent of Government prior to the commencement of production in a Field. Such lifting procedure shall be made available to all the Parties to this Contract.

ARTICLE 19
VALUATION OF PETROLEUM

19.1 For the purpose of this Contract, the value of Crude Oil, Condensate and Natural Gas (refer Article 21) shall be based on the price determined as provided herein.

19.2 A price for Crude Oil shall be determined for each Month or such other period as the Parties may agree (hereinafter referred to as "the Delivery Period") in terms of United States Dollars per Barrel, on import parity basis (with marine freight being determined on the basis of nearest port to the Contract Area) for Crude Oil produced and sold or otherwise disposed of from Contract Area, for each Delivery Period, in accordance with the appropriate basis for that type of sale or disposal specified below. Subject to the provisions of this Article 19, it is clearly understood that the actual prices received by the Company(ies) from the sales will form the basis for the purposes of cost recovery, Profit Petroleum sharing and payment of royalty as provided in the Articles 15, 16 and 17 respectively. The basis of valuation given in this Article for the purpose of Article 15,16 and 17 shall apply only where Government is of the view that sale prices realized by the Company(ies) are not consistent with the price realisable at Arms Length Sales.

19.3 In the event that some or all of a Company's or Contractor's total sales of Crude Oil during a Delivery Period are made to third parties at Arms Length Sales, all sales so made shall be valued at the weighted average of the prices actually received by a Company, calculated by dividing the total receipts from all such sales at the Delivery Point by the total number of Barrels of the Crude Oil sold in such sales.

19.3.1 Each Company constituting the Contractor shall separately submit to the designated nominee of the Government, within fifteen (15) days of the end of each Delivery Period, a report containing the actual prices obtained in their respective Arms Length Sales for any Crude Oil. Such reports shall distinguish between term sales and spot sales and itemize volumes, customers, prices received and credit terms, and a Company shall allow the designated nominee(s) of the Government to examine the relevant sales contracts.

19.4 For the purpose of determining price at Arms Length Sales, the price of the Crude Oil at which sale takes place will generally be based on per Barrel of one or more crude oils which, at the time of calculation, are being freely and actively traded in the international market and are similar in characteristics and quality to the Crude Oil in respect of which the price is being determined, selling price to be ascertained from Platt's Crude Oil Market Wire daily publication ("Platt's"), or the spot market for the same crude oils ascertained in the same manner, whichever price more truly reflects the current value of such crude oils. For any Delivery Period in which sales take place, the price shall be the arithmetic average price per Barrel determined by calculating the average for such Delivery Period of the mean of the high and low FOB prices for each day of the crude oils selected for comparison adjusted for differences in the Crude Oil and the crude oils being compared for quality, transportation costs, delivery time, quantity, payment terms and other contract terms to the extent known and other relevant factors. In the event that Platt's ceases to be published or is not published for a period of thirty (30) consecutive days, the Parties shall agree on an alternative daily publication. The Contractor shall make available all the data pertaining to pricing of Crude to enable Government to decide that the proposed sale price by the Contractor/each constituents of the Contractor reflects a fair market price for the Crude.

19.4.1 In the event that, at the relevant time, no crude oils of similar quality to the Crude Oil to be sold are being actively traded in the international markets where prices can be ascertained by international publication, or the official FOB selling prices and the international spot market price vary widely between producers, the Parties shall meet in good faith to determine an appropriate pricing basis.

19.5 The Contractor shall determine the relevant prices in accordance with this Article and the calculation, basis of calculation and the price determined shall be supplied to the Government and shall be subject to agreement by the Government.

19.6 In the event that the Parties fail to reach agreement on any matter concerning selection of the crude oils for comparison, the calculation, the basis of, or mechanism for the calculation of the prices, the prices arrived at, the adjustment of any price or generally about the manner in which the prices are determined according to the provisions of this Article within thirty (30) days, or such longer period as may be mutually agreed between the Parties, from the date of commencement of Commercial Production or the end of each Delivery Period thereafter, any Party may refer the matter or matters in issue for final determination by a sole expert or arbitrator appointed as provided in Article 33.

19.6.1 If the matter is referred to the sole expert, within ten (10) days of the said appointment, the Parties shall provide the expert with all information they deem necessary or as the expert may reasonably require.

19.6.2 Within fifteen (15) days from the date of his appointment, the expert shall report to the Parties on the issue(s) referred to him for determination, applying the criteria or mechanism set forth herein and indicate his decision thereon to be applicable for the relevant Delivery Period for Crude Oil and such decision shall be accepted as final and binding by the Parties.

19.6.3 Any price or pricing mechanism agreed by the Parties pursuant to the provisions of this Article shall not be changed retroactively.

19.7 In the event that all sales of Crude Oil in a Delivery Period by a Company constituting the Contractor are to be made to an Affiliate, the Parties may agree on an alternative method of valuing the Crude Oil for the purposes of this Contract, provided that such alternative method results in an internationally competitive fair market valuation for that Delivery Period. In case of disagreement, the decision of the Government on determining a Crude price in case of sales to an Affiliate shall be final and binding.

19.8 In the event that in any Delivery Period there is more than one type of sales referred to in Articles 19.3 and 19.7, then, for the purpose of calculating Cost Petroleum and Profit Petroleum entitlement and royalty payments pursuant to Articles 15, 16 and 17 respectively, a single price per Barrel of Crude Oil for all the sales for the relevant Delivery Period shall be used. Such single price shall be the weighted average of the prices determined for each type of sale, weighted by the respective volumes of Crude Oil sold in each type of sale in the relevant Delivery Period.

19.9 The provisions specified above for the determination of the price of sales of Crude Oil shall apply mutatis mutandis to Condensates.

19.10 The price of Natural Gas shall be determined as provided in Article 21.

ARTICLE 20
CURRENCY AND EXCHANGE CONTROL PROVISIONS

20.1 Subject to the provisions herein, and to compliance with the relevant provisions of the laws of general application in India governing currency and foreign exchange and related administrative instructions and procedures issued thereunder on a non-discriminatory basis, each Foreign Company comprising the Contractor shall, during the term of this Contract, have the right to:

(a) repatriate abroad, in United States Dollars or any other freely convertible currency acceptable to the Government and the Foreign Company, the net proceeds of sales of Petroleum in India;

(b) receive, retain and use abroad the proceeds of any export sales of Petroleum under the Contract;

(c) open, maintain and operate bank accounts with reputable banks, both inside and outside India, for the purpose of this Contract;

(d) freely import, through normal banking channels, funds necessary for carrying out the Petroleum Operations;

(e) convert into foreign exchange and repatriate sums imported pursuant to (d) above in excess (if any) of its requirements; and

(f) make payments outside of India for purchases, services and loans obtained abroad without the requirement that funds used in making such payments must come from or originate in India. Provided, however, that repatriation pursuant to subparagraphs (a) and (e) and payments pursuant to subparagraph (f) shall be subject to the provisions of any treaties and bilateral arrangements between the Government and any country with respect to payments to or from that country.

20.2 The rates of exchange for the purchase and sale of currency by the Companies shall be the prevailing rates of general application determined by the Reserve Bank of India or such other financial body as may be mutually agreed by the Parties and, for accounting purposes under this Contract, these rates shall apply as provided in Section 1.6 of Appendix-C.

20.3. A Party other than a Foreign Company comprising the Contractor shall be governed by the relevant currency and foreign exchange laws and related administrative instructions and procedures issued thereunder.

20.4 Indian Companies shall have right to remit their portion of expenditure in foreign currency(ies) in accordance with the exchange control provisions.

ARTICLE 21
NATURAL GAS

21.1 Subject to Article 21.2, the Indian domestic market shall have the first call on the utilisation of Natural Gas discovered and produced from the Contract Area. Accordingly, any proposal by the Contractor relating to Discovery and production of Natural Gas from the Contract Area shall be made in the context of the Government's policy for the utilisation of Natural Gas and shall take into account the objectives of the Government to develop its resources in the most efficient manner and to promote conservation measures.

21.2 The Contractor shall have the right to use Natural Gas produced from the Contract Area for the purpose of Petroleum Operations including reinjection for pressure maintenance in Oil Fields, gas lifting and captive power generation required for Petroleum Operations.

21.3 For the purpose of sales in the domestic market pursuant to this Article 21, the Contractor shall have freedom to market the Gas and sell its entitlement.

21.4 Associated Natural Gas (ANG)

21.4.1 In the event that a Discovery of Crude Oil contains ANG, the Contractor shall declare in the proposal for the declaration of the said Discovery as a Commercial Discovery as specified in Article 10, whether (and by what amount) the estimated production of ANG is anticipated to exceed the quantities of ANG which will be used in accordance with Article 21.2 (such excess being hereinafter referred to as "the Excess ANG"). In such an event the Contractor shall indicate whether, on the basis of the available data and information, it has reasonable grounds for believing that the Excess ANG could be commercially exploited in accordance with the terms of this Contract along with the Commercial Production of the Crude Oil from the Contract Area, and whether the Contractor intends to so exploit the Excess ANG.

21.4.2 Based on the principle of full utilisation and minimum flaring of ANG, a proposed development plan for an Oil Discovery shall, to the extent practicable, include a plan for utilisation of the ANG including estimated quantities to be flared, reinjected, and to be used for Petroleum Operations; and, if the Contractor proposes to commercially exploit the Excess ANG for sale in the domestic market in accordance with Government's policy, or elsewhere, the proposed plans for such exploitation.

21.4.3 If the Contractor wishes to exploit the Excess ANG, subject to Article 21.1, the Contractor shall be free to explore markets for the commercial exploitation of the said Excess ANG and submit its proposals for such exploitation to the Government in accordance with Article 21.4.2.

21.4.4 Where the Contractor is of the view that the Excess ANG cannot be commercially exploited and chooses not to exploit the said Excess ANG, or is unable to find a market for the Excess ANG pursuant to Article 21.4.3, the Government shall be entitled to take and utilize such Excess ANG free of any cost/charge.

21.4.5 If the Government elects to take the Excess ANG as provided in Article 21.4.4:

(a) the Contractor shall deliver such Excess ANG to the Government (or its nominee) free of any cost/charge, at the downstream flange of the Gas/Oil separation facilities;

(b) the Contractor shall, based on sound petroleum engineering practices, install such facilities as would facilitate, insofar as practicable, uninterrupted delivery of such Excess ANG to the Government or its nominee;

(c) the cost of all facilities installed pursuant to paragraph (b) above shall be borne by the Government (or its nominee);

(d) the Government or its nominee shall bear all costs including gathering, treating, processing and transporting costs beyond the downstream flange of the Gas/Oil separation facilities; and

(e) the delivery of such Excess ANG shall be subject to procedures to be agreed between the Government or its nominee and the Contractor prior to such delivery, such procedures to include matters relating to timing of off-take of such Excess ANG. Parties shall endeavour that such procedures do not restrict Oil production.

21.4.6 The Excess ANG which is not commercially exploited by the Contractor, or taken by the Government or its nominee pursuant to this Article 21, shall be returned to the subsurface structure or flared or otherwise disposed off as approved by the Government in the context of the Development Plan, provided that flaring will be resorted to only for small quantities and as a last resort.

21.4.7 As soon as practicable after the submission of the proposed development plan, the Contractor and the Government or its nominee shall meet to discuss the sale and/or disposal of any ANG discovered with a view to giving effect to the provisions of this Article 21 in a timely manner.

21.5 Non Associated Natural Gas (NANG)

21.5.1 In the event of a Discovery of NANG in the Contract Area, the Contractor shall promptly report such Discovery to the Management Committee and the Government and the provisions of Articles 10.1 and 10.2 shall apply. The remaining provisions of Article 10 would apply to the Discovery and development of

NANG only insofar as they are not inconsistent with the provisions of this Article. Notwithstanding the provisions of Article 3, the Contractor shall be entitled to retain the Discovery Area subject to the provisions of this Article 21.

21.5.2 If, pursuant to Article 10.1, the Contractor gives notification that the Discovery is of potential commercial interest, the Contractor shall submit to the Management Committee, within one (1) year from the date of notification of the above said Discovery, the proposed Appraisal Programme, including a Work Programme and Budget to carry out an adequate and effective appraisal of such Discovery, to determine (i) without delay, whether such Discovery is a Commercial Discovery and (ii) with reasonable precision, the boundaries of the area to be delineated as the Development Area. Such proposed Appraisal Programme shall be supported by all relevant data such as Well data, Contractor's best estimate of reserve range and production potential, and shall indicate the date of commencement of the proposed Appraisal Programme.

21.5.3 The proposed Appraisal Programme together with the Work Programme and Budget referred to in Article 21.5.2 shall be reviewed by the Management Committee within sixty (60) days of its submission by the Contractor. The Management Committee shall offer its comments within the said period. The said Appraisal Programme together with the Work Programme and Budget submitted by the Contractor as revised or modified or amended in light of the Management Committee review and advice, shall be adopted as the Appraisal Programme and the Contractor shall promptly proceed with implementation of the said Programme.

21.5.4 If on the basis of the results of the Appraisal Programme, the Contractor is of the opinion that NANG has been discovered in commercial quantities, it shall submit to the Management Committee as soon as practicable but not later than three (3) years from the date of notification of the aforementioned Discovery, a proposal for the declaration of the Discovery as a Commercial Discovery. Such proposal shall take into account the Government's policies on Gas utilization and propose alternative options, if any for use or consumption of the NANG and be accompanied by a report on the Discovery supported by, inter alia, technical and economic data, evaluations, interpretations and analyses of such data and feasibility studies relating to the Discovery prepared by or on behalf of the Contractor, and other relevant information. If no proposal is submitted to the Management Committee by the Contractor within three (3) years from the said Discovery, the Contractor shall relinquish its rights to develop such Discovery and the area relating to such Discovery shall be excluded from the Contract Area.

21.5.5 Where the Contractor has submitted a proposal for the declaration of a Discovery as a Commercial Discovery, the Management Committee shall consider the proposal of the Contractor with reference to commercial utilization or commercial development of the NANG in the domestic market or elsewhere and in the context of Government's policy on Gas utilization and the chain of activities required to bring the NANG from the Delivery Point to potential consumers in the domestic market or elsewhere. The Management Committee may, within eighty five (85) days of the submission of the said proposal, request the Contractor to submit any additional information on the Discovery, the anticipated markets or any other related matter, that may reasonably be required to facilitate a review. The Contractor shall submit the required information within thirty (30) days of the request by the Management Committee. The Management Committee will advise the Contractor of its review within one hundred and thirty five (135) days from the submission of proposal or within fifty five (55) days from the receipt of additional information, as the case may be, on the proposal made by the Contractor to declare the Discovery as a Commercial Discovery.

21.5.6 If the Contractor declares the Discovery a Commercial Discovery after taking into account the advice of the Management Committee as referred to in the Article 21.5.5, the Contractor shall, within one (1) year of the declaration of the Discovery as a Commercial Discovery, submit a development plan for the development of the Discovery to the Management Committee for approval. Such plan shall be supported by all relevant information including, inter alia, the information required in Article 10.7.

21.5.7 Unless otherwise agreed by the Management Committee, it shall consider the proposed development plan and give their approval within one hundred and sixty five (165) days of submission thereof or eighty five (85) days from the receipt of the clarifications/ additional information from the Contractor. Any clarification/additional information required by the Management Committee shall be asked for within eighty five (85) days of receipt of the proposal from the Contractor. The Contractor shall provide such additional information within thirty (30) days from the receipt of request by the Management Committee. If the Management Committee fails to convey its decision within one hundred and sixty five (165) days from the submission of the development plan or eighty five (85) days from the receipt of the clarifications/additional information, whichever is later, the Contractor may submit the development plan for the approval of the Government. Also, where, the Management Committee rejects the development plan of the Contractor, the Contractor can submit the development plan for the approval of the Government.

21.5.8 Where the development plan is submitted to the Government for approval pursuant to Article 21.5.7, the Government shall convey its decision within one hundred and fifteen (115) days from the date of receipt of the proposal from the Contractor. Government, where it considers necessary, may ask clarifications/additional information from the Contractor within eighty five (85) days and shall convey its decision within fifty five (55) days from the date of receipt of such clarifications/additional information.

21.5.9 If the Government has failed to approve or disapproves the Contractor's proposed development plan, within one hundred and fifteen (115) days from receipt or within fifty five (55) days from the receipt of clarifications/ information from the Contractor as mentioned in the Article 21.5.8, the Government shall advise the Contractor, in writing, of the reasons for such failure or disapproval and the Government and the Contractor shall meet to discuss the said development plan and the reasons for the said failure to approve or disapproval, and use their best efforts to agree on appropriate modifications thereto to meet the Government's concerns or objections. Thereafter, the Contractor shall have the right to resubmit, within eighty five (85) days of communication from the Government, the proposed development plan duly amended to meet the Government's concerns. Such right of resubmission of the proposed development plan shall be exercisable by the Contractor only once. The Government will respond to the re-submitted plan within one hundred and fifteen (115) days. If no such plan is submitted to the Government within the above specified period, the Contractor shall relinquish its right to develop such Gas Discovery and such Discovery shall be excluded from the Contract Area.

21.5.10 In the event that the Management Committee or Government, as may be the case, approves the Contractor's development plan for the development of such Commercial Discovery, with such modifications and amendments as the Management Committee or Government, as may be the case, may approve, the said Gas Discovery shall be promptly developed by the Contractor in accordance with the approved plan which shall be the Development Plan.

21.5.11 The Contractor will have a two (2) years period, from the date of approval of the Development Plan by the Management Committee or Government, to tie-up the market(s) for sale of Non-associated Natural Gas.

21.5.12 In the event the Contractor does not commence development of such Discovery within ten (10) years from the date of the first Discovery Well, the Contractor shall relinquish its right to develop such Discovery and the area relating to such Discovery shall be excluded from the Contract Area.

21.6 Valuation of Natural Gas

21.6.1 The Contractor shall endeavour to sell all Natural Gas produced and saved from the Contract Area at arms-length prices to the benefits of Parties to the Contract.

21.6.2 Notwithstanding the provision of Article 21.6.1, Natural Gas produced from the Contract Area shall be valued for the purposes of this Contract as follows:

(a) Gas which is used as per Article 21.2 or flared with the approval of the Government or re-injected or sold to the Government pursuant to Article 21.4.5 shall be ascribed a zero value;

(b) Gas which is sold to the Government or any other Government nominee shall be valued on the terms and conditions actually obtained including pricing formula and delivery; and (Explanation : However, it is clarified that this provision would apply only when the sale is made to the Government or Government nominee under the provisions of the Contract)

(c) Gas which is sold or disposed of otherwise than in accordance with paragraph (a) or (b) shall be valued on the basis of competitive arms length sales in the region for similar sales under similar conditions.

21.7 The formula or basis on which the prices shall be determined pursuant to Article 21.6 shall be approved by the Government prior to the sale of Natural Gas to the consumers/ buyers, within sixty (60) Business Days from the receipt of proposal or from the date of receipt of clarification/additional information, where asked for by the Government. For granting this approval, Government shall take into account the prevailing policy, if any, on pricing of Natural Gas including any linkages with traded liquid fuels, and it may delegate or assign this function to a regulatory authority as and when such an authority is in existence and in place.

ARTICLE 22
EMPLOYMENT, TRAINING AND TRANSFER OF TECHNOLOGY

22.1 Without prejudice to the right of the Contractor to select and employ such number of personnel as, in the opinion of the Contractor, are required for carrying out Petroleum Operations in a safe, cost effective and efficient manner, the Contractor shall, to the maximum extent possible, employ, and require the Operator and Subcontractors to employ, citizens of India having appropriate qualifications and experience, taking into account experience required in the level and nature of the Petroleum Operations.

22.2 The Operator shall offer a mutually agreed number of Indian nationals the opportunity for on-the-job training and practical experience in Petroleum Operations during the Exploration Period. Not later than six (6) months after approval of the Development Plan, the Operator shall, in consultation with the Government, establish and implement training programmes for staff positions in each phase and level of Petroleum Operations including skilled, technical, executive and management positions, with a view to ensuring employment of nationals of India and gradual and progressive reduction of foreign personnel.

22.3 At the request of the Government, the Foreign Companies shall separately endeavour to negotiate, in good faith, technical assistance agreements with the Government or a company nominated by Government for this purpose setting forth the terms by which each Foreign Company constituting the Contractor may render technical assistance and make available commercially proven technical information of a proprietary nature for use in India by the Government or the company nominated by Government. The issues to be addressed in negotiating such technical assistance agreements shall include, but not be limited to, licensing issues, royalty conditions, confidentiality restrictions, liabilities, costs and method of payment.

ARTICLE 23
LOCAL GOODS AND SERVICES

23.1 In the conduct of Petroleum Operations, the Contractor shall:

(a) give preference to the purchase and use of goods manufactured, produced or supplied in India provided that such goods are available on terms equal to or better than imported goods with respect to timing of delivery, quality and quantity required, price and other terms;

(b) employ Indian Subcontractors having the required skills or expertise, to the maximum extent possible, insofar as their services are available on comparable standards with those obtained elsewhere and at competitive prices and on competitive terms; provided that where no such Subcontractors are available, preference shall be given to non-Indian Subcontractors who utilise Indian goods to the maximum extent possible, subject, however, to the proviso in paragraph (a) above; and

(c) ensure that provisions in terms of paragraphs (a) to (b) above are contained in contracts between the Operator and its Subcontractors.

23.2 Subject to Article 8.3(f), the Contractor shall establish appropriate procedures, including tender procedures, for the acquisition of goods and services which shall ensure that suppliers and Subcontractors in India are given adequate opportunity to compete for the supply of goods and services. The tender procedures shall include, inter alia, the financial amounts or value of contracts which will be awarded on the basis of selective bidding or open competitive bidding, the procedures for such bidding, and the exceptions to bidding in cases of emergency, and shall be subject to the approval of the Management Committee.

23.3 Within sixty (60) days after the end of each Year, the Contractor shall provide the Government with a report outlining its achievements in utilising Indian resources during that Year in accordance with Section 10 of Appendix C to this Contract.

23.4 In this Article "goods" means equipment, materials and supplies.

ARTICLE 24
INSURANCE AND INDEMNIFICATION

24.1 **Insurance**

24.1.1 The Contractor shall, during the term of this Contract, maintain and obtain insurance coverage for and in relation to Petroleum Operations for such amounts and against such risks as are customarily or prudently insured in the international petroleum industry in accordance with modern oilfield and petroleum industry practices, and shall furnish to the Government, certificates evidencing that such coverage is in effect. Such insurance policies shall include the Government as additional insured and shall waive subrogation against the Government. The said insurance shall, without prejudice to the generality of the foregoing, cover:

(a) loss or damage to all installations, equipment and other assets for so long as they are used in or in connection with Petroleum Operations; provided, however, that if for any reason the Contractor fails to insure any such installation, equipment or assets, it shall replace any loss thereof or repair any damage caused thereto;

(b) loss, damage or injury caused by pollution in the course of or as a result of Petroleum Operations;

(c) loss of property or damage or bodily injury suffered by any third party in the course of or as a result of Petroleum Operations for which the Contractor may be liable;

(d) any claim for which the Government may be liable relating to the loss of property or damage or bodily injury suffered by any third party in the course of or as a result of Petroleum Operations for which the Contractor is liable to indemnify the Government, or the State Government;

(e) with respect to Petroleum Operations offshore, the cost of removing wrecks and cleaning up operations following any accident in the course of or as a result of Petroleum Operations; and

(f) the Contractor's and/or the Operator's liability to its employees engaged in Petroleum Operations.

24.1.2 The Contractor shall require its Subcontractors to obtain and maintain insurance against the risks referred to in Article 24.1.1 relating mutatis mutandis to such Subcontractors.

24.2 **Indemnity**

Subject to Article 4.7, the Contractor shall indemnify, defend and hold the Government, and the State Government harmless against all claims, losses and damages of any nature whatsoever, including, without limitation, claims for loss or damage to property or injury or death to persons caused by or resulting from any Petroleum Operations conducted by or on behalf of the Contractor.

ARTICLE 25
RECORDS, REPORTS, ACCOUNTS AND AUDIT

25.1 The Contractor shall prepare and maintain in original at an office in India accurate and current books, records, reports and accounts of its activities for and in connection with Petroleum Operations so as to present a fair, clear and accurate record of all its activities, expenditures and receipts.

25.2 Based on generally accepted and recognised accounting principles and modern petroleum industry practices, record, books, accounts and accounting procedures in respect of Petroleum Operations shall be maintained on behalf of the Contractor by the Operator, at its business office in India, in accordance with the Accounting Procedure to this Contract.

25.3 The Contractor shall submit to the Government regular Statements and reports relating to Petroleum Operations as provided in Appendix-C.

25.4.1 The annual audit of accounts shall be carried out on behalf of the Contractor by a qualified, independent firm of recognised chartered accountants, registered in India.

25.4.2 The appointment of auditor and the scope of audit should have prior approval of the Management Committee.

25.4.3 The Contractor shall submit the audited accounts to the Management Committee for approval within sixty (60) days from the end of the Year. The Management Committee shall consider and approve the auditor's report within thirty (30) days after the submission of such report.

25.4.4 Copy of the auditors report shall be submitted to the Government within thirty (30) days after the approval of the Management Committee.

25.5 The Government shall have the right to audit the accounting records of the Contractor in respect of Petroleum Operations as provided in the Accounting Procedure.

25.6 The accounting and auditing provisions and procedures specified in this

Contract are without prejudice to any other requirements imposed by any statute in India, including, without limitation, any specific requirements of the statutes relating to taxation of Companies.

25.7 For the purpose of any audit referred to in Articles 25.5, the Contractor shall make available in original to the auditor all such books, records, accounts and other documents and information as may be reasonably required by the auditor during normal business hours.

ARTICLE 26
INFORMATION, DATA, CONFIDENTIALITY, INSPECTION AND SECURITY

26.1 The Contractor shall, promptly after they become available in India, provide the Government, free of cost, with all data obtained as a result of Petroleum Operations under the Contract including, but not limited to, geological, geophysical, geochemical, petrophysical, engineering, Well logs, maps, magnetic tapes, cores, cuttings and production data as well as all interpretative and derivative data, including reports, analyses, interpretations and evaluation prepared in respect of Petroleum Operations (hereinafter referred to as "Data"). Data shall be the property of the Government, provided, however, that the Contractor shall have the right to make use of such Data, free of cost, for the purpose of Petroleum Operations under this Contract as provided herein.

26.2 The Contractor may, for use in Petroleum Operations, retain copies or samples of material or information constituting the Data and, with the approval of the Government, original material, except that where such material is capable of reproduction and copies have been supplied to the Government, the Contractor may, subject to the right of inspection by the Government, export, subject to any applicable regulations, samples or other original Data for processing or laboratory examination or analysis, provided that representative samples equivalent in quality, size and quantity, or, where such material is capable of reproduction, copies of equivalent quality, have first been delivered to the Government.

26.3 The Contractor shall keep the Government currently advised of all developments taking place during the course of Petroleum Operations and shall furnish the Government with full and accurate information and progress reports relating to Petroleum Operations (on a daily, Monthly, Yearly or other periodic basis) as Government may reasonably require, provided that this obligation shall not extend to proprietary technology. The Contractor shall meet with the Government at a mutually convenient location in India to present the results of all geological and geophysical work carried out as well as the results of all engineering and drilling operations as soon as such Data becomes available to the Contractor.

26.4 All Data, information and reports obtained or prepared by, for or on behalf of, the Contractor pursuant to this Contract shall be treated as confidential and, subject to the provisions herein below, the Parties shall not disclose the contents thereof to any third party without the consent in writing of the other Parties.

26.5 The obligation specified in Article 26.4 shall not operate so as to prevent disclosure:

(a) to Affiliates, contractors, or Subcontractors for the purpose of Petroleum Operations;

(b) to employees, professional consultants, advisers, data processing centres and laboratories, where required, for the performance of functions in connection with Petroleum Operations for any Party comprising the Contractor;

(c) to banks or other financial institutions, in connection with Petroleum Operations;

(d) to bonafide intending assignees or transferees of a Participating Interest of a Party comprising the Contractor or in connection with a sale of the stock or shares of a Party comprising the Contractor;

(e) to the extent required by any applicable law or in connection with any legal proceedings or by the regulations of any stock exchange upon which the shares of a Party comprising the Contractor are quoted;

(f) to Government departments for, or in connection with, the preparation by or on behalf of the Government of statistical reports with respect to Petroleum Operations, or in connection with the administration of this Contract or any relevant law or for any purpose connected with Petroleum Operations; and

(g) by a Party with respect to any Data or information which, without disclosure by such Party, is generally known to the public.

26.6 Any Data, information or reports disclosed by the Parties comprising the Contractor to any other person pursuant to Article 26.5 (a) to (d) shall be disclosed on the terms that such Data, information or reports shall be treated as confidential by the recipient. Prompt notice of disclosures made by Companies pursuant to Article 26.5 shall be given to the Government.

26.7 Any Data, information and reports relating to the Contract Area which, in the opinion of the Government, might have significance in connection with offers by the Government of acreages, may be disclosed by the Government for such purpose. Government may also disclose such Data or information for any exploration programme to be conducted by a third party in adjoining areas with the consent of the Contractor, for better understanding of regional geological set-up and such consent by the Contractor shall not be unreasonably withheld.

26.8 Where an area ceases to be part of the Contract Area, the Contractor shall hand over all the originals and copies of the Data and information with respect to that part to the Government within a period of one (1) year from the date of relinquishment or surrender. The Contractor shall, however, be allowed to retain one copy of the Data in its possession for its own use, where required, and shall not use the Data for sale or any other purposes. Subject to the provisions of this Article, the Contractor shall keep all Data/information confidential.

Explanatory Note : Pursuant to this Article 26 and not withstanding any provisions in the Contract to the contrary, the Government shall have the right to disclose and freely use all data and information at its sole discretion except for data of proprietary nature such as interpretation reports to any party on or after three (3) years from acquisition of such data in order to promote exploration and production activities in the country. For any relinquished areas, the Government shall have the right to disclose and freely use all the data immediately after such relinquishment.

26.9 The Government shall, at all reasonable times, through duly authorised representatives, be entitled to observe Petroleum Operations and to inspect all assets, books, records, reports, accounts, contracts, samples and Data kept by the Contractor or the Operator in respect of Petroleum Operations in the Contract Area, provided, however, that the Contractor shall not be required to disclose any proprietary technology. The duly authorised representatives shall be given reasonable assistance by the Contractor for such functions and the Contractor shall afford such representatives reasonable use of all facilities and privileges afforded to its own personnel in the field including the use of office space and housing for a period not exceeding 30 mandays in a Year and thereafter at the cost of Government. The said representatives shall be entitled to make a reasonable number of surveys, measurements, drawings, tests and copies of documents, take samples, and make reasonable use of the equipment and instruments of the Contractor provided that such functions shall not unduly interfere with the Contractor's Petroleum Operations.

26.10 The Contractor shall give reasonable advance notice to the Government, or to any other authority designated by the Government for such purpose, of its programme of conducting surveys by aircraft or by ships, indicating, inter alia, the name of the survey to be conducted, approximate extent of the area to be covered, the duration of the survey, the commencement date, and the name of the airport or port from which the survey aircraft or ship will commence its voyage.

26.11 The Government, or the authority designated by the Government for such purpose, shall have the right to inspect any aircraft or ship used by the Contractor or a Subcontractor carrying out any survey or other operations in the Contract Area and shall have the right to put on board such aircraft or ship, Government officers in such number as may reasonably be necessary to ensure compliance by the Contractor or the Subcontractor with the security requirements of India.

ARTICLE 27
TITLE TO PETROLEUM, DATA AND ASSETS

27.1 The Government is the sole owner of Petroleum underlying the Contract Area and shall remain the sole owner of Petroleum produced pursuant to the provisions of this Contract except as regards that part of Crude Oil, Condensate or Gas the title whereof has passed to the Contractor or any other person in accordance with the provisions of this Contract.

27.2 Title to Petroleum to which the Contractor is entitled under this Contract, and title to Petroleum sold by the Companies shall pass to the relevant buyer party at the Delivery Point. The Contractor shall be responsible for all costs and risks prior to and including at the Delivery Point and each buyer party shall be responsible for all costs and risks associated with such buyer party's share after the Delivery Point.

27.3 Title to all Data specified in Article 26 shall be vested in the Government and the Contractor shall have the right to use thereof as therein provided.

27.4 Assets purchased by the Contractor for use in Petroleum Operations shall be owned by the Parties comprising the Contractor in proportion to their Participating Interest provided that the Government shall have the right to require vesting of full title and ownership in it, free of charge and encumbrances, of any or all assets, whether fixed or movable, acquired and owned by the Contractor for use in Petroleum Operations inside or outside the Contract Area, such right to be exercisable at the Government's option upon expiry or earlier termination of the Contract.

27.5 The Contractor shall be responsible for proper maintenance, insurance and safety of all assets acquired for Petroleum Operations and for keeping them in good repair, order and working condition at all times, and the costs thereof shall be recoverable as Contract Costs in accordance with Appendix-C.

27.6 So long as this Contract remains in force, subject to Article 27.5, the Contractor shall, free of any charge for the purpose of carrying out Petroleum Operations hereunder, have the exclusive use of assets which have become the property of Government.

27.7 Equipment and assets no longer required for Petroleum Operations during the term of the Contract shall be sold, exchanged or otherwise disposed of by the Contractor, provided however that the proceeds of sale shall be credited to Petroleum Operations as provided in Appendix C, provided that prior written consent of the Management Committee shall be obtained for each transaction in excess of US$ 50,000 (Fifty thousand United States Dollars) or such other value as may be agreed from time to time by the Management Committee. The consent of the Management Committee shall not be unreasonably withheld.

ARTICLE 28
ASSIGNMENT OF PARTICIPATING INTEREST

28.1 Subject to the terms of this Article and other terms of this Contract, any Party omprising the Contractor may assign, or transfer, a part or all of its Participating Interest, with the prior written consent of the Government, which consent shall not be unreasonably withheld, provided that the Government is satisfied that:

(a) the prospective assignee or transferee is of good standing, has the capacity and ability to meet its obligations hereunder, and is willing to provide an unconditional undertaking to the Government to assume its Participating Interest share of obligations and to provide guarantees in respect thereof as provided in the Contract;

(b) the prospective assignee or transferee is not a company incorporated in a country with which the Government, for policy reasons, has restricted trade or business;

(c) the prospective assignor or transferor and assignee or transferee respectively are willing to comply with any reasonable conditions of the Government as may be necessary in the circumstances with a view to ensuring performance under the Contract; and

(d) the assignment or transfer will not adversely affect the performance or obligations under this Contract or be contrary to the interests of India.

28.1.1 Subject to Article 28.7, nothing in this Article 28 shall prevent a Party comprising the Contractor from assigning or transferring a part or all of its Participating Interest to an Affiliate, with the approval of the Management Committee, provided that;

(a) the assignee provides an irrevocable, unconditional bank guarantee from a reputed bank of good standing in India, acceptable to the Government, in favour of the Government, for the amount specified in Article 29.2, in a form provided at Appendix-G;

(b) the assignee provides a parent financial and performance guarantee issued by the guarantor which furnished the guarantee pursuant to Article 29 in respect of the assignor Party's obligations under this Contract in favour of the Government, of the performance of such Affiliate assignee of its obligations under this Contract;

(c) the prospective Affiliate is not a company incorporated in a country with which the Government, for policy reason, has restricted trade or business; and

(d) the assignment will not adversely affect the performance or obligations under this Contract or be contrary to the interest of India.

28.2 In case of any change in the status of a Company or its shareholding resulting in a change in :

(a) the control of the Company; or

(b) its relationship with the company(ies) providing the guarantee under Article 29.1 (a) and 29.1 (b); the Company shall seek the consent of the Government for assigning the Participating Interest under the changed circumstances and the provisions of this Article 28 shall apply, mutatis mutandis, to be obtaining of such consent. For the purpose of this Article 28.2, control has the same meaning as in Article 1.3.

28.3 An application for consent to assign or transfer shall be accompanied by all relevant information concerning the proposed assignment or transfer including detailed information on the proposed assignee or transferee and its shareholding and corporate structure, as was earlier required from the Companies constituting the Contractor, the terms of the proposed assignment or transfer and the unconditional undertaking referred to in Article.

28.4 The applicant shall also submit such information relating to the prospective assignee or transferee of the assignment or transfer as the Government may reasonably require to enable proper consideration and disposal of the application.

28.5 No assignment or transfer shall be effective until the approval of the Government is received or deemed to have been received. Approval may be given by the Government on such terms and conditions as it may deem fit. Provided that such terms and conditions may not increase the obligations of the Parties comprising the Contractor. Upon assignment or transfer of its interest in this Contract, the assignor or transferor shall be released and discharged from its obligations hereunder only to the extent that such obligations are assumed by the assignee or transferee with the approval of the Government.

28.6 In the event that the Government does not give its consent or does not respond to a request for assignment or transfer by a Party comprising the Contractor within one hundred and twenty (120) days of such request and receipt of all information referred to in Article 28.3 above, consent shall be deemed to have been given by the Government.

28.7 An assignment or transfer shall not be made where the Participating Interest to be retained by the proposed assignor or the percentage interest of assignee shall be less than ten per cent (10%) of the total Participating Interest of all the constituents of the Contractor, except where the Government, on the recommendations of the Management Committee may, in special circumstances, so permit.

28.8 Nothing contained in this Article 28, shall prevent a Party comprising the Contractor from mortgaging, pledging, charging or otherwise encumbering at its own risk and cost all or part of its Participating Interest for the purposes of security relating to finance to the extent required for performing its obligation under the Contract, provided that:

(i) such Party shall remain solely liable for all its obligations relating to its Participating Interest to the exclusion of the other participants thereto;

(ii) the encumbrance shall be expressly subordinated to the rights of the other Parties under this Contract. The obligations occurring from the said encumbrance shall be the sole responsibility of the original Party and shall in no manner compromise the rights of other Parties to the Contract;

(iii) such Party has given reasonable notice of such encumbrance and furnishes to all other Parties (including, for the avoidance of doubt, the Government) a certified copy of the executed instrument(s) evidencing the encumbrances;

(iv) keeping in view the national interest of India, prior consent of the Government shall be required (which consent shall not be unreasonably withheld) of the list of potential lenders with whom such Party can consider hypothecation;

(v) the Party creating the charge shall ensure that such charge shall not in any way affect the interest of other Parties or result in interference with joint operations. In the event of any claims or liabilities imposed on other Parties because of the creation of such charges, the Party having created charge on its Participating Interest shall indemnify the other Parties; and

(vi) in case of foreclosure or default by a borrowing Party, the mortgagee shall not be deemed to have acquired a right to carry on either by itself or through an agent, the Petroleum Operation, without the written consent of the Government of India.

28.8.1 The Parties acknowledge that to obtain financing a Party ("Borrower") will be required to secure for a permitted chargee the right to receive a copy of any notice served on the Borrower and the Parties agree that they shall serve a copy of any such notice on any such permitted chargee in accordance with the provisions of Article 37 at the same time as such notice is served on the Borrower. For the purposes of Article 37 the address for service of notices of the permitted chargee shall be that specified in the instrument or instruments referred to in Article 28.8(iii).

28.8.2 In case lender elects to participate directly or through a company other than the Borrower under the financing arrangement referred to above, the same shall be subject to the rights of Government as contained in Article 28.1 of Contract and the pre-emptive rights of the Parties as may be contained in Operating Agreement. Any Party which wishes to exercise the said pre-emptive rights will explicitly assume the obligation on the same terms and conditions as the Borrower.

ARTICLE 29
GUARANTEES

29.1 Subject to Article 29.1 (d), each of the Companies constituting the Contractor shall procure and deliver to the Government within thirty (30) days from the Effective Date of this Contract:

(a) an irrevocable, unconditional bank guarantee from a reputed bank of good standing in India, acceptable to the Government, in favour of the Government, for the amount specified in Article 29.2, in a form provided at Appendix-G;

(b) financial and performance guarantee in favour of the Government from a parent company acceptable to the Government, in the form and substance set out in Appendix-E1, or, where there is no such parent company, the financial and performance guarantee from the Company itself in the form and substance setout in Appendix-E2;

(c) a legal opinion from its legal advisors, in a form satisfactory to the Government, to the effect that the aforesaid guarantees have been duly signed and delivered on behalf of the guarantors with due authority and is legally valid and enforceable and binding upon them;

29.2 The amount of the guarantee referred to in Article 29.1 (a) above shall be an amount equal to thirty five percent (35%) of the Company's Participating Interest share of the total estimated annual expenditure in respect of the Minimum Work Programme to be undertaken by the Contractor in the Contract Area during the relevant Year of a Phase, subject to Article 29.3.

29.3 The guarantee referred to in Article 29.2 shall provide that;

(a) at the end of each Year it shall be automatically renewed for an amount equal to a Company's Participating Interest share of thirty five percent (35%) of the total estimated expenditure in respect of the Minimum Work Programme to be undertaken for the following Year of an Exploration Phase, unless the Contractor has terminated the Contract in accordance with the terms thereof. The guarantee shall be renewed at the end of each Year positively thirty (30) days before the expiry of the guarantee period.

(b) after the completion and due performance of the Minimum Work Programme of a particular Exploration Phase, the guarantee will be released in favour of the Company on presentation to the bank of a certificate from the Government that the obligation of the Contractor has been fulfilled and the guarantee may be released, subject to Article 29.4. Such certificate shall be provided within thirty (30) days from the completion of the Minimum Work Programme and fulfillment of obligations under the Contract to the satisfaction of the Government.

29.4 If the Contractor elects to proceed to the second Exploration Phase of the Exploration Period, a bank guarantee for the Phase in terms of Articles 29.1 (a), 29.2 and 29.3 shall be delivered to the Government with the notice of such election and if such guarantee is not so delivered, the provisions of Article 29.5 shall apply.

29.5 If any of the documents referred to in Article 29.1 is not delivered within the period specified herein, this Contract may be terminated by the Government upon ninety (90) days written notice of its intention to do so.

29.6 Subject to Article 29.7, notwithstanding any change in the composition or shareholding of the parent company furnishing a performance guarantee as provided herein, it shall, not under any circumstances, be absolved of its obligations contained in the guarantees provided pursuant to Article 29.1(b). 29.7 If:

(a) a Party ("Assignor") assigns all or a part of its Participating Interest to a third party ("Assignee") in accordance with Article 28;

(b) the Assignee provides an irrevocable, unconditional bank guarantee from a reputed bank of good standing in India, acceptable to the Government, in favour of the Government, for an amount equal to the assignee's Participating Interest share of the estimated expenditure of the Minimum Work Programme of the Exploration Phase current at the Effective Date of the assignment;

(c) the Assignee provides performance guarantee and legal opinion in terms of this Article; and

(d) the addendum to the Contract giving effect to the assignment of Participating Interest is executed by all Parties; then the Government shall release the guarantee given by the assignor under Article 29.1(a) to the extent of the amount of the guarantee provided by the assignee and where relevant the guarantee under Article 29.1(b).

ARTICLE 30
TERM AND TERMINATION OF THE CONTRACT

30.1 The term of this Contract shall be for the period of the License and any Lease granted thereunder, unless the Contract is terminated earlier in accordance with its terms, and shall be deemed to have been terminated, if for any reason, the Contractor ceases to hold such License or Lease.

30.2 Subject to the provision of Articles 5, 14 and 30.6 and without prejudice to the provisions of Article 30.7 or any other provisions of this Contract, the Contractor shall have the right to terminate this Contract:

(a) with respect to any part of the Contract Area other than a Development Area then producing, or that prior thereto had produced Petroleum, upon giving ninety (90) days written notice of its intention to do so; and

(b) with respect to any Development Area in which Petroleum is being produced, or that prior thereto had produced Petroleum, upon giving at least one hundred and eighty (180) days written notice of its intention to do so.

30.3 This Contract may, subject to the provisions herein below and Article 31, be terminated by the Government upon giving ninety (90) days written notice with reasons to the other Parties of its intention to do so in the following circumstances, namely, that the Contractor or a Party comprising the Contractor ("the Defaulting Party")

(a) has knowingly submitted any false statement to the Government in any manner which was a material consideration in the execution of this Contract; or

(b) has intentionally and knowingly extracted or authorised the extraction of hydrocarbon not authorized to be extracted by the Contract or without the authority of the Government except such extractions as may be unavoidable as a result of operations conducted hereunder in accordance with generally accepted modern oilfield and petroleum industry practices which, when so extracted, were immediately notified to the Government or

(c) is adjudged bankrupt by a competent court or enters into or scheme of composition with its creditors or takes advantage of any law for the benefit of debtors; or

(d) has passed a resolution to apply to a competent court for liquidation of the Company unless the liquidation is for the purpose of amalgamation or reconstruction of which the Government has been given notice and the Government is satisfied that the Company's performance under this Contract would not be adversely affected thereby and has given its approval thereto; or

(e) has assigned any interest in the Contract without the prior consent of the Government as provided in Article 28; or

(f) has failed to make any monetary payment required by law or under this Contract by the due date or within such further period after the due date as may thereafter be specified by the Government; or

(g) has failed to comply with or has contravened the provisions of this Contract in a material particular; or

(h) has failed to comply with any final determination or award made by a sole expert or arbitrators subject to Article 33; or

(i) has failed to carry out or observe any of the terms and conditions of the License or Lease or the provisions of the Acts or Rules in force thereunder, subject however, to Article 31.

(j) on notice of termination as provided in Article 29.5.

PROVIDED THAT where the Contractor comprises two or more Parties, the Government shall not exercise its rights of termination pursuant to Article 30.3, on the occurrence, in relation to one or more, but not all, of the Parties comprising the Contractor, of an event entitling the Government to terminate the Contract,

(a) if any other Party or Parties constituting the Contractor (the non-

Defaulting Party or Parties) satisfies the Government that it, or they, is/are willing and would be able to carry out the obligations of the Contractor.

(b) where the non Defaulting Party or Parties with the consent of the Government has/have acquired the Participating Interest of the Defaulting Party pursuant to the provisions of the Operating Agreement and has/have procured and delivered to the Government a guarantee or guarantees as referred to in Article 29.1 in respect of the Participating Interest of the Defaulting Party acquired by the non Defaulting Party or Parties.

30.4 This Contract may also be terminated by the Government on giving the requisite notice specified above if the events specified in Article 30.3 (c) and (d) occur with respect to a company which has given a performance guarantee pursuant to Article 29 subject however to Article 30.5.

30.5 If the circumstance or circumstances that give rise to the right of termination under Article 30.3(f) or (g) or (i) or Article 30.4 are remedied (whether by the Defaulting Company or by another Party or Parties in its behalf) within the ninety (90) day period, or such extended period as may be granted by the Government, following the notice of the Government's intention to terminate the Contract as aforesaid, such termination shall not become effective.

30.6 On termination of this Contract, for any reason whatsoever, the rights and obligations of the Contractor shall cease but such termination shall not affect any rights of any Party which may have accrued or any obligations undertaken or incurred including obligations under Article 5.7, by the Contractor or any Party comprising the Contractor and not discharged prior to the date of termination.

30.7 In the event of termination pursuant to Articles 30.2, 30.3 or 30.4:

(a) the Government may require the Contractor, for a period not exceeding one eighty (180) days from the date of termination, to continue, for the account and at the cost of the Government, Crude Oil or Natural Gas production activities until the right to continue such production has been transferred to another entity;

(b) a Foreign Company, which is a constituent of the Contractor, shall have to remove and export all its property subject to Article 27 and the provisions hereof provided that in the event that ownership of any property is in doubt, or disputed, such property shall not be exported unless and until the doubt or dispute has been settled in favour of the Foreign Company.

30.8 Within ninety (90) days after the termination of this Contract, pursuant to Article 30.2, 30.3, or 30.4, or such longer period as the Government may agree, the Contractor shall comply with Article 14.9 and any reasonably necessary action as directed by the Government to avoid Environmental Damage or hazards to human life or to the property of others.

ARTICLE 31
FORCE MAJEURE

31.1 Any non-performance or delay in performance by any Party hereto of any of its obligations under this Contract, or in fulfilling any condition of any License or Lease granted to such Party, or in meeting any requirement of the Act, the Rules or any License or Lease, shall, except for the payment of monies due under this Contract or under the Act and the Rules or any law, be excused if, and to the extent that, such non-performance or delay in performance under this Contract is caused by Force Majeure as defined in this Article.

31.2 For the purpose of this Contract, the term Force Majeure means any cause or event, other than the unavailability of funds, whether similar to or different from those enumerated herein, lying beyond the reasonable control of, and unanticipated or unforeseeable by, and not brought about at the instance of, the Party claiming to be affected by such event, or which, if anticipated or foreseeable, could not be avoided or provided for, and which has caused the non-performance or delay in performance. Without limitation to the generality of the foregoing, the term Force Majeure shall include natural phenomena or calamities, earthquakes, typhoons, fires, wars declared or undeclared, hostilities, invasions, blockades, riots, strikes, insurrection and civil disturbances but shall not include the unavailability of funds.

31.3 Where a Party is claiming suspension of its obligations on account of Force Majeure, it shall promptly, but in no case later than seven (7) days after the occurrence of the event of Force Majeure, notify the Management Committee in writing giving full particulars of the Force Majeure, the estimated duration thereof, the obligations affected and the reasons for its suspension.

31.4 A Party claiming Force Majeure shall exercise reasonable diligence to seek to overcome the Force Majeure event and to mitigate the effects thereof on the performance of its obligations under this Contract. The Party affected shall promptly notify the Management Committee as soon as the Force Majeure event has been removed and no longer prevents it from complying with the obligations which have been suspended and shall thereafter resume compliance with such obligations as soon as possible.

31.5 The Party asserting the claim of Force Majeure shall have the burden of proving that the circumstances constitute valid grounds of Force Majeure under this Article and that such Party has exercised reasonable diligence and efforts to remedy the cause of any alleged Force Majeure.

31.6 Where a Party is prevented from exercising any rights or performing any obligations under this Contract due to Force Majeure, the time for the performance of the obligations affected thereby and for performance of any obligation or the exercise of any right dependent thereon, and the term of any Exploration Phase of the Exploration Period or this

Contract, may be extended to the extent of Force Majeure period or by such period as may be agreed by the Management Committee.

31.7 Notwithstanding anything contained herein above, if an event of Force Majeure occurs and is likely to continue for a period in excess of thirty (30) days, the Parties shall meet to discuss the consequences of the Force Majeure and the course of action to be taken to mitigate the effects thereof or to be adopted in the circumstances.

ARTICLE 32
APPLICABLE LAW AND LANGUAGE OF THE CONTRACT

32.1 This Contract shall be governed and interpreted in accordance with the laws of India.

32.2 Nothing in this Contract shall entitle the Contractor to exercise the rights, privileges and powers conferred upon it by this Contract in a manner which will contravene the laws of India.

32.3 The English language shall be the language of this Contract and shall be used in arbitral proceedings. All communications, hearing or visual materials or documents relating to this Contract shall be written or prepared in English.

32.4 The laws will also include amendments, revisions, modifications etc.

ARTICLE 33
SOLE EXPERT, CONCILIATION AND ARBITRATION

33.1 The Parties shall use their best efforts to settle amicably all disputes, differences or claims arising out of or in connection with any of the terms and conditions of this Contract or concerning the interpretation or performance thereof.

33.2 Matters which, by the terms of this Contract, the Parties have agreed to refer to a sole expert and any other matter which the Parties may agree to so refer, may be referred to a sole expert who shall be an independent and impartial person of international standing with relevant qualifications and experience, appointed by a written agreement between the Parties and who shall not, by virtue of nationality, personal connection or commercial interest, have a conflict between his/her own interest and his/her duty as a sole expert. In the event that the Parties fail or are unable to agree on a sole expert within thirty (30) days or such longer period as may be mutually agreed by Parties, the sole expert shall be appointed by a body or an institution or an agency or a person, mutually agreed by Parties. In case, there is no agreement on the body or an institution or an agency or a person for appointing sole expert or such institution or agency or body fails to appoint a sole expert within thirty (30) days or such longer period as may be mutually agreed by Parties, the matter shall be referred to arbitration. Any sole expert appointed shall be acting as an expert and not as an arbitrator and the decision of the sole expert on matters referred to him/her shall be final and binding on the Parties and shall not be subject to arbitration.

33.3 Subject to the provisions of this Contract, the Parties hereby agree that any controversy, difference, disagreement or claim for damages, compensation or otherwise (hereinafter in this Clause referred to as a "dispute") arising between the Parties, which cannot be settled amicably within ninety (90) days after the dispute arises, may (except for those referred to in Article 33.2, which may be referred to a sole expert) be submitted to conciliation or an arbitral tribunal for final decision as hereinafter provided.

33.4 The arbitral tribunal shall consist of three arbitrators. Each Party to the dispute shall appoint one arbitrator and the Party or Parties shall so advise the other Parties. The two arbitrators appointed by the Parties shall appoint the third arbitrator.

33.5 Any Party may, after appointing an arbitrator, request the other Party(ies) in writing to appoint the second arbitrator. If such other Party fails to appoint an arbitrator within thirty (30) days of receipt of the written request to do so, such arbitrator may, at the request of the first Party, be appointed in accordance with Arbitration and Conciliation Act, 1996.

33.6 If the two arbitrators appointed by or on behalf of the Parties fail to agree on the appointment of the third arbitrator within thirty (30) days of the appointment of the second arbitrator and if the Parties do not otherwise agree, at the request of either Party, MPSC Final.p65 81 2/ 20/2006, 12:08 PM 82 the third arbitrator shall be appointed in accordance with Arbitration and Conciliation Act, 1996.

33.7 If any of the arbitrators fails or is unable to act, his successor shall be appointed by the Party or person who originally appointed such in the manner set out in this Article as if he was the first appointment.

33.8 The decision of the arbitral tribunal shall be pronounced within four (4) months unless otherwise extended by the Parties, and, in case of difference among the arbitrators the decision of the majority shall be final and binding on the Parties.

33.9 The arbitration agreement contained in this Article 33 shall be governed by the Arbitration and Conciliation Act, 1996 (Arbitration Act). Arbitration proceedings shall be conducted in accordance with the rules for arbitration provided in Arbitration Act.

33.10 The right to arbitrate disputes under this Contract shall survive expiry or the termination of this Contract.

33.11 Prior to submitting a dispute to arbitration, the Parties may by mutual agreement submit the matter for conciliation in accordance with Part III of the Arbitration and Conciliation Act, 1996. No arbitration proceedings shall be instituted while conciliation proceedings are pending provided that a Party may initiate arbitration proceedings in the event that dispute has not been resolved by conciliation within sixty (60) days of the date of agreement by the Parties to submit such dispute to conciliation.

33.12 The venue of the sole expert, conciliation or arbitration proceedings pursuant to this Article, unless the Parties agree otherwise, shall be New Delhi, India and shall be conducted in the English language. Insofar as practicable, the Parties shall continue to implement the terms of this Contract notwithstanding the initiation of proceedings before a sole expert, conciliator or arbitral tribunal and any pending claim or dispute.

33.13 The fees and expenses of a sole expert or conciliator appointed by the Parties shall be borne equally by the Parties. The cost and expenses of arbitrator appointed by a Party in accordance with the provision of this Article shall be borne by the respective Party and the cost and expenses of third arbitrator and other incidental expenditure in relation to arbitration and liability thereof shall be at the discretion of the arbitrators.

33.14 Notwithstanding anything contrary contained herein above, in the event of dispute among Government Company(ies) and with the Government, such disputes shall be settled in accordance with guidelines issued on the subject by Government from time to time.

ARTICLE 34
CHANGE OF STATUS OF COMPANIES

34.1 The Parties comprising the Contractor shall notify the Government of any change in the management or control of a Company(ies) or the relationship with any guarantor of the Company(ies).

ARTICLE 35
ENTIRE AGREEMENT, AMENDMENTS, WAIVER AND MISCELLANEOUS

35.1 This Contract supersedes and replaces any previous agreement or understanding between the Parties, whether oral or written, on the subject matter hereof, prior to the execution date of this Contract.

35.2 This Contract shall not be amended, modified, varied or supplemented in any respect except by an instrument in writing signed by all the Parties, which shall state the date upon which the amendment or modification shall become effective.

35.3 No waiver by any Party of any one or more obligations or defaults by any other Party in the performance of this Contract shall operate or be construed as a waiver of any other obligations or defaults whether of a like or of a different character.

35.4 The provisions of this Contract shall inure to the benefit of and be binding upon the Parties and their permitted assigns and successors in interest.

35.5 In the event of any conflict between any provisions in the main body of this Contract and any provision in the Appendices, the provision in the main body shall prevail.

35.6 The headings of this Contract are for convenience of reference only and shall not be taken into account in interpreting the terms of this Contract.

35.7 Reference to any law or regulation having the force of law includes a reference to that law or regulation as from time to time may be amended, extended or re-enacted.

35.8 A reference in this Contract to the word "including" shall also mean "including but not limited to".

ARTICLE 36
CERTIFICATES

36.1 A Company shall furnish, prior to execution of this Contract, a duly authorised copy of a resolution properly and legally passed by the Board of Directors of the Company authorising its President or any Vice-President or any other representative to execute this Contract along with a certificate duly signed by the Secretary or an Assistant Secretary of the Company under its seal in this regard and to the effect that the Company has the power and authority to enter into this Contract and to perform its obligations thereunder and has taken all necessary action to authorise the execution, delivery and performance of the Contract.

ARTICLE 37
NOTICES

37.1 All notices, statements, and other communications to be given, submitted or made hereunder by any Party to another shall be sufficiently given if given in writing in English language and sent by registered post, postage paid, or by telegram, telex, facsimile, radio or cable, to the address or addresses of the other Party or Parties as follows:

(a) If to the Government:
Secretary to the Government of India
Ministry of Petroleum and Natural Gas
Shastri Bhavan Dr. Rajendra Prasad Marg,
New Delhi- 110001, INDIA
Facsimile No.: 91 11 23383585

(b) XYZ Company

____

____

Facsimile No. :
Telephone No. :

37.2 Notices when given in terms of Article 37.1 shall be effective when delivered if offered at the address of the other Parties as under Article 37.1 during business hours on working days and, if received outside business hours, on the next following working day. 37.3 Any Party may, by reasonable notice as provided hereunder to the other Parties, change its address and other particulars for notice purpose.

IN WITNESS WHEREOF, the representatives of the Parties to this Contract being duly authorised have hereunto set their hands and have executed these presents this _____ day of __________.

Signed for and on behalf of the President of India

By:_______________

In presence of _____________

Signed for and on behalf
of XYZ Company

By:_______________

In presence of_____________

Signed for and on behalf of XYZ Company

By:_______________

In presence of_____________

APPENDIX A
DESCRIPTION OF THE CONTRACT AREA

The area comprising approximately ____ Sq. Km., Onshore/offshore India identified as Block ____________described herein and shown on the map attached as Appendix B. Longitude and Latitude measurements commencing at points A, B, C, and D are given below:

APPENDIX B
MAP OF THE CONTRACT AREA

APPENDIX C

ACCOUNTING PROCEDURE TO THE CONTRACT BETWEEN THE GOVERNMENT OF INDIA AND XYZ COMPANY(IES) WITH RESPECT TO CONTRACT AREA IDENTIFIED AS BLOCK ___________

TABLE OF CONTENTS

ACCOUNTING PROCEDURE
SECTION 1
GENERAL PROVISIONS

1.1 Purpose

Generally, the purpose of this Accounting Procedure is to set out principles and procedures of accounting which will enable the Government of India to monitor effectively the Contractor's costs, expenditures, production and income so that the Government's entitlement to Profit Petroleum can be accurately determined pursuant to the terms of the Contract. More specifically, the purpose of the Accounting Procedure is to:

- classify costs and expenditures and to define which costs and expenditures shall be allowable for cost recovery and profit sharing and participation purposes;
- specify the manner in which the Contractor's accounts shall be prepared and approved; and
- address numerous other accounting related matters.

This Accounting Procedure is intended to apply to the provisions of the Contract and is without prejudice to the computation of income tax under applicable provisions of the Income-Tax Act, 1961, as amended.

1.2 Definitions

For purposes of this Accounting Procedure, the terms used herein which are defined in the Contract shall have the same meaning when used in this Accounting Procedure.

1.3 Inconsistency

In the event of any inconsistency or conflict between the provisions of this Accounting Procedure and the other provisions of the Contract, the other provisions of the Contract shall prevail.

1.4 Documentation and Statements to be submitted by the Contractor

1.4.1 Within ninety (90) days of the Effective Date of the Contract, the Contractor shall submit to and discuss with the Government a proposed outline of charts of accounts, operating records and reports, which outline shall reflect each of the categories and sub-categories of costs and income specified in Sections 2 and 3 and shall be in accordance with generally accepted standards and recognized accounting systems and consistent with normal petroleum industry practice and procedures for joint venture operations.

Within ninety (90) days of receiving the above submission, the Government shall either provide written notification of its approval of the proposal or request, in writing, revisions to the proposal.

Within one hundred and eighty (180) days from the Effective Date of the Contract, the Contractor and the Government shall agree on the outline of charts of accounts, records and reports which shall also describe the basis of the accounting system and procedures to be developed and used under this Contract. Following such agreement, the Contractor shall expeditiously prepare and provide the Government with formal copies of the comprehensive charts of accounts, records and reports and allow the Government to examine the manuals and to review procedures which are, and shall be, observed under the Contract.

1.4.2 Notwithstanding the generality of the foregoing, the Contractor shall make regular Statements relating to the Petroleum Operations as follows:

(i) Production Statement (see Section 5 of this Accounting Procedure).

(ii) Value of Production and Pricing Statement (see Section 6 of this Accounting Procedure).

(iii) Statement of Costs, Expenditures and Receipts (see Section 7 of this Accounting Procedure).

(iv) Cost Recovery Statement (see Section 8 of this Accounting Procedure).

(v) Profit Sharing Statement (see Section 9 of this Accounting Procedure)

(vi) Local Procurement Statement (see Section 10 of this Accounting Procedure)

(vii) End of Year Statement (see Section 11 of this Accounting Procedure).

(viii) Budget Statement (see Section 12 of this Accounting Procedure).

1.4.3 All reports and Statements shall be prepared in accordance with the Contract and the laws of India and, where there are no relevant provisions in either of these, in accordance with generally accepted practices in the international petroleum industry.

1.4.4 Each of the entities constituting the Contractor shall be responsible for maintaining its own accounting records in order to comply with all legal requirements and to support all returns or any other accounting reports required by any Government authority in relation to the Petroleum Operations. However, for the purposes of giving effect to this Accounting Procedure, the Party constituting the Contractor who is the Operator shall be responsible for maintaining, at its business office in India, on behalf of the Contractor, all the accounts of the Petroleum Operations in accordance with the provisions of the Accounting Procedure and the Contract.

1.5 Language and Units of Account

All accounts, records, books, reports and Statements shall be maintained and prepared in the English language using mercantile basis of accounting. The accounts shall be maintained in United States Dollars, which shall be the controlling currency of account for cost recovery, and profit sharing purposes. Metric units and Barrels shall be employed for measurements required under the Contract. Where necessary for clarification, the Contractor may also maintain accounts and records in other languages, currencies and units.

1.6 Currency Exchange Rates

1.6.1 For conversion purposes between United States Dollars and Indian Rupees or any other currency, the monthly average of the daily mean of the buying and selling rates of exchange as quoted by the Reserve Bank of India (or any other financial body as may be mutually agreed by the Parties) for the Month in which the revenues, costs, expenditures, receipts or income are recorded, shall be used. However, in the case of any single non-US Dollar transaction in excess of the equivalent of fifty thousand (50,000) US Dollars, the conversion into US Dollars shall be performed on the basis of the average of the applicable exchange rates for the day on which the transaction occurred.

1.6.2 Any realized or unrealized gains or losses from the exchange of currency in respect of Petroleum Operations shall be credited or charged to the accounts. A record of the exchange rates used in converting Indian Rupees or any other currencies into United States Dollars as specified in Section 1.6.1 shall be maintained by the Contractor and shall be identified in the relevant Statements required to be submitted by the Contractor in accordance with Section 1.4.2.

1.7 Payments

1.7.1 Subject to Article 20.3 of the Contract and the foreign exchange laws and regulations prevailing from time to time, all payments between the Parties shall, unless otherwise agreed, be in United States Dollars and shall be made through a bank designated by each receiving Party.

1.7.2 Unless otherwise specified, all sums due under the Contract shall be paid within forty five (45) days from the date on which the obligation to pay was incurred.

1.7.3 All sums due by one Party to the other under the Contract during any Month shall, for each day such sums are overdue during such Month, bear interest compounded daily at the applicable LIBOR plus two (2) percentage points.

1.8 Arms Length Transactions Unless otherwise specifically provided for in the Contract, all transactions giving rise to revenues, costs or expenditures which will be credited or charged to the accounts prepared, maintained or submitted hereunder shall be conducted at arms length or on such a basis as will assure that all such revenues, costs or expenditures will not be lower or higher, as the case may be, than would result from a transaction conducted at arms length on a competitive basis with third parties.

1.9 Audit and Inspection Rights of the Government

1.9.1 Without prejudice to statutory rights, the Government, upon at least twenty (20) Business Days advance written notice to the Contractor, shall have the right to inspect and audit, during normal business hours, all records and documents supporting costs, expenditures, expenses, receipts and income, such as the Contractor's accounts, books, records, invoices, cash vouchers, debit notes, price lists or similar documentation with respect to the Petroleum Operations conducted hereunder in each Year, within two (2) years (or such longer period as may be required in exceptional circumstances) from the end of such Year.

1.9.2 The Government may undertake the conduct of the audit either through its own representatives or through a qualified firm of recognised chartered accountants, registered in India or a reputed consulting firm, appointed for the purpose by the Government and the costs of audit in case of Government auditor(s) shall be borne by the Government, where as for outside auditor(s), this shall be borne by the Contractor as a General and Administrative Cost.

1.9.3 In conducting the audit, the Government or its auditors shall be entitled to examine and verify, at reasonable times, all charges and credits relating to the Contractor's activities under the Contract and all books of account, accounting entries, material records and inventories, vouchers, payrolls, invoices and any other documents, correspondence and records considered necessary by the Government to audit and verify the charges and credits. The auditors shall also have the right, in connection with such audit, to visit and inspect, at reasonable times, all sites, plants, facilities, warehouses and offices of the Contractor directly or indirectly serving the Petroleum Operations, and to physically examine other property, facilities and stocks used in Petroleum Operations, wherever located and to question personnel associated with those operations. Where the Government requires verification of charges made by an Affiliate, the Government shall have the right to obtain an audit certificate from an internationally recognized firm of public accountants acceptable to both the Government and the Contractor, which may be the Contractor's statutory auditor. Submission of the audit certificate, shall in no way relieve or diminish the responsibility of the Contractor for the compliance with the obligations under the Contract.

1.9.4 Any audit exceptions shall be made by the Government in writing and notified to the Contractor within one hundred and twenty (120) days following completion of the audit in question.

1.9.5 The Contractor shall answer any notice of exception under Section 1.9.4 within one hundred and twenty (120) days of the receipt of such notice. Where the Contractor has, after the said one hundred and twenty (120) days, failed to answer a notice of exception, the exception shall prevail and deemed to have been agreed to by the Contractor.

1.9.6 All agreed adjustments resulting from an audit and all adjustments required by prevailing exceptions under Section 1.9.5 shall be promptly made in the Contractor's accounts and any consequential adjustments to the Government's entitlement to Petroleum shall be made within thirty (30) days therefrom.

1.9.7 Notwithstanding any reference to a Sole Expert or Arbitration in accordance with the provisions of the Contract, in case any amount is claimed as due to the Government resulting from the audit exception but not accepted or settled by the Contractor, then the Contractor shall deposit such claimed amount in a escrow account to be opened with a financial institution, failing mutually agreed agreement with State Bank of India within thirty (30) days from the date when the amount is disputed by the Contractor. The amount in escrow account along with any interest accumulated thereon shall be appropriated or adjusted in accordance with the decision or award of the Sole Expert or Arbitral Tribunal as may be or otherwise as mutually agreed to between the Parties.

1.9.8 If the Contractor and the Government are unable to reach final agreement on proposed audit adjustments, either Party may refer any dispute thereon to a sole expert as provided for in the Contract. So long as any issues are outstanding with respect to an audit, the Contractor shall maintain the relevant documents and permit inspection thereof until the issue is resolved.

1.10 Revision of the Accounting Procedure By mutual agreement between the Government and the Contractor, this Accounting Procedure may be revised from time to time, in writing, signed by the Parties, stating the date upon which the amendments shall become effective.

SECTION 2

CLASSIFICATION, DEFINITION AND ALLOCATION OF COSTS AND EXPENDITURES

2.1 Segregation of Costs

Costs shall be segregated in accordance with the purposes for which such expenditures are made. All costs and expenditures allowable under Section 3, relating to Petroleum Operations, shall be classified, defined and allocated as set out below in this Section.

2.2 Exploration Costs

Exploration Costs are all direct and allocated indirect expenditures incurred in the search for Petroleum in an area which is, or was at the time when such costs were incurred, part of the Contract Area, including expenditures incurred in respect of:

2.2.1 Aerial, geophysical, geochemical, palaeontological, geological, topographical and seismic surveys, analysis and studies and their interpretation.

2.2.2 Core hole drilling and water Well drilling.

2.2.3 Labour, materials, supplies and services used in drilling Wells with the object of finding Petroleum or in drilling Appraisal Wells provided that if such Wells are completed as producing Wells or injection Well for enhancing Oil recovery, the costs of completion thereof shall be classified as Development Costs.

2.2.4 Facilities used solely in support of the purposes described in Sections 2.2.1, 2.2.2 and 2.2.3 above, including access roads, all separately identified.

2.2.5 Any Service Costs and General and Administrative Costs directly incurred on exploration activities and identifiable as such and a portion of the remaining Service Costs and General and Administrative Costs allocated to Exploration Operations determined by the proportionate share of total Contract Costs (excluding General and Administrative Costs and Service Costs) represented by all other Exploration Costs.

2.2.6 Geological and geophysical information purchased or acquired in connection with Exploration Operations.

2.2.7 Any other expenditures incurred in the search for Petroleum not covered under Sections 2.3 or 2.4.

2.3 Development Costs

Development Costs are all direct and allocated indirect expenditures incurred with respect to the development of discoveries within the Contract Area including expenditures incurred on account of:

2.3.1 Geological and Geophysical information acquired in connection with Development Operations.

2.3.2 Drilling Development Wells, whether these Wells are dry or producing and drilling Wells for the injection of water or Gas to enhance recovery of Petroleum.

2.3.3 Completing of Exploration Wells by way of installation of casing or equipment or otherwise or for the purpose of bringing a Well into use as a producing Well or as a Well for the injection of water or Gas to enhance recovery of Petroleum.

2.3.4 Purchase, installation or construction of production, transport and storage facilities for production of Petroleum, such as pipelines, flow lines, production and treatment units, wellhead equipment, subsurface equipment, enhanced recovery systems, offshore and onshore platforms, export terminals and piers, harbours and related facilities and access roads for production activities.

2.3.5 Engineering and design studies for facilities referred to in Section 2.3.3.

2.3.6 Any Service Costs and General and Administrative Costs directly incurred in Development Operations and identifiable as such and a portion of the remaining Service Costs and General and Administrative Costs allocated to development activities, determined by the proportionate share of total Contract Costs (excluding General and Administrative Costs and Service Costs) represented by all other Development Costs.

2.4 Production Costs

Production Costs are expenditures incurred on Production Operations after the start of production from the Field (which are other than Exploration and Development Costs). The balance of General and Administrative Costs and Service Costs not allocated to Exploration Costs or Development Costs shall be allocated to Production Costs.

2.5 Service Costs

Service Costs are direct and indirect expenditures incurred in support of Petroleum Operations in the Contract Area, including expenditures on warehouses, piers, marine vessels, vehicles, motorized rolling equipment, aircraft, fire and security stations, workshops, water and sewerage plants, power plants, housing, community and recreational facilities and furniture and tools and equipment used in these activities. Service Costs in any Year shall include the costs incurred in such Year to purchase and/ or construct the said facilities as well as the annual costs of maintaining and operating the same, each to be identified separately. All Service Costs shall be regularly allocated as specified in Sections 2.2.5, 2.3.5 and 2.4 to Exploration Costs, Development Costs and Production Costs and shall be separately shown under each of these categories. Where Service Costs are made in respect of shared facilities, the basis of allocation of costs to Petroleum Operations hereunder shall be specified.

2.6 General and Administrative Costs

General and Administrative Costs are expenditures incurred on general administration and management primarily and principally related to Petroleum Operations in or in connection with the Contract Area, and shall include:

2.6.1 main office, field office and general administrative expenditures in India including supervisory, accounting and employee relations services;

2.6.2 an annual overhead charge for services rendered by the parent company or an Affiliate to support and manage Petroleum Operations under the Contract, and for staff advice and assistance including financial, legal, accounting and employee relations services, but excluding any remuneration for services charged separately under this Accounting Procedure, provided that:

(i) for the period from the Effective Date until the date on which the first Development Plan under the Contract is approved by the Government, this annual charge shall be the Contractor's verifiable expenditure but shall in no event be greater than the following percentages of the total Contract Costs incurred during the Contract Year in or in connection with the Contract Area and qualifying for recovery pursuant to Section 3:

Contract costs in any Annual overhead charge
Contract year (in million US$)

0-2 3% Over 2-5 US $ 60,000 + 2% of Contract Costs in excess of US $ 2 million. Over 5 US $ 120,000 + 1% of Contract Costs in excess of US $ 5 million

(ii) from the date on which the first Development Plan is approved, the charge shall be at an amount or rate to be agreed on between the Parties and stated in the Development Plan.

2.6.3 All General and Administrative Costs shall be regularly allocated as specified in Sections 2.2.5, 2.3.5 and 2.4 to Exploration Costs, Development Costs and Production Costs respectively, and shall be separately shown under each of these cost categories.

SECTION 3
COSTS, EXPENSES, EXPENDITURES AND INCIDENTAL INCOME OF THE CONTRACTOR

3.1 Costs Recoverable and Allowable Without Further Approval of the Government Costs incurred by the Contractor on Petroleum Operations as per reviewed or approved Work Programme and Budget by the Management Committee as the case may be, pursuant to the Contract as classified under the headings referred to in Section 2 shall be allowable for the purposes of the Contract except to the extent provided in Section 3.2 or elsewhere in this Accounting Procedure, and subject to audit, as referred to in Articles 25.4.1 to 25.4.4 and Article 25.5, as provided for herein. Further in case of variation in costs over the reviewed/approved Work Programme and Budget, as the case may be, or re-appropriation of costs, shall be submitted to the Management Committee for review/ approval, as the case may be, within thirty (30) days from end of the relevant Financial Year and subject to the audit and other provisions of the Contract, such costs shall be allowable for the purposes of the Contract.

3.1.1 Surface Rights All direct costs necessary for the acquisition, renewal or relinquishment of surface rights acquired and maintained in force for the purposes of the Contract except as provided in Section 3.1.9.

3.1.2 Labour and Associated Labour Costs

(a) Contractor's locally recruited employees based in India Costs of all the Contractor's locally recruited employees who are directly engaged in the conduct of Petroleum Operations under the Contract in India. Such costs shall include the costs of employee benefits and Government benefits for employees and levies imposed on the Contractor as an employer, transportation and relocation costs within India of the employee and such members of the employee's family as per the personnel policy of the employer as required by law or customary practice in India. If such employees are engaged in other activities in India, in addition to Petroleum Operations, the cost of such employees shall be apportioned on a time sheet basis according to sound and acceptable accounting principles.

(b) Assigned Personnel Costs of salaries and wages, including bonuses, of the Contractor's employees directly and necessarily engaged in the conduct of the Petroleum Operations under the Contract, whether temporarily or permanently assigned, irrespective of the location of such employees, it being understood that in the case of those personnel only a portion of whose time is wholly dedicated to Petroleum Operations under the Contract, only that pro rata portion of applicable salaries, wages, and other costs, as specified in Sections 3.1.2(c), (d), (e), (f) and (g), shall be charged and the basis of such pro rata allocation shall be specified.

(c) The Contractor's costs regarding holiday, vacation, sickness and disability benefits and living and housing and other customary allowances applicable to the salaries and wages chargeable under Section 3.1.2(b) above.

(d) Expenses or contributions made pursuant to assessments or obligations imposed under the laws of India which are applicable to the Contractor's cost of salaries and wages chargeable under Section 3.1.2(b) above.

(e) The Contractor's cost of established plans for employees' group life insurance, hospitalization, pension, retirement and other benefit plans of a like nature customarily granted to the Contractor's employees provided, however, that such costs are in accordance with generally accepted standards in the international petroleum industry, applicable to salaries and wages chargeable to Petroleum Operations under Section 3.1.2(b) above.

(f) Personal income taxes where and when they are paid by the Contractor to the Government of India for the employee, in accordance with the Contractor's standard personnel policies.

(g) Reasonable transportation and travel expenses of employees of the Contractor, including those made for travel and relocation of the expatriate employees, including their dependent family and personal effects, assigned to India whose salaries and wages are chargeable to Petroleum Operations under Section 3.1.2(b) above.

Transportation cost as used in this Section shall mean the cost of freight and passenger service and any accountable incidental expenditures related to transfer travel and authorized under the Contractor's standard personnel policies. The Contractor shall ensure that all expenditures related to transportation costs are equitably allocated to the activities which have benefited from the personnel concerned.

3.1.3 Transportation Costs The reasonable cost of transportation of equipment, materials and supplies within India and from outside India to India necessary for the conduct of Petroleum Operations under the Contract, including directly related costs such as unloading charges, dock fees and inland and ocean freight charges.

3.1.4 Charges for Services

(i) Third Parties

The actual costs of contract services, services of professional consultants, utilities and other services necessary for the conduct of Petroleum Operations under the Contract performed by third parties other than an Affiliate of the Contractor, provided that the transactions resulting in such costs are undertaken pursuant to Section 1.8 of this Accounting Procedure.

(ii) Affiliates of Contractor

(a) Professional and Administrative Services and Expenses Cost of professional and administrative services provided by any Affiliate for the direct benefit of Petroleum Operations, including, but not limited to, services provided by the production, exploration, legal, financial, insurance, accounting and computer services divisions other than those covered by

Section 3.1.4 (ii)(b) which the Contractor may use in lieu of having its own employees. Charges shall be equal to the actual cost of providing their services, shall not include any element of profit and shall not be any higher than the most favourable prices charged by the Affiliate to third parties for comparable services under similar terms and conditions elsewhere and will be fair and reasonable in the light of prevailing modern oilfield and petroleum industry practices.

(b) Scientific or Technical Personnel Cost of scientific or technical personnel services provided by any Affiliate of the Contractor for the direct benefit of Petroleum Operations, which cost shall be charged on a cost of service basis. Charges therefor shall not exceed charges for comparable services currently provided by outside technical service organizations of comparable qualifications. Unless the work to be done by such personnel is covered by an Approved Budget and Work Programme, the Contractor shall not authorize work by such personnel without approval of the Management Committee.

(c) Equipment, facilities and property owned and furnished by the Contractor's Affiliates, at rates commensurate with the cost of ownership and operation provided, however, that such rates shall not exceed those currently prevailing for the supply of like equipment, facilities and property on comparable terms in the area where the Petroleum Operations are being conducted. The equipment and facilities referred to herein shall exclude major investment items such as (but not limited to) drilling rigs, producing platforms, oil treating facilities, oil and gas loading and transportation systems, storage and terminal facilities and other major facilities, rates for which shall be subject to separate agreement with the Government.

3.1.5 Communications Cost of acquiring, leasing, installing, operating, repairing and maintaining communication systems including radio, satellite link and microwave facilities between the Contract Area and the Contractor's nearest base facility.

3.1.6 Office, Shore Bases and Miscellaneous Facilities Net cost to the Contractor of establishing, maintaining and operating any office, suboffice, shore base facility, warehouse, housing or other facility directly serving the Petroleum Operations. If any such facility services contract areas other than the Contract Area, or any business other than Petroleum Operations, the net costs thereof shall be allocated on an equitable and consistent basis.

3.1.7 Environmental Studies and Protection Costs incurred in conducting the environmental impact assessment studies for the Contract Area, and in taking environmental protection measures including abandonment cost or contribution to abandonment funds as may be created for abandonment and Site Restoration pursuant to the terms of the Contract.

3.1.8 Materials and equipment

(i) General

So far as is practicable and consistent with efficient and economical operation, only such material shall be purchased or furnished by the Contractor for use in the Petroleum Operations as may be required for use in the reasonably foreseeable future and the accumulation of surplus stocks shall be avoided. Material and equipment held in inventory shall only be charged to the accounts when such material is removed from inventory and used in Petroleum Operations. Costs shall be charged to the accounting records and books based on the "First-in, First-out method".

(ii) Warranty

In the case of defective material or equipment, any adjustment received by the Contractor from the suppliers or manufacturers or their agents in respect of any warranty on material or equipment shall be credited to the accounts under the Contract.

(iii) Value of materials charged to the accounts under the Contract (a) Except as otherwise provided in subparagraph (b) below, materials purchased by the Contractor for use in the Petroleum Operations shall be valued to include invoice price less trade and cash discounts, if any, purchase and procurement fees plus freight and forwarding charges between point of supply and point of shipment, freight to port of destination, insurance, taxes, custom duties, consular fees, other items chargeable against imported material and, where applicable, handling and transportation costs from point of importation to warehouse or operating site, and these costs shall not exceed those currently prevailing in normal arms length transactions on the open market. (b) Material purchased from or sold to Affiliates or transferred to or from activities of the Contractor other than Petroleum Operations under the Contract:

(aa) new material (hereinafter referred to as condition A) shall be valued at the current international price which shall not exceed the price prevailing in normal arms length transactions on the open market;

(bb) used material which is in sound and serviceable condition and is suitable for reuse without reconditioning (hereinafter referred to as condition B) shall be priced at not more than seventy five per cent (75%) of the current price of the above mentioned new materials;

(cc) used material which cannot be classified as condition B, but which, after reconditioning, will be further serviceable for original function as good second-hand condition B material or is serviceable for original function, but substantially not suitable for reconditioning (hereinafter referred to as condition C) shall be priced at not more than fifty per cent (50%) of the current price of the new material referred to above as condition A.

The cost of reconditioning shall be charged to the reconditioned material, provided that the condition C material value plus the cost of reconditioning does not exceed the value of condition B material. Material which cannot be classified as condition B or condition C shall be priced at a value commensurate with its use. Material involving erection expenditure shall be charged at the applicable condition percentage (referred to above) of the current knocked-down price of new material referred to above as condition A. When the use of material is temporary and its service to the Petroleum Operations does not justify the reduction in price in relation to materials referred to above as conditions B and C, such material shall be priced on a basis that will result in a net charge to the accounts under the Contract consistent with the value of the service rendered.

3.1.9 Duties, Fees and Other Charges

Any duties, levies, fees, charges and any other assessments levied by any governmental or taxing authority in connection with the Contractor's activities under the Contract and paid directly by the Contractor except corporate income tax payable by the constituents of the Contractor.

3.1.10 Insurance and Losses

Insurance premia and costs incurred for insurance pursuant to Article 24 of the Contract, provided that such insurance is customary, affords prudent protection against risk and is at a premium no higher than that charged on a competitive basis by insurance companies which are not Affiliates. Except as provided in Sections 3.2 (ix), Section 3.2(x) and Section 3.2(xi), actual costs and losses incurred shall be allowable to the extent not made good by insurance. Such costs may include, but are not limited to, repair and replacement of property in the Contract Area resulting from damages or losses incurred by fire, flood, storm, theft, accident or such other cause.

3.1.11 Legal Expenses

All reasonable costs and expenses, except Section 3.2 (xi) resulting from the handling, investigating, asserting, defending, or settling of any claim or legal action necessary or expedient for the procuring, perfecting, retention and protection of the Contract Area and in defending or prosecuting lawsuits involving the Contract Area or any third party claim arising out of Petroleum Operations under the Contract, or sums paid in respect of legal services necessary for the protection of the joint interest of

Government and the Contractor, shall be allowable. Such expenditures shall include attorney's fees, court costs, costs of investigation and procurement of evidence and amounts paid in settlement or satisfaction of any such litigation and claims provided such costs are not covered elsewhere in the Accounting Procedure. Where legal services are rendered in such matters by salaried or regularly retained lawyers of the Contractor or an Affiliate, such compensation shall be included instead under Section 3.1.2 or 3.1.4 (ii) above as applicable.

3.1.12 Training Costs

All costs and expenses incurred by the Contractor in training as is required under Article 22 of the Contract.

3.1.13 General and Administrative Costs

The costs described in Section 2.6.1 and the charge described in Section 2.6.2 of this Accounting Procedure.

3.1.14 Royalty, License fee, surface rentals etc. Royalty, License fee, surface rentals, dead rents and other levies and taxes paid to the Government of India or State Government or local Government bodies or authority or agency except income tax paid to the Government.

3.2 Costs not recoverable and not allowable under the Contract The following costs and expenses shall not be recoverable or allowable (whether directly as such or indirectly as part of any other charges or expense) for cost recovery and profit sharing purposes under the Contract:

(i) costs and charges incurred before the Effective Date including costs in respect of preparation, signature or ratification of this Contract;

Explanatory Note : It is clarified that costs and expenditures, incurred prior to the Effective Date but after the execution of the Contract, for making statutory payments in connection with the Petroleum Operations such as Petroleum Exploration License (PEL) fee and application fee shall be allowed as Contract Cost and shall be cost recoverable.

(ii) expenditures in respect of any financial transaction to negotiate, float or otherwise obtain or secure funds for Petroleum Operations including, but not limited to, interest, commission, brokerage and fees related to such transactions, as well as exchange losses on loans or other financing, whether between Affiliates or otherwise;

(iii) costs of marketing or transportation of Petroleum beyond the Delivery Point;

(iv) expenditures incurred in obtaining, furnishing and maintaining the guarantees required under the Contract and any other amounts spent on indemnities with regard to non-fulfillment of contractual obligations;

(v) attorney's fees and other costs and charges in connection with arbitration proceedings and sole expert determination pursuant to the Contract;

(vi) fines, interest and penalties imposed by Courts of law of the Republic of India;

(vii) donations and contributions;

(viii) expenditures on creation of any partnership or joint venture arrangement;

(ix) amounts paid with respect to non-fulfillment of contractual obligations;

(x) costs incurred as a result of failure to insure where insurance is required pursuant to the Contract, or of failure to follow procedures laid down by an insurance policy or where the Contractor has elected to self insure, or has under-insured;

(xi) costs and expenditures incurred as a result of misconduct or negligence of the Contractor; and

(xii) expenses of the members of the Management Committee as per Article 6.12.

3.3 Other costs recoverable and allowable only with Management Committee approval Any other costs and expenditures not included in Section 3.1 or 3.2 of this Accounting Procedure but which have been incurred by the Contractor for the necessary and proper conduct of Petroleum Operations shall be allowed to be recovered only with the express prior approval in writing of the Management Committee.

3.4 **Incidental Income and Credits**

All incidental income and proceeds received from Petroleum Operations under the Contract, including but not limited to the items listed below, shall be credited to the accounts under the Contract and shall be taken into account for cost recovery, and Profit Petroleum sharing purposes in the manner described in Articles 15 and 16 of the Contract:-

(i) The proceeds of any insurance or claim or judicial awards in connection with Petroleum Operations under the Contract or any assets charged to the accounts under the Contract where such operations or assets have been insured and the premia charged to the accounts under the Contract;

(ii) Revenue received from third parties for the use of property or assets, the cost of which has been charged to the accounts under the Contract;

(iii) Any adjustment received by the Contractor from the suppliers/ manufacturers or their agents in connection with defective material, the cost of which was previously charged by the Contractor to the accounts under the Contract;

(iv) Rentals, refunds or other credits received by the Contractor which apply to any charge which has been made to the accounts under the Contract;

(v) Prices originally charged to the accounts under the Contract for materials subsequently exported from the Republic of India without being used in Petroleum Operations under the Contract;

(vi) Proceeds from the sale or exchange by the Contractor of assets, plant or facilities, the acquisition costs of which have been charged to the accounts under the Contract;

(vii) Legal costs charged to the accounts under Section 3.1.11 of this Accounting Procedure and subsequently recovered by the Contractor.

3.5 Non-Duplication of Charges and Credits

Notwithstanding any provision to the contrary in this Accounting Procedure, it is the objective of the Parties that there shall be no duplication of charges or credits to the accounts under the Contract.

SECTION 4
RECORDS AND INVENTORIES OF ASSETS

4.1 Records

4.1.1 The Contractor shall keep and maintain detailed records of property and assets in use for or in connection with Petroleum Operations under the Contract in accordance with normal practices in exploration and production activities of the international petroleum industry. Such records shall include information on quantities, location and condition of such property and assets, and whether such property or assets are leased or owned.

4.2 Inventories

4.2.1 The Contractor shall:

(a) not less than once every twelve (12) Months with respect to movable assets; and

(b) not less than once every three (3) Years with respect to immovable assets, take an inventory of the assets used for or in connection with Petroleum Operations in terms of the Contract and address and deliver such inventory to the Government together with a written statement of the principles upon which valuation of the assets mentioned in such inventory has been based.

4.2.2 The Contractor shall give the Government at least thirty (30) days notice in writing in the manner provided for in the Contract of its intention to take the inventory referred to in Section 4.2.1 and the Government shall have the right to be represented when such inventory is taken.

4.2.3 When an assignment of rights under the Contract takes place, a special inventory shall be taken by the Contractor at the request of the assignee provided that the cost of such inventory is borne by the assignee and paid to the Contractor.

4.2.4 In order to give effect to Article 27 of the Contract, the Contractor shall provide the Government with a comprehensive list of all relevant assets when requested by the Government to do so.

SECTION 5
PRODUCTION STATEMENT

5.1 From the date of first production of Petroleum from the Contract Area the Contractor shall submit a monthly Production Statement to Government showing the following information separately of each producing Field and in aggregate for the Contract Area:

5.1.1 The quantity of Crude Oil and Condensate produced and saved.

5.1.2 The quality and characteristics of such Crude Oil and Condensate produced and saved.

5.1.3 The quantity of Associated Natural Gas and Non Associated Natural Gas produced and saved.

5.1.4 The quality, characteristics and composition of such Natural Gas produced and saved separately.

5.1.5 The quantities of Crude Oil, Condensate and Natural Gas used for the purposes of carrying on drilling and production operations and pumping to field storage, as well as quantities re-injected.

5.1.6 The quantities of Crude Oil, Condensate and Natural Gas unavoidably lost.

5.1.7 The quantities of Natural Gas flared and vented.

5.1.8 The size of Petroleum stocks held on the first day of the Month in question.

5.1.9 The size of Petroleum stocks held on the last day of the Month in question.

5.1.10 The quantities of Natural Gas reinjected into the Petroleum Reservoir.

5.1.11 The number of days in the Month during which Petroleum was produced from each Field.

5.1.12 The Gas-Oil ratio for each Reservoir and Field for the relevant Month.

5.1.13 Water production, water injection and Reservoir pressure data for each Reservoir and Field.

5.2 All quantities shown in this Statement shall be expressed in both volumetric terms (barrels of Oil and cubic metres of Gas) and in the case of Oil in weight (metric tonnes).

5.3 For the purpose of reporting Field production quantities pursuant to this Section, the Contractor shall agree with the Management Committee on the exact area to be designated as Development Area.

5.4 The Government may direct in writing that the Contractor include other reasonable particulars relating to the production of Petroleum in its monthly Production Statement, and the Contractor shall comply with such direction.

5.5 The Production Statement for each Month shall be submitted to Government no later than fifteen (15) days after the end of such Month.

SECTION 6
VALUE OF PRODUCTION AND PRICING STATEMENT

6.1 The Contractor shall, for the purposes of Article 19 of the Contract, prepare a Statement providing calculations of the value of Crude Oil and Condensate produced and saved during each Month. This Statement shall contain the following information:

6.1.1 The quantities, prices and receipts realised therefor by the Contractor as a result of sales of Crude Oil and Condensate to third parties made during the Month in question.

6.1.2 The quantities, prices and receipts realised therefor by the Contractor as a result of sales of Crude Oil and Condensate made during the Month in question, other than to third parties, if any.

6.1.3 The quantities of Crude Oil and Condensate appropriated by the Contractor to refining or other processing without otherwise being disposed of in the form of Crude Oil or Condensate.

6.1.4 The value of stocks of Crude Oil and Condensate on the first day of the Month in question.

6.1.5 The value of stocks of Crude Oil and Condensate on the last day of the Month in question.

6.1.6 The percentage volume of total sales of Crude Oil and Condensate made by the Contractor during the Month that are Arms Length Sales to third parties.

6.1.7 Information available to the Contractor, insofar as required for the purposes of Article 19 of the Contract, concerning the prices of competitive crude oils produced by the main petroleum producing and exporting countries including contract prices, discounts and premia, and prices obtained on the spot markets.

6.2 The Contractor shall, for the purpose of Article 21 of the Contract, prepare a Statement providing calculations of the value of Associated Natural Gas and Non Associated Natural Gas produced, flared internally used, saved and sold during each Month. This Statement shall contain all information of the type specified in Section 6.1 for Crude Oil as is applicable to Gas and such other relevant information as may be required by Government.

6.3 The Statements required pursuant to Sections 6.1 and 6.2 shall include a detailed breakdown of the calculation of the prices of Crude Oil, Condensate, Associated Natural Gas and Non Associated Natural Gas pursuant to the provisions of Articles 19 and 21.

6.4 The Value of Production and Pricing Statement for each Month shall be submitted to Government not later than thirty (30) days after the end of such Month.

SECTION 7
STATEMENT OF COSTS, EXPENDITURES AND RECEIPTS

7.1 The Contractor shall prepare with respect to each Quarter a Statement of Costs, Expenditures and Receipts under the Contract using mercantile basis of accounting. The Statement shall distinguish between Exploration Costs, Development Costs and Production Costs and shall separately identify all significant items of costs and expenditure as itemised in Section 3 of this Accounting Procedure within these categories.

The Statement of receipts shall distinguish between income from the sale of Petroleum and incidental income of the sort itemised in Section 3.4 of this Accounting Procedure. If the Government is not satisfied with the degree of disaggregation within the categories, it shall be entitled to request a more detailed breakdown. The Statement shall show the following:

7.1.1 Actual costs, expenditures and receipts for the Quarter in question.

7.1.2 Cumulative costs, expenditures and receipts for the Year in question.

7.1.3 Latest forecast of cumulative costs, expenditures and receipts at the Year end.

7.1.4 Variations between budget forecast and latest forecast and explanations thereof.

7.2 The Statement of Costs, Expenditures and Receipts of each Quarter shall be submitted to Government not later than thirty (30) days after the end of such Quarter.

SECTION 8
COST RECOVERY STATEMENT

8.1 The Contractor shall prepare with respect to each Calendar Quarter a Cost Recovery Statement containing the following information:

8.1.1 Unrecovered Contract Costs carried forward from the previous Quarter, if any.

8.1.2 Contract Costs for the Quarter in question.

8.1.3 Total Contract Costs for the Quarter in question (Section 8.1.1 plus Section 8.1.2).

8.1.4 Quantity and value of Cost Petroleum taken and disposed of by the Contractor for the Quarter in question.

8.1.5 Contract Costs recovered during the Quarter in question as per Article 15.

8.1.6 Total cumulative amount of Contract Costs recovered up to the end of the Quarter in question.

8.1.7 Amount of Contract Costs to be carried forward into the next Quarter.

8.2 The Cost Recovery Statement for each Quarter shall be submitted to Government not later than thirty (30) days after the end of such Quarter.

SECTION 9
PROFIT SHARING STATEMENT

9.1 The Contractor shall prepare with respect to each Quarter a Profit Sharing Statement containing the following information:

9.1.1 The calculation of the applicable net cash flows as defined in Appendix D for the Quarter in question.

9.1.2 The value of the Investment Multiple applicable in the Quarter in question.

9.1.3 Based on Section 9.1.2 and Article 16, the appropriate percentages of Profit Petroleum for the Government and the Contractor in the Quarter in question.

9.1.4 The total amount of Profit Petroleum to be shared between the Government and the Contractor in the Quarter in question.

9.1.5 Based on Sections 9.1.3 and 9.1.4, the amount of Profit Petroleum due to the Government and the Contractor as well as to each constituent of the Contractor in the Quarter in question.

9.1.6 The actual amounts of Petroleum taken or payment received by Government and the Contractor as well as by each constituent of the Contractor during the Quarter in question to satisfy their entitlements pursuant to Section 9.1.5.

9.1.7 Adjustments to be made, if any, in future Quarters in the respective amounts of Profit Petroleum due to the Government and the Contractor as well as to each constituent of the Contractor on account of any differences between the amounts specified in Sections 9.1.5 and 9.1.6, as well as any cumulative adjustments outstanding from previous Quarters. 9.2 The Profit Sharing Statement shall be submitted to Government not later than thirty (30) days after the end of such Quarter. Any amount due or adjustment required in profit sharing among the Parties shall be made within thirty (30) days from the submission of the Statement to the Government.

SECTION 10
LOCAL PROCUREMENT STATEMENT

10.1 In furtherance of the obligation in Article 23 of the Contract for the Contractor to give preference to the procurement of Indian goods and services, the Contractor shall prepare in respect of each Year a local procurement statement, containing the following information:

(a) The amount of expenditure incurred by the Contractor directly, or indirectly through its Subcontractors, on goods supplied, produced or manufactured in India;

(b) the amount of expenditure incurred by the Contractor directly, or indirectly through its Subcontractors, on services provided by Indian entities;

(c) the respective percentages that the expenditures recorded under items (a) and (b) above represent of the Contractor's total expenditures;

(d) a detailed description of the procedures adopted during the Year to identify and purchase goods and services from Indian suppliers; and

(e) a detailed exposition of how the local purchases for the Year as recorded under items (a) and (b) above compared with the projected purchases included in the budget statement for that Year (pursuant to Section 12.1.3), with explanations for any significant variations;

10.2 The local procurement statement shall be submitted to the Government within sixty (60) days after the end of each Year.

SECTION 11
END OF YEAR STATEMENT

11.1 The Contractor shall prepare a definitive End of Year Statement. The Statement shall contain aggregated information in the same format as required in the Production Statement, Value of Production and Pricing Statement, Statement of Costs, Expenditures and Receipts, Cost Recovery Statement and Profit Sharing Statement, but shall be based on actual quantities of Petroleum produced, income received and costs and expenditures incurred. Based upon this Statement, any adjustments that are necessary shall be made to the transactions concerned under the Contract. [Explanation : End of year Statement shall further contain the item wise justification for the variation between the actual costs and expenditure incurred and included in the statement of costs, expenditure and receipts vis-à-vis the Budgets for corresponding line items.]

11.2 The End of Year Statement for each Year shall be submitted to Government within ninety (90) days of the end of such Year.

SECTION 12
BUDGET STATEMENT

12.1 The Contractor shall prepare a Budget Statement for each Year. This Statement shall distinguish between budgeted Exploration Costs, Development Costs and Production Costs and shall show the following:

12.1.1 Forecast costs, expenditures and receipts for the Year in question.

12.1.2 A schedule showing the most important individual items of total costs, expenditures and receipts for the said Year.

12.1.3 Estimated amounts to be spent in the Year on procuring goods and services in India.

12.2 The Budget Statement shall be submitted to Government with respect to each Year not less than ninety (90) days before the start of the said Year provided that in the case of the Year in which the Effective Date falls, the Budget Statement shall be submitted within ninety (90) days of the Effective Date.

APPENDIX-D
CALCULATION OF THE INVESTMENT MULTIPLE FOR PRODUCTION SHARING PURPOSES

1. In accordance with the provisions of Article 16, the share of the Government and the Contractor respectively of Profit Petroleum from the Contract Area in any Year shall be determined by the Investment Multiple earned by the Contractor from the then Petroleum Operations at the end of the preceding Year. These measures of profitability shall be calculated on the basis of the appropriate net cash flows as specified in this Appendix D.

2. The "Net Cash Income" of the Contractor from their Petroleum Operations in any particular Year is the aggregate value for the Year of the following: (i) Cost Petroleum entitlement of the Contractor as provided in Article 15; **plus** (ii) Profit Petroleum entitlement of the Contractor as provided in Article 16; **plus** (iii) the Contractor's all incidental income (of the type specified in section 3.4 of the Accounting Procedure) arising from Petroleum Operations; **less** (iv) the Contractor's Production Costs and royalty payments (Article 17) incurred on or in the Contract Area;

3. The "Investment" made by the Contractor in the Contract Area in any particular Year is the aggregate value for the Year of:

(i) the Contractor's Exploration Costs incurred on or in the Contract Area pursuant to Article 15 **plus**

(ii) the Contractor's Development Costs incurred on or in the Contract Area.

4. For the purposes of the calculation of the Investment Multiple, costs or expenditures which are not allowable as provided in the Accounting Procedure shall be excluded from Contract Costs and be disregarded.

5. The Investment Multiple ratio earned by the Contractor as at the end of any Year shall be calculated by dividing the aggregate value of the addition of each of the annual Net Cash Incomes (accumulated, without interest, up to and including that Year starting from the Year in which Production Costs were first incurred or Production first arose) by the aggregate value of the addition of each of the annual Investments (accumulated, without interest, up to and including that Year starting from the Year in which Exploration and Development Costs were first incurred).

6. Profit Petroleum from the Contract Area in any Year shall be shared between the Government and the Contractor in accordance with the value of the Investment Multiple earned by the Contractor as at the end of the previous Year pursuant to Articles 16.2 to 16.5.

APPENDIX-E1

FORM OF PARENT COMPANY FINANCIAL AND PERFORMANCE GUARANTEE

(to be furnished pursuant to Article 29.1 (b) of the Contract)

WHEREAS______________________________ a company duly organised and existing under the laws of________________ having its registered office at____________________ (hereinafter referred to as 'the Guarantor' which expression shall include its successors and assigns) is [the indirect owner of one hundred percent (100%) of the capital stock of XYZ Company and direct owner of its parent company;] and WHEREAS XYZ Company is signatory to a Production Sharing Contract in respect of an (offshore) (onshore) area identified as Block ____________________ (hereinafter referred to as 'the Contract') made between the Government of India (hereinafter referred to as 'the Government'), and XYZ Company (hereinafter referred to as XYZ which expression shall include its successors and permitted assigns); and WHEREAS the Guarantor wishes to guarantee the performance of XYZ Company or its Affiliate Assignee under the Contract as required by the terms of the Contract;

NOW, THEREFORE this Deed hereby provides as follows:

1. The Guarantor hereby unconditionally and irrevocably guarantees to the Government that it will make available, or cause to be made available, to XYZ Company or any other directly or indirectly owned Affiliate of XYZ Company to which any part or all of XYZ Company's rights or interest under the Contract may subsequently be assigned ('Affiliate Assignee'), financial, technical and other resources required to ensure that XYZ Company or any Affiliate Assignee can carry out its obligations as set forth in the Contract.
2. The Guarantor further unconditionally and irrevocably guarantees to the Government the due and punctual compliance by XYZ Company or any Affiliate Assignee, of any obligations of XYZ Company or any Affiliate Assignee under the Contract.
3. The Guarantor hereby undertakes to the Government that if XYZ Company, or any Affiliate Assignee, shall, in any respect, fail to perform its obligations under the Contract or commit any breach of such obligations, then the Guarantor shall fulfil or cause to be fulfilled the said obligations in place of XYZ Company or any Affiliate Assignee, and will indemnify the Government against all losses, damages, costs, expenses or otherwise which may result directly from such failure to perform or breach on the part of XYZ Company.
4. This guarantee shall take effect from the Effective Date and shall remain in full force and effect for the duration of the said Contract and thereafter until no sum remains payable by XYZ Company, or its Affiliate Assignee, under the Contract or as a result of any decision or award made by any expert or arbitral tribunal thereunder.

5. This guarantee shall not be affected by any change in the articles of association and byelaws of XYZ Company or the Guarantor or in any instrument establishing the Company or Guarantor.
6. The liabilities of the Guarantor shall not be discharged or affected by (a) any time indulgence, waiver or consent given to XYZ Company; (b) any amendment to the Contract or to any security or other guarantee or indemnity to which XYZ Company has agreed; (c) the enforcement or waiver of any terms of the Contract or of any security, other guarantee or indemnity; or (d) the dissolution, amalgamation, reconstruction or reorganisation of XYZ Company.
7. This guarantee shall be governed by and construed in accordance with the laws of India. IN WITNESS WHEREOF the Guarantor, through its duly authorised representatives, has caused its seal to be duly affixed hereto and this guarantee to be duly executed the ____________ day of ________________ 200____.

APPENDIX-E2
FORM OF COMPANY FINANCIAL AND PERFORMANCE GUARANTEE

(to be furnished pursuant to Article 29.1 (b) of the Contract)

WHEREAS___XYZ Company ______________________________ duly organised and existing under the laws of__________________ having its registered office at__________________________ (hereinafter referred to as 'the Guarantor' which expression shall include its successors and assigns) is signatory to a Production Sharing Contract in respect of an (offshore) (onshore) area identified as Block ________________________________ (hereinafter referred to as 'the Contract') made between the Government of India (hereinafter referred to as 'the Government'), and XYZ Company (hereinafter referred to as XYZ which expression shall include its successors and permitted assigns); and WHEREAS the Guarantor wishes to guarantee its performance under the Contract as required by the terms of the Contract;

NOW, THEREFORE this Deed hereby provides as follows:

1. The Guarantor hereby unconditionally and irrevocably guarantees to the Government that it will make available, or cause to be made available, financial, technical and other resources required to ensure that XYZ Company can carry out its obligations as set forth in the Contract.
2. The Guarantor further unconditionally and irrevocably guarantees to the Government the due and punctual compliance by it of any obligations under the Contract.
3. The Guarantor hereby undertakes to the Government that it shall fulfill or cause to be fulfilled all its obligations under the Contract, and if it fails to perform its obligations under the Contract or commits any

breach of such obligations, then it shall indemnify the Government against all losses, damages, costs, expenses or otherwise which may result directly from such failure to perform or breach on its part.

4. This guarantee shall take effect from the Effective Date and shall remain in full force and effect for the duration of the said Contract and thereafter until no sum remains payable by XYZ Company, under the Contract or as a result of any decision or award made by any expert or arbitral tribunal thereunder.
5. This guarantee shall not be affected by any change in the articles of association and byelaws of XYZ Company or in any instrument establishing the Company.
6. The liabilities of the Guarantor shall not be discharged or affected by (a) any time indulgence, waiver or consent given to XYZ Company; (b) any amendment to the Contract or to any security or other guarantee or indemnity to which XYZ Company has agreed; (c) the enforcement or waiver of any terms of the Contract or of any security, other guarantee or indemnity.
7. This guarantee shall be governed by and construed in accordance with the laws of India. IN WITNESS WHEREOF the Guarantor, through its duly authorised representatives, has caused its seal to be duly affixed hereto and this guarantee to be duly executed the ____________ day of ________________ 200__.

APPENDIX-F

PROCEDURE FOR ACQUISITION OF GOODS AND SERVICES I OBJECTIVES

The objectives of these procedures are to:

(a) ensure that the goods and services acquired by the Operator for carrying out the Petroleum Operations are acquired at the optimum cost taking into consideration all relevant factors including price, quality, delivery time and the reliability of potential suppliers.
(b) ensure that goods and services are delivered in a timely manner taking into consideration the consequences of delays in the acquisition of these goods and services on the project as a whole.
(c) ensure that the provisions of Article 23 of the Contract are implemented.

II. PRINCIPLES

The principles upon which these procedures are based are:

(a) The Parties must be satisfied that the Operator is working to an agreed procedure for acquiring goods and services which is auditable and in accordance with the provisions of the Contract.
(b) The Operator must have the ability to acquire goods and services expeditiously so that the project schedules in respect of Approved Work Programmes are maintained.

III. PROCEDURES

The procedures to be adopted by the Operator for the acquisition of goods and services shall be as follows:

Procedure A Procedure B Procedure C

Applicable to $ 50,000 to less $ 200,000 to less Equal to or more Exploration, Appraisal, than $ 200,000 than $ 500,000 than $ 500,000 Development and Production operations.

For contracts valued at less than US$ 5000

The Operator will be at liberty to determine the procurement procedures and methods to procure goods and services valued at less than US Dollars five thousand (US$5000).

For Contracts valued at US$ 5000 and above but less than US$ 50,000

The Operator will be at liberty to determine the preferred method of acquiring goods and services valued at US Dollars five thousand (US$ 5000) and above but less than US Dollars fifty thousand (US$ 50,000) provided that at least three (3) quotations from selected suppliers (including at least one (1) Indian supplier) will be obtained. For items valued at greater than US Dollars twenty thousand (US$ 20,000), Operator is required to report to the Operating Committee if the quote accepted exceed the lowest quote by more than twenty (20) percent. Operator will promptly report to the Operating Committee the Operator's reasons for not selecting the lowest quote.

Procedure A:

Operator shall:

(1) provide the constituents of the Contractor with a list of all the entities approved by the Operating Committee as per Appendix-F (V) for the applicable category of the contract along with other entities, if any, from whom the Operator proposes to invite tender;

(2) add to such list the entities whom other Party requests for adding within five (5) Business Days on receipt of such lists;

(3) if and when any Party so requests, Operator shall evaluate any entity listed in (1) and (2) above to assure that entity is qualified as based on the qualification criteria agreed in accordance with Appendix-F(IV) to perform under the contract;

(4) complete the tendering process within a reasonable period of time;

(5) circulate to all constituents of the Contractor a comparative bid analysis stating Operator's choice of the entity for award of contract. Provide also reasons for such choice in case entity chosen is not the lowest bidder;

(6) inform all the constituents of the Contractor of the entities to whom the contract has been awarded; and

(7) upon the request of a Party, provide such Party with a copy of the final version of the contract awarded.

Procedure B:

Operator shall:

(1) provide the Parties with a list of all the entities approved by the Operating Committee as per Appendix-F (V) for the applicable category of the contract, along with other entities, if any, from whom the Operator proposes to invite tender;
(2) add to such list the entities whom a Party requests for adding within five (5) Business Days on receipt of such list;
(3) if and when any Party so requests, Operator shall evaluate any entity listed in (1) and (2) above to assure that entity is qualified as based on the qualification criteria agreed in accordance with Appendix-F (IV), to perform under the contract;
(4) complete the tendering procedure within a reasonable period of time;
(5) circulate to all constituents of the Contractor a comparative bid analysis stating Operator's choice of the entity for award of contract. Provide also reasons for such choice in case the entity chosen is not the lowest bidder. If the bid selected is not the lowest bid, obtain prior approval of the Operating Committee for award of contract;
(6) award the contract accordingly and inform all the members of the Management Committee of the entities to whom the contract has been awarded; and
(7) upon the request of a Party, provide such Party with a copy of the final version of the contract awarded.

Procedure C:

Operator shall:

(1) publish invitations for parties to pre-qualify for the proposed contract in at least three (3) daily national Indian newspaper. Provide to Non-Operating Companies, a list of responding parties and an analysis of their qualifications for the contract being contemplated to be awarded. Include those who qualify, as per the prequalification criteria approved as per Appendix-F (IV) in the list of entities from whom Operator proposes to invite tender for the said contract;
(2) provide the members of the Management Committee with a total list of all the entities selected as (1) above and all the entities approved by the Operating Committee as per Appendix-F(V) for the applicable category of the contract, along with other entities, if any, from whom the Operator proposes to invite tender;
(3) add to such entities whom a Party requests for adding within five (5) Business Days on receipt of such list;
(4) if and when any Party so requests, Operator shall evaluate any entity listed in (2) and (3) above to assure that entity is qualified as based on the qualification criteria agreed in accordance with Appendix-F(IV), to perform under the contract;
(5) prepare and dispatch the tender documents to the entities as finally listed and to Parties;

(6) after the expiration of the period allowed for tendering, consider and analyse the details of all bids received;

(7) prepare and circulate to the constituents of the Contractor a comparative bid analysis stating Operator's recommendation as to the entity to whom the contract should be awarded, the reasons therefor, and the technical, commercial and contractual terms to be agreed upon;

(8) obtain the approval of the Operating Committee to the recommended bid. However, failing Operating Committee approval any Company may refer the issue to the Management Committee for decision; and

(9) award the contract accordingly and upon the request of a Party, provide such Party with a copy of the final version of the contract; IV. A set of vendor qualifications criteria for each major category contract/supply shall be proposed by the Operator and approved by the Operating Committee within thirty (30) days of its submission. In the event the Operating Committee fails to approve vendor qualification criteria within thirty (30) days of the date the same is first submitted by the Operator, the matter shall be referred to the Management Committee for decision. The Operating Committee may revise the qualification criteria. V. It is anticipated that, in order to expedite joint operations, contracts will be awarded to qualified vendors/contractors who are identified as approved vendors for the specified activities. A list of such approved vendors shall first be established as follows:

Operator shall:

(1) provide the constituents of the Contractor with a list of the entities from whom Operator proposes to invite tender for contracts; and

(2) add to such list entities whom a Company requests for adding within fourteen (14) days on receipt of such list; and

(3) obtain approval of the Operating Committee. Such list shall thereafter be maintained by the Operator. The Operating Committee may add to or delete vendors from such list.

APPENDIX-G

PROFORMA OF BANK GUARANTEE TO BE PROVIDED PURSUANT TO ARTICLE 29

1. In consideration of Government of India (hereinafter referred to as "Government") having entered into a Production Sharing Contract for the block ________dated _________ (hereinafter referred to as "Contract", which expression shall include all the amendments agreed to between the Government and the Contractor, thereto), with M/s ________________ having its registered office at _____________ (hereinafter referred to as __________, which expression unless repugnant to the context or meaning thereof include all its successors, administrators, executors and assigns), which is a constituent of the Contractor, and the Government have agreed that the ___________ Company shall furnish to Government a bank guarantee (hereinafter referred to as "Guarantee") towards its obligations as provided in the Contract for US$(for Foreign Companies)/US$ equivalent in Indian Rupees (for Indian Companies) for the performance of its obligations under the Contract.

2. We _________(name of the Bank) registered under the Law of _________ and having its registered office at ____________ (hereinafter referred to as "the Bank", which expression shall unless repugnant to the context or meaning thereof includes all its successors, administrators, executors and assigns) do hereby guarantee and undertake to pay immediately on the fist demand in writing and any/all money(s) to the extent of Indian Rupees/US$ _______(in figures) and (Indian Rupees/US$ ______ in words) without any demur, reservation, contest or protest and/or without any reference to the Company. Any such demand made by Government on the Bank by serving a written notice shall be conclusive and binding, without any proof, on the Bank as regards the amount due and payable, notwithstanding any dispute(s) pending before any court, tribunal, arbitrator, sole expert, conciliator or any other authority and/or any other matter or thing whatsoever, as liability under these presents being absolute and unequivocal. We agree that the Guarantee herein contained shall be irrevocable and shall continue to be enforceable until it is discharged by Government in writing. This Guarantee shall not be determined, discharged or affected by the liquidation, winding up, dissolution or insolvency of the Contractor and shall remain valid, binding and operative against the Bank.

3. The Bank also agree that Government at its option shall be entitled to enforce this Guarantee against the Bank as a principal debtor, in the first instance, without proceeding against the__________ Company and notwithstanding any security or other guarantee that Government may have in relation to the ___________ Company's liabilities.

4. The Bank further agree that Government shall have fullest liberty without our consent and without affecting in any manner our obligations hereunder to vary any of the terms and conditions of the said Contract or to extend time of performance by the said __________ Company from time to time or to postpone for any time or from time to time exercise of any of the powers vested in Government against the said _________ Company and to forebear or enforce any of the terms and conditions relating to the said Contract and we shall not be relieved from our liability by reason of any such variation, or extension being granted to the said _________ Company or for any forbearance, act or omission on the part of Government or any indulgence by Government to the said _________ Company or any such matter or thing whatsoever which under the law relating to sureties would, but for this provision, have effect of so relieving us.

5. The Bank further agree that the Guarantee herein contained shall remain

in full force during the period that is taken for the performance of the Contract and all dues of Government under or by virtue of this Contract have been fully paid and its claim satisfied or discharged or till Government discharges this Guarantee in writing, whichever is earlier.

6. This Guarantee shall not be discharged by any change in our constitution, in the constitution of ________ Company or that of the Contractor.
7. The Bank confirm that this Guarantee has been issued with observance of appropriate laws of the country of issue.
8. The Bank also agree that this Guarantee shall be governed and construed in accordance with Indian Laws and subject to the exclusive jurisdiction of Indian courts at ________, India.
9. Notwithstanding any thing contained herein above, our liabilities under this Guarantee is limited to Indian Rupees/US$ ___________(in figures) Indian Rupees/US$ ______________ (in words) and our Guarantee shall remain in force upto and including sixty (60) days after the expiry date/ extended date. Any claim under this Guarantee must be received before the expiry of sixty (60) days or before the expiry of sixty (60) days from the extended date if any. If no such claim has been received by us within sixty (60) days after the said date/extended date the Government's right under this will cease. However, if such a claim has been received by us within and upto sixty (60) days after the said date/extended date, all the Government's rights under this Guarantee shall be valid and shall not cease until we have satisfied that claim.

In witness whereof, the Bank through its authorised officers has set its hand and stamp on this ______ day of _________ 200_ at _____________. The seal of__________________ was hereto duly affixed by_____________ this_________ day of ___________200_ in accordance with its bye-laws and this Guarantee was duly signed by_______________ and _______________ as required by the said bye-laws.____________________ ____________________ Secretary President & Director Witness:

MODEL
PRODUCTION SHARING CONTRACT BETWEEN THE GOVERNMENT OF INDIA AND XYZ COMPANY(IES) WITH RESPECT TO THE CONTRACT AREA IDENTIFIED AS BLOCK_______________

TABLE OF CONTENTS

Article Contents

6 Management Committee
7 Operatorship, Joint Operating Agreement and Operating Committee
8 General Rights and Obligations of the Parties
9 Government Assistance
10 Discovery, Development and Production
11 Petroleum Exploration License and Mining Lease
12 Unit Development
13 Measurement of Petroleum
14 Protection of the Environment
15 Recovery of Cost Petroleum
16 Production Sharing of Petroleum
17 Taxes, Royalties, Rentals, Duties etc.
18 Domestic Supply, Sale, Disposal and Export of Crude Oil and Condensate
19 Valuation of Petroleum
20 Currency and Exchange Control Provisions
21 Natural Gas
22 Employment, Training and Transfer of Technology
23 Local Goods and Services
24 Insurance and Indemnification
25 Records, Reports, Accounts and Audit
26 Information, Data, Confidentiality, Inspection and Security
27 Title to Petroleum, Data and Assets
28 Assignment of Participating Interest
29 Guarantees
30 Term and Termination of the Contract
31 Force Majeure
32 Applicable Law and Language of the Contract
33 Sole Expert, Conciliation and Arbitration
34 Change of Status of Companies
35 Entire Agreement, Amendments and Waiver and miscellaneous
36 Certificates
37 Notices

APPENDICES

Appendix A Description of the Contract Area
Appendix B Map of the Contract Area
Appendix C Accounting Procedure to the Contract
Appendix D Calculation of the Investment Multiple for Production Sharing purposes
Appendix E1 Form of Parent company Financial and Performance Guarantee
Appendix E2 Form of Company Financial and Performance Guarantee
Appendix F Procedure for acquisition of goods and services
Appendix G Proforma of Bank Guarantee to be provided pursuant to Article 29

5.2 JOINT OPERATING AGREEMENT (JOA)

Article 7 of the PSC embodies and acknowledges the existence of a Joint Operating Agreement and also prescribes that it must include certain essential features. But this agreement is required only when two or more than two parties constitute the Contractor. Out of which one party is designated as Operator who carries on operations on behalf of the all other parties constituting the Contractor. No change in the operator ship shall be effected without the consent of the Government and such consent shall not be unreasonably withheld. It is often provided in the PSC that the functions required of the Contractor under this Contract shall be performed by the Operator subject to, and in accordance with, the terms and provisions of this Contract and generally accepted Good International Petroleum Industry Practice provided, however, that this provision shall not be construed as relieving the Constituent(s) of Contractor from any of its obligations or liability under the Contract.

Within fifteen (15) days of the Effective Date (or such longer period as may be agreed to by Government), the Companies constituting Contractor are required to execute an Operating Agreement. This agreement should be consistent with the provisions of this Contract. Operator is to provide to the Government a copy of the duly executed Operating Agreement within thirty (30) days of the Effective Date or such longer period as may be agreed to by Government.

The Operator is selected at the time of bid submission. The operator carries on petroleum operations on behalf of all the parties who have entered in to Joint Operating Agreement.

A committee known, as Operating Committee is constituted where all the consortium partners are represented and they take all important decisions in the meetings of the committee.

The procedure to collect money (cash call) from consortium partners gets special treatment and dealt in detail and provision for a penalty which extends to forfeiture of participating interest in the project, is made if a party defaults in the payment of cash call.

Other provisions relating to sole risk operations, assignment with or without a right of pre-emption or first refusal, relinquishment, force-majuere, dispute settlement etc are also provided in the Joint Operating Agreement.

Amongst various regimes in vogue, India has chosen production sharing regime. In this type of regime a contract is entered into between Government of India acting through Ministry of Petroleum and Natural Gas and contracting party or parties as the case may be. In the event of contractor consisting of more then one party, a joint operating agreement is also executed by consortium partners so as to regulate their relationship. These contracts are of long duration extending up to 25 years with provision for extension.

DRAFT
JOINT OPERATING AGREEMENT
BETWEEN
ABCD AND XYZ
FOR

TABLE OF CONTENTS

JOINT OPERATING AGREEMENT

This Joint Operating Agreement is made this ________________ day of ____________________ 2000– by and between

1. ABCD, a body corporate established under the Companies Act 1956, having its registered office at ________________________________ (Hereinafter referred to as "A" and unless the context otherwise require shall include its administrators, successors and permitted assignees), of the FIRST PART.

AND

2. **X** established and existing under the laws of the __________ and having its registered/head office at ________________________ (hereinafter referred to as "________" which expression unless the context otherwise require shall include its administrators, successors and assignees,) of the SECOND PART

AND

3. **Y** established and existing under the laws of the ________________and having its registered/head office at ((hereinafter referred to as "_____" which expression unless the context otherwise require shall include its administrators, successors and assignees,) of the THIRD PART.

All of which are collectively referred herein as Parties, which expression shall include their administrators, successors and such assigns as are permitted under Article 13 hereof.

WITNESSETH

WHEREAS, Parties have entered into a Bidding Agreement (BA) dated __________ which interalia, contains provision for execution of Operating Agreement amongst the Parties.

WHEREAS the Parties have entered in to a Production Sharing Contract (Contract) on __________ with Government of India.

WHEREAS the Parties wish to define their respective rights, interest and obligations with respect to the Petroleum Operations to be conducted under this Joint Operating Agreement (hereinafter referred to as "Agreement").

WHEREAS the Parties have read and understood the rights and obligations contained in the PEL and provisions of the Act and Rules,

NOW THEREFORE, in consideration of the premises and mutual agreements hereinafter contained, the Parties agree as follows:

ARTICLE 1
DEFINITIONS AND INTERPRETATION

Unless the context otherwise requires, for the purposes of this Agreement, the definitions contained in the Contract shall also be applicable herein. Further more, each term listed below shall have the meaning stated therefor, whenever used in this Agreement.

1.1 "Accounting Procedure" means the proceedings accounting to be adopted by the Operator in accordance with the Accounting Procedure attached hereto as Exhibit A.

1.2 "Work Programme" means the work programme as given in the Contract.

1.3 "Advance" means each payment of cash required to be made pursuant to a cash call.

1.4 "Affiliate" means a company that directly or indirectly controls or is controlled by such Party to this Agreement or a company which directly or indirectly controls or is controlled by a company which controls a Party to this Agreement it being understood that "control" means ownership by one company of more than fifty percent (50%) of the voting securities of the other company, or the power to direct, administer and dictate policies of the other company even where the voting securities held by such company exercising such effective control in that other company is less than fifty percent (50%) and the term "controlled" shall have a corresponding meaning.

1.5 "Agreement" means this Joint Operating Agreement including Exhibits.

1.6 "Appraisal Programme" means a programme, carried out following a Discovery in the Contract Area for the purpose of appraising Discovery and delineating the Petroleum Reservoirs to which the Discovery relates in terms of thickness and lateral extent and determining the characteristics thereof and the quantity of recoverable Petroleum therein. "Appraisal Well" means any well drilled as a part of fulfillment of Appraisal Programme.

1.7 "Approved Budget" means a budget that has been approved with respect to the Approved Work Programme.

1.8 "Approved Work Programme" means a Work Programme that has been approved by the Operating Committee.

1.9 "Budget" means a budget formulated in relation to a Work Programme. The term "Budget" shall mean, as the context requires, preliminary, proposed or finally adopted versions thereof, and any revisions or supplements thereto.

1.10 "Business Day" means a day on which the Banks in ______, India and ____________ are customarily open for business.

1.11 "Cash Call" means any request for payment of cash made by the Operator, in accordance with an Approved Work Programme and Budget to the Parties in connection with the Joint Operations (as per the format to be decided by the Operating Committee).

1.12 "Confidential Information" means all legal, financial, commercial, technical and other data, knowledge and information (including any and all information) obtained and/or acquired by any Party(ies) relating to this transaction including but not limited to geological, geophysical, seismic and other data, maps, models, diagrams and modelling exercises any and all reports or other documents created from such data and information and exercises and analyses carried out using such data and information.

1.13 "Contract Area" means at any time the area covered by the PEL,.

1.14 "Consenting Party" means a Party who agrees to participate in and pay its share of the cost of an Exclusive Operation.

1.15 "Delivery Point" means, except as otherwise herein provided or as may be otherwise agreed between the Parties having regard to international practice, the point at which Petroleum reaches the outlet flange of the delivery facility, either offshore or onshore and different Delivery Points may be established by purposes of sales. Delivery point(s) for the purpose of sale(s) of Petroleum from the Contract Area shall be approved by the Management Committee.

1.16 "Development & Production Operations" means any activity connected to exploitation of oil & natural gas from a Commercial Discovery.

1.17 Discovery" means an accumulation of hydrocarbon (oil and/or gas) in the subsurface, the economic viability of development and production of which has been established through exploration and delineation (assessment).

1.18 "Effective Date" shall mean the date on which the Contract is signed.

1.19 "Exploration Phase" means the PEL period of four years as prescribed in the PEL and any extension granted thereto unless otherwise relinquished by the licensee giving two months notice to the GOI before the expiry date. The period may be extended through regrant of PEL by the GOI as per provision of the Act and Rules on application by the licensee atleast two months before the expiry date at the sole discretion of the GOI.

1.20 "Exploration Operations" means operations conducted in the Contract Area pursuant to this Agreement in searching for Petroleum and shall include but not be limited to aerial, geological, geophysical, geochemical, palaeontological investigations relating to the subsurface geology including structural test drilling, stratigraphic test drilling, drilling of Exploration Wells and Appraisal Wells and other related activities such as surveying, drill site preparation and all work necessarily connected therewith that is conducted in connection

1.21 "Joint Account" means the set of accounts maintained by the Operator in accordance with the provisions of this Agreement to record all revenues, expenditure or other transactions conducted on behalf of the Parties as provided herein.

1.21 "Joint Operations" means those operations and activities carried out by the Operator pursuant to this Agreement till the formation of Joint Stock Company, the costs and expenses for which are chargeable to the Joint Account.

1.22 "Joint Property" means, at any point in time, the Contract Area, all wells, facilities, installations, equipment, materials, information, funds and the property held for the Joint Account.

1.23 "Minimum Work Obligations" means the Approved Work Programme related to those items which have been obligated as per the Contract.

1.24 "Operator" means a Party carrying out the operation relating to hydrocarbon exploration in the particular PEL block and who is a partner in the joint venture holding at least 25 % Participating Interest.

1.25 "Non-Operator" means at any time the Party or Parties to this Agreement other than the Operator.

1.26 "Operating Committee" means a committee to evaluate and approve budgets, work programme progress or petroleum operations and many other exigencies as outlined in various Articles and clauses and defined at Article 5 of this Agreement. The headings of the Articles and Clauses in this Agreement are inserted for convenience of reference only and shall not affect the meaning or construction of this Agreement.

1.27 "Party" or "Parties" means Parties to the Agreement and the respective successors and permitted assigns of each.

1.28 "Participating Interest" means an undivided interest (expressed as a percentage) held by a party in and to all rights, privileges, obligations and liabilities under the said PEL and this Agreement which will be converted into share holding on incorporation of the Joint Stock Company.

1.29 "Petroleum Operations" means, as the context may require, Explorations Operations, Development Operations or Production Operations or any combination of two or more such operations, including construction, operation and maintenance of all necessary facilities, plugging and abandonment of Wells, safety, environmental protection, transportation, storage, sale or disposition of Petroleum to the Delivery Point, Site Restoration and any or all other incidental operations or activities as may be necessary.

1.30 "Sub-contractor" means any company or person contracted by the Operator to provide goods or services with respect to Petroleum Operations.

1.31 "Wilful Misconduct" means an intentional and conscious or reckless disregard by the supervisory or management staff or any director of any Party, of the terms of this Agreement or of good oilfield practice but shall not include any act or omission reasonably required to meet emergency conditions, including without limitation the safeguarding of life, property and Joint Operations or, for the avoidance of doubt, any error of judgement or mistake made by any such person in the exercise, in good faith of any function, authority or discretion conferred upon the Party.

Interpretation

1.32 Unless the context otherwise requires, reference to singular shall include a reference to the plural and vice-versa, and reference to any gender shall include a reference to all genders.

1.33 Reference to any Article or Clause shall be an Article or Clause of this Agreement.

1.34 In the event of any inconsistency between the main body of this Agreement and any schedules attached hereto, the provisions of the main body of the Agreement shall prevail.

1.35 Reference to any law or regulation having the force of law includes a reference to that law or regulation from time to time amended, extended or re-enacted.

1.36 In case of any conflict of any term of this Agreement with that of the Contract or Bidding Agreement the provisions of the Contract and this Agreement will prevail.

ARTICLE 2
EFFECTIVE DATE, DURATION AND SCOPE

2.1 This Agreement shall be effective from the Effective Date and shall continue in effect for the term of the license and any lease granted including their extensions and thereafter until a final settlement of all accounts among the Parties has been made or terminated earlier in accordance with the terms of this Agreement or Contract.

2.2 The developing, operating, producing and abandoning the oil and gas fields in the Contract Area, treating and field processing and the transporting of Petroleum produced to the delivery point and appropriate supporting activities for any of the foregoing shall be carried on by a Joint Stock Company formed and incorporated in India for this purpose.

2.3 Except otherwise provided herein, each Party shall share in all costs, obligations and benefits in proportion to its Participating Interest.

ARTICLE 3
PARTICIPATING INTEREST AND JOINT OPERATIONS

3.1 The Participating Interest of the Parties on the Effective Date shall be as follows:

A : ____________ %
X : ____________ %
Y : ____________ %

3.2 The Parties undertake to complete the Work Programme as described in the Contract.

ARTICLE 4
OPERATOR

4.1 All Joint Operations shall be conducted by the Operator in accordance with policies, programmes and budgets approved in accordance with the provisions of this Agreement, the directions of the Operating Committee and in accordance with generally accepted international petroleum industry practices.

4.2 X is designated Operator and agrees to so act until it resigns or ceases to hold a Participating Interest hereunder or until such time as Joint Stock Company assumes the Operatorship as provided in Article 2 of this Agreement. The Parties shall provide the Operator with such powers of attorney or any other documents that it may need in order to carry out operations pursuant to this Agreement.

4.3 Any change in Operator other than PSC formed pursuant Article 4.2 shall be made by resignation of Operator on one hundred eighty (180) days prior written notice to the Parties or by mutual consent of the Parties.

4.4 Subject to Article 4.2, Operator shall be removed upon receipt of notice of occurrence of any of the following events.

(i) An order is made by a court or any effective resolution is passed for dissolution, liquidation, winding up, amalgamation or reorganisation of Operator.

(ii) Operator dissolves, liquidates or terminates its corporate existence.

(iii) Operator becomes insolvent, bankrupt or makes an assignment for the benefit of creditors; or

(iv) A receiver is appointed for a substantial part of Operator's assets.

(v) If Operator together with an Affiliate of Operator is or becomes the holder of a Participating Interest of less than twenty five percent (25%).

(vi) Operator is in substantial breach of its duties or obligations hereunder in any material respects and does not commence to rectify the breach within ten (10) days after written notice from any Non-Operator specifying the breach and requesting Operator to rectify the same and does not rectify the breach within a further thirty (30) days after such ten (10) day period.

4.5 When a change of Operator occurs pursuant to Article 4.3 or Article 4.4:

(i) The Operating Committee shall meet as soon as possible to appoint a successor Operator provided that only the Non-Operators shall have the right to vote to appoint the successor Operator. Such appointment shall be by the unanimous vote of those Parties entitled to vote.

(ii) The Operating Committee shall arrange for the taking of an independent inventory of all Joint Property and an audit of the books and records of the removed Operator. Such inventory and audit shall be completed, if possible, no later than the effective date of the change of the Operator. The liabilities and expenses of such audit shall be charged to the Joint Account unless Operator is removed pursuant to Article 4.4 (except for subparagraph (v) thereof) in which such expenses shall be for the account of the removed Operator.

(iii) The resignation or removal of Operator or replacement by the successor Operator shall not become effective prior to receipt of any approval of the GOI, if same is required.

(iv) Upon the effective date of the resignation, or removal or change of the Operator, the successor Operator shall succeed to all duties, rights and authority of Operator. The former Operator shall transfer to the successor Operator all Joint Property, books of accounts, records and other documents maintained by the outgoing Operator pertaining to the Contract Area and to Joint Operations. Upon delivery of the above described property and data, the former Operator shall be released and discharged from all obligations and liabilities as Operator accruing after such date.

(v) The outgoing Operator shall be liable for all obligations and liabilities as Operator incurred till the effective date of appointment of the successor Operator.

4.6 Subject to Article 4.1 Operator shall have the following powers and authorities and shall perform and carry out the following duties:

4.6.01 Exert its bonafide efforts to explore the Contract Area in accordance with generally accepted international petroleum industry practice to discover commercial accumulations of Petroleum.

4.6.02 Subject to the provisions of this Agreement, exercise control and exclusively manage and carry out the Joint Operations.

4.6.03 Comply with the respective obligations imposed upon the Parties by the laws of India including the timely filing of reports and payment of all fees, levies, taxes (except income taxes payable by the Parties), and charges of every nature payable by the Parties under the said laws.

4.6.04 Take all appropriate steps to obtain, maintain in force, renew or extend, consents, approvals, authorisations in accordance with the terms of this Agreement.

4.6.05 Represent the Parties before the GOI and any and all courts, administrative departments, bureaux and agencies whatsoever in India in matters related to the Joint Operations, and in so doing prepare and sign, file and receive any affidavits, undertakings, contracts, petitions, requests, certificates, authorisations, reports or other documents in connection therewith; advise the Parties in a timely fashion of any such acts which in the Operator's opinion may significantly affect the Parties; and perform all other acts of a similar nature necessary and proper in connection therewith, except to the extent any Party has indicated its desire to represent itself in connection with any of the foregoing. Non-Operators shall be entitled to attend for the purposes of observation, meetings between the Operator and GOI and any court appearance of the Operator where such meetings or appearances pertain to matters relating to this Agreement.

4.6.06 Establish and maintain such relationship with GOI, local authorities and with the public as shall be necessary or appropriate to assure that the Joint Operations are conducted and carried out to the best advantage of the Parties.

4.6.07 Prepare and submit to the Operating Committee, programs and budgets as provided in Article 6.

4.6.08 In accordance with the procedure outlined by Operating Committee establish and maintain such offices, installations and facilities in India as are necessary, adequate and appropriate for the efficient performance of Joint Operations and service hereunder.

4.6.09 Establish and maintain such books, records and accounts as are required by this Agreement together with such additional books, records and accounts as from time to time may be reasonably specified by the Parties. Maintain such production records, well and reservoir data, field reserves and rate of production studies and estimates, and the like as will reflect a thorough and accurate history of Petroleum Operations, together with such additional similar records and data as from time to time may be specified by the Operating Committee or required under the Agreement.

4.6.10 Employ, administer, use and have sole responsibility for all the selection, employment, administration and compensation for all personnel reasonably required for the Joint Operations. Also obtain when appropriate, such occasional or part-time services of experts, loan employees and others as operational requirements may, in Operator's opinion, dictate, while always keeping in mind the Parties desire to conduct Joint Operations in an efficient and cost effective manner.

4.6.11 Subject to Articles 5 and 7 herein, acquire on behalf of the Parties and utilise in the Joint Operations all equipment, materials and supplies necessary or desirable for carrying on the Joint Operations.

4.6.12 Acquire such lands, leases, surfaces rights, easements and rights-of-way as may be necessary or desirable for the Joint Operations; provided, however, that Operator shall not acquire for the Joint Operations any lands, surface rights, easements or rights-of-way involving a total commitment in excess of the equivalent of fifty thousand United States Dollars (U.S. $50,000) for each such acquisition without prior approval of Operating Committee; and provided, further, that Operator shall not acquire for the Joint Operations any lease for a longer period of time than ten (10) years at an annual rental in excess of the equivalent of twenty five thousand United States Dollars (U.S. $25,000) without prior approval of the Operating Committee.

4.6.13 Enter into contracts with others in its own name as principal or as agent for the Parties, as appropriate and in the interest of the

Parties following the procedure, for the performance of services or supply of facilities, equipment, materials, or supplies by such others, provided that there is a reasonable basis to believe that such others shall be competent and capable technically and financially, to perform properly their obligations under such contracts, and that the selection of such contractors is to be made on the basis of the Procedures for Acquisition of Goods and Services contained in Appendix VI to the Agreement.

4.6.14 In cases of emergency, Operator may award sub-contracts without following the procedure set out in the Procedure for Acquisition of Goods and Services contained in Appendix B subject, however, to such transactions being reported with written detailed reason for awarding such subcontracts along with the financial implications thereof to the Operating Committee at its first meeting following such award.

4.6.15 Promptly pay and duly discharge all costs and expenses incurred in connection with the Joint Operations and keep and maintain the Contract Area and any property acquired for the Joint Operations free of all liens, charges and encumbrances arising out of the Joint Operations or in connection therewith.

4.6.15.1 The Operator shall obtain and maintain in respect of the Joint Operations and the Joint Property, insurance as required under this Agreement or any applicable laws and such other insurance as the Operating Committee may from time to time determine. All such required insurance shall name all the Parties having a Participating Interest as additional insured with appropriate waivers of subrogation. No other insurance shall be carried by the Operator for the Joint Operations. Each Party individually may maintain such additional insurance or self-insurance as it deems proper to protect its own interests in the Joint Operations, the costs of which shall not be charged to the Joint Account.

4.6.15.2 The policies of any such insurance shall be endorsed with waivers of all explicit or implicit rights of subrogation to eventual rights against the Non Operators and, alternatively, to the extent possible, the Operator shall have and the Non-Operators named as additional insured. The Operator shall take care that Sub-contractors while taking insurance shall have the Parties named as additional insured.

4.6.15.3 Each Party shall be solely liable for any loss or damage or liability of whatsoever nature when such loss, damage or liability is caused by such Party's Gross Negligence or Wilful Misconduct and in such event such Party shall indemnify the other Parties against all claims in respect of any loss or damage so arising.

4.6.15.4 Subject to Article 4.6.15.3 and to Article 4.9.01, all damage, loss and liability incurred in the Joint Operations which are not completely covered by insurance policies procured or furnished under Article 4.6.15.1 shall be borne by the Parties in proportion to their respective Participating Interests.

4.6.16 Open and maintain a bank account or accounts as may be appropriate for the Joint Operations.

4.6.17 Keep the Parties informed regarding the Joint Operations and prepare and submit reports, technical studies and production forecasts, including significant changes in production forecasts, and other data and information as required under this Agreement.

4.6.18 Subject to Article 10.3, allow authorised representatives of the Parties full access to Joint Operations, data and information pertaining to Joint Operations including derivative maps, reports, books, records, accounts at all reasonable times and conduct audits as provided for in the JOA Accounting Procedure.

4.6.19 Give timely notice to the Parties prior to physically demarcating the location of an Exploration Well (including co-ordinates, and all other relevant information), as well as reasonable advance notice of the commencement of any exploratory drilling operations and give notice of proposed testing.

4.6.20 Take prompt and necessary action and measures to protect and safeguard life, health, the environment and the property and interests of the Parties and the Joint Operations.

4.6.21 Make such recommendations from time to time to the Operating Committee for the efficient carrying out of the Joint Operations as Operator may consider advisable.

4.6.22 Perform duties for the Operating Committee set out in Article 5, submit Work Programmes, Budgets and AERs as provided in Article 6 and perform such additional operations and services as Operating Committee may from time to time direct.

4.7 Operator undertakes to carry out each Joint Operations programme within the limits of the related Work Programme and budgets in accordance with the generally accepted international petroleum industry practice and shall not undertake any operations hereunder not included in the Joint Operations programme or make any expenditure as to a financial year in excess of the amounts budgeted in the related Budgets except as follows:

4.7.01 If necessary to carry out an Approved Work Programme, Operator is authorised to make expenditure in excess of the budget adopted therefor up to but not exceeding ten percent (10%) of each budget item, provided that such excess expenditures shall not exceed five percent (5%) of the total Budget and be reported promptly to the Operating Committee by Operator.

4.7.02 Operator in a financial year is also authorised to make expenditures for operations in the Contract Area not included in an Approved Work Programme or not provided in an adopted Budget, limited, however, to a total not exceeding fifty thousand United States Dollars (U.S. $50,000) provided that the expenditures are not for purposes therefore rejected by the Operating Committee and provided such expenditures shall be reported promptly to the Operating Committee by the Operator. Once the itemised report of such expenditures has been approved by the Operating Committee, the amount shall be increased back to fifty thousand United States Dollars (U.S. $50,000).

4.7.03 In case of emergency, Operator may make such immediate expenditures as it deems necessary for the protection of life, property or environment, and such emergency expenditures shall be reported promptly to the Operating Committee.

4.7.04 Operator shall annually submit a forecast of operating expenses (also called operating expenditure budget) to Operating Committee. Should actual or anticipated operating expenses for any annual period be in excess of such forecast by ten percent (10%) or one hundred thousand United States Dollars (U.S. $100,000), whichever is less, appropriate explanation of the over expenditure shall be presented to the Operating Committee as soon as such overrun is evident and Operating Committee shall determine corrective actions to be taken, if any. Operator shall furnish to the Parties quarterly statements of actual versus budgeted expenditures with appropriate explanations for variances.

4.8 Operator shall promptly provide the Parties all duplicate samples, properly packed, of rocks and fluids collected during operations. Operator shall always be mindful in the conduct of Petroleum Operations of the rights and interest of India.

4.9 If a claim is made against any Non-Operator or if any Party is sued on account of any matter arising from operations hereunder over which such Party has no control because of the rights given to the Operator by this Agreement, such Party shall immediately notify all other Parties and the claim or suit shall be treated as any other claim or suit involving operations hereunder. Operator shall (except in the case of class actions) without admission of liability, compromise, settle or defend claims and litigation (to the extent not covered by insurance) arising out of Joint Operations provided, however, that Operator shall not pay or waive rights to more than the equivalent of twenty five thousand United States Dollars (US$25,000) in full settlement of any claim or suit without prior written approval of the Operating Committee. However, the aggregate value payable by Operator without prior approval of Operating Committee towards settlement of all claims or suits in a year shall not be more than equivalent of one hundred thousand United States Dollars

(US$100,000). Any claim or litigation involving an amount in excess of the equivalent of twenty five thousand United States Dollars (US$25,000) shall be reported promptly to the parties and the Parties shall have the right to be represented by its own counsel at its own cost and expense in the compromise, settlement or defence of such claims or litigation.

4.9.01 The Operator shall not be liable to the Parties or any of them for any loss or damage arising in connection with the conduct by the Operator of Joint Operations hereunder, and, further each Party shall, to the extent of its Participating Interest, indemnify the Operator against any action, claim or demand arising in connection therewith except in the case of any loss or damage or any action, claim or demand, aforesaid arising directly out of Gross Negligence or Wilful Misconduct on the part of the Operator.

4.10 Claims and demands made on Non-Operators

In the event a claim, demand, action is made by the Government or any third party or arising out of or liability is made against or incurred by or demanded from any party who is not the Operator under the terms of this Agreement, then such Party shall upon receipt of any such claim or demand or upon receipt of notification of such suit or action, promptly report the same to the Operator and all of the other Parties and provide therewith details of such claim, demand, suit or action. Any compromise, settlement or waiver of any rights in respect of any such claim, demand, suit or action made by such party without the prior approval of the Operating Committee, shall be solely to the account of such party and the other Parties hereto shall not be responsible or liable therefore or any part thereof. The aforesaid is without prejudice to legal rights as may be available to the Parties to be represented by their own counsel at their expense in the participation or defence of any such claim, demand, suit or action.

4.11 Each of the Parties shall, obtain and maintain, in respect of its Percentage Interest share of any liability to third parties which may arise in connection with the Joint Operations, such insurance or other evidence of ability to meet any such liability as shall from time to time be determined by the Operating Committee. Each of the Parties shall, as and when required by the Operating Committee, produce to it such evidence as it shall reasonably require to establish that such insurance or such ability exists and is being maintained.

ARTICLE 5
OPERATING COMMITTEE

5.1 Subject to Article 2, to provide for the orderly supervision and direction of Joint Operations, there shall be established an Operating Committee consisting of one (1) representative of each of the Parties to this

Agreement. The nominee of the Operator will be the convenor and chairman of the Operating Committee.

Each Party shall within two weeks of the Effective Date notify the other Party in writing of the name and address of its representative and may at any time and from time to time replace its representative by notice to the other Party and may designate one or more alternate representatives to act in the absence of its representative. A representative shall be deemed authorised to bind the Party nominating such representative for the purposes of this Agreement.

The Operating Committee is the co-ordinating body for the direction, control and administration of the Joint Operations. The principal functions of the Operating Committee shall be:

5.1.01 To establish policies from time to time governing various aspects or activities of the Joint Operations.

5.1.02 To review, approve and revise annual exploration Work Programmes and corresponding Budgets, as proposed by the Operator.

5.1.03 To appoint such technical, financial, accounting, legal or other sub-committees as the representatives may deem appropriate for studies, analyses, reports, etc., on matters pertaining to the Joint Operations.

5.1.04 To review reports on Joint Operations conducted in the Contract Area.

5.1.05 To review and approve any proposal for the appraisal of an area and to delineate any Discovery.

5.1.06 To review, revise and approve Work Programmes and Budgets for Exploration Operation as defined herein and as proposed by the Operator.

5.1.07 To review and approve Exploration Wells and locations (including locations for Wells required for any purposes whatsoever).

5.1.07A To review and approve delineation plan including release of assessment wells for establishing a Discovery as a Commercial Discovery or otherwise.

5.1.08 To review and approve well stimulation programmes.

5.1.09 To review and determine the area to be relinquished

5.1.10 To approve appointment of subcontractors for carrying out any Petroleum Operations by Operator beyond the authority vested in the Operator under this Agreement.

5.1.11 To review and approve the declaration of a Discovery as a Commercial Discovery.

5.1.12 To review and approve any proposed plan for joint development

5.1.12A To finalise proposal for formation of a joint stock company for developing a Commercial discovery.

5.1.13 To review and determine the area that may be voluntarily relinquished.

5.1.14 To review and approve such other matters with respect to Joint

Operations in the Contract Area as may be referred to the Operating Committee by any member of the Operating Committee.

5.1.15 Any other matter which is required by the terms of this Agreement to be approved by the Operating Committee.

5.2 Unless otherwise agreed, the Operating Committee shall meet at least once every six months at the office of the Operator or elsewhere as the Parties shall unanimously decide. Additional meetings may be requested and convened by any Party. A Party which wishes to hold an additional meeting shall give notice at least twenty (20) days before the proposed meeting date, specifying the proposed time, place and agenda of the meeting. In case of emergency, a meeting shall be convened whenever necessary, upon such notice as shall be reasonable in the circumstances. Matters not included in the agenda may be taken up with the unanimous consent of all Parties (whether present at the meeting or otherwise).

5.3 The Chairman of the Operating Committee, who will direct the meeting and prepare the minutes, shall be a representative of the Operator. The minutes shall include the names of the representatives present, the Parties they represent and any formal action taken by the Operating Committee. The minutes shall be deemed correct and approved when the Operator receives written, telex or facsimile approval from the Parties. The Chairman shall maintain a permanent file of all Operating Committee determinations.

5.4 Any of the Parties may submit matters for determination without holding a meeting provided such matters are submitted by facsimile, telex or acknowledged telegram to the other Parties. In such event, the other Parties shall vote by giving advice by facsimile, telex or acknowledge telegram of such vote to the submitting Party within thirty (30) days with copies to all the other Parties unless the communication relates to the deepening, completing, side-tracking, plugging back, reworking, coring, testing or abandoning of a well on which drilling equipment is located, in which case each Party will vote within seventy two (72) hours. A failure to vote shall be deemed to be a negative response.

5.5 Each representative or alternate representative at the cost and expense of the Party it represents shall be entitled to have present at any Operating Committee meeting such reasonable number of advisors as they may desire and which are appropriate to the Joint Operations matters under consideration.

5.6 Subject to Article 8 of this Agreement all determinations of the Operating Committee shall be made by seventy five percent (75%) vote of the Participating Interests with each Party voting according to its Participating Interest and such determinations shall be binding on the Parties.

5.7 The lodging, food, transportation and incidental costs incurred by representatives on the Operating Committee to attend meetings of the

Operating Committee shall be borne by the respective Parties they represent.

ARTICLE 6
PROGRAMMES AND BUDGETS

6.1 On or before the 1st day of November of each Year, Operator shall submit to the Parties a recommended operating programme and budget for the Contract Area for the subsequent financial year, which operating programme and budget shall include as a minimum the work and expenditure required to be performed or incurred during such financial year. At the same time as that financial year's program and budget is submitted, a provisional program for the next succeeding financial year shall be presented by the Operator. Operator shall submit an operating program and budget as soon as possible after the execution of this Agreement for the remainder of the Year in which this Agreement is executed.

6.2 On or before the 30th of November of each Year, the Operating Committee shall agree upon and adopt an operating programme and budget for the subsequent year which shall include as a minimum the work and expenditure required to be performed or incurred under the Contract during such year. At the time of agreeing upon and adopting an operating programme and budget, the Operating Committee shall provisionally consider, but not act upon or adopt, an operating programme for the next succeeding Year. As soon as possible after the adoption of an operating programme and budget, Operator shall provide a copy thereof to each Party.

6.3 Each budget shall be subdivided into the two major functional categories, i.e. exploration and appraisal, and within each of those by sub-categories further broken down by programmed activity and individual projects. Purchases of materials and supply inventory not specifically made for an approved budget project will be budgeted as a separate item. Each individual project costing more than one hundred thousand United States Dollars (U.S. $100,000) per program will be either labelled as "Firm" or "To Mature" depending upon the degree of complete details furnished at the time of budget presentation.

6.4 For a project to be considered "Firm" within the budget it will require program description, objectives, cost estimates and economic justification sufficiently complete and in such detail as to allow thorough evaluation of the project.

6.5 Projects that do not meet this presentation standard at the time of budget formulation may also be included in the budget to receive approval in principle, and will be labelled as "To Mature". Such projects must be subsequently fully justified to the Operating Committee and receive its specific approval prior to the physical commencement of the project.

6.6 Operator shall submit to the Parties holding a Participating Interest for prior approval Authorisation Expenditure Requests ("AER's") covering each individual project within the budget in the following categories and shall obtain approval prior to undertaking the following projects,(A sample AER is Exhibit C)

(a) Licences – any initial cash payment for a licence.

(b) Exploration – quarterly projects for geological and geophysical work, core drilling and concession rentals after any initial payment, and overhead charges applicable to geological and geophysical operations. Under special circumstances Operator shall obtain separate approval for major geological or geophysical projects, which are in themselves in excess of fifty thousand United States Dollars (U.S. $50,000).

(c) Drilling:

   (i) Each Exploration and

   (ii) delineation well.

   (iii) Deepening of any well below original total depth, involving exploratory footage.

   (iv) Workovers in excess of one hundred thousand United States Dollars (U.S. $100,000) for any well, including deepening in to development zones.

(d) Plant and equipment.

   (i) Individual construction projects and equipment purchases exceeding fifty thousand United States Dollars (U.S. $50,000) each.

   (ii) Equipping of wells exceeding one hundred thousand United States Dollars (U.S. $100,000). Equipping of wells includes generally the purchase and installation of equipment and material for lifting, heating, storing and otherwise handling production.

   (iii) Projects and purchases costing less than fifty thousand United States Dollars (U.S. $50,000) each shall be approved by Operator but shall be included for information purposes in approval requests submitted monthly to the Parties.

   (iv) Storehouse stocks – quarterly commitments for purchases of advance materials for projects not yet approved.

   (v) Unusual commitments of any kind.

(e) The AER's shall be based on the best current estimates in hand and shall be developed prior to the availability of firm contract rates. Operator shall also submit supplements for approval when it is anticipated that AER's will be overextended by more than ten percent (10%).

(f) Approval of an AER shall be confirmed by returning a signed copy of the AER to the Operator.

ARTICLE 7
COSTS AND EXPENSES

7.1 All costs and expenses incurred by the Operator on behalf of the Parties in carrying out Joint Operations under this Agreement (whether or not specifically mentioned herein) shall be borne and paid by the Parties in proportion to their Participating Interests and shall be determined and/ or allocated in accordance with current accounting procedures as prescribed in the Accounting Procedure, attached hereto and made a part hereof.

7.2 In the event of any conflict between this Agreement and the applicable provisions contained in the Accounting Procedure or in any supplemental accounting instructions approved by the Operating Committee, the applicable provisions in the body of this Agreement shall govern to the extent of such conflict.

7.3 The terms and provisions of the Accounting Procedure may be changed or modified as agreed by the Parties from time to time.

7.4 Audits of Operator's accounts and records relating to the accounting made under this Agreement shall be conducted as provided for in the Accounting Procedure.

7.5 Inventories of Joint Operations assets shall be taken as provided for in the Accounting Procedure.

7.6 **Default**

i) If any Party (hereinafter called the "Defaulting Party") fails to pay in full its share of any Cash Call within 30 days from the date of cash call ("Due Date") as provided in the body of this Agreement or in the Accounting Procedure :

a) The Operator shall notify all Parties of such default no later than the fifth (5th) Business Day from the Due Date.

b) The Parties other than the Defaulting Party (hereinafter called the "Non-Defaulting Parties") shall contribute, as hereinafter provided, the amount in default.

c) Within three (3) Business Days following the notification by the Operator under (a) above, the Operator shall notify the Non-Defaulting Parties of the amount in Default and shall make a further Cash Call to the Non-Defaulting Parties, with a copy to the Defaulting Party.

d) If such default continues, each of the Non-Defaulting Parties, in the proportion that its then current Participating Interest bears to the total of the then current Participating Interests of all Non-Defaulting Parties shall pay the amount specified in the Cash Call issued pursuant to (c) above no later than ten (10) Business Days from date of receipt of such notice, and shall continue to pay, in addition to its Participating Interest share of subsequent Cash Calls, the same proportion of that part of all subsequent Cash Calls attributable to the Defaulting Party until such time as the Defaulting Party has remedied its default as hereinafter provided or until, pursuant to Article 7.7 the Defaulting Party's Participating Interest has been diluted or the Non-Defaulting Party has withdrawn. Any Non-Defaulting Party that fails to make payment of the amount specified in the Cash Call as issued to it pursuant to paragraph (c) above and any subsequent Cash Calls shall itself become a Defaulting Party.

e) The Operator shall promptly notify the Parties whenever a default has been remedied.

ii) The Defaulting Party shall have the right to remedy its default up to the time of reduction of its Participating Interest by payment of the amount due together with interest thereon calculated at the LIBOR rate on the Due Date plus two percent (2%) per annum, such interest being compounded on a monthly basis throughout the period of the delinquency. All interest received by the Operator shall be paid to the Non-Defaulting Parties as well as any funds advanced in respect of the amount in default refunded, in the proportions specified in Article 7.6 (i) (d).

iii) The Non-Defaulting Parties shall at all times have a lien on the Defaulting Party's Participating Interest share of Petroleum to secure the payment, in full, of the amount in default together with interest as provided in Article 7.6 (ii) and, for the purpose of enforcing such lien, shall have the right to require the purchaser of the Defaulting Party's share of Petroleum to make payment in respect thereof to the Operator or whomsoever they nominate until the Non-Defaulting Parties have recovered, in full, the amount in default together with interest as provided in Article 7.6(ii). Such purchaser shall be entitled to rely upon information given to it by the Operator for and on behalf of the Parties as to the amount in default and interest due, and the Operator or whomsoever it nominates, shall be empowered to give such purchaser a full and complete receipt in respect of such payment.

iv) All sums which may become payable to a Non-Defaulting Party under Article 7.6(iii) shall be paid by the Operator to it (or to a designated bank account of the Non-Defaulting Party's choice) as soon as possible and shall be applied first to accrued interest and then to reduce the principal amount of the default, and any residual balance shall be paid promptly to the Defaulting Party. Any foreign exchange gains or losses as calculated by the Operator shall be for the account of the Defaulting Party.

7.7 Continuation of Default.

i) After any default has continued for thirty (30) Business Days from the date of written notice of Default under Article 7.6 and for as long thereafter as the Defaulting Party remains in default on any payment due under this Agreement, the Defaulting Party shall not be entitled to vote on any matter coming before the Operating Committee during the period such default continues. Unless agreed otherwise by the Non-Defaulting Parties, the voting interest of each Non-Defaulting Party shall be in the proportion which its Participating Interest bears to the total of the Participating Interest of all the Non-Defaulting Parties. Any matter requiring unanimous vote of the Parties shall be deemed to exclude the Defaulting Party. Notwithstanding the foregoing, the Defaulting Party shall be deemed to have approved, and shall join with the Non-Defaulting Parties in taking any action to maintain and preserve the Contract.

ii) In the event that the default continues for more than Ninety (90) days (the "Default Period") and the Defaulting Party does not pay the amount in default plus accrued interest by the end of such time, a proportion of the Participating Interest of such Defaulting Party shall, at the sole election of the Non-Defaulting Parties be forfeited to the Non-Defaulting Parties to reflect the ratio that the Participating Interest of the Defaulting Party bears to the cumulative contribution made by the Defaulting Party to the cumulative contribution made plus amount in default of the Defaulting Party.

Following such forfeiture the reduced Participating Interest of the Defaulting Party shall be in accordance of the following formula, except that in the case of a default on the first Cash Call made on a Party, ten percent (10%) of the Participating Interest of the Defaulting Party shall stand forfeited:

$$A = (B/(B+C)) \times D$$

where :

A = the reduced Participating Interest of the Defaulting Party.

B = the total contributions to the Joint Operations costs of the Defaulting Party up to but not including the amount in default.

C = the amount in default.

D = Participating Interest of the Defaulting Party.

Such forfeiture will not restore the Defaulting Party's powers and rights forfeited under Article 7.7(i) until such Defaulting Party has paid the first Cash Call following the date of such forfeiture. If the Defaulting Party forfeits fifty percent (50%) or more of its Participating Interest as at the Due Date (fifty percent times Participating Interest) to the Non-Defaulting Parties under this Article, such Defaulting Party "Operating Committee" means a committee to evaluate and approve budgets, work programme progress or petroleum operations and many other exigencies as outlined in various Articles and clauses and defined at Article 5 of this Agreement. The headings of the Articles and Clauses in this Agreement are inserted for convenience of reference only and shall not affect the meaning or construction of this Agreement. shall, at the election of the Willing Parties, forfeit all of its Participating Interest to such Non-Defaulting Parties who shall assume such forfeited Participating Interest, including the obligations relative thereto, in proportion to their Participating Interests or in such other proportion as unanimously agreed by them. The Defaulting Party shall execute such documents as are necessary to transfer its Participating Interest at its sole cost.

iii) Notwithstanding the provisions of article 7.7(ii) in the event that as a result of a forfeiture by the Defaulting Party of a part of its Participating Interest pursuant to the provisions of Article 7.7(ii), the remaining Participating Interest of the Defaulting Party falls below ten percent (10%) the Non-Defaulting Parties shall assume such Participating Interest of the Defaulting Party in proportion to their Participating Interest or in such other proportion as unanimously agreed by them.

7.8 Forfeiture to be Without Prejudice to other remedies and no right of Set Off.

i) The foregoing provisions, including but not limited to the transfer of the Defaulting Party's Participating Interest are without prejudice to any other remedies or rights which the Non-Defaulting Parties may have against the Defaulting Party with respect to a default.

ii) Each Party acknowledges and accepts that a fundamental principle of this Agreement is that each Party must pay its Participating Interest share of all amounts due under this Agreement as and when required. Accordingly, any Party which becomes a Defaulting Party undertakes that, in respect of either any exercise by the Non-Defaulting Parties of any rights under or the application of any of the provisions of Articles 7.6 or 7.7 such Party shall not raise by way of set off or invoke as defence, whether in law or equity, any failure to pay amounts due and owning under this Agreement or any alleged or unliquidated claim that such Party may have against Operator or any Non-Operator, whether such claim arises under this Agreement or otherwise. Such Party further undertakes not to raise by defence, whether in law or in equity, that the nature or the amount of the remedies granted to the Non-Defaulting Parties is unreasonable or excessive.

7.10 Any Party may make an additional contribution on behalf of another Party so as to satisfy a Cash Call of such other Party in accordance with this Article.

ARTICLE 8
OWNERSHIP OF ASSETS

8.1 Subject to Article 2 and 8.3 hereunder, all assets acquired by Operator for the Joint Operations shall be owned by the Parties in proportion to their respective contributions to the costs of acquisition of such assets.

8.2 Unless the Operating Committee previously authorises otherwise, Operator shall acquire for the Joint Operations only such assets as are reasonably estimated to be required in carrying out the operations provided for in approved Joint Operations programs and budgets, or revisions thereof, it being intended that Operator shall not unreasonably stockpile materials and equipment for future use without the approval of the Operating Committee.

8.3 Whenever Operator considers that any Joint Operations asset is no longer needed in carrying out of the Operations, Operator shall notify the Parties and such surplus assets shall be disposed of as provided in Article 9.4

8.4 Each Party shall separately sell and dispose of its interest in any asset no longer required for the Joint Operations; provided that any Party may delegate to Operator for periods of not more than one (1) year authority to sell and dispose of such asset and the proceeds of any such sale shall reflect the percentage ownership which the Parties have in such assets.

8.5 Operator may also utilise in the operations equipment owned and made available by any Party and charges to the Joint Operations for use of such separately-owned equipment shall be made as provided in the Accounting Procedure.

8.6 For projects involving expenditures for equipment, supplies and services of one million United States dollars (U.S.$ 1,000,000) or more, any Party may furnish in kind (in whole or in Part) its Participating Interest share of the necessary equipment, supplies and services, provided such furnishing in kind does not interfere with or add to the cost of the programs and projects being carried out for the Joint Account as reasonably determined by Operator nor inhibit Operator's ability to discharge its obligation to conduct the operations under this Agreement prudently, efficiently and in a timely manner.

ARTICLE 9
INFORMATION, REPORTS AND CONFIDENTIALITY

9.1 Operator shall furnish to the Parties data, information and following reports:

9.1.1 Daily well progress report (for Exploration and Appraisal Wells) which shall include for the day concerned, a brief description of the work performed, the number of feet drilled, the type and depth of the formation or formations penetrated, the size and landed depth of any casing landed, the type and results of any tests made, and such other well information or data as the Parties from time to time may reasonably specify.

9.1.2 During drilling, daily reports of cumulative Well Costs.

9.1.3 Monthly report of Joint Operations which shall include a summary of all services and all Exploration, Appraisal, drilling and testing Operations performed during the period covered by such report.

9.1.4 Immediate reports on well developments of significance, such as blow-out or Discovery, and immediate notice of special events of importance to the Joint Operations, such as: fire, accident, sabotage or acts of God involving loss of life or serious property damage; strikes and riots; significant competitor activities; or GOI actions threatening or adversely affecting the rights and interests of the Parties. Such immediate reports or notices shall be given by telex, telegraph, telephone or equivalent means and confirmed in writing.

9.1.5 A comprehensive final report upon the completion of each significant phase in the operations.

9.1.6 Such other reports as the Parties may from time to time request as to all lands, concessions, surface right, easements, right-of-way, facilities, equipment, materials and supplies acquired by or disposed of by Operator pursuant to this Agreement.

9.2 Subject to Article 10.3, the Parties shall have full access at all reasonable times to all of the Joint Operations and may:

9.2.1 Inspect all records, books and accounts, data studies, logs, surveys, files, maps and other information kept by or available to Operator;

9.2.2 Receive samples of cores, well cuttings and Petroleum;

9.2.3 Have copies made (first copy at Joint Operations expense) of all data or information, including laboratory analyses and maps;

9.2.4 Have special studies performed by Operator (at such requesting Party's expense, unless approved by Operating Committee) to the extent and when studies do not interfere with or delay Operator's performance of its duties under this Agreement; and

9.2.5 Request and receive from Operator forecasts, schedules, statements or status reports where reasonable and where not otherwise provided for in this Agreement, preparation costs of which shall be for such Party's account unless approved by the Operating Committee.

9.3 Nothing in this Agreement shall require any Party to divulge proprietary technology to the other Parties, and subject to Article 22 of the Contract the following principles as to utilisation and protection of proprietary technology in the Joint Operations shall apply:

9.3.1 Operator may from time to time utilise its proven proprietary technology at Joint Operations expense in the conduct of Joint Operations and subject to Article 10.3.2, may take such reasonable precautions as Operator deems necessary or desirable (including agreement with or commitments from its employees, contractors and agents against such disclosure) to prevent disclosure of or entitlement to its proprietary technology including improvements and inventions resulting from use hereunder) to others.

9.3.2 Non-Operators shall have access to basic field data obtained through Operator's utilisation of proprietary technology and to final maps, data, information and conclusions resulting from any such utilisation, with entitlement to copies of such basic and final data, interpretations, maps, information and conclusions as provided for in this Agreement.

9.3.3 Any Party may at its own expense utilise its proprietary technology in developing from Joint Operations field data whatever information or conclusions it desires without the obligations to disclose such information, its result or conclusions to the Parties without adversely affecting Joint Operation.

9.3.4 Each Party utilising proprietary technology as mentioned in this Article 10.3 shall hold all other Parties free and harmless from all third party claims, charges and damages arising out of such utilisation, including but not limited to claims based on patent infringement.

9.3.5 Should future difficulties or controversies between the Parties occur regarding application of the principles of this Article 10.3, the Parties shall consult in good faith with the view to best possible utilisation of proprietary technology in the Joint Operations, giving due regard to each party's right to protect the proprietary nature of its technology and the Parties, right and need for access to data and information as well as to final results for justifiable determinations of programmes and budgets.

9.4 Except as otherwise provided herein, each Party undertakes to treat as confidential, and prevent disclosure to any third party of any and all information and data obtained in the Joint Operations and relating to the Contract Area or to Operations undertaken pursuant to this Agreement.

9.4.1 In consideration a Party (Disclosing Party) allowing the receiving Party (Receiving Party) access to the Confidential Information, the Receiving Party agrees to keep the Confidential Information confidential and acknowledges that the Disclosing Party considers the information disclosed to the Receiving Party to be proprietary. The Receiving Party shall not disclose the Confidential Information to any one except in accordance with the following terms:

a) The Receiving Party agrees to keep the Confidential Information strictly confidential and subject to sub-clause (b) to (d) inclusive, below, agrees not to disclose such Confidential Information to any person in any manner or mode whatsoever (including by means of photocopy or reproduction) without the prior written consent of the Disclosing Party, such consent not being unreasonably withheld.

b) The Receiving Party may disclose the Confidential Information without the prior written consent of the Disclosing Party to the extent such information

i) is already in the public domain or comes into the public domain other than through a breach of the terms of this Agreement by the Receiving Party or by any person to whom disclosure of Confidential Information by the Receiving Party is permitted pursuant to Sub-Clause (c) or (d) below ; or

ii) is required to be disclosed under applicable law or by a governmental order, decree, regulation or rule binding upon the Receiving Party provided that, as soon as practicable, the Receiving Party shall give prior written notice to the Disclosing Party that such disclosure is required and consults with the Disclosing Party on whether, and if so what action should be taken to resist the requirement.

c) The Receiving Party may disclose the Confidential Information without the prior written consent of the Disclosing Party to the Receiving Party's Affiliate.

d) The Receiving Party may also disclose the Confidential Information without the prior written consent of the Disclosing Party to such of the following persons who have a clear need to access it for the purpose of evaluating, negotiating or advising on matters arising out of this Agreement e.g. employees, officers and directors of the Receiving Party.

e) The Receiving Party may also disclose the Confidential Information with prior intimation to any professional adviser or financial institution retained by the Receiving Party for the purpose of evaluating the Confidential Information or evaluating financing, negotiating or advising on matters arising out of this Agreement.

f) Prior to making any disclosure of Confidential Information as permitted under Sub-Clause (c) or (d) above, the Receiving Party will procure that the proposed recipient of such Confidential Information is made aware of the terms of this Article and the Receiving Party will procure that each such proposed recipient adheres to and be bound by the terms of this Article as if they were a party to the extent of agreement contained in this Article.

9.4.2 Notwithstanding anything contained in hereinabove, the provisions of this Article shall continue and bind the Parties for a period not exceeding Five(5) Years from the date of termination of the Agreement.

ARTICLE 10
WORK PROGRAMME COMMITMENT

The Parties shall carry on the Work Programme as given in the Contract. In case of Minimum Work Programme, the Operator is authorised to take all actions necessary as per the commitment made in the Contract notwithstanding any difference of opinion in Operating Committee. The decision taken by the Operator for meeting the contractual commitment with respect to Minimum Work Programme shall be final and binding on the Parties.

ARTICLE 11
RELINQUISHMENT

11.1 When it becomes necessary pursuant to the terms of the PEL to relinquish portions of the Contract Area:

11.1.1 The Parties holding a Participating Interest shall make reasonable best efforts to reach agreement on the area proposed to be relinquished one hundred twenty (120) days before such relinquishment is required to be made. If such Parties are unable to reach unanimous agreement ninety (90) days prior to the date on which the area or areas must be relinquished, the following procedure shall be observed provided all selections of the area or areas to be relinquished are within the requirements and restrictions of the Contract:

a) In the event Parties holding at least sixty percent (60%) of the total Participating Interests agree on an area or areas to be relinquished, such area or areas shall be relinquished.

b) In respect to any area or areas to be relinquished on which the Parties holding at least sixty percent (60%) of the total Participating Interests cannot agree, each Party shall select a proportionate part of the remaining area or areas to be retained equal to its Participating Interest.

11.1.2 The Operator shall proceed in a timely manner to relinquish the area or areas determined in accordance with the foregoing procedures and pursuant to Act and Rules.

ARTICLE 12
WITHDRAWAL

12.1 Any Party may withdraw from this Agreement at any time provided that during the Exploration Period no Party shall withdraw until all the Exploration Operations obligations have been met and completed.

12.2 The Party desiring to withdraw (hereinafter called the "Withdrawing Party") shall give notice of intention to withdraw to the other Parties at least forty five (45) days in advance of the date of withdrawal and such notice shall contain:

12.2.1 Designation of the effective date for withdrawal, which shall be the last day of a Calendar Month;

12.2.2 An offer to assign, without compensation, to the other Parties having a Participating Interest all of the Withdrawing Party's right, title and interest in the Contract Area, and all Joint Property, such offer to be conditioned on the receiving Parties assumption, subject to Article 13.4, of all future obligation related to the right, title and interest offered.

12.2.3 The other Parties shall have thirty (30) days from date of receipt of such notice to notify the withdrawing Party whether they accept the offer and elect to receive the assignment provided for in Article 13.2.2.

(a) If one or more of the other Parties having a Participating Interest gives notice of acceptance and election to receive such an assignment within such thirty (30) day period, the Parties shall promptly execute and deliver and do and perform all acts and things necessary and appropriate to effect such assignment. If more than one of the other Parties elects to receive the assignment, the Withdrawing Party's interest shall be apportioned among the assignees in the proportion that the Participating Interest of each (prior to assignment) bears to the total of the Participating Interest of all such assignees unless another apportionment is agreed to by all the assignees.

(b) If none of the other Parties having a Participating Interest gives notice of its acceptance and election to receive such an assignment within the thirty (30) day period, the Parties shall promptly cause Operator to wind up the Joint Operations and the Parties shall do and perform all acts and things necessary and appropriate to this end, including termination of the Contract, subject to approval of Operating Committee. In such event all Parties, including the Withdrawing Party, shall continue to be responsible for their respective Participating Interest share of all financial and other obligations hereunder and under the Contract until the Contract is terminated without further obligation of the Parties and final settlement has been made among the Parties.

12.2.4 In the event of an assignment under Article 12.2.3 (a), the Withdrawing Party shall be responsible for its financial obligations accrued under the Contract and this Agreement up to the date of giving notice of intention to withdraw, including but not limited to payment of its full share of costs and expenses as to any particular well, geophysical survey or other individual project investment for which they were committed either under the terms of the Contract or through an approved program or budget which has been commenced (or funds therefor committed) before such date of notice, even though completion thereof may post date such date of notice, except to the extent such obligations shall be assumed or paid by any third party which acquired from the remaining Party or Parties within three (3) months of the effective date of withdrawal all or part of the interest assigned by the Withdrawing Party. The Withdrawing Party shall not be obligated for any obligations or commitments undertaken by the remaining

Parties after the date of giving notice of withdrawal, but the Withdrawing Party shall be obligated for its Participating Interest share of any obligations which may accrue under the Contract prior to the date of notice.

12.3 The Withdrawing Party shall remain liable and obligated for its Participating Interest share of all expenditure accruing to the Joint Account under any relevant Programme and budget prior to the date on which the notice of withdrawal was given and also to pay all costs and expenses incurred by the other Parties in connection with such withdrawal.

12.4 A Withdrawing Party shall not be allowed to withdraw from the Contract and this Agreement if its Participating Interest is subject to any lien, charges or encumbrance other than those arising under the Contract unless the other Parties are willing to accept the assignment subject to the additional lien, charges or encumbrance.

12.5 In the event a Party or Parties elect to receive an assignment of the Withdrawing Party's interest, all cost of the assignment, including notarial and registration fees and transfer taxes, shall be the sole obligation of such Party or Parties in the proportion of the Participating Interest each receives to the total Participating Interest assigned and such Party or Parties shall indemnify the Withdrawing Party against and hold it harmless from any further costs, expenses or liabilities incurred under the Contract from the date of notice.

ARTICLE 13
SALE, TRANSFER AND ASSIGNMENT

13.1 The right to sell, assign or transfer a Participating Interest, in whole or in part, shall be subject to the consent of other Parties and to the terms of the Contract and Article 11 and 12 and the following provisions of Article 13.

13.2 Any Party shall have the right at any time, with or without compensation, to sell, transfer and assign to an Affiliate all or part of its right, title and interest in the Contract Area and in and under this Agreement provided the assignee assumes all of the Party's obligation with respect to such interest and such Affiliate is considered by other Parties comparable technically and financially and assignor provides a bank guarantee to the extent of committed and additional work programme.

13.3 **Rights of First Refusal**

13.3.1 If any Party ("Transferring Party") wishes to sell, assign or transfer all or any part of its Participating Interest ("Sale Interest") other than to an Affiliate it shall notify each of the other Parties in writing of its desire to do so and if it wishes it may state the price and other material commercial terms upon which it is willing to dispose of the Sale Interest or it may request offers for the Sale

Interest. Each of the other Parties shall then have a period of forty-five (45) days after receipt of such notice ("JV Negotiating Period") in which to make a firm offer in writing to, and negotiate with, the Transferring Party for all or part of the Sale Interest. If the Transferring Party accepts any such offer the related sale of all or part of the Sale Interest shall be completed (other than for receipt of GOI approval) within forty five (45) days after the end of the JV Negotiating Period failing which the provisions of this Article 13.3.1 shall be re-applied. If the Transferring Party does not receive any such offer(s) from the other Parties or does not accept such offer(s) from the other Parties (whether for all or part of the Sale Interest) it may transfer all or part of the Sale Interest to a third party (subject always to the other provisions of this Article 13 provided that:

13.3.1.1 such transfer shall be at a price and on other commercial terms no more favourable to the third party than the final price and terms on which the Transferring Party and any other Party offer or failed to agree for all or part of the Sale Interest; and

13.3.1.2 such transfer to a third party shall have been completed (other than for receipt of GOI approval) within ninety days from the end of the JV Negotiating Period failing which the provisions of this Article 13.3.1 shall be re-applied.

13.3.2 The Parties intend to apply the principles of Article 13.3.1 in good faith and be bound by the spirit as well as the terms thereof and, in particular but without limitation:

13.3.2.1 a fair cash equivalent shall be placed on any non-cash consideration offered to the Transferee by an other Party or a third party, supported, if requested by any Party, by a report from an internationally recognised independent consultant; and

13.3.2.2 if requested by any other Party, the Transferring Party shall produce a report from its external auditors confirming that it has complied with the provisions of this Article 13.

13.4 No assignment, sale or transfer shall be effective unless the assignee shall have agreed in writing to be bound by all the terms and provisions contained in this Agreement and the Contract which shall include the assumption of all the assigning Party's rights and obligations under this Agreement and the PEL.

13.5 All costs of sales, transfers and assignments provided for in this Article 13, including notarial and registration fees, tax or other expense of whatsoever kind and nature, shall be the sole obligation of the assignor Party(ies).

13.6 The Parties shall execute and deliver all documents do and perform all acts and things necessary and appropriate to validly effect the sales and assignments provided in this Article 13.

13.7 Nothing contained in this Article 13 shall prevent a Party from mortgaging, pledging, charging or otherwise encumbering all or part of its interest in the Contract Area and in and under this Agreement for the purpose of security relating to raising of finance so as to meet its obligations under this Agreement provided that:

(1) such Party shall remain liable for all obligations relating to such interest;

(2) the encumbrance shall be subject to any necessary approval of the GOI and be expressly subordinated to the rights of the other Parties under this Agreement; and

(3) such Party shall ensure that any such mortgage, pledge, charge or encumbrance shall be expressed to be without prejudice and subject to the provisions of this Agreement.

ARTICLE 14
FORCE MAJEURE

14.1 Any non-performance or delay in performance by any Party hereto of any of its obligations under this Agreement or in fulfilling any condition of any licence or lease granted to such Party, or in meeting any requirement of the Act, the Rules or any licence or lease issued thereunder, shall, except for the payment of moneys due under this Agreement or under the Act and the Rules or any law, be excused if, and to the extent that, such non-performance or delay in performance is caused by Force Majeure as defined in this Article.

14.2 For the purpose of this Agreement, the term Force Majeure means any cause or event, other than the non-availability of funds, whether similar to or different from those enumerated herein, lying beyond the reasonable control of, and unanticipated or enforceable by, and not brought about at the instance of, the Party claiming to be affected by such event, or which, if anticipated or foreseeable, could not be avoided or provided for, and which has caused the non-performance or delay in performance. Without limitation to the generality of the foregoing, the term Force Majeure shall include natural phenomena or calamities, earthquakes, typhoons, fires, wars declared or undeclared, hostilities, invasions, blockades, riots, strikes, insurrection and civil disturbance.

14.3 Where a Party is claiming suspension of its obligations on account of Force Majeure, it shall promptly, but in no case later than seven (7) days after the occurrence of the event of Force Majeure, notify the other Parties in writing giving full particulars of the Force Majeure, the estimated duration thereof, the obligations affected and the reasons for its suspension.

14.4 A Party claiming Force Majeure shall exercise reasonable diligence to seek to overcome the Force Majeure event and to mitigate the effects thereof on the performance of its obligations under this Agreement provided, however, that the settlement of strikes or differences with employees shall be within the discretion of the Party having the difficulty. The Party affected shall promptly notify the other Parties as soon as the Force Majeure event has been removed and no longer prevents it from complying with the obligations which have been suspended and shall thereafter resume compliance with such obligations as soon as possible.

14.5 The Party asserting the claim of Force Majeure shall have the burden of proving that the circumstances constitute valid grounds of Force Majeure under this Article and that such Party has exercised reasonable diligence and efforts to remedy the cause of any alleged Force Majeure.

14.6 Where a Party is prevented from exercising any rights or performing any obligations under this Agreement due to Force Majeure, the time for the performance of the obligations affected thereby and for performance of any obligation or the exercise of any right dependent thereon, and the term of any stage of the exploration Phase or this Agreement, may be extended by such additional period as may be agreed between the Parties.

Notwithstanding anything contained hereinabove, if an event of Force Majeure occurs and is likely to continue for a period of thirty (30) days, the Parties shall meet to discuss the consequences of the Force Majeure and the course of action to be taken to mitigate the effects thereof or to be adopted in the circumstances

ARTICLE 15
RELATIONSHIP OF PARTIES, MUTUAL INDEMNITIES AND NO PARTITION

15.1 Several Liability

a) The rights, covenants, obligations and responsibilities of the Parties pursuant to or arising under this Agreement shall be several to the extent of their respective Participating Interests and not joint or joint and several.

b) Each Party shall be individually responsible for its own covenants, obligations and responsibilities to the extent of its Participating Interest except as herein expressly provided. Subject to Articles 16.2 and 4.9.01, in the event a claim, demand, action or liability is made against or incurred by or demanded from any Party under the terms of or pursuant to the Contract or this Agreement in excess of its Participating Interest, such Party shall have the right of recovery of such excess by contribution from each of the Parties in proportion to their respective Participating Interests.

c) Further, and except as expressly provided in this Agreement, nothing herein contained shall be construed to create an association, trust or partnership between the Parties.

d) No Party shall be under the control of, or be the agent of or have a right or power to bind any other Party without its express written consent, except as herein expressly provided.

15.2 Mutual Indemnities

Subject to the provisions of this Agreement, each Party shall indemnify and keep indemnified each of the other Parties from every claim, demand, action, liability or loss resulting from each and every breach or default by the indemnifying Party of any of its obligations under the Agreement.

15.3 No Partition

Unless unanimously agreed by the Parties no Party shall seek to partition any Joint Property, whether by court order or otherwise.

15.4 ONGC hereby warrants, and represents to Co-venturer as follows :

15.4.1 ONGC hereby warrant and represent to the Parties that the PEL has been granted and is on the Effective Date valid and subsisting in respect of Block ....and is the only agreement which materially affects or is capable of materially affecting the Farmout Interests.

15.4.2 The Operator hereby warrants and represents to other Party(s) that the best industry practices in operations and safety will be employed at all times and state of art of technology will be used during Exploration Phase.

15.6 Each party shall indemnify and hold harmless other Parties from and against :

15.6.1 any reasonable loss or damage arising from any action, litigation or arbitration arising out of the exploration operations, affecting the PEL Area or affecting the Farmout Interests where and to the extent that the circumstances giving rise to such action, litigation or arbitration arise on or after the Effective Date, and

15.6.2 any reasonable loss or damage which ONGC and/or Co-venturer suffer(s) as a direct result of breaches of or inaccuracies in any of the respective warranties or respective representations contained in this Article.

15.7 Operator shall furnish to ONGC with guarantee (in case of Affiliates a parent company guarantee) in respect of warranties and representations as contained in Article 15.5

15.8 Co-venturer(s) undertake(s) that differences or disputes between them shall not under any circumstances affect the due performance of their obligations

ARTICLE 16
NOTICES

16.1 Any notice or other communication provided for in this Agreement shall be in writing and shall be deemed to have been properly given or delivered when delivered in person to an authorised representative of the Party to whom it is addressed, or when sent by airmail, electronic mail, facsimile, telex, telegram or cable to the Party at its address hereinafter specified:

The Company Secretary
Telex No. : 031-65184/66262
Fax No. : +91-11-3316413

M/s
____
____
____

Telephone:
Fax:

and shall be effective from the date of receipt thereof.

16.2 Any Party may from time to time change its address under this Article on fifteen (15) days notice to the other Party (ies).

ARTICLE 17
TERMINATION

17.1 The Agreement may be terminated in accordance with terms and conditions described hereunder.

17.2 This Contract may, subject to the provisions herein below, be terminated by ONGC upon giving ninety (90) days written notice of to the other Parties of its intention to do so in the following circumstances, namely, that the Co-venturer or a Party comprising the Co-venturer ("the Defaulting Party").

a) has knowingly submitted any false statement to ONGC in any manner which was a material consideration in the execution of Bidding Agreement and this Agreement; or

b) is adjudged bankrupt by a competent court or enters into or scheme or composition with its creditors or takes advantages of any law for the benefit of debtors; or

c) has passed a resolution to apply to a competent court for its liquidation unless the liquidation is for the purpose of amalgamation or reconstruction of which the ONGC has given notice and the ONGC is satisfied that the performance under this Agreement would not be adversely affected thereby and has given its approval thereto; or

d) has assigned any interest in the Agreement without the prior consent of the Other Parties; or

e) has failed to make any monetary payment required by law or under this Agreement by the due date or within such further period after the due date as may thereunder be specified by the Government or any of its authorities; or

f) has failed to comply with or has contravened the provisions of this Agreement in a material particular;

PROVIDED THAT

Where the Contractor comprises two or more Parties, ONGC shall not exercise its rights of termination pursuant to Article 17.2, on the occurrence, in relation to one or more, but not all, of the Parties comprising the Contractor, of an event entitling to terminate the Agreement, if any other Party or Parties constituting the Contractor (the Non-Defaulting Party or Parties) satisfies ONGC that it, or they, is/are willing and would be able to carry out the obligations of the Agreement.

17.3 If the circumstance or circumstances that give rise to the right of termination under Article 17.3(f) or (g) or (i) are remedied (whether by the Defaulting Company or by another Party or Parties in its behalf) within the ninety (90) days period, or such extended period as may be granted by the ONGC following the notice of the intention to terminate the Agreement as aforesaid, such termination shall not become effective.

17.4 If the circumstance or circumstances that would otherwise result in termination are the subject matter of proceedings under Article 19, then termination shall not take place so long as such proceedings continue and thereafter may only take place when and if consistent with the arbitral award.

17.5 On termination of this Agreement, for any reason whatsoever, the rights and obligations under this Agreement shall cease but such termination shall not affect any rights of any Party which may have accrued or any obligations undertaken or incurred by the Co-venturer(s) or any Party comprising the Co-venturer(s) and not discharged prior to the date of termination.

17.6 In the event of termination a Foreign Company, which is a constituent of the Co-venturer(s), shall have to remove and export all its property and the provisions hereof provided that in the event that ownership of any property is in doubt, or disputed, such property shall not be exported unless and until the doubt or dispute has been settled in favour of the Foreign Company.

17.7 Within ninety (90) days after the termination of this Agreement, or such longer period as ONGC may agree, the Parties shall comply with directions of the GOI to avoid Environmental Damage or hazards to human life or to the property of others.

ARTICLE 18
ARBITRATION

18.1 All differences relating to, disputes arising out of in connection with this Agreement, which cannot be amicably resolved by the Parties shall be referred to Arbitration in accordance with Arbitration and Conciliation Act, 1996.

18.2 The Arbitral Tribunal shall consist of three Arbitrators. Each party shall appoint one Arbitrator notifying the differences / disputes to be adjudicated by the Arbitrators. The two Arbitrators appointed by the Parties shall appoint the Third Arbitrator i.e. Presiding Arbitrator. In case the parties or arbitrators fail to appoint the third arbitrator, the same shall be appointed in accordance with Arbitration and Conciliation Act, 1996.

18.3 The venue of Arbitration shall be New Delhi (India).

18.4 A reasoned decision in writing by the arbitrators shall be issued not more than 30 days following the conclusion of the arbitration hearing. Such decision shall be binding upon the Parties. To the extent the decision involves construction of the agreement, such construction shall thereafter be deemed a part of the Agreement and shall be used as appropriate in future interpretations hereof. The arbitrators shall assess the costs of the arbitration based on the positions of the Parties, the evidence submitted and the decisions rendered. The arbitration proceedings and decisions shall be in the English language.

ARTICLE 19
APPLICABLE LAW

This Agreement shall be governed by and interpreted in accordance with the Laws of India, both substantive and procedural. Delhi Courts in India shall have exclusive jurisdiction in all matters arising under this Agreement.

ARTICLE 20
GENERAL

20.1 No waiver by any Party of any provision of this Agreement, no consent to or departure therefrom shall be binding unless made expressly and confirmed in writing by authorised representatives of the Parties. Further, any such waiver or consent shall relate only to such matter, non-compliance or breach as it expressly relates to and for the purpose for which it is given and shall not apply to any subsequent or other matter, non compliance or breach. No default or delay on the part of either Party in exercising any rights, power or privilege hereunder shall operate as a waiver thereof or of any rights or remedies hereunder.

20.2 This Agreement may be amended or varied only by an instrument in writing executed by all of the Parties.

20.3 The English language shall be the language of this Agreement and shall be used in all communications between the Parties as well as in the Arbitral proceedings.

20.4 Conflict of Interest

(A) Each Party undertakes that it shall avoid any conflict of interest between its own interests (including the interests of Affiliates) and the interests of the other Parties in dealing with suppliers, customers and all other organisations or individuals doing or seeking to do business with the Parties in connection with activities contemplated under this Agreement.

(B) The provisions of the proceedings paragraph shall not apply to:

(1) A Party's performance which is in accordance with the local preference laws or policies of the host government; or

(2) A Party's acquisition of products or services from an Affiliate, or the sale thereof to an Affiliate, made in accordance with rules and procedures provided in this Agreement

20.5 Public Announcements

(A) Operator shall be responsible for the preparation and release of all public announcements and statements regarding this Agreement or the Joint Operations; provided that, no public announcement or statement shall be issued or made unless prior to its release all the Parties have been furnished with a copy of such statement or announcement and the approval of Parties has been obtained. Where a public announcement or statement becomes necessary or desirable because of danger to or loss of life, damage to property or pollution as a result of activities arising under this Agreement, Operator is authorised to issue and make such announcement or statement without prior approval of the Parties, but shall promptly furnish all the Parties with a copy of such announcement or statement.

(B) If a Party wishes to issue or make any public announcement or statement regarding this Agreement or the Joint Operations, it shall not do so unless prior to its release, such Party furnishes all the Parties with a copy of such announcement or statement, and obtains the approval of at least two (2) non-affiliated Parties holding fifty percent (50%) or more of the Participating Interests; provided that, notwithstanding any failure to obtain such approval, no party shall be prohibited from issuing or making any such public announcement or statement if it is necessary to do so in order to comply with the applicable laws, rules or regulations, legal proceedings or stock exchange having jurisdiction over such party.

20.6 Severance of Invalid Provisions

If and for so long as any provision of this Agreement is deemed or declared to be invalid for any reason whatsoever, such invalidity shall not affect the validity or operation of any other provision of this Agreement except only so far as shall be necessary to give effect to the construction of such invalidity, and any such invalid provision shall be deemed severed from this Agreement without affecting the validity of the balance of this Agreement.

20.7 Modifications

There shall be no modification of this Agreement except by written consent of authorised representative of all Parties.

IN WITNESS WHEREOF the Parties have caused this Agreement to be executed by their duly authorised officers and representatives as of the day and year first above written.

Signed by

for and on behalf of
Oil & Natural Gas Corporation Limited

In presence of
for and on behalf of

M/s x

In Presence of

EXIHIBIT "A"
ACCOUNTING PROCEDURE

ARTICLE I
GENERAL PROVISIONS

1.1. DEFINITIONS

The definitions contained in Article 1 of the Operating Agreement to which this Accounting Procedure is attached shall also be applicable herein. Furthermore, as and when used in this Accounting Procedure the terms listed below shall have the following meaning :

A. The "said Agreement" means the Operating Agreement of which this Exhibit is a part.
B. "Material" shall mean all material, supplies and equipment acquired for or held for use for Joint Operations.
C. "Outsider" or "Outsiders" means parties other than Operator and Non-Operators.
D. "Net Expenditures" as used herein means one hundred percent (100%) of Total Expenditures less credits applicable to the Joint Account.

E. Other terms used in this Accounting Procedure which are defined or described in said Agreement shall have the same meaning given them in said Agreement.

1.2 Inconsistency

In the event of any inconsistency or conflict between the provisions of this Accounting Procedure and the other provisions of the Agreement, the other provisions of the Agreement shall prevail.

1.3. Documentation and Statements to be submitted by the Operator

1.36.1 Within ninety (90) days of the signing of this Agreement, the Operator shall submit to and discuss with Non-Operator a proposed outline of charts of accounts, operating records and reports, which outline shall reflect each of the categories and sub-categories of costs and income specified in Sections 2 and 3 and shall be in accordance with generally accepted standards and recognised accounting systems and consistent with normal petroleum industry practice and procedures for joint venture operations.

Within ninety (90) days of receiving the above submission, the Non-Operators shall either provide written notification of its approval of the proposal or request, in writing, revisions to the proposal.

Within one hundred and eighty (180) days from the signing of the Agreement, the Operator and the Non-Operator shall agree on the outline of charts of accounts, records and reports which shall also describe the basis of the accounting system and procedures to be developed and used. Following such agreement, the Operator shall expeditiously prepare and provide the non-operator with formal copies of the comprehensive charts of accounts, records and reports and allow the Non-operator to examine the manuals and to review procedures which are, and shall be, observed.

1.4 Arms Length Transactions

Unless otherwise specifically provided, all transactions giving rise to revenues, costs or expenditures which will be credited or charged to the accounts prepared, maintained or submitted hereunder shall be conducted at arms length or on such a basis as will assure that all such revenues, costs or expenditures will not be lower or higher, as the case may be, than would result from a transaction conducted at arms length on a competitive basis with third parties.

1.5 Revision of the Accounting Procedure

By mutual agreement between the Operator and the Non-Operator, this Accounting Procedure may be revised from time to time, in writing, signed by the Parties, stating the date upon which the amendments shall become effective.

2. JOINT ACCOUNT RECORDS AND CURRENCY EXCHANGE

Operator shall maintain accounting records for the Joint Account in accordance with generally accepted accounting practices used in the petroleum industry and in such a manner that all expenditures will be segregated or can be allocated to appropriate areas in connection with the Petroleum Operations Joint Account records shall be maintained in the Operator's offices in India, and stated in United States Dollars and Indian Rupees. Expenditures other than United States Dollars or Indian Rupees shall be translated into United State Dollars and charged as United States Dollar expenditures. Unless mutually agreed otherwise by the Parties, translation of non United States Dollars expenditures into United States Dollars and United States Dollar expenditures into Indian Rupees will be as follows :

(A) Expenditures of currencies purchased by the Operator for the Joint Account with United States Dollars will be translated back into United States Dollars at the rate of exchange at which the currency was purchased with individual currency purchased deemed to be utilised on a first in/first out basis to provide the funds expended.

(B) Indian Rupees advanced pursuant to Indian Rupee calls shall be translated into United States Dollars at the monthly average of the daily mean of the buying and selling rates of exchange as quoted by State Bank of India (or any other financial body as may be mutually agreed by the Parties) for the Month. However, in case of any single non-US Dollar transaction in excess of the equivalent of fifty thousand United States Dollar (US $ 50,000), the conversion into US Dollars shall be performed on the basis of the average of the applicable rates for the day on which the transaction occurred.

(C) Expenditures of United States Dollars advanced pursuant to United States Dollar Cash Calls shall be translated into Indian Rupees using the applicable exchange rate for the month of disbursement based on the current procedure that the Operator uses to translate United States Dollars.

The intent of the procedures outlined in Paragraph 2 of this Article-I is to minimise any gain or loss on exchange chargeable to the Joint Account.

It is agreed, however, that any loss or gain resulting from the exchange of currency required for the use of the Joint Operations or from the translation above listed, shall be charged or credited to the Joint Account. The matter of translation rates will be reconsidered if it is determined that the above methods result in inequities, or burdensome administration, or do not provide the proper value for stating expenditures under the terms of the Agreement.

3. ADVANCES AND CASH CALLS

3.1 (a) Upon approval of Work Programmes and Budgets and subject to Article 4.7 of the Agreement, Operator shall have the right at its option to require the Parties to advance their share of Joint Operations net cash requirements by submitting to the Parties, on or before the 1st day of any month an itemised estimate by Budget categories of such requirements for the succeeding month.

(b) Each such Cash Call shall be equal to the Operator's estimate of the money to be spent to perform its duties under the Approved Work Programme and Budget for the month concerned and shall specify the currency or currencies in which the money shall be expended. The Parties shall pay their Participating Interest share of the cash requirements within thirty (30) days after receipt of the estimates or by the first Business Day of the month for which advances are required, whichever is the later. A reconciliation between estimates and actual cash requirements shall be made by Operator at the close of each Calendar Month and any differences will be adjusted in the subsequent cash requirement request.

(c) A Party shall have the right to make payment in the currency required to be expended by the Operator where such Party's share of the payment required is US$100,000 (United States Dollars one hundred thousand) or more. The principles referred to in Para 2 of this Article shall apply in relation to the exchange rate to be used by the Operator where a Party exercises its right pursuant to this subparagraph (c).

(d) For information purposes the Cash Call shall contain an estimate of the funds required for the succeeding 2 months.

(e) In the event that Operator subsequently determines that due to unanticipated expenditures the estimated monthly requirements submitted to the Parties are inadequate, Operator shall have the right to require the Parties to advance a portion of additional requirements by submitting an itemised estimate by Budget categories of such requirements. Each Party shall pay its proportional share of such advance within thirty(30) Business Days after receipt of such notice or the date of requirement specified by the Operator whichever is later.

(f) Each Cash Call shall indicate the AER pursuant to which it is made. If a Party believes that the Cash Call exceeds the AER to which it relates such Party may within three (3) days of receipt of such Cash Call send notice to all other Parties including the Operator advising that it believes that all or a part of the Cash Call exceeds the applicable AER and providing its reasons therefor. If within a further three (3) days all Non-Operators agree that all or a part of the Cash Call does not fall within an AER then that part of Cash Call on which the Non-Operators have unanimously agreed exceeds the AER shall be invalidated. The Parties shall not be subject to Articles 7.6 and 7.7 for failing to pay within the time required that part of a Cash Call invalidated pursuant to this provision and in such event within a further five (5) days the Operating Committee shall meet to resolve the issue.

(g) Where separate Joint Operations bank accounts are maintained as provided in Article I.3.1 (j) of this Accounting Procedure, any interest earned from deposits of cash advances will be for the benefit of the individual Parties' account in proportion to their contributions.

However, any Party may require that funds it has advanced in excess of its proportionate share of actual cash required by the Joint Operations, should said excess funds exceed twenty five thousand United States Dollars (U.S.$25,000) or their equivalent in Indian Rupees, be returned in the currency of their original denomination within a period of fifteen (15) Business Days from the date of receipt of the request at the Party's expense. If the Operator does not make arrangements in good faith to return such funds in fifteen (15) Business Days then the Operator shall pay the Party requesting the refund the difference between the interest specified in Article 7.6 (ii) of the Agreement and the interest earned on the bank account of the Joint Account.

(h) If the Operator does not request advance funds from the Parties, then the monthly billing referred to in Paragraph 4(A) of Article I shall be paid by the Parties in accordance with their Participating Interest within thirty(30) days after receipt thereof.

(i) Cash requirements shall be specified by the Operator in United States Dollars and/or in Indian Rupees as required for the Joint Operations and the Parties shall advance their share of United States Dollars and Indian Rupees as so specified.

(j) Subject to paragraph 3.1(e) of this Article I if any Party fails to pay in full its share of any Cash Call by the due date as provided above, it shall be in default, and Article 7.6 and 7.7 of the Agreement shall apply.

(k) Funds under the said Agreement held by Operator outside of India shall be segregated into a separate bank account or accounts, and shall not be commingled with Operator's other funds. Funds under the Agreement held by the Operator within India shall be held in separate bank accounts maintained solely for the respective purposes of the Joint Account. A monthly statement summarising receipts, disbursements, transfers to each Joint Account and beginning and ending balances thereof shall be provided by the Operator to the Parties.

(l) The Operator shall endeavour to maintain funds in bank accounts for the Joint Account at a level consistent with that required for the prudent conduct of Joint Operations.

4. STATEMENTS AND BILLINGS

(A) Unless otherwise agreed by the Parties, Operator shall render monthly to all the Parties by the twenty fifth (25th) day of each Month a billing statement of the costs and expenditure incurred during the prior Month, indicating by appropriate classification the nature thereof and the portion of such costs charged to each of the Parties.

These statements shall contain the following information :

(a) advances of funds setting forth the currencies received from each Party,

(b) the share of each Party in total expenditures on a cash and accrual basis.

(c) the current cash balance of each Party, and

(d) a summary of costs, credits and expenditure on current month, year to date, and inceptions to date basis or other periodic basis, as agreed by the Parties for each activity of the Approved Work Programme and Budget.

(B) Each Party shall be responsible for preparing its own accounting and tax reports to meet the requirements of India and other applicable countries. Operator to the extent that the information is reasonably available from the Joint Account records, will provide in a timely manner, Non-Operators with the necessary statements to facilitate the discharge of such responsibility.

(C) The "cash" rather than "accrual" basis for accounting shall be used in charging the Joint Account. "Cash" basis as used herein means that expenditures for the Joint Operations are regarded as applicable to the period in which cash disbursements are made, whereas "accrual" basis means that expenses are regardless of when paid. For Non-Operator's internal accounting purposes, the Operator shall reflect accruals applicable to the Joint Account as memorandum items.

(D) The billing statement is to be accompanied by billing schedules which shall be schedules dividing such expenditures and income into main classifications of expenditure as indicated by Approved Budget and AER's. The billing schedule shall also show cumulative totals of all payments linked to AER's Budget categories and Programmes.

5. ADJUSTMENTS

Payments of any such bills shall not prejudice the right of Non-Operators to protest or question the correctness thereof. Subject to the exception noted in paragraph 6 of this Article I, all statements rendered to Non-Operators by Operator during any Financial Year shall be presumed to be true and correct after twenty seven (27) months following the end of such Financial Year, unless within said twenty seven (27) month period any Non-Operator requests for adjustment. Failure on the part of Non-Operators to make claim on Operator for adjustment within such period shall establish the correctness thereof and preclude the filing of exceptions thereto or making of claims for adjustment thereon. No adjustment favourable to the Operator shall be made unless it is made within the same prescribed period. The provisions of this paragraph shall not prevent any adjustments resulting from amongst others, physical inventory of property as provided for in Article IV, INVENTORIES, hereof, and exchange gains or losses as contemplated by Article I (2).

6. AUDITS

Non-Operators, upon written notice to the Operator, shall have the right to audit Operator's accounts and records relating to the accounting hereunder for any Financial Year within the twenty four (24) month period following the end of such Financial Year; provided, however, that Non-Operators must take written exception to and make claim upon the Operator for all discrepancies disclosed by said audit within the said twenty four (24) month period. Non-Operators shall make every reasonable effort to conduct audits in a manner which will result in a minimum of inconvenience to the Operator. Audits by Non-Operators shall be at Non-Operator's expense.

At the conclusion of each such audit, the Non-Operators and the Operator shall endeavour to settle outstanding matters and a written report shall be circulated to all the Non-Operators and the Operator within three (3) months of the conclusion of the audit and shall also be placed before the Operating Committee. In addition, a written statement of settlement reached between the Non-Operator and the Operator after the outstanding matters have been settled between the Non-Operators and the Operators shall be placed before the Operating Committee. The report shall include all claims arising from such audit together with comments pertinent to the operations of the accounts and records. The Operator shall reply to the report in writing as soon as possible and in any event not later than three (3) months following the receipt of such report.

Should the Non-Operator consider that the report or reply requires further investigation of any item therein, the Non-Operator shall have the right to conduct further investigation in relation to such matter notwithstanding that such period of twenty four (24) months has expired. Such further investigations shall be commenced within thirty (30) days and be concluded within sixty (60) days of receipt of such reply or report.

Notwithstanding that the said period of twenty four (24) months may have expired, if evidence exists that the Operator has been guilty of Wilful Misconduct, the Non-Operator shall have the right to conduct further audits in respect of any earlier period.

All adjustments resulting from audit agreed between Operator and Non-Operators conducting the audit shall be rectified promptly in the Joint Account by the Operator and reported to other Non-Operators as well as to the Operating Committee. If any dispute shall arise in connection with an audit, it shall be discussed by the Parties and if a settlement between the Operator and the Non-Operators or among the Parties is not unanimously agreed, the item or items in dispute shall be referred to arbitration.

7. If there should be any conflict between the provisions of this Accounting Procedure and the said Agreement, the latter shall control.

ARTICLE II
BASIS OF CHARGES TO THE JOINT ACCOUNT

1. The Operator shall fully charge the Joint Account with all direct costs and expenses incurred in accordance with all Approved Work Programmes and Budgets and the provisions of this Agreement, Materials and Joint Operations. Without in any way limiting the generality of the foregoing, chargeable direct and indirect costs shall include all costs set forth herein

the Accounting Procedure and any other costs approved by the Operating Committee.

2. CLASSIFICATION, DEFINITION AND ALLOCATION OF COSTS AND EXPENDITURES

2.1 *Segregation of Costs*

Costs shall be segregated in accordance with the purposes for which such expenditures are made. All costs and expenditures allowable under Section 3, relating to Petroleum Operations, shall be classified, defined and allocated as set out below in this Section.

2.2 *Exploration Costs*

Exploration Costs are all direct and allocated indirect expenditures incurred in the search for Petroleum in an area which is, or was at the time when such costs were incurred, part of the Contract Area, including expenditures incurred in respect of:

2.2.1 Aerial, geophysical, geochemical, palaeontological, geological, topographical and seismic surveys, analyses and studies and their interpretation.

2.2.2 Core hole drilling and water Well drilling.

2.2.3 Labour, materials, supplies and services used in drilling Wells with the object of finding Petroleum or in drilling Appraisal Wells provided that if such Wells are completed as producing Wells or injection well for enhancing oil recovery, the costs of completion thereof shall be classified as Development Costs.

2.2.4 Facilities used solely in support of the purposes described in Sections 2.2.1, 2.2.2 and 2.2.3 above, including access roads, all separately identified.

2.2.5 Any Service Costs and General and Administrative Costs directly incurred on exploration activities and identifiable as such and a portion of the remaining Service Costs and General and Administrative Costs allocated to Exploration Operations determined by the proportionate share of total Contract Costs (excluding General and Administrative Costs and Service Costs) represented by all other Exploration Costs.

2.2.6 Geological and geophysical information purchased or acquired in connection with Exploration Operations.

2.2.7 Any other expenditures incurred in the search for Petroleum not covered under Sections 2.3 or 2.4.

2.3 *Development Costs*

Development Costs are all direct and allocated indirect expenditures incurred with respect to the development of discoveries within the Contract Area including expenditures incurred on account of:

2.3.1 Drilling Development Wells, whether these Wells are dry or producing and drilling Wells for the injection of water or Gas to enhance recovery of Petroleum.

2.3.2 Completing of Exploration Wells by way of installation of casing or equipment or otherwise or for the purpose of bringing a Well into use as a producing Well or as a Well for the injection of water or Gas to enhance recovery of Petroleum.

2.3.3 Purchase, installation or construction of production, transport and storage facilities for production of Petroleum, such as pipelines, flow lines, production and treatment units, wellhead equipment, subsurface equipment, enhanced recovery systems, offshore and onshore platforms, export terminals and piers, harbours and related facilities and access roads for production activities.

2.3.4 Engineering and design studies for facilities referred to in Section 2.3.3.

2.3.5 Any Service Costs and General and Administrative Costs directly incurred in Development Operations and identifiable as such and a portion of the remaining Service Costs and General and Administrative Costs allocated to development activities, determined by the proportionate share of total Contract Costs (excluding General and Administrative Costs and Service Costs) represented by all other Development Costs.

2.4 *Production Costs*

Production Costs are expenditures incurred on Production Operations after the start of production from the Field (which are other than Exploration and Development Costs). The balance of General and Administrative Costs and Service Costs not allocated to Exploration Costs or Development Costs shall be allocated to Production Costs.

2.5 *Service Costs*

Service Costs are direct and indirect expenditures incurred in support of Petroleum Operations in the Contract Area, including expenditures on warehouses, piers, marine vessels, vehicles, motorized rolling equipment, aircraft, fire and security stations, workshops, water and sewerage plants, power plants, housing, community and recreational facilities and furniture and tools and equipment used in these activities. Service Costs in any Year shall include the costs incurred in such Year to purchase and/or construct the said facilities as well as the annual costs of maintaining and operating the same, each to be identified separately. All Service Costs shall be regularly allocated as specified in Sections 2.2.5, 2.3.5 and 2.4 to Exploration Costs, Development Costs and Production Costs and shall be separately shown under each of these categories. Where Service Costs are made in respect of shared facilities, the basis of allocation of costs to Petroleum Operations hereunder shall be specified.

2.6 *General and Administrative Costs*

General and Administrative Costs are expenditures incurred on general administration and management primarily and principally related to Petroleum Operations in or in connection with the Contract Area, and shall include:

2.6.1 main office, field office and general administrative expenditures in India including supervisory, accounting and employee relations services;

2.6.2 an annual overhead charge for services rendered by the parent company or an Affiliate to support and manage Petroleum Operations under the Contract, and for staff advice and assistance including financial, legal, accounting and employee relations services, but excluding any remuneration for services charged separately under this Accounting Procedure, provided that:

(i) for the period from the Effective Date until the date on which the first Development Plan is approved by the Government, this annual charge shall be the Contractor's verifiable expenditure but shall in no event be greater than the following percentages of the total Contract Costs incurred during the Calendar Year in or in connection with the Contract Area and qualifying for recovery pursuant to Section 3:

Contract costs in calendar year	*Annual overhead charge*
0-$2 mm	3% plus
over $2 mm – $5 mm	2% plus
over $5 mm	1%

(ii) from the date on which the first Development Plan is approved, the charge shall be at an amount or rate to be agreed on between the Parties and stated in the Development Plan.

2.6.3 All General and Administrative Costs shall be regularly allocated as specified in Sections 2.2.5, 2.3.5 and 2.4 to Exploration Costs, Development Costs and Production Costs respectively, and shall be separately shown under each of these cost categories.

ARTICLE III
COSTS, EXPENSES, EXPENDITURES AND INCIDENTAL INCOME OF THE CONTRACTOR

3.1 *Costs Recoverable and Allowable Without Further Approval of the Operating Committee*

Costs incurred on Petroleum Operations pursuant to the Contract as classified under the headings referred to in Section 2 shall be allowable for the purposes of the Contract except to the extent provided in Section 3.2 or elsewhere in this Accounting Procedure, and subject to audit as provided for herein.

3.1.1 *Surface Rights*

All direct costs necessary for the acquisition, renewal or relinquishment of surface rights acquired and maintained in force for the purposes of the Contract except as provided in Section 3.1.9.

3.1.2 *Labour and Associated Labour Costs Operator's locally recruited employees based in India*

(a) Costs of all Operator's locally recruited employees who are directly engaged in the conduct of Petroleum Operations under the Agreement in India. Such costs shall include the costs of employee benefits and Government benefits for employees and levies imposed on the Contractor as an employer, transportation and relocation costs within India of the employee and such members of the employee's family as per the personnel policy of the employer as required by law or customary practice in India. If such employees are engaged in other activities in India, in addition to Petroleum Operations, the cost of such employees shall be apportioned on a time sheet basis according to sound and acceptable accounting principles.

(b) *Assigned Personnel*

Costs of salaries and wages, including bonuses, of the Contractor's employees directly and necessarily engaged in the conduct of the Petroleum Operations under the Contract, whether temporarily or permanently assigned, irrespective of the location of such employees, it being understood that in the case of those personnel only a portion of whose time is wholly dedicated to Petroleum Operations, only that pro rata portion of applicable salaries, wages, and other costs, as specified in Sections 3.1.2(c), (d), (e), (f) and (g), shall be charged and the basis of such pro rata allocation shall be specified.

(c) The Contractor's costs regarding holiday, vacation, sickness and disability benefits and living and housing and other customary allowances applicable to the salaries and wages chargeable under Section 3.1.2(b) above.

(d) Expenses or contributions made pursuant to assessments or obligations imposed under the laws of India which are applicable to the Contractor's cost of salaries and wages chargeable under Section 3.1.2(b) above.

(e) The Contractor's cost of established plans for employees' group life insurance, hospitalization, pension, retirement and other benefit plans of a like nature customarily granted to the Contractor's employees provided, however, that such costs are in accordance with generally accepted standards in the international petroleum industry, applicable to salaries and wages chargeable to Petroleum Operations under Section 3.1.2(b) above.

(f) Personal income taxes where and when they are paid by the Contractor to the Government of India for the employee, in accordance with the Contractor's standard personnel policies.

(g) Reasonable transportation and travel expenses of employees of the Contractor, including those made for travel and relocation of the expatriate employees, including their dependent family and personal effects, assigned to India whose salaries and wages are chargeable to Petroleum Operations under Section 3.1.2(b) above.

Transportation cost as used in this Section shall mean the cost of freight and passenger service and any accountable incidental expenditures related to transfer travel and authorized under Contractor's standard personnel policies. Contractor shall ensure that all expenditures related to transportation costs are equitably allocated to the activities which have benefited from the personnel concerned.

3.1.3 *Transportation Costs*

The reasonable cost of transportation of equipment, materials and supplies within India and from outside India to India necessary for the conduct of Petroleum Operations under the Contract, including directly related costs such as unloading charges, dock fees and inland and ocean freight charges.

3.1.4 *Charges for Services*

(i) *Third Parties*

The actual costs of contract services, services of professional consultants, utilities and other services necessary for the conduct of Petroleum Operations under the Contract performed by third parties other than an Affiliate of the Contractor, provided that the transactions resulting in such costs are undertaken pursuant to Section 1.8 of this Accounting Procedure.

(ii) *Affiliates of Contractor*

(a) *Professional and Administrative Services and Expenses*

Cost of professional and administrative services provided by any Affiliate for the direct benefit of Petroleum Operations, including, but not limited to, services provided by the production, exploration, legal, financial, insurance, accounting and computer services divisions other than those covered by Section 3.1.4 (ii)(b) which Contractor may use in lieu of having its own employees. Charges shall be equal to the actual cost of providing their services, shall not include any element of profit and shall not be any higher than the most favourable prices charged by the Affiliate to third parties for comparable services under similar terms and conditions elsewhere and will be fair and reasonable in the light of prevailing international oil industry practice and experience.

(b) *Scientific or Technical Personnel*

Cost of scientific or technical personnel services provided by any Affiliate of Contractor for the direct benefit of Petroleum Operations, which cost shall be charged on a cost of service basis. Charges therefor shall not exceed charges for comparable services currently provided by outside technical service organizations of comparable qualifications. Unless the work to be done by such personnel is covered by an approved budget and Work Programme, Contractor shall not authorize work by such personnel without approval of the Management Committee.

(c) Equipment, facilities and property owned and furnished by the Contractor's Affiliates, at rates commensurate with the cost of ownership and operation provided, however, that such rates shall not exceed those currently prevailing for the supply of like equipment, facilities and property on comparable terms in the area where the Petroleum Operations are being conducted. The equipment and facilities referred to herein shall exclude major investment items such as (but not limited to) drilling rigs, producing platforms, oil treating facilities, oil and gas loading and transportation systems, storage and terminal facilities and other major facilities, rates for which shall be subject to separate agreement with the Government.

3.1.5 Communications

Cost of acquiring, leasing, installing, operating, repairing and maintaining communication systems including radio and microwave facilities between the Contract Area and the Contractor's nearest base facility.

3.1.6 *Office, Shore Bases and Miscellaneous Facilities*

Net cost to Contractor of establishing, maintaining and operating any office, sub-office, shore base facility, warehouse, housing or other facility directly serving the Petroleum Operations. If any such facility services contract areas other than the Contract Area, or any business other than Petroleum Operations, the net costs thereof shall be allocated on an equitable and consistent basis.

3.1.7 *Environmental Studies and Protection*

Costs incurred in conducting the environmental impact assessment studies for the Contract Area, and in taking environmental protection measures including abandonment cost or contribution to abandonment funds as may be created for abandonment and Site Restoration pursuant to the terms of the Contract.

3.1.8 *Materials and equipment*

(i) *General*

So far as is practicable and consistent with efficient and economical operation, only such material shall be purchased or furnished by the Contractor for use in the Petroleum Operations as may be required for use in the reasonably foreseeable future and the accumulation of surplus stocks shall be avoided. Material and equipment held in inventory shall only be charged to the accounts when such material is removed from inventory and used in Petroleum Operations.

(ii) *Warranty*

In the case of defective material or equipment, any adjustment received by the Contractor from the suppliers or manufacturers or their agents in respect of any warranty on material or equipment shall be credited to the accounts under the Contract.

(iii) *Value of materials charged to the accounts under the Contract*

(a) Except as otherwise provided in subparagraph (b) below, materials purchased by the Contractor for use in the Petroleum Operations shall be valued to include invoice price less trade and cash discounts, if any, purchase and procurement fees plus freight and forwarding charges between point of supply and point of shipment, freight to port of destination, insurance, taxes, custom duties, consular fees, other items chargeable against imported material and, where applicable, handling and transportation costs from point of importation to warehouse or operating site, and these costs shall not exceed those currently prevailing in normal arms length transactions on the open market.

(b) Material purchased from or sold to Affiliates or transferred to or from activities of the Contractor other than Petroleum Operations under the Contract:

(aa) new material (hereinafter referred to as condition A) shall be valued at the current international price which shall not exceed the price prevailing in normal arms length transactions on the open market;

(bb) used material which is in sound and serviceable condition and is suitable for reuse without reconditioning (hereinafter referred to as condition B) shall be priced at not more than seventy five per cent (75%) of the current price of the above mentioned new materials;

(cc) used material which cannot be classified as condition B, but which, after reconditioning, will be further serviceable for original function as good second-hand condition B material or is serviceable for original function, but substantially not suitable for reconditioning (hereinafter referred to as condition C) shall be priced at not more than fifty per cent (50%) of the current price of the new material referred to above as condition A.

The cost of reconditioning shall be charged to the reconditioned material, provided that the condition C material value plus the cost of reconditioning does not exceed the value of condition B material.

Material which cannot be classified as condition B or condition C shall be priced at a value commensurate with its use.

Material involving erection expenditure shall be charged at the applicable condition percentage (referred to above) of the current knocked-down price of new material referred to above as condition A.

When the use of material is temporary and its service to the Petroleum Operations does not justify the reduction in price in relation to materials referred to above as conditions B and C, such material shall be priced on a basis that will result in a net charge to the accounts under the Contract consistent with the value of the service rendered.

3.1.9 *Duties, Fees and Other Charges*

Any duties, levies, fees, charges and any other assessments levied by any governmental or taxing authority in connection with the Contractor's activities under the Contract and paid directly by the Contractor except corporate income tax payable by the constituents of the Contractor.

3.1.10 *Insurance and Losses*

Insurance premia and costs incurred for insurance pursuant to Article 24 of the Contract, provided that such insurance is customary, affords prudent protection against risk and is at a premium no higher than that charged on a competitive basis by insurance companies which are not Affiliates. Except as provided in Sections 3.2 (ix), Section 3.2(x) and Section 3.2(xi), actual costs and losses incurred shall be allowable to the extent not made good by insurance. Such costs may include, but are not limited to, repair and replacement of property in the Contract Area resulting from damages or losses incurred by fire, flood, storm, theft, accident or such other cause.

3.1.11 *Legal Expenses*

All reasonable costs and expenses, except Section 3.2 (xi) resulting from the handling, investigating, asserting, defending, or settling of any claim or legal action necessary or expedient for the procuring, perfecting, retention and protection of the Contract Area and in defending or prosecuting lawsuits involving the Contract Area or any third party claim arising out of Petroleum Operations under the Contract, or sums paid in respect of legal services necessary for the protection of the joint interest of Government and the Contractor, shall be allowable. Such expenditures shall include attorney's fees, court costs, costs of investigation and procurement of evidence and amounts paid in settlement or satisfaction of any such litigation and claims provided such costs are not covered elsewhere in the Accounting

Procedure. Where legal services are rendered in such matters by salaried or regularly retained lawyers of the Contractor or an Affiliate, such compensation shall be included instead under Section 3.1.2 or 3.1.4 (ii) above as applicable.

3.1.12 *Training Costs*

All costs and expenses incurred by the Contractor in training as is required under Article 22 of the Contract.

3.1.13 *General and Administrative Costs*

The costs described in Section 2.6.1 and the charge described in Section 2.6.2 of this Accounting Procedure.

3.1.14 Royalty, License fee and other levies and taxes (except Income Tax) paid to the Government of India.

3.2 *Costs not recoverable and not allowable under the Contract*

The following costs and expenses shall not be recoverable or allowable (whether directly as such or indirectly as part of any other charges or expense) unless specifically agreed by all Parties to the contract.

(i) expenditures in respect of any financial transaction to negotiate, float or otherwise obtain or secure funds for Petroleum Operations including, but not limited to, interest, commission, brokerage and fees related to such transactions, as well as exchange losses on loans or other financing, whether between Affiliates or otherwise;

(ii) expenditures incurred in obtaining, furnishing and maintaining the guarantees required under the Contract and any other amounts spent on indemnities with regard to non- fulfilment of contractual obligations;

(iii) attorney's fees and other costs and charges in connection with arbitration proceedings and sole expert determination pursuant to the Contract;

(iv) fines, interest and penalties imposed by Courts of law of the Republic of India;

(v) donations and contributions;

(vi) amounts paid with respect to non-fulfilment of contractual obligations;

(vii) costs incurred as a result of failure to insure where insurance is required pursuant to the Agreement, or of failure to follow procedures laid down by an insurance policy or where Contractor has elected to self insure, or has under-insured;

(viii) costs and expenditures incurred as a result of misconduct or negligence of the Contractor;

3.3 *Other costs recoverable and allowable only with Operating Committee approval*

Any other costs and expenditures not included in Section 3.1 or 3.2 of this Accounting Procedure but which have been incurred by the Contractor for the necessary and proper conduct of Petroleum Operations shall be allowed to be recovered only with the express prior approval in writing of the Management Committee.

3.4 *Incidental Income and Credits*

All incidental income and proceeds received from Petroleum Operations under the Contract, including but not limited to the items listed below, shall be credited to the accounts under the Contract and shall be taken into account for cost recovery, and profit production sharing purposes in the manner described in Articles 15 and 16 of the Contract:

(i) The proceeds of any insurance or claim or judicial awards in connection with Petroleum Operations under the Contract or any assets charged to the accounts under the Contract where such operations or assets have been insured and the premia charged to the accounts under the Contract;

(ii) Revenue received from third parties for the use of property or assets, the cost of which has been charged to the accounts under the Contract;

(iii) Any adjustment received by the Contractor from the suppliers/ manufacturers or their agents in connection with defective material, the cost of which was previously charged by the Contractor to the accounts under the Contract;

(iv) Rentals, refunds or other credits received by the Contractor which apply to any charge which has been made to the accounts under the Contract;

(v) Prices originally charged to the accounts under the Contract for materials subsequently exported from the Republic of India without being used in Petroleum Operations under the Contract;

(vi) Proceeds from the sale or exchange by the Contractor of assets, plant or facilities, the acquisition costs of which have been charged to the accounts under the Contract;

(vii) Legal costs charged to the accounts under Section 3.1.11 of this Accounting Procedure and subsequently recovered by the Contractor.

3.5 *Non-Duplication of Charges and Credits*

Notwithstanding any provision to the contrary in this Accounting Procedure, it is the objective of the Parties that there shall be no duplication of charges or credits to the accounts under the Contract.

ARTICLE IV
DISPOSAL OF MATERIALS

The Operator shall be under no obligation to purchase for its own account surplus, new or second hand Material. The disposition of major items of surplus equipment and Material, such as derricks, tanks, engines, pumping units and tubular goods shall be subject to determination by the Operating Committee, provided the Operator shall have the right to dispose of normal accumuiations of junk and scrap material either by transfer or sale for the Joint Properties, **subject to relevant provisions of the Agreement.**

1. MATERIAL PURCHASED BY OPERATOR OR NON-OPERATOR

Material purchased by either Operator or Non-Operators for use in operations outside this Agreement shall be credited by Operator to the Joint Account for the month in which the Material is removed by the purchaser.

2. DIVISION IN KIND

Division of Material in kind, if made between Operator and Non-Operators, shall be in proportion to their respective interests in such Material. Each Party will thereupon be charged individually with its share of the agreed value of Material received or receivable by each Party, and corresponding credits will be made by Operator to the Joint Account. Such credits shall appear in the monthly statement of Joint Operations.

3. SALE TO OUTSIDERS

Sale to Outsiders of Material from the Joint Properties shall be credited by the Operator to the Joint Account at the net amount collected by the Operator from vendee. If such sale shall be priced lower than the basis stipulated in Article III.3, then such sale shall be approved by the Operating Committee prior to such sale. Any claims by vendee for defective Materials or otherwise shall be charged back to the Joint Account if and when paid by the Operator.

ARTICLE V
INVENTORIES

1. PERIODIC INVENTORIES, NOTICE AND REPRESENTATION

Unless otherwise agreed by the Parties, the Operator shall take physical inventory of Joint Account assets as are ordinarily considered controllable by operators of oil and gas properties at the following intervals:

A. Storehouse stocks - annually.

B. Other assets - cyclical basis so that all controllable assets shall be inventoried at intervals of not more than five (5) years.

C. Completed construction projects - within sixty (60) days following physical completion of project.

Written notice of intention to take inventory shall be given by the Operator at least thirty (30) days before any inventory is to begin so that the Non-Operators may be represented when any inventory is taken.

Failure of the Non-Operators to be represented at an inventory shall bind the Non-Operators to accept the inventory taken by the Operator.

2. RECONCILIATION AND ADJUSTMENT OF INVENTORIES

Reconciliation of inventory with charges to the Joint Account shall be made by the Operator, and the Operator shall furnish the Non-Operators a copy of the inventory and a priced list of overages and shortages. Inventory adjustments shall be made by the Operator to the Joint Account for overages and shortages except that any particular item exceeding five thousand United

States Dollars (U.S.$ 5,000) or class of Material adjustment exceeding twenty five thousand United States Dollars (U.S.$ 25,000) or net total adjustments of both overages and shortages for any particular inventory exceeding in the aggregate an amount equivalent to forty thousand United States Dollars (U.S.$ 40,000) shall require the approval of the Non-Operator(s). The Operator shall be held accountable to the Non-Operator(s) only for shortages due to lack of reasonable diligence.

3. SPECIAL INVENTORIES

Special inventories may be taken, at the expenses of the purchaser, whenever there is any sale or change of interest in the Joint Properties; and it shall be the duty of the Party selling to notify all other Parties hereto as quickly as possible after the transfer of interest takes place. In such cases, both the seller and the purchaser shall be represented and shall be governed by the inventory so taken.

ARTICLE VI
FORM IN WHICH ACCOUNTS SHALL BE MAINTAINED

The form of the accounts as required for the purpose of complying with provisions of the Agreement will include:

1. Accounts recorded through Operator's Uniform Chart of Accounts or an equivalent system of Accounts. A manual of Operator's accounts will be maintained at Operator's office in India.
2. Statements produced from the Operator's accounts of income and expenditure. Such statements produced from the Operator's accounts must be developed over time to reflect actual progress and circumstances.

EXHIBIT "B"
PROCEDURE FOR ACQUISITION OF GOODS AND SERVICES

I. OBJECTIVES

The objectives of these procedures are to:

(a) ensure the goods and services acquired by the Operator for the carrying out of the Petroleum Operations are acquired at the optimum cost taking into consideration all relevant factors including price, quality, delivery times and the reliability of potential suppliers.

(b) ensure that goods and services are delivered in a timely taking into consideration the consequences of delays in the acquisition of these goods and services on the project as a whole.

(c) Ensure that the relevant provisions of the Agreement are implemented.

II. PRINCIPLES

The principles upon which these procedures are based are:

(a) The parties must be satisfied that the Operator is working in accordance with agreed procedure for acquiring goods and services which is auditable and in accordance with the provisions of the Agreement.

(b) The Operator must have the ability to acquire goods and services expeditiously so that the project schedules in respect of Approved Work Programmes are maintained.

III. PROCEDURE

The procedure to be adopted by the Operator for the acquisition of goods and services shall be as follows:

Sl.No.	Procedure A	Procedure B	Procedure C
Applicable to Exploration, Appraisal	$ 50,000 to $200,00	$200,000 to $500,000	more than or equal to $500,000

For Contract valued at less than US $50,000

The Operator will be at liberty to determine the preferred method of acquiring goods and services valued at less than US$ 50,000 with the understanding that at least three (3) quotations from selected suppliers (including at least one (1) Indian supplier will be obtained if manufactured in India. For items valued at greater than US $20,000, Operator is required to report to the Operating Committee if the quote accepted exceed the lowest quote by more than 20 percent. Operator will promptly report to the Operating Committee the Operator's reasons for not selecting the lowest quote.

Procedure A:

Operator shall:

(1) provide the Non-Operators with a list of all the entities approved by the Operating Committee as per para (v) for the applicable category of the contract along with other entities, if any, from whom the Operator proposes to invite tender;

(2) Add to such list the entities whom other Party requests for adding within five (5) Business Days on receipt of such lists;

(3) If and when any party so requests, Operator shall evaluate any entity listed in (1) and (2) above to assure that entity is qualified as based on the qualification criteria agreed in accordance with para (IV) of Exhibit B to perform under the contract;

(4) Complete the tendering process within a reasonable period of time;

(5) Circulate to Non-Operators a comparative bid analysis stating Operator's choice of the entity for award of contract. Provide also reasons for such choice in case entity chosen is not the lowest bidder;

(6) Inform Non-Operators of the entities to whom the contract has been awarded; and

(7) Upon the request of a Non-Operator, provide with a copy of the final version of the contract awarded.

Procedure B:

(1) Provide the Non-Operators with a list of all the entities approved by the Operating Committee as per para (V) of Exhibit B for the applicable category of the contract, along with other entities, if any, from whom the Operator proposes to invite tender;
(2) Add to such list the entities whom a Party requests for adding within five (5) Business Days on receipt of such list;
(3) If and when any Party so requests, Operator shall evaluate any entity listed in (1) and (2) above to assure that entity is qualified as based on the qualification criteria agreed in accordance with para (IV) of Exhibit B, to perform under the contract;
(4) Complete the tendering procedure within a reasonable period of time;
(5) Circulate to Non-Operators a comparative bid analysis stating Operator's choice of the entity for award of contract. Provide also reasons for such choice in case the entity chosen is not the lowest bidder. If the bid selected is not the lowest bid, obtain prior approval of the Operating Committee for award of contract;
(6) Award the contract accordingly and inform all the members of Operating Committee of the entities to whom the contract has been awarded; and
(7) Upon the request of a Party, provide with a copy of the final version of the contract awarded.

Procedure C:

Operator shall:

(1) Publish invitations for parties to pre-qualify for the proposed contract in at least three (3) daily national newspapers. Provide to Non-Operators, a list of responding parties and an analysis of their qualifications for the contract being contemplated to be awarded. Include those who qualify, as per the prequalification criteria approved as per para (IV) of Exhibit B in the list of entities from whom Operator proposes to invite tender for the said contract.
(2) Provide the members of Operating Committee with a total list of all the entities selected as (1) above all the entities approved by the Operating Committee as per para (V) of Exhibit B for the applicable category of the contract, along with entities, if any, from whom the Operator proposes to invite tender.;
(3) Add to such entities whom a Party requests for adding within five (5) Business Days on receipt of such list;
(4) If and when any Party so requests, Operator shall evaluate any entity listed in (2) and (3) above to assure that entity is qualified as based on the qualification criteria agreed in accordance with para (IV) of Exhibit B, to perform under the Contract;
(5) Prepare and despatch the tender documents to the entities as finally

listed and to Non- Operators;

(6) After the expiration of the period allowed for tendering, consider and analyse the details of all bids received.

(7) Prepare and circulate to the Non-Operators a comparative bid analysis stating Operator's recommendations as to the entity to whom the contract should be awarded, the reasons therefor, and the technical, commercial and contractual terms to be agreed upon.

(8) Obtain the approval of the Operating Committee to the recommended bid.

(9) Award the contract accordingly and upon the request of the Non-Operator, provide with a copy of the final version of the contract;

IV. A set of vendor qualifications criteria for each major category contract/ supply shall be proposed by the Operator and approved by the Operating Committee within thirty (30) days of its submission. In the event the Operating Committee fails to approve vendor qualification criteria within (30) days of the date of same is first submitted by the Operator, the matter shall be referred to the Operating Committee for decision. The Operating Committee may revise the qualification criteria.

V. It is anticipated that, in order to expedite Joint Operations, contracts will be awarded to qualified vendors/ contractors who are identified as approved vendors for the specified activities. A list of such approved vendors shall first be established as follows:

Operator shall:

(1) Provide the Non- Operator(s) with a list of the entities from whom Operator proposes to invite tender for contracts; and

(2) Add to such list entities whom a Non -Operator requests for adding within fourteen (14) days on receipt of such list; and

Obtain approval of the Operating Committee by the Operator. Such list shall thereafter be maintained by the Operator. The Operating Committee may add to or delete vendors from such list.

EXHIBIT "C"

Performa for Authorisation Expenditure Request

AUTHORISATION EXPENDITURE REQUEST

AER Title

Block **AER No.**

Description

Cost Breakdown **As per detail attached**

PHASING	**Quarter - 1**	**Quarter - 2**	**Quarter - 3**	**Quarter - 4**	**Total**
Accrued					
Cash					

BUDGET STATUS

TOTAL AER REQUEST
PRIOR AER REQUESTS
BALANCE REMAINING
TOTAL BUDGET

COMMENCEMENT DATE **COMPLETION DATE**

OPERATOR'S APPROVALS	**POSITION**	**SIGNATURE**	**DATE**
PREPARED BY			
RECOMMENDED BY			
APPROVED BY			

AUTHORISATION	INTEREST	$000'S SIGNATURE	DATE
M/s	%		
M/s	%		
M/s	%		
Total	%		

EXHIBIT "D"

PROFORMA FOR CASH CALL

Name & Address of Operator:				Block: Date: Cash Call No.	
AER No.	Description	Approved Value US Dollars	Estimated Expenditure of the month	Estimated Expenditure of the month	Estimated Expenditure of the month
					OVERHEAD %

TOTAL PARTNER ALLOCATION	INTEREST	US DOLLARS
M/S A	%	
M/S B	%	
M/S C	%	
TOTAL	%	
PARTNER ADVANCES	DATE	US DOLLARS
M/s A		
M/s B		
M/s C		
TOTAL		
PARTNER NET PAYABLE		US DOLLARS
M/s A		
M/S B		
M/s C		
TOTAL		
M/s Operator		Bank Details:

ENERGY CONTRACTS AND DISPUTE RESOLUTION

Energy contracts are generally of long duration, high value and complicated in nature. For example in petroleum sector the basic production sharing contracts are 25 to 30 years duration with a provision for extension as well. Consultants have a major role in international commercial arbitration arising out of such long duration and complicated contracts. Even fixation of price of crude oil or gas in itself a specialized job and can taken up only by an expert.

- **Sole expert**

  Technical and specialized matters
  Not referred to arbitration

Decision final and binding
Generally single

- **Mediation**

  Mediator

- **Conciliation**

  More than one
  Decision is not binding
  Parties to Agree to make finding binding
  Conciliation got legislative recognition

- **Arbitration**

  One or more arbitrators
  Decision is binding

- **Arbitration and Conciliation Act, 1996**
- **International Centre For Alternate Dispute Resolution**
- **International Chamber of Commerce, Paris**
- **Indian Council of Arbitration**
- **UNCITRAL Conciliation Rules**
- **UNCITRAL Arbitration Rules**
- **Applicable Law**

Expert Witness

The consultants are brought in an arbitration proceeding as expert witness on a topic which is of technical and specialized nature. For example in America there are various societies of consultants for providing specialized consultancy services to arbitral tribunals and courts.

American Society of Questioned Documents Examiners

The purposes of the Society and of its members are to foster education, sponsor scientific research, establish standards, exchange experience, and provide instruction in the field of questioned document examination, and to promote justice in matters that involve questions about documents. To accomplish these purposes, the Society and its members subscribe to the following objectives:

To establish and maintain high professional standards for ethics, for education and training, and for excellence in work performance.

To engage upon and to encourage scientific research and development in document examination and related matters and to disseminate the results by presentation at annual meetings and by publication.

To prepare, maintain, and administer a comprehensive course of study in document examination; to record, preserve, and correlate significant experiences in solving questioned document problems.

To improve knowledge and understanding of the work of the document examiner by the public, the bar, the judiciary and among the forensic sciences.

To discover prospective document examiners of good character and good potential for development and to assist in their training.

In a commercial arbitration technical, and financial consultants may be required from time to time but legal consultants are required on a regular basis as case is fought from one stage to another. The legal consultant chalk out strategy and fix milestones in consultation with the client

- **Study of a Typical Sole Expert, Conciliation and Arbitration Clause in an Energy Contract**

ARTICLE 32and 33 of the Production Sharing Contract in India
Article 32: APPLICABLE LAW AND LANGUAGE OF THE CONTRACT

32.1 This Contract shall be governed and interpreted in accordance with the laws of India.

32.2 Nothing in this Contract shall entitle the Contractor to exercise the rights, privileges and powers conferred upon it by this Contract in a manner which will contravene the laws of India.

32.3 The English language shall be the language of this Contract and shall be used in arbitral proceedings. All communications, hearing or visual materials or documents relating to this Contract shall be written or prepared in English.

32.4 The laws will also include amendments, revisions, modifications etc.

ARTICLE 33
SOLE EXPERT, CONCILIATION AND ARBITRATION

33.1 The Parties shall use their best efforts to settle amicably all disputes, differences or claims arising out of or in connection with any of the terms and conditions of this Contract or concerning the interpretation or performance thereof.

33.2 Matters which, by the terms of this Contract, the Parties have agreed to refer to a sole expert and any other matters which the Parties may agree to so refer, may be referred to a sole expert who shall be an independent and impartial person of international standing with relevant qualifications and experience, appointed by agreement between the Parties and who shall not, by virtue of nationality, personal connection or commercial interest, have a conflict between his/her own interest and his/her duty as a sole expert. In the event that the Parties fail or are unable, to agree on a sole expert, the sole expert shall be appointed by the Chief Justice of India pursuant to this Article. Any sole expert appointed shall be acting as an expert and not as an arbitrator and the decision of the sole expert on matters referred to him/her shall be final and binding on the Parties and not subject to arbitration.

33.3 Subject to the provisions of this Contract, the Parties hereby agree that any controversy difference, disagreement or claim for damages, compensation or otherwise (hereinafter in this Clause referred to as a "dispute") arising between the Parties, which cannot be settled amicably within ninety (90) days after the dispute arises, may (except for those referred to in Article 33.2, which may be referred to a sole expert) be submitted to an arbitral tribunal for final decision as hereinafter provided.

33.4 The arbitral tribunal shall consist of three arbitrators. Each Party to the dispute shall appoint one arbitrator and the Party or Parties shall so advise the other Parties. The two arbitrators appointed by the Parties shall appoint the third arbitrator.

33.5 Any Party may, after appointing an arbitrator, request the other Party(ies) in writing to appoint the second arbitrator. If such other Party fails to appoint an arbitrator within thirty (30) days of receipt of the written request to do so, such arbitrator may, at the request of the first Party, be appointed by the Chief Justice of India or by a person authorised by him or within thirty (30) days of the date of receipt of such request, from amongst persons who are not nationals of the country of any of the Parties to the arbitration proceedings.

33.6 If the two arbitrators appointed by on behalf of the Parties fail to agree on the appointment of the third arbitrator within thirty (30) days of the appointment of the second arbitrator and if the Parties do not otherwise agree, at the request of either Party, the third arbitrator shall be appointed in accordance with Conciliation and Arbitration Act, 1996.

33.7 If any of the arbitrators fails or is unable to act, his successor shall be appointed by the Party or person who originally appointed such in the manner set out in this Article as if he was the first appointment.

33.8 The decision of the arbitral tribunal, and, in the case of difference among the arbitrators, the decision of the majority, shall be final and binding upon the Parties.

33.9 The arbitration agreement contained in this Article 33 shall be governed by the Arbitration and Conciliation Act, 1996 (Arbitration Act). Arbitration proceedings shall be conducted in accordance with the rules for arbitration provided in Arbitration Act and the United Nations Commission on International Trade Law (UNCITRAL) rules may apply to the extent where corresponding rules are not provided in the Act. In case of any conflict between the Arbitration Act and the provisions of this Article 33, the provisions of the Arbitration Act shall prevail.

33.10 The right to arbitrate disputes under this Contract shall survive expiry or the termination of this Contract.

33.11 Prior to submitting a dispute to arbitration, the Parties may by mutual agreement submit the matter for conciliation in accordance with Part III of the Arbitration and Conciliation Act, 1996. No arbitration proceedings shall be instituted while conciliation proceedings are pending provided that a Party may initiate arbitration proceedings in the event that dispute has not been resolved by conciliation within twenty one (21) days of the date of agreement by the Parties to submit such dispute to conciliation.

33.12 The venue of the sole expert, conciliation or arbitration proceedings pursuant to this Article, unless the Parties agree otherwise, shall be _______, India and shall be conducted in the English language. Insofar as practicable, the Parties shall continue to implement the terms of this Contract notwithstanding the initiation of arbitral proceedings before a sole expert, conciliator or arbitral tribunal and any pending claim or dispute.

33.13 The fees and expenses of a Sole Expert or Conciliator appointed by the Parties shall be borne equally by the Parties. The cost and expenses of arbitrator appointed by a Party in accordance with the provision of this Article shall be borne by the respective Party and the cost and expenses of third arbitrator and liability thereof shall be at the discretion of the arbitrators.

5.3 FARM-IN AGREEMENTS

A farm-in agreement is an arrangement under which a holder of an exploration or mining right agrees to assign a portion of its interest to another party (who "farms-in") in exchange for a certain amount of expenditure or the performance of specific activities.

Farm out (in) arrangement: An arrangement, used primarily in the oil and gas industry, in which the owner or lessee of mineral rights (the first party) assigns a working interest to an operator (the second party), the consideration for which is specified exploration and/or development activities. The first party retains an overriding royalty or other type of economic interest in the mineral production. The arrangement from the viewpoint of the second party is termed a "farm-in arrangement."

Shell and Cairn conclude a Farmout/Farmin Agreement in Rajasthan, India

Shell India Production and Development B.V. ("Shell") and Cairn Energy PLC announce today that they have concluded an agreement whereby Shell will immediately assign to Cairn a 27.5% interest in Block RJ-ON-90/1 in Rajasthan. In addition Cairn has been granted options to increase its stake in the Block to 40% on a promoted basis. The Rajasthan Production Sharing Contract area, awarded to Shell in 1995, is located 100 kms north of the proven oil and gas province of the Cambay Basin, and covers an area of 8,856 sq. kms. The assignment is subject to obtaining all necessary Government approvals.

Shell will assign a 27.5% interest to Cairn. In return Cairn will pay Shell 100% of the first well costs to a cap of $10 million and historic costs of approximately $1.75 million. To date Shell has spent approximately $10 million on 2D and 3D seismic. Shell intends spudding its first exploration well on RJ-ON-90/1 at the end of 1998. ONGC, the Indian Government owned oil company, has a 30% back-in right.

Bill Gammell, Cairn's Chief Executive, said:

> "We are delighted to have acquired a material early entry position in an exploration Block in one of our target growth areas in India. The fact that RJ-ON-90/1 is operated by Shell, our joint venture partner in Bangladesh, is doubly pleasing."

Tony Wildig, Shell's Vice President for New Business, Middle East and South Asia, said:

> "The geology of this Block is fascinating, potentially a mirror-image of the Cambay Basin. It represents an excellent opportunity for ourselves and Cairn to take forward in the spirit of our developing relationship."

Notes

"Cairn" as referred to above means Cairn Energy PLC and/or its subsidiaries as appropriate.

ONGC-CARIN FARM-IN Agreement

ONGC announces that it has signed a Heads of Agreement with Cairn for farm-in in certain assets of Cairn, the participation of Cairn in two exploration blocks of ONGC, and the formation of a strategic alliance for future opportunities.

The proposed transaction with Cairn is subject, inter alia, to finalisation of mutually acceptable sale and purchase agreements between the parties and approval by the Government of India.

The proposed transaction comprises the following:

(i) The farm-out by Cairn to ONGC of a 90% operated interest in deep water exploration Block KG- DWN-98/2 with an effective economic date of 30th September 2003. Cairn will retain operatorship of Block KG-DWN-98/2 until the closing date, following which ONGC will assume operatorship under the terms of transfer of operatorship agreements to be agreed between Cairn and ONGC.

(ii) The farm-out by Cairn to ONGC of a 15% exploration interest in Block CB/OS-2 and a 10% interest in the Lakshmi and Gauri Development Areas with an effective economic date of 1st January, 2003. (together "the Farm-out Interests")

(iii) The farm-in interest by ONGC to Cairn of a 30% interest in each of Block GV-ONN-97/1 onshore Northern India and CB-ONN-2001/1 onshore Gujarat Western India, with an effective economic date of 30th September, 2003 ("the farm-in in Interests")

ONGC will pay to Cairn a cash consideration of US$135 million ("the Cash Consideration") for the farm-out interests. The cash consideration will be adjusted to reflect net working capital movements between the respective effective economic dates and the closing date.

Cairn will pay to ONGC a cash consideration equivalent to economic monetary value of the blocks as assessed by ONGC at the time of bidding for its farm-in participative interest in two ONGC blocks (to be evaluated by Cairn).

In addition, ONGC and Cairn yesterday submitted joint bids for three blocks onshore Gujarat in the NELP-IV licensing round. If successful, Cairn will retain a 30% interest in these blocks and ONGC will be operator.

Further information on the Farm-out Interests and the Farm-in Interests is included in the Appendix to this announcement.

Bill Gammell, Chief Executive, commented :

"We are delighted to extend our relationship with ONGC and to create a strategic alliance to look at opportunities both within India and elsewhere. The farm-out of discovered and producing interests is consistent with our strategy of both creating and realising value from exploration success. The transaction also significantly augments our onshore acreage position in India."

Subir Raha, Chairman & Managing Director of ONGC commented:

We are delighted to reinforce the equation with Cairn through a strategic alliance, to add value from opportunities in Indiaand abroad. These transactions will synergise the competencies of ONGC and CAIRN, and enhance our cost- efficiencies.

APPENDIX

Block KG-DWN-98/2

Cairn currently holds a 100% operated Interest in Block KG-DWN-98/2, which was awarded pursuant to the first round in India's New Exploration Licensing Policy. It is a deep water block located in the Krishna-Godavari basin offshore Eastern Indian which water depths range from approximately 300 to 3,000 metres.

Cairn signed a PSC for the block in April 2000 following which it conducted an extensive exploration campaign resulting in several hydrocarbon discoveries during 2001. During 2002 Cairn completed a detailed subsurface technical review of the results of the exploration programme with a view to seeking a partner to optimise plans for future exploration, appraisal and development activity on the block.

Block CB/OS-2

Cairn signed a PSC for Block CB/OS-2, which is located in the Gulf of Cambayoffshore Western India, in June 1998. Cairn currently holds a 75% exploration interest in the block and is operator for the joint venture, which also includes TATA Petrodyne Limited ("TATA") (15%) and ONGC (10%).

ONGC has a right to increase its stake by 30% in the event of a commercial discovery on the block and has exercised this right in respect of the Lakshmi and Gauri Development Areas. The current equity interests for these development areas are therefore Cairn 50%, ONGC 40% and TATA 10%.

The Lakshmi gas field commenced production in November 2002 and average sales from the field during the first half of 2003 were 106 million standard cubic feet of gas per day. Development of the neighbouring satellite Gauri gas field is currently underway with production expected to commence in the first half of 2004.

Block CB-ONN-2001/1

Block CB-ONN-2001/1 occupies an area of 210 square kilometres in the heart of productive Cambay onshore basin and was awarded to ONGC in the NELP-III bid round. Cairn also bid for this block in NELP-III. ONGC Currently holds 100% operated interest in the block & is also the operator.

Block GV-ONN-97/1

Block GV-ONN-97/1 occupies an areas of 36,750 square kilometres in the virtually unexplored Ganga basin adjacent to the large Delhi market. The block was awarded to ONGC under NELP-I bid round. ONGC currently holds 70% exploration interest in the block & is operator for the block. Indian Oil Corporation (IOC) holds 30% participating interest.

LAK RANCH FARMIN AGREEMENT

THIS FARMIN AGREEMENT (the "Agreement") is dated and effective as of January 15, 2004, by and between IVANHOE ENERGY (USA) INC. ("Ivanhoe") and DEREK RESOURCES (U.S.A.) INC. ("Derek");

WHEREAS Derek is the lessee of the LAK Ranch Leases as set forth in Exhibit "A", attached hereto and incorporated herein by this reference;

AND WHEREAS pursuant to the SEC Agreement, as defined below, Derek holds an undivided 5% Working Interest in the LAK Ranch Area in trust for SEC until such time as Derek has expended the Funds provided by SEC to Derek pursuant to the SEC Agreement and transfers the aforementioned Working Interest to SEC;

AND WHEREAS Ivanhoe desires to earn a Working Interest in the LAK Ranch Leases and LAK Ranch Area, subject to the conditions hereof and the requirements and obligations to be performed by Ivanhoe set out in this Agreement;

AND WHEREAS the Parties desire to possibly further explore and develop the lands subject to the LAK Ranch Leases and the LAK Ranch Area for the production of oil, gas and other hydrocarbon substances;

NOW THEREFORE, in consideration of the premises and the mutual covenants and agreements contained herein, it is understood and agreed by and between the Parties, as follows:

ARTICLE I
DEFINITIONS

In this Agreement, including the recitals, this Section and the Exhibits, unless there is something in the context or subject matter inconsistent therewith, the following terms shall have the meanings hereinafter set forth:

1.1 "Affiliate" - of a Party means any company or other entity that (i) controls, (ii) is controlled by or (iii) is under common control with such Party. For the purpose of this definition, control shall mean the ownership, directly or indirectly, of 50% or more of the stock or other units of ownership having the right to vote for the election of directors of such company or other entity.

1.2 "Area of Mutual Interest" or "AMI" - shall have the meaning ascribed in Article III, Section 3.1.

1.3 "Capital Expenditures" - means expenditures for capital items or purposes as distinguished from operating costs and expenditures, in accordance with generally accepted accounting principles generally applied in the United States oil and gas industry and as defined in COPAS.

1.4 "Dollars" or the symbol "$" - means legal tender of the United States of America.

1.5 "Effective Date" - means the effective date of this Agreement, being 12:00 p.m. Central Standard Time on January 15, 2004.

1.6 "Funds" - means the monies provided by SEC to Derek pursuant to the SEC Agreement which Derek agreed to spend on operating and development expenses on The LAK Ranch Area and which upon its complete expenditure, SEC will earn an undivided 5% working interest in the LAK Ranch Area.

1.7 "Gross Negligence or Willful Misconduct" - means any act or failure to act (whether sole, joint or concurrent) by any person or entity which was intended to cause, or which was in reckless disregard of or wanton indifference to, harmful consequences such person or entity knew, or should have known, such act or failure would have on the safety or property of another person or entity.

1.8 "Joint Account" - means the account established and maintained by Operator to record all expenditures, receipts, materials and other assets acquired or used in the conduct of Joint Operations.

1.9 "Joint Operating Agreement" or "JOA" - means the Joint Operating Agreement referred to in Article VIII, a form of which is attached hereto and made a part hereof as Exhibit "C".

1.10 "Joint Operations" - mean operations, whether for the Pilot Phase, Phase 1 or subsequently, conducted by Operator for the benefit and account of the Parties pursuant to any or all this Agreement, the Joint Operating Agreement, the LAK Ranch Lease and any New Lease.

1.11 "LAK Ranch Area" - refers to the area outlined on the plat attached hereto and made a part hereof as Exhibit "B."

1.12 "LAK Ranch Leases" - means the Leases listed on Exhibit "A" attached hereto and made a part hereof and covering lands located within the confines of the LAK Ranch Area.

1.13 "LAK Ranch Program" - means the program summarized on Exhibit "E" to this Agreement summarizing the work to be done during the Pilot Phase and Phase 1.

1.14 "New Lease" - has the meaning ascribed to it in Section 3.2.

1.15 "Non-Operator" - means a Party other than the Operator. Derek is initially the Non-Operator, but should SEC be assigned a Working Interest by Derek, SEC shall also be a Non-Operator.

1.16 "Operator" - means at any time the Party appointed to conduct Joint Operations on behalf of the Parties, and initially shall mean Ivanhoe.

1.17 "Other Costs" - mean costs other than Capital Expenditures, for which the Parties will be liable and pay in accordance with their Working Interests, as provided in Section 4.4.

1.18 "Lease" - means an oil and gas lease and any oil, gas, and mineral lease, mineral interest, or any other instrument granting the right to explore for, drill, and take oil, gas, and other minerals.

1.19 "Party" and "Parties" - means initially Ivanhoe and Derek, individually or collectively, as the case may be, and/or their successors and assigns. If Derek spends the Funds, and assigns a 5% Working Interest to SEC, then SEC shall be a Party or one of the Parties, and assume its Working Interest share of all obligations and liabilities as a Party to and under this Agreement, including with respect to representations and warranties and under the Joint Operating Agreement and the LAK Ranch Lease.

1.20 "Phase 1" - means the work and activities described as Phase 1 in the LAK Ranch Program Summary attached as Exhibit "E".

1.21 "Pilot Phase" - means the work described as the Pilot Phase in the LAK Ranch Program Summary attached as Exhibit "E".

1.22 "SEC" - means SEC Oil and Gas Partnership.

1.23 "SEC Agreement" - means the document entitled "LAK Ranch Oil Project -Amended Heads of Agreement", dated October 30, 2003, between Derek and SEC and any and all agreements that may be substituted therefore.

1.24 "Substances" - means all severed crude oil, natural gas, casinghead gas, gasoline, natural gasoline, petroleum, natural gas liquids, condensate, products, liquids and other hydrocarbons and other minerals of every kind and description produced to be produced under the LAK Ranch Leases from the LAK Ranch Area.

1.25 "Working Interest" - means the undivided participating interest of a Party in the benefits and obligations under this Agreement, the Joint Operating Agreement, the LAK Ranch Lease and any other Lease that may be taken by the Parties in the LAK Ranch Area. Expression of working interests as a percentage means the indicated percentage of 100%.

ARTICLE II
ATTACHMENT AND EXHIBITS

2.1 Exhibits - The following Exhibits, attached hereto, are incorporated herein and made a part hereof for all purposes:
Exhibit "A" LAK Ranch Leases and Lands
Exhibit "B LAK Ranch Area/ AMI
Exhibit "C" Joint Operating Agreement
Exhibit "D" Environmental Compliance
Exhibit "E" LAK Ranch Program Summary

ARTICLE III
AREA OF MUTUAL INTEREST

3.1 Area of Mutual Interest - The Parties hereby establish an Area of Mutual Interest" or "AMI", covering the lands located within the area outlined on the plat attached hereto and made a part hereof as Exhibit "B" and within two (2) miles of that area. The AMI shall continue in effect as long as the Parties jointly own Leases within the AMI and shall continue in effect under the JOA and until one year after the JOA terminates.

3.2 Acquisition/Option - Should any Party directly or indirectly, acquire a

Lease ("New Lease"), in addition to or in substitution for all or any part of the LAK Ranch Lease, affecting any lands located, fully or partially within the AMI, such Party (the "Acquiring Party") shall immediately give written notice thereof to each other Party (a "Non-Acquiring Party"), together with all pertinent details and information, including copies of all instruments of conveyance (including but not limited to, copies of the New Lease, assignments, subleases, farmouts, and other contracts affecting the LAK Ranch Lease and or the LAK Ranch Area), copies of paid drafts or checks and itemized invoices of the actual costs expended incurred and to be incurred by the Acquiring Party, including any bonus, administrative fees, brokerage, legal and recordation costs, and any other direct costs (said actual costs are referred to as "Acquisition Costs"). A Non-Acquiring Party shall have 30 days (or 48 hours if a rig capable of performing a drilling operation is on location in preparation for drilling), from receipt of such notice to elect in writing to acquire its proportionate interest in the New Lease by paying its proportionate share of the cost and assuming its proportionate share of the obligations. Failure to respond by the deadlines stated herein shall be deemed an election not to participate in the acquisition described in said notice. Failure to pay said proportionate share of costs within 30 days following an election to participate shall result in a forfeiture of a Non-Acquiring Party's interest with the same effect as if said Party had initially elected not to participate. If less than all Parties elect to participate in said acquisition, the lands and/or depths covered by any New Lease shall be excluded from the AMI and the New Lease acquired shall not be subject to this Agreement, but shall belong entirely to the Parties participating in the acquisition.

3.3 Payment/Assignment - If a Non-Acquiring Party elects to participate in such acquisition it shall pay to the Acquiring Party its proportionate share of the Acquisition Costs for such interest within 30 days after receipt of an invoice for said costs. Upon receipt of said payment, the Acquiring Party shall execute and deliver to the Non-Acquiring Party an assignment of the interest due the Non-Acquiring Party, pursuant to which the Non-Acquiring Party shall bear and assume its proportionate share of all obligations, covenants, conditions, requirements, and terms associated with the acquisition.

3.4 Encumbrances - Any assignment made pursuant to this Article III shall be free and clear of any encumbrances placed on the assigned New Lease in favor of or by the Acquiring Party, but otherwise shall be made without warranty of title, either express or implied, except by, through, and under the Acquiring Party. The assignment shall be made and accepted subject to, and assignee shall expressly assume its proportionate share of, all obligations of the assignor pertaining to the New Lease.

3.5 Merger - The provisions of this Article III shall not apply to Lease acquisitions resulting from mergers, consolidations or reorganizations involving all or substantially all of the properties or assets of the Acquiring Party.

ARTICLE IV
IVANHOE CASH PAYMENT AND PARTIES' WORKING INTERESTS

4.1 Ivanhoe Cash Payment. Contemporaneously with the signing of this Agreement, Ivanhoe will make a one-time cash payment to Derek of $100,000, in consideration of which Derek will assign to Ivanhoe a 30% Working Interest and Ivanhoe shall undertake to complete the Pilot Phase as described in Section 4.5.

4.2 Ivanhoe Earning. Ivanhoe is entitled to earn up to an additional 30% Working Interest for a total Working Interest of 60%, by participating in the Pilot Phase and Phase 1, and paying Capital Expenditures and Other Costs, as provided in Sections 4.3 through 4.6 below. The following are the initial Working Interests of the Parties:
Derek................................70%
Ivanhoe............................. 30%

4.3 Capital Expenditures. If Ivanhoe participates in the Pilot Phase and Phase 1, Ivanhoe shall make Capital Expenditures in accordance with work programs and budgets prepared by Ivanhoe in consultation with the Parties. Ivanhoe's initial 30% Working Interest shall be increased by a proportional 6% for every $1,000,000, or part thereof, it incurs on a 100% basis for Capital Expenditures (that is without contribution by the other Parties) for the LAK Ranch Program during the Pilot Phase and Phase 1, up to a maximum of 60% Working Interest. For further clarity, at the time of determination and adjustment of Ivanhoe's interest for incurred Capital Expenditures, for any portion of the Capital Expenditures less than a full multiple of one million ($1,000,000) dollars (the "Portion"), the increase in interest shall be determined based upon the following formula:

$$\frac{[\text{Portion}] \times 6\%}{1{,}000{,}000}$$

Ivanhoe's Working Interest will be determined and adjusted at the end of each quarter based on cumulative Capital Expenditures it has incurred on a 100% basis through the quarter. If Ivanhoe participates in Phase I and, upon completion of the work program for Phase I, Ivanhoe has incurred less than $5,000,000 on Capital Expenditures on a 100% basis, additional Capital Expenditures for work on the LAK Ranch Lease agreed by the Parties to be performed, shall be paid 100% by Ivanhoe until it has incurred an aggregate of $5,000,000 between the Pilot Phase and Phase I. When Ivanhoe has incurred a total of $5,000,000 on Capital Expenditures on a 100% basis during the Pilot Phase and Phase 1 and subsequently, as appropriate, additional LAK Ranch Program costs, including Capital Expenditures, will be paid by the Parties in proportion to their respective Working Interests at the time, which will be as follows:

Ivanhoe..................................60%
Derek....................................40%

4.4 OTHER COSTS. Ivanhoe's commitments to make Capital Expenditures on a 100% basis shall apply only to Capital Expenditures of the LAK Ranch Program. Lease operating costs, Operator's overhead costs agreed in the JOA and all such similar costs during the Pilot Phase and Phase 1 ("Other Costs") will be paid and shared by the Parties in proportion to their respective Working Interests from time to time during the Pilot Phase, Phase 1 and subsequently. Ivanhoe shall have no liability for Other Costs incurred prior to the Effective Date. Other Costs from and after the Effective Date will be paid by the Parties pursuant to cash calls or advance provisions of the JOA. If one or more of the Parties fails to pay their share of Other Costs for the Pilot Phase and/or Phase 1, and Ivanhoe pays Other Costs to continue operations, then such delinquent amounts shall bear interest at the rate provided in the JOA and shall be subject to the remedies for collection provided in the JOA.

4.5 Pilot Phase. Ivanhoe undertakes the work program for the Pilot Phase, which consists of the work program described in the LAK Ranch Program details attached as Exhibit "E" to this Agreement. Ivanhoe will pay 100% of the Capital Expenditures of the Pilot Phase. The Capital Expenditures budgeted to be spent in performing such work does not constitute an obligation to spend that amount.

Operations being conducted on the LAK Ranch Lease by Derek were ordered stopped by the Wyoming Oil and Gas Commission. Before Ivanhoe can commence work on the Pilot Phase, it must obtain approval of the Commission and possibly other Wyoming and Federal agencies in order to resume operations. Promptly after the Effective Date, Ivanhoe, with the other Parties' cooperation and assistance, will undertake the necessary studies and work in order to obtain all the approvals to resume operations. Such work will be pursued diligently in order to obtain the approval as quickly as possible, but in the event that notwithstanding such diligent efforts such approvals are not obtained within one year after the Effective Date, Ivanhoe will have the option to terminate this Agreement and will have no further obligation to the other Parties. The costs of obtaining the above approvals to resume operations shall be Other Costs to be shared by the Parties in proportion to their respective Working Interests, i.e., Ivanhoe 30% and Derek 70%.

4.6 Phase 1. Phase 1 consists of the work program described in the LAK Ranch Program details attached as Exhibit "E" to this Agreement. Ivanhoe shall have the option to elect to participate in Phase 1, within 30 days after completion and evaluation of the results of the Pilot Phase. If Ivanhoe does not elect to participate in Phase 1, Ivanhoe shall retain a 15% Working Interest and shall reassign to the Parties the remaining

15% of its initial Working Interest.

4.7 SEC Interest. If the Funds have been spent by Derek and a Working Interest has been assigned to SEC, then Derek's Working Interest shall be reduced by 5% and SEC shall have a 5% Working Interest. Ivanhoe's Working Interest shall not be reduced or affected by the assignment of a Working Interest to SEC.

4.8 Delivery of Assignments - Unless otherwise provided herein, any assignment due any Party, shall be delivered in a form suitable for recording and within 15 days of the date on which it was executed.

4.9 Assignment of Real and Personal Property - Any assignment of a Working Interest hereunder shall include an assignment of an equal undivided interest in all real and personal property and which is owned by the assigning party, located on or off the LAK Ranch Leases and used in connection with the operations of the Leases, including without limitation, wells, well equipment, casing, tubing, tanks, gathering lines, separators, buildings, pumps, compressors, gas plants, motors, fixtures, machinery, power lines, telegraph and telephone lines, roads, field processing plants, and other improvements used in the operation thereof.

ARTICLE V
JOINT OPERATING AGREEMENT

5.1 Operations - Except as otherwise provided herein, or in the Joint Operating Agreement, the costs, risk, and expenses shall be based on each Party's Working Interest. All operations hereunder shall be conducted and governed in accordance with the terms and conditions of the Joint Operating Agreement. The Parties shall execute the Joint Operating Agreement signed simultaneously with the execution of this Agreement. Ivanhoe shall be designated Operator under the Joint Operating Agreement, and Derek or Derek and SEC, as the case may be, shall be Non-Operators. In the event of a conflict between the terms and conditions of this Agreement and the Joint Operating Agreement, the terms and conditions of this Agreement shall prevail.

ARTICLE VI
INFORMATION, TESTS, AND NOTICES

6.1 Access to Operations and Furnishing of Data - The Non-Operators shall, at all times, and at their sole cost and risk, have full and free access to the LAK Ranch Area. Ivanhoe shall obtain and furnish to Non-Operators the following data reports as they are currently produced or compiled so far as they relate to Joint Operations:

(a) copies of all logs or surveys, including in digitally recorded format if exists;

b) daily drilling reports;

(c) copies of all tests and core data and analysis reports;
(d) final well recap report;
(e) copies of plugging reports;
(f) copies of final geological and geophysical maps, seismic sections and shot point location maps;
(g) engineering studies, development schedules and quarterly progress reports relating to the LAK Ranch Area;
(h) field and well performance reports, including reservoir studies and reserve estimates;
(i) production data;
(j) as requested by Non-Operators (i) copies of all material reports relating to the LAK Ranch Area furnished to Ivanhoe by any governmental or regulatory body or authority; and (ii) other material studies and reports relating to LAK Ranch Area; and
(k) such other information as Non-Operators may reasonably request, provided that the requesting Party or Parties pay the costs of preparation of such information and that the preparation of such information will not unduly burden Operator's administrative and technical personnel.

6.2 Ivanhoe shall test any wells encountered which Ivanhoe wishes to test in accordance with good and prudent petroleum industry practices and field conservation principles as are generally followed by the United States petroleum industry, and shall further test all formations indicated by such testing to be productive and considered to have the potential for commercial production in the LAK Ranch Area, provided that they are subject to the LAK Ranch Lease or other Leases granting the Parties the right to do so. Ivanhoe shall keep an accurate and detailed log of each well drilled hereunder.

ARTICLE VII
CONDUCT OF OPERATIONS/ABANDONMENT

7.1 Diligent Operations and Abandonment - Ivanhoe as Operator shall conduct all operations for the Pilot Phase, Phase 1 and thereafter, in a diligent, safe and efficient manner in accordance with good and prudent petroleum industry practices and field conservation principles as are generally followed by the United States petroleum industry under similar circumstances and at all times in compliance with all applicable laws, ordinances, rules and regulations. When any programmed well drilled hereunder, reaches the contract depth and has been tested as agreed, Ivanhoe shall either complete the same as a producer, or plug and abandon the same in accordance with all applicable laws and regulations; provided however, that if Ivanhoe desires to abandon the well as a dry hole, Ivanhoe shall so notify Non-Operators who shall then have 48 hours to elect to take over such well and attempt to complete it at their sole cost, risk, and expense. Should any Non-Operator take over the well ("Takeover Parties"), Ivanhoe shall relinquish all rights thereto, including the materials and equipment used in connection therewith free and clear of all liens and encumbrances not existing as of the date of this Agreement, other than ordinary and usual liens and encumbrances arising from the drilling of the well. The Takeover Parties shall pay to Ivanhoe the fair market value of such material and of the salvageable in-hole equipment, less the estimated cost of salvage, and the Takeover Parties shall become responsible for the rig time costs. Upon any such takeover, all rights of Ivanhoe pertaining to the well shall automatically terminate and the Takeover Parties shall be solely responsible for the completion and/or abandonment of the well, and shall indemnify and hold Ivanhoe harmless from any cost or loss relating thereto.

ARTICLE VIII
OPERATOR

8.1 Status of Ivanhoe as Operator. – Notwithstanding the provisions of the JOA, in the absence of Gross Negligence or Willful Misconduct on the part of management level or supervisory personnel of Ivanhoe directly and causally connected with the subject action, Ivanhoe shall not be liable as Operator for damages or losses or otherwise to any other Party for anything done or omitted to be done as Operator in carrying out any Joint Operations, which shall be for the Joint Account, subject however to Ivanhoe's obligation to pay Capital Expenditures on a 100% basis. Each Party, including Ivanhoe as a Party, shall bear any losses or liabilities in proportion to their Working Interests; provided, however, that Ivanhoe as a Party or as Operator shall not be liable for losses or liabilities resulting from operations on or related to the LAK Ranch Lease or LAK Ranch Area prior the Effective Date, the liability for which shall be for the account of the other Parties.

8.2 No liability for Losses. In no case shall the Operator be liable to any Party, or to any third party, for any loss or spillage of oil or gas or other hydrocarbon substances or for any consequential loss, including loss of profit, arising out of its conduct of Joint Operations, or for any damages or losses resulting therefrom. Each Party shall bear any such losses or liability in proportion to their Working Interests.

8.3 Government Approvals. In the conduct of the Pilot Phase, Phase 1 and Joint Operations, the Operator shall:

(a) Pay to the government of the State of Wyoming or the Federal Government, as applicable, (collectively, the "Government") for the Joint Account, within the periods and in the manner prescribed by all applicable laws, ordinances, rules, and regulations, all periodic payments, royalties, taxes, fees and other payments pertaining to the Joint Operations (such amounts to be treated under this Agreement as Capital Expenditures or Other Costs, as the case may be, in accordance with generally accepted accounting principles) but excluding any taxes measured by the incomes of any of the Parties; and

(b) Have the exclusive right and obligation to represent the Parties in all dealings with the Government with respect to matters arising under the Joint Operations. The Operator shall notify the Non-Operators as soon as possible of such meetings. Subject to any necessary Government approvals, Non-Operators shall have the right to attend any meetings with the Government with respect to such matters, but only in the capacity of observers. Nothing contained in this Agreement shall restrict any Party from holding discussions with the Government with respect to any issue peculiar to its particular business interests arising under the LAK Ranch Leases or this Agreement, but in such event, such Party shall promptly advise the other Parties, if possible, before and in any event promptly after such discussions, provided that such Party shall not be required to divulge to the other Parties any matters discussed to the extent the same involve proprietary information or matters not affecting the other Parties.

ARTICLE IX
TITLE AND OTHER REPRESENTATIONS AND WARRANTIES

9.1 Title and Curative Work. Except for those encumbrances and royalties created by, through or under Derek or SEC including but not limited to those set out in Exhibit "A", Derek does not warrant title, express or implied, to the LAK Ranch Leases. Derek shall not be obligated to perform any curative work or to furnish any materials with respect thereto, other than copies of such title opinions and other relevant documents as they may have in their possession. Nevertheless, during the term of this Agreement or the JOA, any curative work by Derek shall enure to the benefit of Ivanhoe and Ivanhoe shall be promptly furnished with copies of all opinions and curative instruments pertaining to the LAK Ranch Leases and the LAK Ranch Area. Derek shall make available to Ivanhoe, all records and lease files pertaining to the LAK Ranch Leases. To the extent that title to one or more of the Leases included in the drilling unit for any programmed well in the LAK Ranch Program is such that a reasonable and prudent Operator would require in Ivanhoe's opinion curative action to be taken, except as set out above, Ivanhoe may undertake such curative action, in which case, the time in which operations are to be started, shall be extended by the period of time necessary for Ivanhoe to complete such curative work.

9.2 Other Representations and Warranties of Derek. Derek, on its own behalf and for SEC to the extent that SEC will earn a Working Interest in the LAK Ranch Area, represents and warrants to Ivanhoe as follows:

9.2.1 Encumbrances on the LAK Ranch Area. Exhibit "A" sets forth a description of (a) all overriding royalties, production payments or preferential rights of any nature of third parties with respect to the LAK Ranch Leases and the Substances, (b) all rights-of-way and other agreements and other rights relating to the use or ownership of the surface or surface properties that are used or held for use in connection with the production of Substances from the LAK Ranch Area, (c) all contracts, licenses, permits, approvals, consents, certificates and other rights granted by governmental authorities and all certificates of convenience or necessity, immunities, privileges, grants and other rights that relate to the LAK Ranch Area or the ownership or operation of the LAK Ranch Area and the LAK Ranch Leases. Ivanhoe and the interests of Ivanhoe in the LAK Ranch Leases and the LAK Ranch Area shall at all times be free and clear of any obligation under the SEC Agreement. Derek and SEC agree that Derek and its interests in the LAK Ranch Leases and the LAK Ranch Area, alone shall be responsible for performance of all obligations to SEC under the SEC Agreement and/or any other agreement or arrangement between Derek and SEC.

9.2.2 Litigation and Claims. There are no claims, demands, filings, investigations, administrative proceedings, actions, suits or other legal proceedings pending or, to the best of their knowledge, threatened, with respect the LAK Ranch Leases or LAK Ranch Area or the ownership or operation thereof, other than proceedings relating to the oil and gas industry generally and as to which neither Derek and or SEC is a named party; and no written notice from any governmental authority or any other person has been received by either Derek or SEC claiming any violation of any law, rule, regulation, ordinance, order, decision or decree of any governmental authority (including, without limitation, any such law, rule, regulation, ordinance, order, decision or decree concerning the conservation of natural resources) except for funds owed to Bateman Engineering Inc. ("Bateman") in the approximate amount of $285,000, which funds shall be paid by Derek to Bateman within ten (10) days of the Effective Date.

9.2.3 Approval and Preferential Rights. There are no approvals or consents required to be obtained by Derek for consummation of the transactions contemplated by this Agreement and no preferential purchase rights that affect the transactions contemplated by this Agreement.

9.2.4 Compliance with Laws and Permits. The operations conducted by Derek on the LAK Ranch Area and LAK Ranch Leases have been and currently are, in compliance with the provisions and requirements of all laws, rules, regulations, ordinances, orders, decisions and decrees of all governmental authorities having jurisdiction with respect thereto, except where noncompliance would not reasonably be expected to have a material adverse effect on the operation thereof. All necessary government permits, licenses and other authorizations would not reasonably be expected to have a material adverse effect on the ownership or operation of the LAK Ranch Leases. Neither has Derek received a demand from a governmental authority to plug and abandon any well on the LAK Ranch Leases that has not been plugged and abandoned.

9.2.5 Environmental Compliance. No pollutant, waste, contaminant, or hazardous, extremely hazardous, or toxic material, substance, chemical or waste identified, defined or regulated as such under any law relating to health, safety or environmental matters is present, or has been handled, managed, stored, transported, processed, treated, disposed of, released, migrated or has escaped on, in, from, under or in connection with the LAK Ranch Lease or LAK Ranch Area or the ownership or operation thereof such as to cause a condition or circumstance that would reasonably be expected to result in a violation of any law relating to health and safety or environmental matters or in a remediation, removal, response, restoration, abatement, investigative or monitoring obligation, other than violations or obligations that would not reasonably be expected to have a material adverse effect on the ownership or operation of the LAK Ranch Leases or LAK Ranch Area.

9.2.6 Production Burdens, Taxes and Expenses. All payments due under or with respect to the LAK Ranch Leases have been properly and timely paid. All ad valorem, property, production, severance and other taxes based on or measured by the ownership of the LAK Ranch Leases and the LAK Ranch Area based on or measured by the ownership of such properties or the production of Substances therefrom have been properly and timely paid.

ARTICLE X
RELATIONSHIP/TAXATION

10.1 Relationship of the Parties – This Agreement does not create, and shall not be construed to create, a partnership, association, joint venture or fiduciary relationship of any kind or character between the Parties, and shall not be construed to imposeany duty, obligation or liability arising from such a relationship by or with respect to any Party. Notwithstanding any provisions herein that the rights and liabilities hereunder are several and not joint or collective or that this Agreement and the operations hereunder shall not constitute a partnership, each Party elects to be excluded from the application of all or any part of the provisions of Subchapter "K", Chapter 1, Subtitle "A", U.S. Internal Revenue Code of 1986, as amended, or similar provisions of applicable State laws.

ARTICLE XI
FORCE MAJEURE

11.1 Force Majeure – If a Party is rendered unable, wholly or in part, by Force Majeure, to carry out its obligations under this Agreement, other than the obligations to pay money, that Party shall give the other Parties prompt written notice of the Force Majeure with reasonably full particulars concerning it; thereupon the obligations of the Party giving the notice, so far as they are affected by the Force Majeure, shall be suspended during the continuance of the Force Majeure, and such reasonable time thereafter as the affected Party reasonably requires under the circumstances to carry out its obligations. The affected Party shall use all reasonable diligence to remove the Force Majeure situation and resume the performance of its obligations as quickly as practicable. The requirement that any Force Majeure shall be remedied with all reasonable dispatch, shall not require the settlement of strikes, lockouts, or other labor difficulty by the Parties involved, contrary to its wishes; how all such difficulties shall be handled shall be entirely within the discretion of the Party concerned.

11.2 Claims. The term "Force Majeure" shall mean any circumstances not within the reasonable control, directly or indirectly, of the Party affected including but not limited to an act of God, strike, lockout or other industrial disturbance, terrorist act, act of the public enemy, war, blockade, public riot, lightning, fire, storm, flood, explosion, governmental delay, restraint or inaction, unavailability of equipment, but only if and to the extent that any act to which Force Majeure is claimed:

(a) Such circumstances, despite the exercise of reasonable diligence, cannot be or be caused to be prevented, avoided or removed by such Party; and

(b) In the case of claims by such Party, such event materially adversely affects (in case and/or time) the ability of the Party to perform its obligations under this Agreement, the Joint Operating Agreement or any Lease, and the Party has taken all reasonable precautions, exercised due care and has taken all reasonable alternative measures in order to avoid the effect of such event on the Party's ability to perform its obligations under this Agreement and to mitigate the consequences thereof.

ARTICLE XII
ASSIGNMENTS
OPTION TO PURCHASE INTERESTS

12.1 Right of First Refusal – Other than as required and allowed as between one Party to this Agreement to another Party or third party elsewhere in this Agreement and subject to this Section, a Party shall not dispose of any of its Working Interest, whether by assignment, sale, trade, lease, sublease or farmout, without first complying with the following provisions:

(a) The Party wishing to make the disposition (in this Section called the "Disposing Party") shall give notice to each other Party (in this Section called an "Offeree") of its intention to make the disposition, including in such notice a description of the Working Interest proposed to be disposed of, the identity of the proposed assignee, the price or other consideration for which the Disposing Party is prepared to make such disposition, the proposed effective date and closing date of the transaction and any other information respecting the transaction which would be material to the exercise of the Offerees' rights hereunder (such notice in this Article called the "Disposition Notice").

(b) In the event the consideration described in the Disposition Notice cannot be matched in kind and the Disposition Notice does not include the Disposing Party's bona fide estimate of the value, in cash, of such consideration, an Offeree may, within seven (7) days of the receipt by the Offerees of the Disposition Notice, request the Disposing Party to provide such estimate to the Offerees, whereupon the Disposing Party shall provide such estimate in a timely manner, and in no event in excess of 15 business days, and the election period provided herein to the Offerees shall be suspended until such estimate is received by the Offerees. Failure to provide such cash equivalent estimate shall be grounds for refusal by any or all of the Offerees of consent to the disposition.

(c) In the event of a dispute as to the reasonableness of an estimate of the cash value of the consideration described in the Disposition Notice or provided pursuant to paragraph (b) herein, as the case may be, the matter shall be referred to arbitration as provided in Section 16.8.

(d) Within the later of: (i) thirty (30) days from the receipt of the Disposition Notice, as modified by any suspension pursuant to paragraph (b) herein; or ii) if applicable, fifteen (15) days from receipt of notice of the arbitrated value determined pursuant to Section 16.8, an Offeree may give notice to the Disposing Party that it elects to purchase the Working Interest described in the Disposition Notice for the applicable price (in this Section called a "Notice of Acceptance"). A Notice of Acceptance shall create a binding contractual obligation upon the Disposing Party to sell, and upon an Offeree giving a Notice of Acceptance to purchase, for the applicable price, all of the Working Interest included in such Disposition Notice on the terms and conditions set forth in the Disposition Notice. However, if more than one Offeree gives a otice of Acceptance, each such Offeree shall purchase the Working Interest to which such Notice of Acceptance pertains in the proportion its

Working Interest bears to the total Working Interests of such Offerees.

(e) In the event that the Working Interest described in the Disposition Notice is not disposed of to one or more of the Offerees pursuant to the preceding paragraph, the disposition to the proposed assignee shall be subject to the consent of the Offerees. Such consent shall not be unreasonably withheld, and it shall be reasonable for an Offeree to withhold its consent to the disposition if it reasonably believes that the disposition would be likely to have a material adverse effect on it, its Working Interest or operations to be conducted hereunder, including, without limiting the generality of all or any part of the foregoing, a reasonable belief that the proposed assignee does not have the financial capability to meet prospective obligations arising out of this Agreement. However, an Offeree shall be deemed to have consented to the disposition to the proposed assignee, unless, within the time period prescribed in paragraph (d) herein, the Offeree advises the other parties, by notice, that it is not prepared to consent to such disposition.

(f) If the Working Interest described in the Disposition Notice is not disposed of to one or more the Offerees pursuant to paragraph (d) herein, the Disposing party may, subject to the obtaining the consents prescribed by the preceding paragraph, dispose of such Working Interest at any time within one hundred and fifty (150) days from the issuance of such Disposition Notice, provided that such disposition is not on terms that are more favorable to such purchaser than those offered in the Disposition Notice.

(g) Following a disposition herein or one hundred fifty (150) days following the issuance of a Disposition Notice from which a disposition did not result, as the case may be, the provisions of this Section shall once again apply to the Working Interest described in the Disposition Notice.

12.2 Section 12.1 shall not apply in the following instances, namely:

(a) An assignment made by way of security for the assignor's present or future indebtedness, or liabilities (whether contingent, direct or indirect and whether financial or otherwise), the issuance of the bonds or debentures of a corporation, or the performance of the obligations of the assignor as a guarantor under a guarantee, provided that in the event the security is enforced by sale or foreclosure, Section 12.1 shall apply.

(b) A disposition to an Affiliate of the assignor, or in consequence of a merger or amalgamation of the assignor with another corporation or pursuant to an assignment, sale or disposition made by a Party of its entire Working Interest to a corporation in return for shares in that corporation or to a registered partnership in return for an interest in that partnership.

(c) A disposition made by the assignor of all, or substantially all, or of an undivided interest in all or substantially all, of its petroleum and natural gas rights; provided, however, if the LAK Ranch Leases and any other Lease interests of Derek and/or SEC are all or substantially all of the net assets held by them, then Section 12.1 shall apply.

(d) A disposition by Derek to SEC pursuant to the terms of the SEC Agreement. However, a Party making such a disposition pursuant to paragraph (b), (c) or (d) of this Clause shall advise the other Parties of such disposition in a timely manner.

12.3 If any assignment of Working Interest is made to multiple assignees so as to increase the expense or duties of the Operator, the Operator may require the assignees (and the assignor if it retains a Working Interest) to appoint one of their number as representing all of them for the purposes of this Agreement, unless arrangements satisfactory to the Operator are made to compensate the Operator for the increased expenses or duties.

ARTICLE XIII
INSURANCE

13.1 Insurance Requirements – Before commencing operations within the LAK Ranch Area, pursuant to this Agreement, Ivanhoe shall procure and shall thereafter maintain in force during the term of this Agreement, the insurance hereinafter described, covering the operations to be performed hereunder by Ivanhoe as Operator. The costs of all insurance carried by the Operator for the benefit of the Parties and the Joint Operations shall be for the Joint Account of the Parties.

(a) Workers' Compensation and Occupational Disease Insurance, satisfying the legal requirements of the State of Wyoming and Employers Liability Insurance with a minimum of $1,000,000.00 per occurrence.

(b) Comprehensive general liability insurance, including personal injury blanket contractual liability and property damage, with coverage of at least $1,000,000 combined single limit or the equivalent; owned and non-owned automobile liability, including personal injury and property damage, with coverage of at least $1,000,000, combined single limit or the equivalent; products liability and completed operations coverage of at least $1,000,000, combined single limit or the equivalent

(c) Umbrella liability insurance, with at least $4,000,000 combined single limit per occurrence and at least $10,000,000 in the aggregate;

(d) Control of well seepage and pollution, and well Restoration/Redrill Insurance with a combined single limit of at least $3,000,000 per occurrence.

(e) Each and every policy required hereinabove in (b), (c) and (d) shall name Non-Operators as additional insured.

(f) With the exception of Paragraph 13.1 (a), the Parties agree to mutually waive subrogation in favor of each other on all insurances carried by each Party and/or obtain such waiver from the insurance carrier if so required by the insurance contract. If such waiver is not obtained, the Party failing to do so shall indemnify the other Parties for any claim by an insurance carrier arising out of subrogation.

13.2 Certificates of Insurance. Before commencing any work hereunder, Ivanhoe shall furnish to Non-Operators Certificates of Insurance evidencing the issuance to Ivanhoe of the policies of insurance providing the types of coverage and limits of liability prescribed above, and certifying that Non-Operators shall be given not less than 30 days written notice prior to any material change, substitution or cancellation prior to the normal expiration dates. Cancellation or expiration of any of said policies of insurance shall not preclude Ivanhoe from recovery thereunder for any liability arising under this Agreement.

ARTICLE XIV
FAILURE TO DRILL

14.1 Failure to Comply – Under no circumstances shall Ivanhoe be liable for its failure to drill or complete any programmed well in the LAK Ranch Program, or the manner in which it is drilled, tested or completed.

ARTICLE XV
NOTICES

15.1 Notices – All notices between the Parties authorized or required by any of the provisions of this Agreement, unless otherwise specifically provided, shall be given in writing and delivered in person by mail, courier service, facsimile, or e-mail, postage or charges prepaid or by telex or telecopier and addressed to the Party to whom the notice is given as follows.

Derek: Derek Resources (U.S.A.) Inc.
Street Address: Suite 1550, 355 Burrard Street
Vancouver B. C., Canada V6C 2G8
Attention: Barry C. J. Ehrl, President & CEO
Telephone: 604-331-1757
Facsimile: 604-669-5193
E- mail: info@derekresources.com
Ivanhoe: Ivanhoe Energy (USA) Inc.
Street Address: 5060 California Avenue, Suite 400
Bakersfield, California 93309
Mailing Address: P.O. Box 9279
Bakersfield, CA 93389-9279

Attention:	Robert Coffey, Vice-President, Operations
Telephone:	661-869-8332
Facsimile:	661-869-2820
E-mail	rcoffey@ivanhoeenergy.com
Copy to:	Oscar Blake, General Counsel

The originating notice given under any provision hereof shall be deemed given only when received by the Party to whom such notice is directed, and the time for such Party to give any notice in response thereto shall run from the date the originating notice is received. The second or any responsive notice shall be deemed given when deposited in the mail or with the courier service, with postage or charges prepaid, or upon transmission by facsimile or telecopier. Each Party shall have the right to change its address, or the person to receive such notice, at any time, and from time to time by giving written notice thereof to the other Party.

ARTICLE XVI
ADDITIONAL TERMS, PROVISIONS, AND COVENANTS

16.1 Time and Specific Performance of Essence - Time and specific performance are of the essence of this Agreement.

16.2 Laws and Regulations; Applicable Law - Ivanhoe as Operator shall comply with and conduct its operations hereunder in accordance with all applicable laws, ordinances, rules, and regulations. This Agreement shall be construed and interpreted under and in accordance with the laws of the State of Wyoming.

16.3 Headings and Interpretation - The section and other headings used throughout this Agreement are for administrative convenience only and shall be disregarded for purposes of construing this Agreement. Words importing the singular shall include the plural and vice versa, and the use of one gender includes any other gender, except where the context otherwise requires.

16.4 Waiver - Any of the terms, provisions, covenants, representations or conditions hereof may be waived only by a written instrument executed by the Party waiving compliance. The failure of any Party at any time or times to require performance of any provisions hereof shall in no manner affect the right to enforce the same. No waiver by any Party of any condition or of the breach of any term, provision, covenant, representation or warranty contained in this Agreement, whether by conduct or otherwise, in any one or more instances, shall be deemed to be or construed as a further or continuing waiver of any such condition or breach, or a waiver of any other condition or of the breach of any other term, provision, covenant, representation or warranty.

16.5 Press Releases - No Party shall distribute any information or photographs concerning operations hereunder to the press or other media without the approval of the other Parties. In the event of a requirement by a regulatory agency [including the Securities Exchange Commission or the U.S. or securities regulatory authorities of Canada, or any stock exchange on which the securities of a Party or its Affiliate (to the extent the Affiliate either owns a majority interest in a Party, or a Party owns a majority interest in the Affiliate) are listed], or of an emergency involving extensive property damage, operations failure, loss of human life or other clear emergency, the Parties are deemed authorized to furnish such minimum, strictly factual information as is necessary to satisfy requirements of appropriate regulatory authorities, or the legitimate public interest on the part of the press, other media, and duly constituted authorities if time does not permit the obtaining of prior approval by the other Party, but the Party furnishing such information shall promptly advise the other Party of the information so furnished.

16.6 Amendments - This Agreement may be amended, modified, changed, altered or supplemented only by written instrument duly executed by the Parties specifically for such purpose and which specifically refers to this Agreement.

16.7 Conflicts - In the event of a conflict between provisions of this Agreement and those contained in any of the agreements attached hereto, the provisions of this Agreement shall prevail.

16.8 Arbitration - Unless otherwise expressly agreed in writing by all of the Parties to any claim, dispute or controversy (the "Dispute"), the following procedures shall be implemented for resolving any Dispute:

(a) Any Party may invoke these dispute resolution procedures (the "Procedures") by providing written notice of any Dispute to the other Parties. Within thirty (30) days after such notice is given, all the Parties shall attempt to resolve the Dispute through negotiations at a meeting, which shall be attended by representatives of each of the Parties having decision-making authority as well as by management-level personnel of each of the Parties who have not previously been directly engaged in directing or responding to the Dispute;

(b) If a Dispute is not resolved after negotiations in a meeting as set out above, the Dispute shall be submitted to mandatory and binding arbitration at the election of any Party by sending a notice of arbitration to the other Parties;

(c) The Arbitration proceedings shall administered and conducted in accordance with the Commercial Rules of Arbitration of the American Arbitration Association.

(d) The Dispute shall be decided by three independent arbitrators ("Arbitrators") (that is having no substantial economic or other material relationship with any or all of the Parties). The Parties shall attempt to agree on two of the independent arbitrators (the "Agreed Arbitrators") with the third being appointed by the Agreed

Arbitrators. Should any Party refuse or neglect to join in the timely appointment of the Agreed Arbitrators, the Agreed Arbitrators shall be appointed by the American Arbitration Association, and the Arbitration shall proceed notwithstanding the non-appointment and non-attendance of such Party in the appointment process.

(e) The Arbitrators shall have 30 days after their appointment within which to allow discovery, hear evidence, and issue its decision or award and shall in good faith attempt to comply with such time limits; provided, however, if the Arbitrators, at their sole and reasonable discretion, believe additional time is necessary to reach a decision, they may notify the Parties and extend the time to reach a decision in 30 day increments, but in no event to exceed an additional 60 days.

(f) The decision or award of the Arbitrators shall be by majority vote. shall be in writing and shall state its detailed reasoning for the award. Discovery of evidence shall be conducted expeditiously by the Parties, bearing in mind the Parties desire to limit discovery and to expedite the decision or award of the Arbitrators at the most reasonable cost and expense of the Parties. Judgment upon an award rendered pursuant to such arbitration may be entered in any court having jurisdiction or application may be made to such court for a judicial acceptance of the award, and an order of enforcement, as the case may be. The place of Arbitration shall be Denver, Colorado. All questions arising out of this Agreement and their rights and obligations created herein, or its validity, existence, interpretation, performance or breach shall be governed by the laws of the State of Wyoming, without regard to conflict of laws principles. The Arbitrators shall not award consequential or punitive damages to any Party. The costs and expenses of the arbitration proceeding, including the fees of the arbitrators and all costs and expenses, including legal fees and witness fees, incurred by the prevailing Party, shall be borne by the losing Party, such award of costs to be decided and specified by the Arbitrators.

16.9 Entire Agreement – When executed by the duly authorized representatives of the Parties, this Agreement shall constitute the entire agreement between the Parties regarding the LAK Ranch Leases, and the LAK Ranch Area and Area of Mutual Interest and shall supersede and replace any and all other writings, understandings or memoranda of understanding entered into or discussed prior to the Effective Date.

16.10 Savings Clause – If any part or portion of this Agreement is held to be invalid, such invalidity of any such part or portion shall not affect any remaining part or portion hereof.

16.11 Corporate Authority – The Parties represent that, as of the date of the execution hereof, they are duly authorized, validly existing, and in good standing under the laws of the jurisdictions of organization and are qualified and authorized to do business in the State of Wyoming and that all requisite corporate power and authority to execute, deliver, and effectuate this Agreement have been duly obtained.

16.12 Further Assurances – Each Party shall promptly execute, acknowledge, and deliver to the other Party such other instruments and take such other action as may be necessary or convenient in order to effect the transactions contemplated in this Agreement.

IN WITNESS WHEREOF, the Parties hereto have executed this Agreement to be effective as of the Effective Date.

IVANHOE ENERGY (USA) INC.

BY: ______________________________

E. L. DANIEL, President

DEREK RESOURCES (U.S.A.) INC.

BY: ______________________________

BARRY C. J. EHRL, President & CEO

EXHIBIT "A"

Attached to and made a part of that certain Farmin Agreement dated January 15, 2004, by and between Ivanhoe Energy (USA Inc., and Derek Resources (USA) Inc. All lands are in Weston County, Wyoming

Description of Lands

Tract A

Township 44 North, Range 60 West, 6th P.M.

Section 6: Southwest Quarter, Northeast Quarter, East Half of the Northwest Quarter, and West Half of the Southeast Quarter (SW/4, NE/4, E/2 NW/4, W/2 SE/4).
Section 7: West Half, and West Half of the East Half (W/2, W/2 E/2).
Section 18: South Half of the Southwest Quarter, and the Northwest Quarter (S/2 SW/4, NW/4).
Section 19: West Half of West Half, Northeast Quarter of the Northwest Quarter (W/2 W/2, NE/4 NW/4).

Township 44 North, Range 61 West, 6th P.M.

Section 1: Southeast Quarter (SE/4).
Section 11: Southeast Quarter, East Half of the Northeast Quarter, and South Half of the Southwest Quarter (SE/4, E/2 NE/4, S/2 SW/4).
Section 12: All.
Section 13: West Half, and South Half of the Southeast Quarter (W/2, S/2 SE/4).
Section 14: All.
Section 22: Northeast Quarter (NE/4).
Section 23: North Half, Southeast Quarter, East Half of the Southwest Quarter (N/2, SE, E/2 SW/4).
Section 24: North Half, Southeast Quarter, and East Half of the Southwest Quarter (N/2, SE/4, E/2 SW/4).
Section 25: Northwest Quarter of the Northeast Quarter, North Half of the Northwest Quarter, Southwest Quarter of the Northwest Quarter, and Northwest Quarter of the Southwest Quarter (NW/4 NE/4, N/2 NW/4, SW/4 NW/4, NW/4 SW/4.
Section 26: East Half of the Northeast Quarter, and Northeast Quarter of the Southeast Quarter (E/2 NE/4, NE/4 SE/4). Section 30: West Half of the Northwest Quarter (W/4 NW/4).

Tract B

Township 44 North, Range 61 West, 6th P.M. ________________
Section 24: Southeast Quarter of the Southeast Quarter (SE/4 SE/4).

Section 24: Northeast Quarter of the Southeast Quarter (NE/4 SE/4).
Section 25: South Half of the Southwest Quarter, the Northeast Quarter of the Southwest Quarter, Northwest Quarter of the Southeast Quarter, East Half of the Northeast Quarter, Southwest Quarter of the Northeast Quarter, and Southeast Quarter of the Northwest Quarter (S/2 SW/4, NE/4 SW/4, NW/4 SE/4, E/4 NE/4, SW/4 NE/4, SE/4 NW/4).

Tract C

Township 44 North, Range 60 West, 6th P.M.

Section 18: North Half of the Southwest Quarter (N/2 SW/4).

Township 44 North, Range 61 West, 6th P.M.

Section 13: Northeast Quarter, and North Half of the Southeast Quarter (NE/4, N/2 SE/4).

Tract D

Township 44 North, Range 61 West, 6th P.M.

Section 22: Southeast Quarter (SE/4).
Section 27: West Half of the East Half (W/2 E/2).

Tract E

Township 44 North, Range 60 West, 6th P.M.

Section 24: Southeast Quarter of the Southeast Quarter (SE/4,SE/4).

Additional Property Leased from the Federal Government

Township 44 North, Range 61 West, 6th P.M.

Section 1: South Half of the Southwest Quarter (S/2 SW/4).
Section 10: Southwest Quarter of the Northeast Quarter (SW/4 NE/4).
Section 10: West Half of the Southeast Quarter, Southeast Quarter of the Southeast Quarter (W/2 SE/4, SE/4 SE/4), excluding 13.77 acres in railroad right-of-way WyW0119068.
Section 15: West Half of the East Half, East Half of the West Half, Northwest Quarter of the Northwest Quarter, Southwest Quarter of the Southwest Quarter (W/2 E/2, E/2 W/2, NW/4 NW/4, SW/4 SW/4).
Section 15: East Half of the East Half (E/2 E/2), excluding 23.78 acres in railroad right-of-way WyW0119068.
Section 25: Northeast Quarter of the Southeast Quarter, South half of the

Southeast Quarter (NE/4 SE/4, S/2 SE/4).

Section 26: Northwest Quarter of the Northeast Quarter, East Half of the Southwest Quarter (NW/4 NE/4, E/2 SW/4).

Section 26: East Half of the Northwest Quarter, Southeast Quarter of the Southeast Quarter (E/2 NW/4, SE/4 SE/4), excluding 14.23 acres in railroad right-of-way WyW0119068.

Oil and Gas Leases subject to this Agreement

A. Oil and Gas Lease, dated January 10, 1981, by and between Donald B. Roberts, B.M. Stewart, and Sheri Vineyard, as Lessors, and Exoil Services, Inc., as Lessee, recorded in Book 87, Page 189, Official Records of Weston County, Wyoming.
B. Oil and Gas Lease, dated effective August 31, 1958, by and between the U.S. Department of Interior, Bureau of Land Management, as Lessor, and Edel P. Smith, as Lessee, being identified as Lease Serial Number W-022180-A.
C. Oil and Gas Lease, dated effective December 1, 1999, by and between the U.S. Department of Interior, Bureau of Land Management, as Lessor, and Cove Energy, as Lessee, being identified as Lease Serial Number W-149512.

Burdens on Production

Tract A	19.9501%
Tract B	19.9501%
Tract C	11.7875% (Based on 50% leased)
Tract D	11.7875% (Based on 50% leased)
Tract E	19.9501%

EXHIBIT "B" TO FARMIN AGREEMENT
LAK RANCH AREA/AMI

[GRAPHIC - FULL PAGE MAP OF DEREK RESOURCES CORPORATION LAK RANCH PROPERTY WESTON COUNTY WYOMING]

Original LAK Property and Adjoining Property

DESCRIPTION OF PROPERTY

TRACT A

Township 44 North, Range 60 West, 6th P.M., Weston County, Wyoming

Section 6: SW1/4, NE1/4, E1/2NW1/4, W1/2SE1/4
Section 7: W1/2, W1/2E1/2
Section 18: S1/2S W1/4, NW1/4
Section 19: W1/2 W1/2, NE1/4NW1/4

Township 44 North, Range 61 West, 6th P.M., Weston County, Wyoming
Section 1: SE1/4
Section 11: SE1/4, E1/2NE1/4, S1/2SW1/4
Section 12: All
Section 13: W1/2, S1/2SE1/4
Section 14: All
Section 22: NE1/4
Section 23: N1/2, SE, E1/2SW1/4
Section 24: N1/2, SE1/4, E1/2SW1/4
Section 25: NW1/4NE1/4, N1/2NW1/4, SW1/4NW1/4, NW1/4SW1/4
Section 26: E1/2NE1/4, NE1/4SE1/4
Section 30: W1/2NW1/4

TRACT B

Township 44 North, Range 61 West, 6th P.M., Weston County, Wyoming

Section 24: SE1/4SE1/4
Section 24: NE1/4SE1/4
Section 25: S1/2SW1/4, NE1/4SW1/4, NW1/4SE1/4, E1/2NE1/4, SW1/4NE1/4, SE1/4NW1/4

TRACT C

Township 44 North, Range 60 West, 6th P.M., Weston County, Wyoming Section 18: N1/2SW1/4 Township 44 North, Range 61 West, 6th P.M., Weston County, Wyoming Section 13: NE1/4, N1/2SE1/4

TRACT D

Township 44 North, Range 61 West, 6th P.M., Weston County, Wyoming
Section 22: SE1/4 Section 27: W1/2E1/2

Additional Property Leased from the Federal Government

DESCRIPTION OF PROPERTY

Township 44 North, Range 61 West, 6th P.M., Weston County, Wyoming

Section 10: SW1/4NE1/4
Section 10: W1/2SE1/4, SE1/4SE1/4, (Excluding 13.77 acres in railroad right-of-way WyW0119068)
Section 15: W1/2E1/2, E1/2W1/2, NW1/4NW1/4, SW1/4SW1/4
Section 15: E1/2E1/2(Excluding 23.87 acres in railroad right-of-way WyW0119068)
Section 25: NE1/4SE1/4, S1/2S1/4
Section 26: NW1/4NE1/4, E1/2SW1/4
Section 26: E1/2NW1/4, SE1/4SE1/4(Excluding 14.23 acres in railroad right-of-way WyW0119068)

Additional Property Leased from Edel P. Smith and Accidental Oil Company

DESCRIPTION OF PROPERTY

Township 44 North, Range 61 West, 6th P.M., Weston County, Wyoming
Section 1: SW1/2 SW1/4

EXHIBIT "D"
TO LAK RANCH FARMIN AGREEMENT LAK RANCH PROGRAM SUMMARY

ENVIRONMENTAL COMPLIANCE

1. Oil spill occurred on December 20, 2001 as a result of high winds breaking the wire that held the bypass loop on the Jacuzzi pump. Approximately 9 to 10 barrels was displaced on the surface. Chuck Brothers were retained to handle the oil waste disposal which was completed on or about December 26, 2001.
2. On December 31, 2001, Mr. Jack Eslinger, a landowner with property near the LAK Ranch Project called the Derek LAK Ranch site office regarding water surfacing on his property. Dr. Paul Trost was consulted regarding the water seepage and informed Derek site staff that the water was ground water being pressured to the surface and that the water was not contaminated so that there need not be any concern in this regard. Derek did not shut down its LAK Ranch Project operation due to the water seepage as Derek had previously decided to shut down operations for financial reasons. In addition, Derek was in the process of applying for a surface water discharge permit, which was subsequently granted. There was no further contact with Mr. Eslinger with regard to this matter beyond his original call to the LAK Ranch site office until mid-January of 2004, when Mr. Eslinger spoke with the Derek site manager, Mr. Art Huckins. At this time, Mr. Eslinger expressed further concerns about the water damage to his property. Derek acknowledges that it is Derek's responsibility to resolve this matter with Mr. Eslinger and Derek intends to do so without admitting any liability for the water damage.

EXHIBIT "E"

ATTACHED TO AND MADE A PART OF THAT CERTAIN FARMIN AGREEMENT DATED JANUA RY 15, 2004, BY AND BETWEEN IVANHOE ENERGY (USA) INC. AND DEREK RESOURCES (U.S.A.) INC.

LAK RANCH PROGRAM DETAILS

PILOT PHASE

PROGRAM SUMMARY

- Cycle steam existing horizontal producer
- Drill approx. 5 vertical slim hole steam injection wells
- Test response from continuous steam injection
- Acquire and process high resolution 3-D seismic survey

Estimated Pilot Cost – 1.1 MM$ (US$)
Timeframe – 1.5 years starting in 1/1/2004

DISCUSSION

The intent of this phase is to test the technical feasibility of using thermal recovery methods to enhance the production of the horizontal producing well completed in the Newcastle Sand on the LAK Ranch.

The existing horizontal producing well will be tested in a cyclic steam mode using a conventional beam pump to produce the well. This portion of the program is estimated to take approximately six months. After evaluating the cyclic results, a continuous steam injection program will be evaluated using five vertical steam injection wells completed up dip of the horizontal well. After one year of continuous injection there should be enough data to support the go-forward decision.

During this phase, and depending on performance of the above injection program, a 3-D seismic program will be conducted over prospective acreage on the LAK Ranch in order to more accurately map the geologic features and support the exploitation of oil reserves on this project. In order to resolve the stratigraphic variability, and successfully identify the productive zones, an ultra-high resolution survey will be obtained. We estimate that a project of 4-5 square miles of 3-D seismic data may be acquired and processed for $500,000 US. At a minimum, Ivanhoe will spend 250 M$US acquiring and processing seismic data during the Pilot Phase. It is estimated that the geophysical program will take a minimum of 6 months and is scheduled for the later half of 2004.

PHASE I

PROGRAM SUMMARY

- Drill three horizontal producers
- Drill approx. 15 vertical slim hole steam injection wells
- Drill six temperature observation wells
- Drill six delineation wells
- Expand existing facility infrastructure (well testing, water treatment and steam generation).
- Drill approximately six delineation wells to test 3-D geological model

Estimated Phase I Cost – 5.8 MM$ (US$)
Estimated Cumulative (Pilot +Phase I) Cost = 6.9MM$ (US$)
Timeframe – 2 years starting on 7/1/2005

DISCUSSION

During Phase I three additional horizontal producers will be drilled along with 15 vertical steam injection wells and one water disposal well. The total drilling cost is estimated at 3.65 MM$ and will occur during 2005 (2.4 MM$) and 2006 (1.25 MM$). A detailed reservoir simulation model will be constructed and used to plan the Phase I resèrvoir development.

The existing facility infrastructure should be maximized with a total of four horizontal producers with some modifications. The modifications that are anticipated are purchasing a 50 MMBTU/hr steam generator, building a water plant to process produced water for generator feed water, adding well test equipment, modifying the heater treater with fire tubes, adding a free water knockout and installing a LACT. Facility costs are anticipated to total 1.6 MM$ and occur mainly during 2005 (1.5 MM$).

EXHIBIT "C"

ATTACHED TO AND MADE A PART OF THAT CERTAIN FARMIN AGREEMENT DATED JANUARY 15, 2004, BY AND BETWEEN IVANHOE ENERGY (USA) INC. (IVAN), AS OPERATOR AND DEREK RESOURCES (U.S.A.) INC., AS NON-OPERATOR.

A.A.P.L. FORM 610 – 1989

MODEL FORM OPERATING AGREEMENT

OPERATING AGREEMENT

DATED
JANUARY 15, 2004,
year

OPERATOR IVANHOE ENERGY (USA) INC.

CONTRACT AREA PURSUANT TO THE FARMIN AGREEMENT REFERENCED ABOVE

____

____

____

COUNTY OR PARISH OF WESTON, STATE OF WYOMING

A.A.P.L. NO. 610 – 1989

TABLE OF CONTENTS

OPERATING AGREEMENT

THIS AGREEMENT, entered into by and between IVANHOE ENERGY (USA) INC. ______________________________, hereinafter designated and referred to as "Operator," and the signatory party or parties other than Operator, sometimes hereinafter referred to individually as "Non-Operator," and collectively as "Non-Operators."

WITNESSETH:

WHEREAS, the parties to this agreement are owners of Oil and Gas Leases and/or Oil and Gas Interests in the land identified in Exhibit "A," and the parties hereto have reached an agreement to explore and develop these Leases and/or Oil and Gas Interests for the production of Oil and Gas to the extent and as hereinafter provided, NOW, THEREFORE, it is agreed as follows:

ARTICLE I
DEFINITIONS

As used in this agreement, the following words and terms shall have the meanings here ascribed to them:

A. The term "AFE" shall mean an Authority for Expenditure prepared by a party to this agreement for the purpose of estimating the costs to be incurred in conducting an operation hereunder.

B. The term "Completion" or "Complete" shall mean a single operation intended to complete a well as a producer of Oil and Gas in one or more Zones, including, but not limited to, the setting of production casing, perforating, well stimulation and production testing conducted in such operation.

C. The term "Contract Area" shall mean all of the lands, Oil and Gas Leases and/or Oil and Gas Interests intended to be developed and operated for Oil and Gas purposes under this agreement. Such lands, Oil and Gas Leases and Oil and Gas Interests are described in Exhibit "A."

D. The term "Deepen" shall mean a single operation whereby a well is drilled to an objective Zone below the deepest Zone in which the well was previously drilled, or below the Deepest Zone proposed in the associated AFE, whichever is the lesser.

E. The terms "Drilling Party" and "Consenting Party" shall mean a party who agrees to join in and pay its share of the cost of any operation conducted under the provisions of this agreement.

F. The term "Drilling Unit" shall mean the area fixed for the drilling of one well by order or rule of any state or federal body having authority. If a Drilling Unit is not fixed by any such rule or order, a Drilling Unit shall be the drilling unit as established by the pattern of drilling in the Contract Area unless fixed by express agreement of the Drilling Parties.

G. The term "Drillsite" shall mean the Oil and Gas Lease or Oil and Gas Interest on which a proposed well is to be located.

H. The term "Initial Well" shall mean the well required to be drilled by the parties hereto as provided in Article VI.A.

I. The term "Non-Consent Well" shall mean a well in which less than all parties have conducted an operation as provided in Article VI.B.2.

J. The terms "Non-Drilling Party" and "Non-Consenting Party" shall mean a party who elects not to participate in a proposed operation.

K. The term "Oil and Gas" shall mean oil, gas, casinghead gas, gas condensate, and/or all other liquid or gaseous hydrocarbons and other marketable substances produced therewith, unless an intent to limit the inclusiveness of this term is specifically stated.

L. The term "Oil and Gas Interests" or "Interests" shall mean unleased fee and mineral interests in Oil and Gas in tracts of land lying within the Contract Area which are owned by parties to this agreement.

M. The terms "Oil and Gas Lease," "Lease" and "Leasehold" shall mean the oil and gas leases or interests therein covering tracts of land lying within the Contract Area which are owned by the parties to this agreement.

N. The term "Plug Back" shall mean a single operation whereby a deeper Zone is abandoned in order to attempt a Completion in a shallower Zone.

O. The term "Recompletion" or "Recomplete" shall mean an operation whereby a Completion in one Zone is abandoned in order to attempt a Completion in a different Zone within the existing wellbore.

P. The term "Rework" shall mean an operation conducted in the wellbore of a well after it is Completed to secure, restore, or improve production in a Zone which is currently open to production in the wellbore. Such operations include, but are not limited to, well stimulation operations but exclude any routine repair or maintenance work or drilling, Sidetracking, Deepening, Completing, Recompleting, or Plugging Back of a well.

Q. The term "Sidetrack" shall mean the directional control and intentional deviation of a well from vertical so as to change the bottom hole location unless done to straighten the hole or drill around junk in the hole to overcome other mechanical difficulties.

R. The term "Zone" shall mean a stratum of earth containing or thought to contain a common accumulation of Oil and Gas separately producible from any other common accumulation of Oil and Gas. Unless the context otherwise clearly indicates, words used in the singular include the plural, the word "person" includes natural and artificial persons, the plural includes the singular, and any gender includes the masculine, feminine, and neuter.

ARTICLE II
EXHIBITS

The following exhibits, as indicated below and attached hereto, are incorporated in and made a part hereof:

A. Exhibit "A," shall include the following information:
   (1) Description of lands subject to this agreement,
   (2) Restrictions, if any, as to depths, formations, or substances,
   (3) Parties to agreement with addresses and telephone numbers for notice purposes,
   (4) Percentages or fractional interests of parties to this agreement,
   (5) Oil and Gas Leases and/or Oil and Gas Interests subject to this agreement,
   (6) Burdens on production.

C. Exhibit "C," Accounting Procedure.
D. Exhibit "D," Insurance.
E. Exhibit "E," Gas Balancing Agreement.
F. Exhibit "F," Non-Discrimination and Certification of Non-Segregated Facilities.
H. Other: MEMORANDUM OF OPERATING AGREEMENT

If any provision of any exhibit, except Exhibits "E," "F" and "G," is inconsistent with any provision contained in the body of this agreement, the provisions in the body of this agreement shall prevail.

ARTICLE III
INTERESTS OF PARTIES

B. INTERESTS OF PARTIES IN COSTS AND PRODUCTION

Unless changed by other provisions, all costs and liabilities incurred in operations under this agreement shall be borne and paid, and all equipment and materials acquired in operations on the Contract Area shall be owned, by the parties as their interests are set forth in Exhibit "A." In the same manner, the parties shall also own all production of Oil and Gas from the Contract Area subject, however, to the payment of royalties and other burdens on production as described hereafter.

Regardless of which party has contributed any Oil and Gas Lease or Oil and Gas Interest on which royalty or other burdens may be payable and except as otherwise expressly provided in this agreement, each party shall pay or deliver, or cause to be paid or delivered, all burdens on its share of the production from the Contract Area up to, but not in excess of, 20% and shall indemnify, defend and hold the other parties free from any liability therefor. Except as otherwise expressly provided in this agreement, if any party has contributed hereto any Lease or Interest which is burdened with any royalty, overriding royalty, production payment or other burden on production in excess of the amounts stipulated above, such party so burdened shall assume and alone bear all such excess obligations and shall indemnify, defend and hold the other parties hereto harmless from any and all claims attributable to such excess burden.

However, so long as the Drilling Unit for the productive Zone(s) is identical with the Contract Area, each party shall pay or deliver, or cause to be paid or delivered, all burdens on production from the Contract Area due under the terms of the Oil and Gas Lease(s) which such party has contributed to this agreement, and shall indemnify, defend and hold the other parties free from any liability therefor.

No party shall ever be responsible, on a price basis higher than the price received by such party, to any other party's lessor or royalty owner, and if such other party's lessor or royalty owner should demand and receive settlement on a higher price basis, the party contributing the affected Lease shall bear the additional royalty burden attributable to such higher price.

Nothing contained in this Article III.B. shall be deemed an assignment or cross-assignment of interests covered hereby, and in the event two or more parties contribute to this agreement jointly owned Leases, the parties' undivided interests in said Leaseholds shall be deemed separate leasehold interests for the purposes of this agreement.

C. SUBSEQUENTLY CREATED INTERESTS

If any party has contributed hereto a Lease or Interest that is burdened with an assignment of production given as security for the payment of money, or if, after the date of this agreement, any party creates an overriding royalty, production payment, net profits interest, assignment of production or other burden payable out of production attributable to its working interest hereunder, such burden shall be deemed a "Subsequently Created Interest." Further, if any party has contributed hereto a Lease or Interest burdened with an overriding royalty, production payment, net profits interests, or other burden payable out of production created prior to the date of this agreement, and such burden is not shown on Exhibit "A," such burden also shall be deemed a Subsequently Created Interest to the extent such burden causes the burdens on such party's Lease or Interest to exceed the amount stipulated in Article III.B. above.

The party whose interest is burdened with the Subsequently Created Interest (the "Burdened Party") shall assume and alone bear, pay and discharge the Subsequently Created Interest and shall indemnify, defend and hold harmless the other parties from and against any liability therefor. Further, if the Burdened Party fails to pay, when due, its share of expenses chargeable hereunder, all provisions of Article VII.B. shall be enforceable against the Subsequently Created Interest in the same manner as they are enforceable against the working interest of the Burdened Party. If the Burdened Party is required under this agreement to assign or relinquish to any other party, or parties, all or a portion of its working interest and/or the production attributable thereto, said other party, or parties, shall receive said assignment and/or production free and clear of said Subsequently Created Interest, and the Burdened Party shall indemnify, defend and hold harmless said other party, or parties, from any and all claims and demands for payment asserted by owners of the Subsequently Created Interest.

ARTICLE IV
TITLES

A. TITLE EXAMINATION

Title examination shall be made on the Drillsite of any proposed well prior to commencement of drilling operations and, if a majority in interest of the Drilling Parties so request or Operator so elects, title examination shall be made on the entire Drilling Unit, or maximum anticipated Drilling Unit, of the well. The opinion will include the ownership of the working interest, minerals, royalty, overriding royalty and production payments under the applicable Leases. Each party contributing Leases and/or Oil and Gas Interests to be included in the Drillsite or Drilling Unit, if appropriate, shall furnish to Operator all abstracts (including federal lease status reports), title opinions, title papers and curative material in its possession free of charge. All such information not in the possession of or made available to Operator by the parties, but necessary for the examination of the title, shall be obtained by Operator. Operator shall cause title to be examined by attorneys on its staff or by outside attorneys. Copies of all title opinions shall be furnished to each Drilling Party. Costs incurred by Operator in procuring abstracts, fees paid outside attorneys for title examination (including preliminary, supplemental, shut-in royalty opinions and division order title opinions) and other direct charges as provided in Exhibit "C" shall be borne by the Drilling Parties in the proportion that the interest of each Drilling Party bears to the total interest of all Drilling Parties as such interests appear in Exhibit "A." Operator shall make no charge for services rendered by its staff attorneys or other personnel in the performance of the above functions.

Each party shall be responsible for securing curative matter and pooling amendments or agreements required in connection with Leases or Oil and Gas Interests contributed by such party. Operator shall be responsible for the preparation and recording of pooling designations or declarations and communitization agreements as well as the conduct of hearings before governmental agencies for the securing of spacing or pooling orders or any other orders necessary or appropriate to the conduct of operations hereunder. This shall not prevent any party from appearing on its own behalf at such hearings.

Costs incurred by Operator, including fees paid to outside attorneys, which are associated with hearings before governmental agencies, and which costs are necessary and proper for the activities contemplated under this agreement, shall be direct charges to the joint account and shall not be covered by the administrative overhead charges as provided in Exhibit "C."

Operator shall make no charge for services rendered by its staff attorneys or other personnel in the performance of the above functions.

No well shall be drilled on the Contract Area until after (1) the title to the Drillsite or Drilling Unit, if appropriate, has been examined as above provided, and (2) the title has been approved by the examining attorney or title has been accepted by all of the Drilling Parties in such well.

B. LOSS OR FAILURE OF TITLE

3. Losses: All losses of Leases or Interests committed to this agreement, shall be joint losses and shall be borne by all parties in proportion to their interests shown on Exhibit "A." This shall include but not be limited to the loss of any Lease or Interest through failure to develop or because express or implied covenants have not been performed (other than performance which requires only the payment of money), and the loss of any Lease by expiration at the end of its primary term if it is not renewed or extended. There shall be no readjustment of interests in the remaining portion of the Contract Area on account of any joint loss.

ARTICLE V
OPERATOR

A. DESIGNATION AND RESPONSIBILITIES OF OPERATOR

IVANHOE ENERGY (USA) INC. shall be the Operator of the Contract Area, and shall conduct and direct and have full control of all operations on the Contract Area as permitted and required by, and within the limits of this agreement. In its performance of services hereunder for the Non-Operators, Operator shall be an independent contractor not subject to the control or direction of the Non-Operators except as to the type of operation to be undertaken in accordance with the election procedures contained in this agreement. Operator shall not be deemed, or hold itself out as, the agent of the Non-Operators with authority to bind them to any obligation or liability assumed or incurred by Operator as to any third party. Operator shall conduct its activities under this agreement as a reasonable prudent operator, in a good and workmanlike manner, with due diligence and dispatch, in accordance with good oilfield practice, and in compliance with applicable law and regulation, but in no event shall it have any liability as Operator to the other parties for losses sustained or liabilities incurred except such as may result from gross negligence or willful misconduct.

B. RESIGNATION OR REMOVAL OF OPERATOR AND SELECTION OF SUCCESSOR

1. Resignation or Removal of Operator: Operator may resign at any time by giving written notice thereof to Non-Operators. If Operator terminates its legal existence, no longer owns an interest hereunder in the Contract Area, or is no longer capable of serving as Operator, Operator shall be deemed to have

resigned without any action by Non-Operators, except the selection of a successor. Operator may be removed only for good cause by the affirmative vote of Non-Operators owning a majority interest based on ownership as shown on Exhibit "A" remaining after excluding the voting interest of Operator; such vote shall not be deemed effective until a written notice has been delivered to the Operator by a Non-Operator detailing the alleged default and Operator has failed to cure the default within thirty (30) days from its receipt of the notice or, if the default concerns an operation then being conducted, within forty-eight (48) hours of its receipt of the notice. For purposes hereof, "good cause" shall mean not only gross negligence or willful misconduct but also the material breach of or inability to meet the standards of operation contained in Article V.A. or material failure or inability to perform its obligations under this agreement.

Subject to Article VII.D.1., such resignation or removal shall not become effective until 7:00 o'clock A.M. on the first day of the calendar month following the expiration of ninety (90) days after the giving of notice of resignation by Operator or action by the Non-Operators to remove Operator, unless a successor Operator has been selected and assumes the duties of Operator at an earlier date. Operator, after effective date of resignation or removal, shall be bound by the terms hereof as a Non-Operator. A change of a corporate name or structure of Operator or transfer of Operator's interest to any single subsidiary, parent or successor corporation shall not be the basis for removal of Operator.

2. Selection of Successor Operator: Upon the resignation or removal of Operator under any provision of this agreement, a successor Operator shall be selected by the parties. The successor Operator shall be selected from the parties owning an interest in the Contract Area at the time such successor Operator is selected. The successor Operator shall be selected by the affirmative vote of two (2) or more parties owning a majority interest based on ownership as shown on Exhibit "A"; provided, however, if an Operator which has been removed or is deemed to have resigned fails to vote or votes only to succeed itself, the successor Operator shall be selected by the affirmative vote of the party or parties owning a majority interest based on ownership as shown on Exhibit "A" remaining after excluding the voting interest of the Operator that was removed or resigned. The former Operator shall promptly deliver to the successor Operator all records and data relating to the operations conducted by the former Operator to the extent such records and data are not already in the possession of the successor operator. Any cost of obtaining or copying the former Operator's records and data shall be charged to the joint account.

3. Effect of Bankruptcy: If Operator becomes insolvent, bankrupt or is placed in receivership, it shall be deemed to have resigned without any action by Non-Operators, except the selection of a successor. If a petition for relief under the federal bankruptcy laws is filed by or against Operator, and the removal of Operator is prevented by the federal bankruptcy court, all Non-

Operators and Operator shall comprise an interim operating committee to serve until Operator has elected to reject or assume this agreement pursuant to the Bankruptcy Code, and an election to reject this agreement by Operator as a debtor in possession, or by a trustee in bankruptcy, shall be deemed a resignation as Operator without any action by Non-Operators, except the selection of a successor. During the period of time the operating committee controls operations, all actions shall require the approval of two (2) or more parties owning a majority interest based on ownership as shown on Exhibit "A." In the event there are only two (2) parties to this agreement, during the period of time the operating committee controls operations, a third party acceptable to Operator, Non-Operator and the federal bankruptcy court shall be selected as a member of the operating committee, and all actions shall require the approval of two (2) members of the operating committee without regard for their interest in the Contract Area based on Exhibit "A."

C. EMPLOYEES AND CONTRACTORS

The number of employees or contractors used by Operator in conducting operations hereunder, their selection, and the hours of labor and the compensation for services performed shall be determined Operator, and all such employees or contractors shall be the employees or contractors of Operator.

D. RIGHTS AND DUTIES OF OPERATOR

1. Competitive Rates and Use of Affiliates: All wells drilled on the Contract Area shall be drilled on a competitive contract basis at the usual rates prevailing in the area. If it so desires, Operator may employ its own tools and equipment in the drilling of wells, but its charges therefor shall not exceed the prevailing rates in the area and the rate of such charges shall be agreed upon by the parties in writing before drilling operations are commenced, and such work shall be performed by Operator under the same terms and conditions as are customary and usual in the area in contracts of independent contractors who are doing work of a similar nature. All work performed or materials supplied by affiliates or related parties of Operator shall be performed or supplied at competitive rates, pursuant to written agreement, and in accordance with customs and standards prevailing in the industry.

2. Discharge of Joint Account Obligations: Except as herein otherwise specifically provided, Operator shall promptly pay and discharge expenses incurred in the development and operation of the Contract Area pursuant to this agreement and shall charge each of the parties hereto with their respective proportionate shares upon the expense basis provided in Exhibit "C." Operator shall keep an accurate record of the joint account hereunder, showing expenses incurred and charges and credits made and received.

3. Protection from Liens: Operator shall pay, or cause to be paid, as and when they become due and payable, all accounts of contractors and suppliers and wages and salaries for services rendered or performed, and for materials supplied on, to or in respect of the Contract Area or any operations for the joint account thereof, and shall keep the Contract Area free from liens and encumbrances resulting therefrom except for those resulting from a bona fide dispute as to services rendered or materials supplied.

4. Custody of Funds: Operator shall hold for the account of the Non-Operators any funds of the Non-Operators advanced or paid to the Operator, either for the conduct of operations hereunder or as a result of the sale of production from the Contract Area, and such funds shall remain the funds of the Non-Operators on whose account they are advanced or paid until used for their intended purpose or otherwise delivered to the Non-Operators or applied toward the payment of debts as provided in Article VII.B. Nothing in this paragraph shall be construed to establish a fiduciary relationship between Operator and Non-Operators for any purpose other than to account for Non-Operator funds as herein specifically provided. Nothing in this paragraph shall require the maintenance by Operator of separate accounts for the funds of Non-Operators unless the parties otherwise specifically agree.

5. Access to Contract Area and Records: Operator shall, except as otherwise provided herein, permit each Non-Operator or its duly authorized representative, at the Non-Operator's sole risk and cost, full and free access at all reasonable times to all operations of every kind and character being conducted for the joint account on the Contract Area and to the records of operations conducted thereon or production therefrom, including Operator's books and records relating thereto. Such access rights shall not be exercised in a manner interfering with Operator's conduct of an operation hereunder and shall not obligate Operator to furnish any geologic or geophysical data of an interpretive nature unless the cost of preparation of such interpretive data was charged to the joint account. Operator will furnish to each Non-Operator upon request copies of any and all reports and information obtained by Operator in connection with production and related items, including, without limitation, meter and chart reports, production purchaser statements, run tickets and monthly gauge reports, but excluding purchase contracts and pricing information to the extent not applicable to the production of the Non-Operator seeking the information. Any audit of Operator's records relating to amounts expended and the appropriateness of such expenditures shall be conducted in accordance with the audit protocol specified in Exhibit "C."

6. Filing and Furnishing Governmental Reports: Operator will file, and upon written request promptly furnish copies to each requesting Non-Operator not in default of its payment obligations, all operational notices, reports or applications required to be filed by local, State, Federal or Indian agencies or authorities having jurisdiction over operations hereunder. Each Non-Operator shall provide to Operator on a timely basis all information necessary to Operator to make such filings.

7. Drilling and Testing Operations: The following provisions shall apply to each well drilled hereunder, including but not limited to the Initial Well:

(a) Operator will promptly advise Non-Operators of the date on which the well is spudded, or the date on which drilling operations are commenced.

(b) Operator will send to Non-Operators such reports, test results and notices regarding the progress of operations on the well as the Non-Operators shall reasonably request, including, but not limited to, daily drilling reports, completion reports, and well logs.

(c) Operator shall adequately test all Zones encountered which may reasonably be expected to be capable of producing Oil and Gas in paying quantities as a result of examination of the electric log or any other logs or cores or tests conducted hereunder.

8. Cost Estimates: Upon request of any Consenting Party, Operator shall furnish estimates of current and cumulative costs incurred for the joint account at reasonable intervals during the conduct of any operation pursuant to this agreement. Operator shall not be held liable for errors in such estimates so long as the estimates are made in good faith.

9. Insurance: At all times while operations are conducted hereunder, Operator shall comply with the workers compensation law of the state where the operations are being conducted; provided, however, that Operator may be a self-insurer for liability under said compensation laws in which event the only charge that shall be made to the joint account shall be as provided in Exhibit "C." Operator shall also carry or provide insurance for the benefit of the joint account of the parties as outlined in Exhibit "D" attached hereto and made a part hereof. Operator shall require all contractors engaged in work on or for the Contract Area to comply with the workers compensation law of the state where the operations are being conducted and to maintain such other insurance as Operator may require.

In the event automobile liability insurance is specified in said Exhibit "D," or subsequently receives the approval of the parties, no direct charge shall be made by Operator for premiums paid for such insurance for Operator's automotive equipment.

ARTICLE VI
DRILLING AND DEVELOPMENT

Pursuant to the Farmin Agreement to which this Operating Agreement is attached.

1. Proposed Operations: If any party hereto should desire to drill any well on the Contract Area other than the Initial Well, or if any party should desire to Rework, Sidetrack, Deepen, Recomplete or Plug Back a dry hole or a well no longer capable of producing in paying quantities in which such party has not otherwise relinquished its interest in the proposed objective Zone under this agreement, the party desiring to drill, Rework, Sidetrack, Deepen, Recomplete or Plug Back such a well shall give written notice of the proposed operation to the parties who have not otherwise relinquished their interest in such objective Zone under this agreement and to all other parties in the case of a proposal for Sidetracking or Deepening, specifying the work to be performed, the location, proposed depth, objective Zone and the estimated cost of the operation. The parties to whom such a notice is delivered shall have thirty (30) days after receipt of the notice within which to notify the party proposing to do the work whether they elect to participate in the cost of the proposed operation. If a drilling rig is on location, notice of a proposal to Rework, Sidetrack, Recomplete, Plug Back or Deepen may be given by telephone and the response period shall be limited to forty-eight (48) hours, exclusive of Saturday, Sunday and legal holidays. Failure of a party to whom such notice is delivered to reply within the period above fixed shall constitute an election by that party not to participate in the cost of the proposed operation. Any proposal by a party to conduct an operation conflicting with the operation initially proposed shall be delivered to all parties within the time and in the manner provided in Article VI.B.6.

If all parties to whom such notice is delivered elect to participate in such a proposed operation, the parties shall be contractually committed to participate therein provided such operations are commenced within the time period hereafter set forth, and Operator shall, no later than ninety (90) days after expiration of the notice period of thirty (30) days (or as promptly as practicable after the expiration of the forty-eight (48) hour period when a drilling rig is on location, as the case may be), actually commence the proposed operation and thereafter complete it with due diligence at the risk and expense of the parties participating therein; provided, however, said commencement date may be extended upon written notice of same by Operator to the other parties, for a period of up to thirty (30) additional days if, in the sole opinion of Operator, such additional time is reasonably necessary to obtain permits from governmental authorities, surface rights (including rights-of-way) or appropriate drilling equipment, or to complete title examination or curative matter required for title approval or acceptance. If the actual operation has not been commenced within the time provided (including any extension thereof as specifically permitted herein or in the force majeure provisions of Article XI) and if any party hereto still desires to conduct said operation, written notice proposing same must be resubmitted to the other parties in accordance herewith as if no prior proposal had been made. Those parties that did not participate in the drilling of a well for which a proposal to Deepen or Sidetrack is made hereunder shall, if such parties desire to participate in the proposed Deepening or Sidetracking operation, reimburse the Drilling Parties in accordance with Article VI.B.4. in the event of a Deepening operation and in accordance with Article VI.B.5. in the event of a Sidetracking operation.

2. Operations by Less Than All Parties: (a) Determination of Participation. If any party to whom such notice is delivered as provided in Article VI.B.1. or VI.C.1. (Option No. 2) elects not to participate in the proposed operation, then, in order to be entitled to the benefits of this Article, the party or parties giving the notice and such other parties as shall elect to participate in the operation shall, no later than ninety (90) days after the expiration of the notice period of thirty (30) days (or as promptly as practicable after the expiration of the forty-eight (48) hour period when a drilling rig is on location, as the case may be) actually commence the proposed operation and complete it with due diligence. Operator shall perform all work for the account of the Consenting Parties; provided, however, if no drilling rig or other equipment is on location, and if Operator is a Non-Consenting Party, the Consenting Parties shall either: (i) request Operator to perform the work required by such proposed operation for the account of the Consenting Parties, or (ii) designate one of the Consenting Parties as Operator to perform such work. The rights and duties granted to and imposed upon the Operator under this agreement are granted to and imposed upon the party designated as Operator for an operation in which the original Operator is a Non-Consenting Party. Consenting Parties, when conducting operations on the Contract Area pursuant to this Article VI.B.2., shall comply with all terms and conditions of this agreement.

If less than all parties approve any proposed operation, the proposing party, immediately after the expiration of the applicable notice period, shall advise all Parties of the total interest of the parties approving such operation and its recommendation as to whether the Consenting Parties should proceed with the operation as proposed. Each Consenting Party, within forty-eight (48) hours (exclusive of Saturday, Sunday and legal holidays) after receipt of such notice, shall advise the proposing party of its desire to (i) limit participation to such party's interest as shown on Exhibit "A" or (ii) carry only its proportionate part (determined by dividing such party's interest in the Contract Area by the interests of all Consenting Parties in the Contract Area) of Non-Consenting Parties' interests, or (iii) carry its proportionate part (determined as provided in (ii)) of Non-Consenting Parties' interests together with all or a portion of its proportionate part of any Non-Consenting Parties' interests that any Consenting Party did not elect to take. Any interest of Non-Consenting Parties that is not carried by a Consenting Party shall be deemed to be carried by the party proposing the operation if such party does not withdraw its proposal. Failure to advise the proposing party within the time required shall be deemed an election under (i). In the event a drilling rig is on location, notice may be given by telephone, and the time permitted for such a response shall not exceed a total of forty-eight (48) hours (exclusive of Saturday, Sunday and legal holidays). The proposing party, at its election, may withdraw such proposal if there is less than 100% participation and shall notify all parties of such decision within ten (10) days, or within twenty-four (24) hours if a drilling rig is on location, following expiration of the applicable response period. If 100% subscription to the proposed operation is obtained, the proposing party shall promptly notify the Consenting Parties of their proportionate interests in the operation and the party serving as Operator shall commence such operation within the period provided in Article VI.B.1., subject to the same extension right as provided therein.

(b) Relinquishment of Interest for Non-Participation. The entire cost and risk of conducting such operations shall be borne by the Consenting Parties in the proportions they have elected to bear same under the terms of the preceding paragraph. Consenting Parties shall keep the leasehold estates involved in such operations free and clear of all liens and encumbrances of every kind created by or arising from the operations of the Consenting Parties. If such an operation results in a dry hole, then subject to Articles VI.B.6. and VI.E.3., the Consenting Parties shall plug and abandon the well and restore the surface location at their sole cost, risk and expense; provided, however, that those Non-Consenting Parties that participated in the drilling, Deepening or Sidetracking of the well shall remain liable for, and shall pay, their proportionate shares of the cost of plugging and abandoning the well and restoring the surface location insofar only as those costs were not increased by the subsequent operations of the Consenting Parties. If any well drilled, Reworked, Sidetracked, Deepened, Recompleted or Plugged Back under the provisions of this Article results in a well capable of producing Oil and/or Gas in paying quantities, the Consenting Parties shall Complete and equip the well to produce at their sole cost and risk, and the well shall then be turned over to Operator (if the Operator did not conduct the operation) and shall be operated by it at the expense and for the account of the Consenting Parties. Upon commencement of operations for the drilling, Reworking, Sidetracking, Recompleting, Deepening or Plugging Back of any such well by Consenting Parties in accordance with the provisions of this Article, each Non-Consenting Party shall be deemed to have relinquished to Consenting Parties, and the Consenting Parties shall own and be entitled to receive, in proportion to their respective interests, all of such Non-Consenting Party's interest in the well and share of production therefrom or, in the case of a Reworking, Sidetracking, Deepening, Recompleting or Plugging Back, or a Completion pursuant to Article VI.C.1. Option No. 2, all of such Non-Consenting Party's interest in the production obtained from the operation in which the Non-Consenting Party did not elect to participate. Such relinquishment shall be effective until the proceeds of the sale of such share, calculated at the well, or market value thereof if such share is not sold (after deducting applicable ad valorem, production, severance, and excise taxes, royalty, overriding royalty and other interests not excepted by Article III.C. payable out of or measured by the production from such well accruing with respect to such interest until it reverts), shall equal the total of the following:

(i) 100% of each such Non-Consenting Party's share of the cost of any newly acquired surface equipment beyond the wellhead connections (including but not limited to stock tanks, separators, treaters, pumping equipment and piping), plus 100% of each such Non-Consenting Party's share of the cost of operation of the well commencing with first production and continuing until each such Non-Consenting Party's relinquished interest shall revert to it under other provisions of this Article, it being agreed that each Non-Consenting Party's share of such costs and equipment will be that interest which would have been chargeable to such Non-Consenting Party had it participated in the well from the beginning of the operations; and

(ii) 300% of (a) that portion of the costs and expenses of drilling, Reworking, Sidetracking, Deepening, Plugging Back, testing, Completing, and Recompleting, after deducting any cash contributions received under Article VIII.C., and of (b) that portion of the cost of newly acquired equipment in the well (to and including the wellhead connections), which would have been chargeable to such Non-Consenting Party if it had participated therein.

Notwithstanding anything to the contrary in this Article VI.B., if the well does not reach the deepest objective Zone described in the notice proposing the well for reasons other than the encountering of granite or practically impenetrable substance or other condition in the hole rendering further operations impracticable, Operator shall give notice thereof to each Non-Consenting Party who submitted or voted for an alternative proposal under Article VI.B.6. to drill the well to a shallower Zone than the deepest objective Zone proposed in the notice under which the well was drilled, and each such Non- Consenting Party shall have the option to participate in the initial proposed Completion of the well by paying its share of the cost of drilling the well to its actual depth, calculated in the manner provided in Article VI.B.4. (a). If any such Non-Consenting Party does not elect to participate in the first Completion proposed for such well, the relinquishment provisions of this Article VI.B.2. (b) shall apply to such party's interest.

(c) Reworking, Recompleting or Plugging Back. An election not to participate in the drilling, Sidetracking or Deepening of a well shall be deemed an election not to participate in any Reworking, Recompletion or Plugging Back operation proposed in such a well, or portion thereof, to which the initial non-consent election applied that is conducted at any time prior to full recovery by the Consenting Parties of the Non-Consenting Party's recoupment amount. Similarly, an election not to participate in the Completing or Recompleting of a well shall be deemed an election not to participate in any Reworking operation proposed in such a well, or portion thereof, to which the initial non-consent election applied that is conducted at any time prior to full recovery by the Consenting Parties of the Non-Consenting Party's recoupment amount. Any such Reworking, Recompleting or Plugging Back operation conducted during the recoupment period shall be deemed part of the cost of operation of said well and there shall be added to the sums to be recouped by the Consenting Parties 200% of that portion of the costs of the Reworking, Recompleting or Plugging Back operation which would have been chargeable to such Non-Consenting Party had it participated therein. If such a Reworking, Recompleting or Plugging Back operation is proposed during such recoupment period, the provisions of this Article VI.B. shall be applicable as between said Consenting Parties in said well.

(d) Recoupment Matters. During the period of time Consenting Parties are entitled to receive Non-Consenting Party's share of production, or the proceeds therefrom, Consenting Parties shall be responsible for the payment of all ad valorem, production, severance, excise, gathering and other taxes, and all royalty, overriding royalty and other burdens applicable to Non-Consenting Party's share of production not excepted by Article III.C.

In the case of any Reworking, Sidetracking, Plugging Back, Recompleting or Deepening operation, the Consenting Parties shall be permitted to use, free of cost, all casing, tubing and other equipment in the well, but the ownership of all such equipment shall remain unchanged; and upon abandonment of a well after such Reworking, Sidetracking, Plugging Back, Recompleting or Deepening, the Consenting Parties shall account for all such equipment to the owners thereof, with each party receiving its proportionate part in kind or in value, less cost of salvage. Within ninety (90) days after the completion of any operation under this Article, the party conducting the operations for the Consenting Parties shall furnish each Non-Consenting Party with an inventory of the equipment in and connected to the well, and an itemized statement of the cost of drilling, Sidetracking, Deepening, Plugging Back, testing, Completing, Recompleting, and equipping the well for production; or, at its option, the operating party, in lieu of an itemized statement of such costs of operation, may submit a detailed statement of monthly billings. Each month thereafter, during the time the Consenting Parties are being reimbursed as provided above, the party conducting the operations for the Consenting Parties shall furnish the Non-Consenting Parties with an itemized statement of all costs and liabilities incurred in the operation of the well, together with a statement of the quantity of Oil and Gas produced from it and the amount of proceeds realized from the sale of the well's working interest production during the preceding month. In determining the quantity of Oil and Gas produced during any month, Consenting Parties shall use industry accepted methods such as but not limited to metering or periodic well tests. Any amount realized from the sale or other disposition of equipment newly acquired in connection with any such operation which would have been owned by a Non-Consenting Party had it participated therein shall be credited against the total unreturned costs of the work done and of the equipment purchased in determining when the interest of such Non-Consenting Party shall revert to it as above provided; and if there is a credit balance, it shall be paid to such Non-Consenting Party.

If and when the Consenting Parties recover from a Non-Consenting Party's relinquished interest the amounts provided for above, the relinquished interests of such Non-Consenting Party shall automatically revert to it as of 7:00 a.m. on the day following the day on which such recoupment occurs, and, from and after such reversion, such Non-Consenting Party shall own the same interest in such well, the material and equipment in or pertaining thereto, and the production therefrom as such Non-Consenting Party would have been entitled to had it participated in the drilling, Sidetracking, Reworking, Deepening, Recompleting or Plugging Back of said well. Thereafter, such Non-Consenting Party shall be charged with and shall pay its proportionate part of the further costs of the operation of said well in accordance with the terms of this agreement and Exhibit "C" attached hereto.

3. Stand-By Costs: When a well which has been drilled or Deepened has reached its authorized depth and all tests have been completed and the results thereof furnished to the parties, or when operations on the well have been otherwise terminated pursuant to Article VI.F., stand-by costs incurred pending response to a party's notice proposing a Reworking, Sidetracking, Deepening, Recompleting, Plugging Back or Completing operation in such a well (including the period required under Article VI.B.6. to resolve competing proposals) shall be charged and borne as part of the drilling or Deepening operation just completed. Stand-by costs subsequent to all parties responding, or expiration of the response time permitted, whichever first occurs, and prior to agreement as to the participating interests of all

Consenting Parties pursuant to the terms of the second grammatical paragraph of Article VI.B.2. (a), shall be charged to and borne as part of the proposed operation, but if the proposal is subsequently withdrawn because of insufficient participation, such stand-by costs shall be allocated between the Consenting Parties in the proportion each Consenting Party's interest as shown on Exhibit "A" bears to the total interest as shown on Exhibit "A" of all Consenting Parties.

In the event that notice for a Sidetracking operation is given while the drilling rig to be utilized is on location, any party may request and receive up to five (5) additional days after expiration of the forty-eight hour response period specified in Article VI.B.1. within which to respond by paying for all stand-by costs and other costs incurred during such extended response period; Operator may require such party to pay the estimated stand-by time in advance as a condition to extending the response period. If more than one party elects to take such additional time to respond to the notice, standby costs shall be allocated between the parties taking additional time to respond on a day-to-day basis in the proportion each electing party's interest as shown on Exhibit "A" bears to the total interest as shown on Exhibit "A" of all the electing parties.

4. Deepening: If less than all parties elect to participate in a drilling, Sidetracking, or Deepening operation proposed pursuant to Article VI.B.1., the interest relinquished by the Non-Consenting Parties to the Consenting Parties under Article VI.B.2. shall relate only and be limited to the lesser of (i) the total depth actually drilled or (ii) the objective depth or Zone of which the parties were given notice under Article VI.B.1. ("Initial Objective"). Such well shall not be Deepened beyond the Initial Objective without first complying with this Article to afford the Non-Consenting Parties the opportunity to participate in the Deepening operation.

In the event any Consenting Party desires to drill or Deepen a Non-Consent Well to a depth below the Initial Objective, such party shall give notice thereof, complying with the requirements of Article VI.B.1., to all parties (including Non-Consenting Parties). Thereupon, Articles VI.B.1. and 2. shall apply and all parties receiving such notice shall have the right to participate or not participate in the Deepening of such well pursuant to said Articles VI.B.1. and 2. If a Deepening operation is approved pursuant to such provisions, and if any Non-Consenting Party elects to participate in the Deepening operation, such Non-Consenting party shall pay or make reimbursement (as the case may be) of the following costs and expenses.

(a) If the proposal to Deepen is made prior to the Completion of such well as a well capable of producing in paying quantities, such Non-Consenting Party shall pay (or reimburse Consenting Parties for, as the case may be) that share of costs and expenses incurred in connection with the drilling of said well from the surface to the Initial Objective which Non-Consenting Party would have paid had such Non-Consenting Party agreed to participate therein, plus the Non-Consenting Party's share of the cost of Deepening and of participating in any further operations on the well in accordance with the other provisions of this Agreement; provided, however, all costs for testing and Completion or attempted Completion of the well incurred by Consenting Parties prior to the point of actual operations to Deepen beyond the Initial Objective shall be for the sole account of Consenting Parties.

(b) If the proposal is made for a Non-Consent Well that has been previously Completed as a well capable of producing in paying quantities, but is no longer capable of producing in paying quantities, such Non-Consenting Party shall pay (or reimburse Consenting Parties for, as the case may be) its proportionate share of all costs of drilling, Completing, and equipping said well from the surface to the Initial Objective, calculated in the manner provided in paragraph (a) above, less those costs recouped by the Consenting Parties from the sale of production from the well. The Non-Consenting Party shall also pay its proportionate share of all costs of re-entering said well. The Non-Consenting Parties' proportionate part (based on the percentage of such well Non-Consenting Party would have owned had it previously participated in such Non-Consent Well) of the costs of salvable materials and equipment remaining in the hole and salvable surface equipment used in connection with such well shall be determined in accordance with Exhibit "C." If the Consenting Parties have recouped the cost of drilling, Completing, and equipping the well at the time such Deepening operation is conducted, then a Non-Consenting Party may participate in the Deepening of the well with no payment for costs incurred prior to re-entering the well for Deepening The foregoing shall not imply a right of any Consenting Party to propose any Deepening for a Non-Consent Well prior to the drilling of such well to its Initial Objective without the consent of the other Consenting Parties as provided in Article VI.F.

5. Sidetracking: Any party having the right to participate in a proposed

Sidetracking operation that does not own an interest in the affected wellbore at the time of the notice shall, upon electing to participate, tender to the wellbore owners its proportionate share (equal to its interest in the Sidetracking operation) of the value of that portion of the existing wellbore to be utilized as follows:

(a) If the proposal is for Sidetracking an existing dry hole, reimbursement shall be on the basis of the actual costs incurred in the initial drilling of the well down to the depth at which the Sidetracking operation is initiated.

(b) If the proposal is for Sidetracking a well which has previously produced, reimbursement shall be on the basis of such party's proportionate share of drilling and equipping costs incurred in the initial drilling of the well down to the depth at which the Sidetracking operation is conducted, calculated in the manner described in Article VI.B.4(b) above. Such party's proportionate share of the cost of the well's salvable materials and equipment down to the depth at which the Sidetracking operation is initiated shall be determined in accordance with the provisions of Exhibit "C."

6. Order of Preference of Operations. Except as otherwise specifically provided in this agreement, if any party desires to propose the conduct of an operation that conflicts with a proposal that has been made by a party under this Article VI, such party shall have fifteen (15) days from delivery of the initial proposal, in the case of a proposal to drill a well or to perform an operation on a well where no drilling rig is on location, or twenty-four (24) hours, (exclusive of Saturday, Sunday and legal holidays), from delivery of the initial proposal, if a drilling rig is on location for the well on which such operation is to be conducted, to deliver to all parties entitled to participate in the proposed operation such party's alternative proposal, such alternate proposal to contain the same information required to be included in the initial proposal. Each party receiving such proposals shall elect by delivery of notice to Operator within five (5) days after expiration of the proposal period, or within twenty-four (24) hours (exclusive of Saturday, Sunday and legal holidays) if a drilling rig is on location for the well that is the subject of the proposals, to participate in one of the competing proposals. Any party not electing within the time required shall be deemed not to have voted. The proposal receiving the vote of parties owning the largest aggregate percentage interest of the parties voting shall have priority over all other competing proposals; in the case of a tie vote, the initial proposal shall prevail. Operator shall deliver notice of such result to all parties entitled to participate in the operation within five (5) days after expiration of the election period (or within twenty-four (24) hours, exclusive of Saturday, Sunday and legal holidays, if a drilling rig is on location). Each party shall then have two (2) days (or twenty-four (24) hours if a rig is on location) from receipt of such notice to elect by delivery of notice to Operator to participate in such operation or to relinquish interest in the affected well pursuant to the provisions of Article VI.B.2.; failure by a party to deliver notice within such period shall be deemed an election not to participate in the prevailing proposal.

7. Paying Wells. No party shall conduct any Reworking, Deepening, Plugging Back, Completion, Recompletion, or Sidetracking operation under this agreement with respect to any well then capable of producing in paying quantities except with the consent of parties with majority interest in percentage of that have not relinquished interests in the well at the time of such operation.

C. COMPLETION OF WELLS; REWORKING AND PLUGGING BACK

1. Completion: Without the consent of all parties, no well shall be drilled, Deepened or Sidetracked, except any well drilled, Deepened or Sidetracked pursuant to the provisions of Article VI.B.2. of this agreement. Consent to the drilling, Deepening or Sidetracking shall include:

Option No. 2: All necessary expenditures for the drilling, Deepening or Sidetracking and testing of the well. When such well has reached its authorized depth, and all logs, cores and other tests have been completed, and the results thereof furnished to the parties, Operator shall give immediate notice to the Non-Operators having the right to participate in a Completion attempt whether or not Operator recommends attempting to Complete the well, together with Operator's AFE for Completion costs if not previously provided. The parties receiving such notice shall have forty-eight (48) hours (exclusive of Saturday, Sunday and legal holidays) in which to elect by delivery of notice to Operator to participate in a recommended Completion attempt or to make a Completion proposal with an accompanying AFE. Operator shall deliver any such Completion proposal, or any Completion proposal conflicting with Operator's proposal, to the other parties entitled to participate in such Completion in accordance with the procedures specified in Article VI.B.6. Election to participate in a Completion attempt shall include consent to all necessary expenditures for the Completing and equipping of such well, including necessary tankage and/or surface facilities but excluding any stimulation operation not contained on the Completion AFE. Failure of any party receiving such notice to reply within the period above fixed shall constitute an election by that party not to participate in the cost of the Completion attempt; provided, that Article VI.B.6. shall control in the case of conflicting Completion proposals. If one or more, but less than all of the parties, elect to attempt a Completion, the provision of Article VI.B.2. hereof (the phrase "Reworking, Sidetracking, Deepening, Recompleting or Plugging Back" as contained in Article VI.B.2. shall be deemed to include "Completing") shall apply to the operations thereafter conducted by less than all parties; provided, however, that Article VI.B.2. shall apply separately to each separate Completion or Recompletion attempt undertaken hereunder, and an election to become a Non-Consenting Party as to one Completion or Recompletion attempt shall not prevent a party from becoming a Consenting Party in subsequent Completion or Recompletion attempts regardless whether the Consenting Parties as to earlier Completions or Recompletion have recouped their costs pursuant to Article VI.B.2.; provided further, that any recoupment of costs by a Consenting Party shall be made solely from the production attributable to the Zone in which the Completion attempt is made. Election by a previous Non-Consenting party to participate in a subsequent Completion or Recompletion attempt shall require such party to pay its proportionate share of the cost of salvable materials and equipment installed in the well pursuant to the previous Completion or Recompletion attempt, insofar and only insofar as such materials and equipment benefit the Zone in which such party participates in a Completion attempt.

2. Rework, Recomplete or Plug Back: No well shall be Reworked, Recompleted or Plugged Back except a well Reworked, Recompleted, or Plugged Back pursuant to the provisions of Article VI.B.2. of this agreement. Consent to the Reworking, Recompleting or Plugging Back of a well shall include all necessary expenditures in conducting such operations and Completing and equipping of said well, including necessary tankage and/or surface facilities.

D. OTHER OPERATIONS

Operator shall not undertake any single project reasonably estimated to require an expenditure in excess of TWENTY-FIVE THOUSAND Dollars ($25,000.00) except in connection with the drilling, Sidetracking, Reworking, Deepening, Completing, Recompleting or Plugging Back of a well that has been previously authorized by or pursuant to this agreement; provided, however, that, in case of explosion, fire, flood or other sudden emergency, whether of the same or different nature, Operator may take such steps and incur such expenses as in its opinion are required to deal with the emergency to safeguard life and property but Operator, as promptly as possible, shall report the emergency to the other parties. If Operator prepares an AFE for its own use, Operator shall furnish any Non-Operator so requesting an information copy thereof for any single project costing in excess of TWENTY-FIVE THOUSAND Dollars ($25,000.00). Any party who has not relinquished its interest in a well shall have the right to propose that Operator perform repair work or undertake the installation of artificial lift equipment or ancillary production facilities such as salt water disposal wells or to conduct additional work with respect to a well drilled hereunder or other similar project (but not including the installation of gathering lines or other transportation or marketing facilities, the installation of which shall be governed by separate agreement between the parties) reasonably estimated to require an expenditure in excess of the amount first set forth above in this Article VI.D. (except in connection with an operation required to be proposed under Articles VI.B.1. or VI.C.1. Option No. 2, which shall be governed exclusively be those Articles). Operator shall deliver such proposal to all parties entitled to participate therein. If within thirty (30) days thereof Operator secures the written consent of any party or parties owning at least 51% of the interests of the parties entitled to participate in such operation, each party having the right to participate in such project shall be bound by the terms of such proposal and shall be obligated to pay its proportionate share of the costs of the proposed project as if it had consented to such project pursuant to the terms of the proposal.

E. ABANDONMENT OF WELLS

1. Abandonment of Dry Holes: Except for any well drilled or Deepened pursuant to Article VI.B.2., any well which has been drilled or Deepened under the terms of this agreement and is proposed to be completed as a dry hole shall not be plugged and abandoned without the consent of all parties. Should Operator, after diligent effort, be unable to contact any party, or should any party fail to reply within forty-eight (48) hours (exclusive of Saturday, Sunday and legal holidays) after delivery of notice of the proposal to plug and abandon such well, such party shall be deemed to have consented to the proposed abandonment. All such wells shall be plugged and abandoned in accordance with applicable regulations and at the cost, risk and expense of the parties who participated in the cost of drilling or Deepening such well. Any party who objects to plugging and abandoning such well by notice delivered to Operator within forty-eight (48) hours (exclusive of Saturday, Sunday and legal holidays) after delivery of notice of the proposed plugging shall take over the well as of the end of such forty-eight (48) hour notice period and conduct further operations in search of Oil and/or Gas subject to the provisions of Article VI.B.; failure of such party to provide proof reasonably satisfactory to Operator of its financial capability to conduct such operations or to take over the well within such period or thereafter to conduct operations on such well or plug and abandon such well shall entitle Operator to retain or take possession of the well and plug and abandon the well. The party taking over the well shall indemnify Operator (if Operator is an abandoning party) and the other abandoning parties against liability for any further operations conducted on such well except for the costs of plugging and abandoning the well and restoring the surface, for which the abandoning parties shall remain proportionately liable.

2. Abandonment of Wells. That Have Produced: Except for any well in which a Non-Consent operation has been conducted hereunder for which the Consenting Parties have not been fully reimbursed as herein provided, any well which has been completed as capable of production, whether or not such well has produced, shall not be plugged and abandoned without the consent of all parties. If all parties consent to such abandonment, the well shall be plugged and abandoned in accordance with applicable regulations and at the cost, risk and expense of all the parties hereto. Failure of a party to reply within sixty (60) days of delivery of notice of proposed abandonment shall be deemed an election to consent to the proposal. If, within sixty (60) days after delivery of notice of the proposed abandonment of any well, all parties do not agree to the abandonment of such well, those wishing to continue its operation from the Zone then open to production shall be obligated to take over the well as of the expiration of the applicable notice period and shall indemnify Operator (if Operator is an abandoning party) and the other abandoning parties against liability for any further operations on the well conducted by such parties. Failure of such party or parties to provide proof reasonably satisfactory to Operator of their financial capability to conduct such operations or to take over the well within the required period or thereafter to conduct operations on such well shall entitle operator to retain or take possession of such well and plug and abandon the well.

Parties taking over a well as provided herein shall tender to each of the other parties its proportionate share of the value of the well's salvable material and equipment, determined in accordance with the provisions of Exhibit "C," less the estimated cost of salvaging and the estimated cost of plugging and abandoning and restoring the surface; provided, however, that in the event the estimated plugging and abandoning and surface restoration costs and the estimated cost of salvaging are higher than the value of the well's salvable material and equipment, each of the abandoning parties shall tender to the parties continuing operations their proportionate shares of the estimated excess cost. Each abandoning party shall assign to the non-abandoning parties, without warranty, express or implied, as to title or as to quantity, or fitness for use of the equipment and material, all of its interest in the wellbore of the well and related equipment, together with its interest in the Leasehold insofar and only insofar as such Leasehold covers the right to obtain production from that wellbore in the Zone then open to production. If the interest of the abandoning party is or includes and Oil and Gas Interest, such party shall execute and deliver to the non-abandoning party or parties an oil and gas lease, limited to the wellbore and the Zone then open to production, for a term of one (1) year and so long thereafter as Oil and/or Gas is produced from the Zone covered thereby, such lease to be on the form attached as Exhibit "B." The assignments or leases so limited shall encompass the Drilling Unit upon which the well is located. The payments by, and the assignments or leases to, the assignees shall be in a ratio based upon the relationship of their respective percentage of participation in the Contract Area to the aggregate of the percentages of participation in the Contract Area of all assignees. There shall be no readjustment of interests in the remaining portions of the Contract Area.

Thereafter, abandoning parties shall have no further responsibility, liability, or interest in the operation of or production from the well in the Zone then open other than the royalties retained in any lease made under the terms of this Article. Upon request, Operator shall continue to operate the assigned well for the account of the non-abandoning parties at the rates and charges contemplated by this agreement, plus any additional cost and charges which may arise as the result of the separate ownership of the assigned well. Upon proposed abandonment of the producing Zone assigned or leased, the assignor or lessor shall then have the option to repurchase its prior interest in the well (using the same valuation formula) and participate in further operations therein subject to the provisions hereof.

3. Abandonment of Non-Consent Operations: The provisions of Article VI.E.1. or VI.E.2. above shall be applicable as between Consenting Parties in the event of the proposed abandonment of any well excepted from said Articles; provided, however, no well shall be permanently plugged and abandoned unless and until all parties having the right to conduct further operations therein have been notified of the proposed abandonment and afforded the opportunity to elect to take over the well in accordance with the provisions of this Article VI.E.; and provided further, that Non-Consenting Parties who own an interest in a portion of the well shall pay their proportionate shares of abandonment and surface restoration cost for such well as provided in Article VI.B.2.(b).

F. TERMINATION OF OPERATIONS

Upon the commencement of an operation for the drilling, Reworking, Sidetracking, Plugging Back, Deepening, testing, Completion or plugging of a well, including but not limited to the Initial Well, such operation shall not be terminated without consent of parties bearing 75% of the costs of such operation; provided, however, that in the event granite or other practically impenetrable substance or condition in the hole is encountered which renders further operations impractical, Operator may discontinue operations and give notice of such condition in the manner provided in Article VI.B.1, and the provisions of Article VI.B. or VI.E. shall thereafter apply to such operation, as appropriate.

G. TAKING PRODUCTION IN KIND

Option No. 1: Gas Balancing Agreement Attached

Each party shall have the right to take in kind or separately dispose of its proportionate share of all Oil and Gas produced from the Contract Area, exclusive of production which may be used in development and producing operations and in preparing and treating Oil and Gas for marketing purposes and production unavoidably lost. Any extra expenditure incurred in the taking in kind or separate disposition by any party of its proportionate share of the production shall be borne by such party. Any party taking its share of production in kind shall be required to pay for only its proportionate share of such part of Operator's surface facilities which it uses.

Each party shall execute such division orders and contracts as may be necessary for the sale of its interest in production from the Contract Area, and, except as provided in Article VII.B., shall be entitled to receive payment directly from the purchaser thereof for its share of all production.

If any party fails to make the arrangements necessary to take in kind or separately dispose of its proportionate share of the Oil produced from the Contract Area, Operator shall have the right, subject to the revocation at will by the party owning it, but not the obligation, to purchase such Oil and/or Gas or sell it to others at any time and from time to time, for the account of the non-taking party. Any such purchase or sale by Operator may be terminated by Operator upon at least ten (10) days written notice to the owner of said production and shall be subject always to the right of the owner of the production upon at least ten (10) days written notice to Operator to exercise at any time its right to take in kind, or separately dispose of, its share of all Oil and/or Gas not previously delivered to a purchaser. Any purchase or sale by Operator of any other party's share of Oil and/or Gas shall be only for such reasonable periods of time as are consistent with the minimum needs of the industry under the particular circumstances, but in no event for a period in excess of one (1) year.

Any such sale by Operator shall be in a manner commercially reasonable under the circumstances but Operator shall have no duty to share any existing market or to obtain a price equal to that received under any existing market. The sale or delivery by Operator of a non-taking party's share of Oil under the terms of any existing contract of Operator shall not give the non-taking party any interest in or make the non-taking party a party to said contract. No purchase shall be made by Operator without first giving the non-taking party at least ten (10) days written notice of such intended purchase and the price to be paid or the pricing basis to be used.

All parties shall give timely written notice to Operator of their Gas marketing arrangements for the following month, excluding price, and shall notify Operator immediately in the event of a change in such arrangements. Operator shall maintain records of all marketing arrangements, and of volumes actually sold or transported, which records shall be made available to Non-Operators upon reasonable request. In the event one or more parties' separate disposition of its share of the Gas causes split-stream deliveries to separate pipelines and/or deliveries which on a day-to-day basis for any reason are not exactly equal to a party's respective proportion- ate share of total Gas sales to be allocated to it, the balancing or accounting between the parties shall be in accordance with any Gas balancing agreement between the parties hereto, whether such an agreement is attached as Exhibit "E" or is a separate agreement. Operator shall give notice to all parties of the first sales of Gas from any well under this agreement.

ARTICLE VII
EXPENDITURES AND LIABILITY OF PARTIES

A. LIABILITY OF PARTIES

The liability of the parties shall be several, not joint or collective. Each party shall be responsible only for its obligations, and shall be liable only for its proportionate share of the costs of developing and operating the Contract Area. Accordingly, the liens granted among the parties in Article VII.B. are given to secure only the debts of each severally, and no party shall have any liability to third parties hereunder to satisfy the default of any other party in the payment of any expense or obligation hereunder. It is not the intention of the parties to create, nor shall this agreement be construed as creating, a mining or other partnership, joint venture, agency relationship or association, or to render the parties liable as partners, co-venturers, or principals. In their relations with each other under this agreement, the parties shall not be considered fiduciaries or to have established a confidential relationship but rather shall be free to act on an arm's-length basis in accordance with their own respective self-interest, subject, however, to the obligation of the parties to act in good faith in their dealings with each other with respect to activities hereunder.

B. LIENS AND SECURITY INTERESTS

Each party grants to the other parties hereto a lien upon any interest it now owns or hereafter acquires in Oil and Gas Leases and Oil and Gas Interests in the Contract Area, and a security interest and/or purchase money security interest in any interest it now owns or hereafter acquires in the personal property and fixtures on or used or obtained for use in connection therewith, to secure performance of all of its obligations under this agreement including but not limited to payment of expense, interest and fees, the proper disbursement of all monies paid hereunder, the assignment or relinquishment of interest in Oil and Gas Leases as required hereunder, and the proper performance of operations hereunder. Such lien and security interest granted by each party hereto shall include such party's leasehold interests, working interests, operating rights, and royalty and overriding royalty interests in the Contract Area now owned or hereafter acquired and in lands pooled or unitized therewith or otherwise becoming subject to this agreement, the Oil and Gas when extracted therefrom and equipment situated thereon or used or obtained for use in connection therewith (including, without limitation, all wells, tools, and tubular goods), and accounts (including, without limitation, accounts arising from gas imbalances or from the sale of Oil and/or Gas at the wellhead), contract rights, inventory and general intangibles relating thereto or arising therefrom, and all proceeds and products of the foregoing.

To perfect the lien and security agreement provided herein, each party hereto shall execute and acknowledge the recording supplement and/or any financing statement prepared and submitted by any party hereto in conjunction herewith or at any time following execution hereof, and Operator is authorized to file this agreement or the recording supplement executed herewith as a lien or mortgage in the applicable real estate records and as a financing statement with the proper officer under the Uniform Commercial Code in the state in which the Contract Area is situated and such other states as Operator shall deem appropriate to perfect the security interest granted hereunder. Any party may file this agreement, the recording supplement executed herewith, or such other documents as it deems necessary as a lien or mortgage in the applicable real estate records and/or a financing statement with the proper officer under the Uniform Commercial Code.

Each party represents and warrants to the other parties hereto that the lien and security interest granted by such party to the other parties shall be a first and prior lien, and each party hereby agrees to maintain the priority of said lien and security interest against all persons acquiring an interest in Oil and Gas Leases and Interests covered by this agreement by, through or under such party. All parties acquiring an interest in Oil and Gas Leases and Oil and Gas Interests covered by this agreement, whether by assignment, merger, mortgage, operation of law, or otherwise, shall be deemed to have taken subject to the lien and security interest granted by this Article VII.B. as to all obligations attributable to such interest hereunder whether or not such obligations arise before or after such interest is acquired.

To the extent that parties have a security interest under the Uniform Commercial Code of the state in which the Contract Area is situated, they shall be entitled to exercise the rights and remedies of a secured party under the Code. The bringing of a suit and the obtaining of judgment by a party for the secured indebtedness shall not be deemed an election of remedies or otherwise affect the lien rights or security interest as security for the payment thereof. In addition, upon default by any party in the payment of its share of expenses, interests or fees, or upon the improper use of funds by the Operator, the other parties shall have the right, without prejudice to other rights or remedies, to collect from the purchaser the proceeds from the sale of such defaulting party's share of Oil and Gas until the amount owed by such party, plus interest as provided in "Exhibit C," has been received, and shall have the right to offset the amount owed against the proceeds from the sale of such defaulting party's share of Oil and Gas. All purchasers of production may rely on a notification of default from the non-defaulting party or parties stating the amount due as a result of the default, and all parties waive any recourse available against purchasers for releasing production proceeds as provided in this paragraph.

If any party fails to pay its share of cost within one hundred twenty (120) days after rendition of a statement therefor by Operator, the non-defaulting parties, including Operator, shall upon request by Operator, pay the unpaid amount in the proportion that the interest of each such party bears to the interest of all such parties. The amount paid by each party so paying its share of the unpaid amount shall be secured by the liens and security rights described in Article VII.B., and each paying party may independently pursue any remedy available hereunder or otherwise.

If any party does not perform all of its obligations hereunder, and the failure to perform subjects such party to foreclosure or execution proceedings pursuant to the provisions of this agreement, to the extent allowed by governing law, the defaulting party waives any available right of redemption from and after the date of judgment, any required valuation or appraisement of the mortgaged or secured property prior to sale, any available right to stay execution or to require a marshaling of assets and any required bond in the event a receiver is appointed. In addition, to the extent permitted by applicable law, each party hereby grants to the other parties a power of sale as to any property that is subject to the lien and security rights granted hereunder, such power to be exercised in the manner provided by applicable law or otherwise in a commercially reasonable manner and upon reasonable notice.

Each party agrees that the other parties shall be entitled to utilize the provisions of Oil and Gas lien law or other lien law of any state in which the Contract Area is situated to enforce the obligations of each party hereunder. Without limiting the generality of the foregoing, to the extent permitted by applicable law, Non-Operators agree that Operator may invoke or utilize the mechanics' or materialmen's lien law of the state in which the Contract Area is situated in order to secure the payment to Operator of any sum due hereunder for services performed or materials supplied by Operator.

C. ADVANCES

Operator, at its election, shall have the right from time to time to demand and receive from one or more of the other parties payment in advance of their respective shares of the estimated amount of the expense to be incurred in operations hereunder during the next succeeding month, which right may be exercised only by submission to each such party of an itemized statement of such estimated expense, together with an invoice for its share thereof. Each such statement and invoice for the payment in advance of estimated expense shall be submitted on or before the 20th day of the next preceding month. Each party shall pay to Operator its proportionate share of such estimate within fifteen (15) days after such estimate and invoice is received. If any party fails to pay its share of said estimate within said time, the amount due shall bear interest as provided in Exhibit "C" until paid. Proper adjustment shall be made monthly between advances and actual expense to the end that each party shall bear and pay its proportionate share of actual expenses incurred, and no more.

D. DEFAULTS AND REMEDIES

If any party fails to discharge any financial obligation under this agreement, including without limitation the failure to make any advance under the preceding Article VII.C. or any other provision of this agreement, within the period required for such payment hereunder, then in addition to the remedies provided in Article VII.B. or elsewhere in this agreement, the remedies specified below shall be applicable. For purposes of this Article VII.D., all notices and elections shall be delivered only by Operator, except that Operator shall deliver any such notice and election requested by a non-defaulting Non-Operator, and when Operator is the party in default, the applicable notices and elections can be delivered by any Non-Operator. Election of any one or more of the following remedies shall not preclude the subsequent use of any other remedy specified below or otherwise available to a non-defaulting party.

1. Suspension of Rights: Any party may deliver to the party in default a Notice of Default, which shall specify the default, specify the action to be taken to cure the default, and specify that failure to take such action will result in the exercise of one or more of the remedies provided in this Article. If the default is not cured within thirty (30) days of the delivery of such Notice of Default, all of the rights of the defaulting party granted by this agreement may upon notice be suspended until the default is cured, without prejudice to the right of the non-defaulting party or parties to continue to enforce the obligations of the defaulting party previously accrued or thereafter accruing under this agreement. If Operator is the party in default, the Non-Operators shall have in addition the right, by vote of Non-Operators owning a majority in interest in the Contract Area after excluding the voting interest of Operator, to appoint a new Operator effective immediately. The rights of a defaulting party that may be suspended hereunder at the election of the non-defaulting parties shall include, without limitation, the right to receive information as to any operation conducted hereunder during the period of such default, the right to elect to participate in an operation proposed under Article VI.B. of this agreement, the right to participate in an operation being conducted under this agreement even if the party has previously elected to participate in such operation, and the right to receive proceeds of production from any well subject to this agreement.

2. Suit for Damages: Non-defaulting parties or Operator for the benefit of non-defaulting parties may sue (at joint account expense) to collect the amounts in default, plus interest accruing on the amounts recovered from the date of default until the date of collection at the rate specified in Exhibit "C" attached hereto. Nothing herein shall prevent any party from suing any defaulting party to collect consequential damages accruing to such party as a result of the default.

3. Deemed Non-Consent: The non-defaulting party may deliver a written Notice of Non-Consent Election to the defaulting party at any time after the expiration of the thirty-day cure period following delivery of the Notice of Default, in which event if the billing is for the drilling a new well or the Plugging Back, Sidetracking, Reworking or Deepening of a well which is to be or has been plugged as a dry hole, or for the Completion or Recompletion of any well, the defaulting party will be conclusively deemed to have elected not to participate in the operation and to be a Non-Consenting Party with respect thereto under Article VI.B. or VI.C., as the case may be, to the extent of the costs unpaid by such party, notwithstanding any election to participate theretofore made.

Until the delivery of such Notice of Non-Consent Election to the defaulting party, such party shall have the right to cure its default by paying its unpaid share of costs plus interest at the rate set forth in Exhibit "C," provided, however, such payment shall not prejudice the rights of the non-defaulting parties to pursue remedies for damages incurred by the non-defaulting parties as a result of the default. Any interest relinquished pursuant to this Article VII.D.3. shall be offered to the non-defaulting parties in proportion to their interests, and the non-defaulting parties electing to participate in the ownership of such interest shall be required to contribute their shares of the defaulted amount upon their election to participate therein.

4. Advance Payment: If a default is not cured within thirty (30) days of the delivery of a Notice of Default, Operator, or Non-Operators if Operator is the defaulting party, may thereafter require advance payment from the defaulting party of such defaulting party's anticipated share of any item of expense for which Operator, or Non-Operators, as the case may be, would be entitled to reimbursement under any provision of this agreement, whether or not such expense was the subject of the previous default. Such right includes, but is not limited to, the right to require advance payment for the estimated costs of drilling a well or Completion of a well as to which an election to participate in drilling or Completion has been made. If the defaulting party fails to pay the required advance payment, the non-defaulting parties may pursue any of the remedies provided in the Article VII.D. or any other default remedy provided elsewhere in this agreement. Any excess of funds advanced remaining when the operation is completed and all costs have been paid shall be promptly returned to the advancing party.

5. Costs and Attorneys' Fees: In the event any party is required to bring legal proceedings to enforce any financial obligation of a party hereunder, the prevailing party in such action shall be entitled to recover all court costs, costs of collection, and a reasonable attorney's fee, which the lien provided for herein shall also secure.

E. RENTALS, SHUT-IN WELL PAYMENTS AND MINIMUM ROYALTIES

Rentals, shut-in well payments and minimum royalties which may be required under the terms of any lease shall be paid by the party or parties who subjected such lease to this agreement at its or their expense. In the event two or more parties own and have contributed interests in the same lease to this agreement, such parties may designate one of such parties to make said payments for and on behalf of all such parties. Any party may request, and shall be entitled to receive, proper evidence of all such payments. In the event of failure to make proper payment of any rental, shut-in well payment or minimum royalty through mistake or oversight where such payment is required to continue the lease in force, any loss which results from such non-payment shall be borne in accordance with the provisions of Article IV.B.2.

Operator shall notify Non-Operators of the anticipated completion of a shut-in well, or the shutting in or return to production of a producing well, at least five (5) days (excluding Saturday, Sunday, and legal holidays) prior to taking such action, or at the earliest opportunity permitted by circumstances, but assumes no liability for failure to do so. In the event of failure by Operator to so notify Non-Operators, the loss of any lease contributed hereto by Non-Operators for failure to make timely payments of any shut-in well payment shall be borne jointly by the parties hereto under the provisions of Article IV.B.3.

F. TAXES

Beginning with the first calendar year after the effective date hereof, Operator shall render for ad valorem taxation all property subject to this agreement which by law should be rendered for such taxes, and it shall pay all such taxes assessed thereon before they become delinquent. Prior to the rendition date, each Non-Operator shall furnish Operator information as to burdens (to include, but not be limited to, royalties, overriding royalties and production payments) on Leases and Oil and Gas Interests contributed by such Non-Operator. If the assessed valuation of any Lease is reduced by reason of its being subject to outstanding excess royalties, overriding royalties or production payments, the reduction in ad valorem taxes resulting therefrom shall inure to the benefit of the owner or owners of such Lease, and Operator shall adjust the charge to such owner or owners so as to reflect the benefit of such reduction. If the ad valorem taxes are based in whole or in part upon separate valuations of each party's working interest, then notwithstanding anything to the contrary herein, charges to the joint account shall be made and paid by the parties hereto in accordance with the tax value generated by each party's working interest. Operator shall bill the other parties for their proportionate shares of all tax payments in the manner provided in Exhibit "C."

If Operator considers any tax assessment improper, Operator may, at its discretion, protest within the time and manner prescribed by law, and prosecute the protest to a final determination, unless all parties agree to abandon the protest prior to final determination. During the pendency of administrative or judicial proceedings, Operator may elect to pay, under protest, all such taxes and any interest and penalty. When any such protested assessment shall have been finally determined, Operator shall pay the tax for the joint account, together with any interest and penalty accrued, and the total cost shall then be assessed against the parties, and be paid by them, as provided in Exhibit "C."

Each party shall pay or cause to be paid all production, severance, excise, gathering and other taxes imposed upon or with respect to the production or handling of such party's share of Oil and Gas produced under the terms of this agreement.

ARTICLE VIII
ACQUISITION, MAINTENANCE OR TRANSFER OF INTEREST

A. SURRENDER OF LEASES

The Leases covered by this agreement, insofar as they embrace acreage in the Contract Area, shall not be surrendered in whole or in part unless all parties consent thereto.

However, should any party desire to surrender its interest in any Lease or in any portion thereof, such party shall give written notice of the proposed surrender to all parties, and the parties to whom such notice is delivered shall have thirty (30) days after delivery of the notice within which to notify the party proposing the surrender whether they elect to consent thereto. Failure of a party to whom such notice is delivered to reply within said 30-day period shall constitute a consent to the surrender of the Leases described in the notice. If all parties do not agree or consent thereto, the party desiring to surrender shall assign, without express or implied warranty of title, all of its interest in such Lease, or portion thereof, and any well, material and equipment which may be located thereon and any rights in production thereafter secured, to the parties not consenting to such surrender. If the interest of the assigning party is or includes an Oil and Gas Interest, the assigning party shall execute and deliver to the party or parties not consenting to such surrender an oil and gas lease covering such Oil and Gas Interest for a term of one (1) year and so long thereafter as Oil and/or Gas is produced from the land covered thereby. Upon such assignment or lease, the assigning party shall be relieved from all obligations thereafter accruing, but not theretofore accrued, with respect to the interest assigned or leased and the operation of any well attributable thereto, and the assigning party shall have no further interest in the assigned or leased premises and its equipment and production other than the royalties retained in any lease made under the terms of this Article. The party assignee or lessee shall pay to the party assignor or lessor the reasonable salvage value of the latter's interest in any well's salvable materials and equipment attributable to the assigned or leased acreage. The value of all salvable materials and equipment shall be determined in accordance with the provisions of Exhibit "C", less the estimated cost of salvaging and the estimated cost of plugging and abandoning and restoring the surface. If such value is less than such costs, then the party assignor or lessor shall pay to the party assignee or lessee the amount of such deficit. If the assignment or lease is in favor of more than one party, the interest shall be shared by such parties in the proportions that the interest of each bears to the total interest of all such parties. If the interest of the parties to whom the assignment is to be made varies according to depth, then the interest assigned shall similarly reflect such variances.

Any assignment, lease or surrender made under this provision shall not reduce or change the assignor's, lessor's or surrendering party's interest as it was immediately before the assignment, lease or surrender in the balance of the Contract Area; and the acreage assigned, leased or surrendered, and subsequent operations thereon, shall not thereafter be subject to the terms and provisions of this agreement but shall be deemed subject to an Operating Agreement in the form of this agreement.

B. RENEWAL OR EXTENSION OF LEASES

If any party secures a renewal or replacement of an Oil and Gas Lease or Interest subject to this agreement, then all other parties shall be notified promptly upon such acquisition or, in the case of a replacement Lease taken before expiration of an existing Lease, promptly upon expiration of the existing Lease. The parties notified shall have the right for a period of thirty (30) days following delivery of such notice in which to elect to participate in the ownership of the renewal or replacement Lease, insofar as such Lease affects lands within the Contract Area, by paying to the party who acquired it their proportionate shares of the acquisition cost allocated to that part of such Lease within the Contract Area, which shall be in proportion to the interest held at that time by the parties in the Contract Area. Each party who participates in the purchase of a renewal or replacement Lease shall be given an assignment of its proportionate interest therein by the acquiring party.

If some, but less than all, of the parties elect to participate in the purchase of a renewal or replacement Lease, it shall be owned by the parties who elect to participate therein, in a ratio based upon the relationship of their respective percentage of participation in the Contract Area to the aggregate of the percentages of participation in the Contract Area of all parties participating in the purchase of such renewal or replacement Lease. The acquisition of a renewal or replacement Lease by any or all of the parties hereto shall not cause a readjustment of the interests of the parties stated in Exhibit "A," but any renewal or replacement Lease in which less than all parties elect to participate shall not be subject to this agreement but shall be deemed subject to a separate Operating Agreement in the form of this agreement.

If the interests of the parties in the Contract Area vary according to depth, then their right to participate proportionately in renewal or replacement Leases and their right to receive an assignment of interest shall also reflect such depth variances.

The provisions of this Article shall apply to renewal or replacement Leases whether they are for the entire interest covered by the expiring Lease or cover only a portion of its area or an interest therein. Any renewal or replacement Lease taken before the expiration of its predecessor Lease, or taken or contracted for or becoming effective within six (6) months after the expiration of the existing Lease, shall be subject to this provision so long as this agreement is in effect at the time of such acquisition or at the time the renewal or replacement Lease becomes effective; but any Lease taken or contracted for more than six (6) months after the expiration of an existing Lease shall not be deemed a renewal or replacement Lease and shall not be subject to the provisions of this agreement. The provisions in this Article shall also be applicable to extensions of Oil and Gas Leases.

C. ACREAGE OR CASH CONTRIBUTIONS

While this agreement is in force, if any party contracts for a contribution of cash towards the drilling of a well or any other operation on the Contract Area, such contribution shall be paid to the party who conducted the drilling or other operation and shall be applied by it against the cost of such drilling or other operation. If the contribution be in the form of acreage, the party to whom the contribution is made shall promptly tender an assignment of the acreage, without warranty of title, to the Drilling Parties in the proportions said Drilling Parties shared the cost of drilling the well. Such acreage shall become a separate Contract Area and, to the extent possible, be governed by provisions identical to this agreement. Each party shall promptly notify all other parties of any acreage or cash contributions it may obtain in support of any well or any other operation on the Contract Area. The above provisions shall also be applicable to optional rights to earn acreage outside the Contract Area which are in support of well drilled inside Contract Area.

If any party contracts for any consideration relating to disposition of such party's share of substances produced hereunder, such consideration shall not be deemed a contribution as contemplated in this Article VIII.C.

D. ASSIGNMENT; MAINTENANCE OF UNIFORM INTEREST

For the purpose of maintaining uniformity of ownership in the Contract Area in the Oil and Gas Leases, Oil and Gas Interests, wells, equipment and production covered by this agreement no party shall sell, encumber, transfer or make other disposition of its interest in the Oil and Gas Leases and Oil and Gas Interests embraced within the Contract Area or in wells, equipment and production unless such disposition covers either:

1. the entire interest of the party in all Oil and Gas Leases, Oil and Gas Interests, wells, equipment and production; or
2. an equal undivided percent of the party's present interest in all Oil and Gas Leases, Oil and Gas Interests, wells, equipment and production in the Contract Area.

Every sale, encumbrance, transfer or other disposition made by any party shall be made expressly subject to this agreement and shall be made without prejudice to the right of the other parties, and any transferee of an ownership interest in any Oil and Gas Lease or Interest shall be deemed a party to this agreement as to the interest conveyed from and after the effective date of the transfer of ownership; provided, however, that the other parties shall not be required to recognize any such sale, encumbrance, transfer or other disposition for any purpose hereunder until thirty (30) days after they have received a copy of the instrument of transfer or other satisfactory evidence thereof in writing from the transferor or transferee. No assignment or other disposition of interest by a party shall relieve such party of obligations previously incurred by such party hereunder with respect to the interest transferred, including without limitation the obligation of a party to pay all costs attributable to an operation conducted hereunder in which such party has agreed to participate prior to making such assignment, and the lien and security interest granted by Article VII.B. shall continue to burden the interest transferred to secure payment of any such obligations.

If, at any time the interest of any party is divided among and owned by four or more co-owners, Operator, at its discretion, may require such co-owners to appoint a single trustee or agent with full authority to receive notices, approve expenditures, receive billings for and approve and pay such party's share of the joint expenses, and to deal generally with, and with power to bind, the co-owners of such party's interest within the scope of the operations embraced in this agreement; however, all such co-owners shall have the right to enter into and execute all contracts or agreements for the disposition of their respective shares of the Oil and Gas produced from the Contract Area and they shall have the right to receive, separately, payment of the sale proceeds thereof.

E. WAIVER OF RIGHTS TO PARTITION

If permitted by the laws of the state or states in which the property covered hereby is located, each party hereto owning an undivided interest in the Contract Area waives any and all rights it may have to partition and have set aside to it in severalty its undivided interest therein.

F. PREFERENTIAL RIGHT TO PURCHASE
(Optional; Check if applicable.)

Should any party desire to sell all or any part of its interests under this agreement, or its rights and interests in the Contract Area, it shall promptly give written notice to the other parties, with full information concerning its proposed disposition, which shall include the name and address of the prospective transferee (who must be ready, willing and able to purchase), the purchase price, a legal description sufficient to identify the property, and all other terms of the offer. The other parties shall then have an optional prior right, for a period of ten (10) days after the notice is delivered, to purchase for the stated consideration on the same terms and conditions the interest which the other party proposes to sell; and, if this optional right is exercised, the purcasing parties shall share the purchased interest in the proportions that the interest of each bears to the total interest of all purchasing parties. However, there shall be no preferential right to purchase in those cases where any party wishes to mortgage its interests, or to transfer title to its interests to its mortgagee in lieu of or pursuant to foreclosure of a mortgage of its interests, or to dispose of its interests by merger, reorganization, consolidation, or by sale of all or substantially all of its Oil and Gas assets to any party, or by transfer of its interests to a subsidiary or parent company or to a subsidiary of a parent company, or to any company in which such party owns a majority of the stock.

ARTICLE IX
INTERNAL REVENUE CODE ELECTION

If, for federal income tax purposes, this agreement and the operations hereunder are regarded as a partnership, and if the parties have not otherwise agreed to form a tax partnership pursuant to Exhibit "G" or other agreement between them, each party thereby affected elects to be excluded from the application of all of the provisions of Subchapter "K," Chapter 1, Subtitle "A," of the Internal Revenue Code of 1986, as amended ("Code"), as permitted and authorized by Section 761 of the Code and the regulations promulgated thereunder. Operator is authorized and directed to execute on behalf of each party hereby affected such evidence of this election as may be required by the Secretary of the Treasury of the United States or the Federal Internal Revenue Service, including specifically, but not by way of limitation, all of the returns, statements, and the data required by Treasury Regulation ss.1.761. Should there be any requirement that each party hereby affected give further evidence of this election, each such party shall execute such documents and furnish such other evidence as may be required by the Federal Internal Revenue Service or as may be necessary to evidence this election. No such party shall give any notices or take any other action inconsistent with the election made hereby. If any present or future income tax laws of the state or states in which the Contract Area is located or any future income tax laws of the United States contain provisions similar to those in Subchapter "K," Chapter 1, Subtitle "A," of the Code, under which an election similar to that provided by Section 761 of the Code is permitted, each party hereby affected shall make such election as may be permitted or required by such laws. In making the foregoing election, each such party states that the income derived by such party from operations hereunder can be adequately determined without the computation of partnership taxable income.

ARTICLE X
CLAIMS AND LAWSUITS

Operator may settle any single uninsured third party damage claim or suit arising from operations hereunder if the expenditure does not exceed TWENTY-FIVE THOUSAND Dollars ($25,000.00) and if the payment is in complete settlement of such claim or suit. If the amount required for settlement exceeds the above amount, the parties hereto shall assume and take over the further handling of the claim or suit, unless such authority is delegated to Operator. All costs and expenses of handling settling, or otherwise discharging such claim or suit shall be at the joint expense of the parties participating in the operation from which the claim or suit arises. If a claim is made against any party or if any party is sued on account of any matter arising from operations hereunder over which such individual has no control because of the rights given Operator by this agreement, such party shall immediately notify all other parties, and the claim or suit shall be treated as any other claim or suit involving operations hereunder.

ARTICLE XI
FORCE MAJEURE

If any party is rendered unable, wholly or in part, by force majeure to carry out its obligations under this agreement, other than the obligation to indemnify or make money payments or furnish security, that party shall give to all other parties prompt written notice of the force majeure with reasonably full particulars concerning it; thereupon, the obligations of the party giving the notice, so far as they are affected by the force majeure, shall be suspended during, but no longer than, the continuance of the force majeure. The term "force majeure," as here employed, shall mean an act of God, strike, lockout, or other industrial disturbance, act of the public enemy, war, blockade, public riot, lightening, fire, storm, flood or other act of nature, explosion, governmental action, governmental delay, restraint or inaction, unavailability of equipment, and any other cause, whether of the kind specifically enumerated above or otherwise, which is not reasonably within the control of the party claiming suspension.

The affected party shall use all reasonable diligence to remove the force majeure situation as quickly as practicable. The requirement that any force majeure shall be remedied with all reasonable dispatch shall not require the settlement of strikes, lockouts, or other labor difficulty by the party involved, contrary to its wishes; how all such difficulties shall be handled shall be entirely within the discretion of the party concerned.

ARTICLE XII
NOTICES

All notices authorized or required between the parties by any of the provisions of this agreement, unless otherwise specifically provided, shall be in writing and delivered in person or by United States mail, courier service, telegram, telex, telecopier or any other form of facsimile, postage or charges prepaid, and addressed to such parties at the addresses listed on Exhibit "A." All telephone or oral notices permitted by this agreement shall be confirmed immediately thereafter by written notice. The originating notice given under any provision hereof shall be deemed delivered only when received by the party to whom such notice is directed, and the time for such party to deliver any notice in response thereto shall run from the date the originating notice is received. "Receipt" for purposes of this agreement with respect to written notice delivered hereunder shall be actual delivery of the notice to the address of the party to be notified specified in accordance with this agreement, or to the telecopy, facsimile or telex machine of such party. The second or any responsive notice shall be deemed delivered when deposited in the United States mail or at the office of the courier or telegraph service, or upon transmittal by telex, telecopy or facsimile, or when personally delivered to the party to be notified, provided, that when response is required within 24 or 48 hours, such response shall be given orally or by telephone, telex, telecopy or other facsimile within such period. Each party shall have the right to change its address at any time, and from time to time, by giving written notice thereof to all other parties. If a party is not available to receive notice orally or by telephone when a party attempts to deliver a notice required to be delivered within 24 or 48 hours, the notice may be delivered in writing by any other method specified herein and shall be deemed delivered in the same manner provided above for any responsive notice.

ARTICLE XIII
TERM OF AGREEMENT

This agreement shall remain in full force and effect as to the Oil and Gas Leases and/or Oil and Gas Interests subject hereto for the period of time selected below; provided, however, no party hereto shall ever be construed as having any right, title or interest in or to any Lease or Oil and Gas Interest contributed by any other party beyond the term of this agreement.

[X] Option No. 1: So long as any of the Oil and Gas Leases subject to this agreement remain or are continued in force as to any part of the Contract Area, whether by production, extension, renewal or otherwise. The termination of this agreement shall not relieve any party hereto from any expense, liability or other obligation or any remedy therefor which has accrued or attached prior to the date of such termination. Upon termination of this agreement and the satisfaction of all obligations hereunder, in the event a memorandum of this Operating Agreement has been filed of record, Operator is authorized to file of record in all necessary recording offices a notice of termination, and each party hereto agrees to execute such a notice of termination as to Operator's interest, upon request of Operator, if Operator has satisfied all its financial obligations.

ARTICLE XIV
COMPLIANCE WITH LAWS AND REGULATIONS

A. LAWS, REGULATIONS AND ORDERS

This agreement shall be subject to the applicable laws of the state in which the Contract Area is located, to the valid rules, regulations, and orders of any duly constituted regulatory body of said state; and to all other applicable federal, state, and local laws, ordinances, rules, regulations and orders.

B. GOVERNING LAW

This agreement and all matters pertaining hereto, including but not limited to matters of performance, non-performance, breach, remedies, procedures, rights, duties, and interpretation or construction, shall be governed and determined by the law of the state in which the contract area is located. If the contract area is in two or more states, the law of the state of wyoming shall govern.

C. REGULATORY AGENCIES

Nothing herein contained shall grant, or be construed to grant, Operator the right or authority to waive or release any rights, privileges, or obligations which Non-Operators may have under federal or state laws or under rules, regulations or orders promulgated under such laws in reference to oil, gas and mineral operations, including the location, operation, or production of wells, on tracts offsetting or adjacent to the Contract Area. With respect to the operations hereunder, Non-Operators agree to release Operator from any and all losses, damages, injuries, claims and causes of action arising out of, incident to or resulting directly or indirectly from Operator's interpretation or application of rules, rulings, regulations or orders of the Department of Energy or Federal Energy Regulatory Commission or predecessor or successor agencies to the extent such interpretation or application was made in good faith and does not constitute gross negligence. Each Non-Operator further agrees to reimburse Operator for such Non-Operator's share of production or any refund, fine, levy or other governmental sanction that Operator may be required to pay as a result of such an incorrect interpretation or application, together with interest and penalties thereon owing by Operator as a result of such incorrect interpretation or application.

ARTICLE XV
MISCELLANEOUS

A. EXECUTION

This agreement shall be binding upon each Non-Operator when this agreement or a counterpart thereof has been executed by such Non-Operator and Operator notwithstanding that this agreement is not then or thereafter executed by all of the parties to which it is tendered or which are listed on Exhibit "A" as owning an interest in the Contract Area or which own, in fact, an interest in the Contract Area. Operator may, however, by written notice to all Non-Operators who have become bound by this agreement as aforesaid, given at any time prior to the actual spud date of the Initial Well but in no event later than five days prior to the date specified in Article VI.A. for commencement of the Initial Well, terminate this agreement if Operator in its sole discretion determines that there is insufficient participation to justify commencement of drilling operations. In the event of such a termination by Operator, all further obligations of the parties hereunder shall cease as of such termination. In the event any Non-Operator has advanced or prepaid any share of drilling or other costs hereunder, all sums so advanced shall be returned to such Non-Operator without interest. In the event Operator proceeds with drilling operations for the Initial Well without the execution hereof by all persons listed on Exhibit "A" as having a current working interest in such well, Operator shall indemnify Non-Operators with respect to all costs incurred for the Initial Well which would have been charged to such person under this agreement if such person had executed the same and Operator shall receive all revenues which would have been received by such person under this agreement if such person had executed the same.

B. SUCCESSORS AND ASSIGNS

This agreement shall be binding upon and shall inure to the benefit of the parties hereto and their respective heirs, devisees, legal representatives, successors and assigns, and the terms hereof shall be deemed to run with the Leases or Interests included within the Contract Area.

C. COUNTERPARTS

This instrument may be executed in any number of counterparts, each of which shall be considered an original for all purposes.

D. SEVERABILITY

For the purposes of assuming or rejecting this agreement as an executory contract pursuant to federal bankruptcy laws, this agreement shall not be severable, but rather must be assumed or rejected in its entirety, and the failure of any party to this agreement to comply with all of its financial obligations provided herein shall be a material default.

ARTICLE XVI

OTHER PROVISIONS

Notwithstanding any other provision contained in Articles I through XV to the contrary,

A. ADDITIONAL PARTIES

There shall be no obligation on the Operator hereof to perform to a multiplicity of parties succeeding to the interest of the parties hereto. Therefore, in exercising any option, election, consent, notice, acceptance, declination or other right of performance, the respective successors in interest shall act as a unit and shall, moreover, designate one of their number to whom all notices and performance may be tendered and from whom all notices and performance may be obtained and exacted, respectively.

B. BILLING ADDITIONAL INTERESTS

The parties further agree that in no event during the term of this agreement shall Operator be required to make more than one (1) billing for the entire interest credited to each Party hereunder. It is further agreed that if any Party to this Agreement, hereinafter referred to as "Selling Party", disposes of part of the interest credited to it hereunder, the Selling Party will be solely responsible for billing its assignee(s), and shall remain primarily liable to the other party for the interest(s) assigned and shall make prompt payment to Operator for the entire amount of statements and billings rendered to it. It is further understood and agreed that if Selling Party disposes of all its interest as set out hereunder, whether to one or several assignees, Operator shall continue to issue statements and billings to the Selling Party for the interest conveyed until such time as Selling Party has designated and qualified one assignee to receive the billing for the entire interest. In order to qualify one assignee to receive the billing for the entire interest credited to the Selling Party hereunder, Selling Party shall furnish to Operator the following:

1. Written notice of the conveyance and photostatic or certified copies of the recorded assignments by which the transfer was made.
2. The name of the assignee to be billed and a written statement signed by the assigned to be billed in which it consents to receive statements and billings for the entire interest credited to Selling Party on Exhibit "A" hereunder.
3. Written consent from all assignees evidencing their consent to have the designee act on their behalf as hereinabove provided.

C. WORKOVER OPERATIONS

It is agreed that without the mutual consent of all parties, no workover or other operations will be conducted under the provisions of Article VI so long as any completion in the well proposed to be worked over is producing in paying quantities.

D. SEQUENCE OF FURTHER OPERATIONS

It is agreed that where a well, which has been authorized under the terms of this agreement, by all parties, or by one or more, but less than all parties under Paragraph VI.B. (1) or (2), shall have been drilled to the objective depth or the objective formation, and the parties participating in the well cannot mutually agree upon the sequence and timing of further operations regarding said well, the following elections shall control in the order enumerated hereafter, to wit:

1. An election to do additional logging, coring or testing;
2. An election to attempt to complete the well at either the objective depth or objective formation;
3. An election to plug back and attempt to complete said well;
4. An election to sidetrack the well;
5. An election to deepen said well; and
6. An election to plug and abandon said well.

If at the time said participating parties are considering any of the above elections, the hole is in such a condition that a reasonably prudent Operator would not conduct the operations contemplated by the particular election involved for fear of placing the hole in jeopardy or losing the same prior to completing the well in the objective depth or objective formation, such election shall not be given the priority hereinabove set forth.

E. MEMORANDUM OF OPERATING AGREEMENT

Parties agree to execute a memorandum of this Agreement simultaneously with the execution of this Agreement, which shall be recorded by Operator in each County in which lands comprising all or any part of the Contract Area hereunder are situated. Operator will prepare the memorandum of this Agreement and circulate same to the Non-Operator for execution. Upon recording, Operator will provide recordation information to Non-Operator for its records. A form of said Memorandum, which is attached as Exhibit "H" hereto, shall include a reference to the pertinent substantive provisions of this Agreement, including the Gas Balancing Agreement.

F. HEADINGS

All headings in this agreement are for reference purposes only and have no binding effect on the terms, conditions or provisions of this Agreement.

G. CONFIDENTIALITY OF INFORMATION

All geophysical, geological or engineering data acquired by the parties under this Agreement, as a result of joint operations conducted hereunder, shall be kept CONFIDENTIAL by the parties unless the release of such information to a third party is agreed upon by the parties or is required by law; however, it is agreed that such data may be released to a third party if such third party is a consultant of any party or provided such third party is a prospective partner of such party or is a prospective purchaser of all or any portion of such parties interest. The term during which information is to be kept SECRET AND CONFIDENTIAL shall coincide with the term of this Agreement or for a period of one (1) years from the date of this Agreement, whichever is later. Unless otherwise provided above, any release of information to a third party must have the prior written consent of all parties hereto, but in any event, said third party must agree in writing to be bound by the provisions of this paragraph.

Nothing herein shall prohibit any party from disclosing necessary information to its affiliate company or whatever information in such manner as may be required by statute, rule or regulation, including the rules and regulations of any stock exchange on which any securities of such party or any affiliate are traded; nor shall any party be prohibited by the terms hereof form disclosing information acquired under this agreement to any financial institution providing or proposing to provide financing to the disclosing party.

H. NEWS RELEASES

Any party hereto or any related party desiring to issue a news release concerning operations conducted within the Contract Area, shall provide the other party(ies) hereto with copies of the proposed release and no such news release shall be issued without first obtaining the written consent of all parties hereto, which consent shall not be unreasonably withheld. Notwithstanding the immediately preceding sentence, no prior consent shall be required for any news release required by law and/or the Securities and Exchange Commission. The only exception to the foregoing shall be that in the event of an emergency involving extensive property damage, operations failure, loss of human life or other clear emergency, the Operator is authorized to furnish such minimum, strictly factual information as shall be necessary to satisfy the legitimate public interest on the part of the press and duly constituted authorities if time does not permit the obtaining of prior approval by the other parties hereto. The

Operator shall thereupon promptly advise the other parties hereto of the information so furnished.

I. LANDS EXCLUDED FROM THIS AGREEMENT

In the event that lands covered by a lease or a portion thereof which were initially subject to this Agreement, but in accordance with the terms hereof are no longer subject to this Agreement, and such lands are acquired by one or more parties to this Agreement (provided that at such time two (2) or more parties are signatories to this Agreement), such leasehold or contractual rights and the leasehold covered thereby shall be governed by an operating agreement identical to this Agreement, except that such operating agreement shall be revised to reflect the parties thereto and their percentage ownership, the Contract Area and the Operator, if different.

J. CONFLICT OF TERMS

This Agreement is made expressly subject to the Farmin agreement dated January, 2004, by and between Ivanhoe Energy (USA) Inc., Derek Resources (U.S.A.), Inc., and the parties hereto shall be bound by the terms and conditions contained in said Agreement. In the event that any of the terms of this Operating Agreement conflict with the terms of the Farmin Agreement, the parties hereto agree that the terms of the Farmin Agreement shall prevail.

K. METERING OF PRODUCTION

If a diversity of the working interest ownership in production from a lease subject to this Agreement occurs as a result of operations by less than all parties pursuant to the provisions of this Agreement, it is agreed that the oil, gas or other hydrocarbons produced from the well or wells completed by the consenting party or parties shall be separately measured either by well test or by standard metering equipment (to be properly tested periodically for accuracy), at the option of the Operator, and that the setting of a separate tank battery will not be required unless the purchaser of the production or a governmental regulatory body having jurisdiction will not approve metering for separately measuring the production.

L. COVENANT RUNNING WITH THE LAND/EXTENT OF OBLIGATIONS

Should any party hereto sell or transfer any or all of its leasehold estate committed to this Agreement, the obligations, terms and covenants hereof shall be considered covenants running with the land and shall inure to and be binding upon the parties hereto, their respective heirs, devisees, legal representatives, successors and assigns. The transferring party shall remain bound to and liable for the performance of obligations and covenants of this Agreement until the transferee executes and agrees to become a party to this Agreement. Should any such transferee fail to assume all of the obligations and covenants of this Agreement, then the transferring party remains bound to the other parties to this agreement for the performance of all obligations, covenants and indemnifications hereof.

M. AUTHORITY FOR EXPENDITURE REQUIREMENTS

Notwithstanding anything herein to the contrary, the proposing party will furnish both a detailed Authority For Expenditure and a detailed program for each proposed operation that is estimated to exceed Twenty-Five Thousand Dollars ($25,000.00) For purposes of this provision, "detailed" shall mean information understood by the proposing party to be reasonably necessary for the receiving party to make an informed decision, but in no event shall the proposing party be required to furnish information beyond what is deemed reasonable and prudent in the industry.

IN WITNESS WHEREOF, this agreement shall be effective as of the 15th day of January, 2004.______________________________, who has prepared and circulated this form for execution, represents and warrants that the form was printed from and, with the exception(s) listed below, is identical to the AAPL Form 610-1989 Model Form Operating Agreement, as published in computerized form by Forms On-A-Disk, Inc. No changes, alterations, or modifications, other than those made by strikethrough and/or insertion and that are clearly recognizable as changes in Articles ______________________________, have been made to the form.

ATTEST OR WITNESS: OPERATOR

IVANHOE ENERGY (USA) INC.

By ______________________________

____

Type or print name

Title ______________________________

Date ______________________________

Tax ID or S.S. No. 77-0492835

NON-OPERATORS

DEREK RESOURCES (U.S.A.) INC.

By ______________________________

____

Type or print name

Title ______________________

Date ______________________

Tax ID or S.S. No. ______________________

ACKNOWLEDGMENTS

Note: The following forms of acknowledgment are the short forms approved by the Uniform Law on Notarial Acts.

The validity and effect of these forms in any state will depend upon the statutes of that state.

Individual acknowledgment:

State of ______________) ss.

County of) ______________

This instrument was acknowledged before me on

by ______________________________

(Seal, if any) ______________________

Title (and Rank) ______________________

My commission expires:______________________

Acknowledgment in representative capacity:

State of) ______________________

) ss.
County of) ______________________

This instrument was acknowledged before me on

by as

of .

(Seal, if any)

Title (and Rank)

My commission expires:

EXHIBIT "A"

Attached to and made a part of that certain Operating Agreement dated January 15, 2004, by and between Ivanhoe Energy (USA Inc., as Operator, and Derek Resources (USA) Inc. as Non-Operators.

All lands are in Weston County, Wyoming

1. Description of Lands

Tract A

Township 44 North, Range 60 West, 6th P.M.

Section 6: Southwest Quarter, Northeast Quarter, East Half of the Northwest Quarter, and West Half of the Southeast Quarter (SW/4, NE/4, E/2 NW/4, W/2 SE/4).

Section 7: West Half, and West Half of the East Half (W/2, W/2 E/2).

Section 18: South Half of the Southwest Quarter, and the Northwest Quarter (S/2 SW/4, NW/4).

Section 19: West Half of West Half, Northeast Quarter of the Northwest Quarter (W/2 W/2, NE/4 NW/4).

Township 44 North, Range 61 West, 6th P.M.

Section 1: Southeast Quarter (SE/4).

Section 11: Southeast Quarter, East Half of the Northeast Quarter, and South Half of the Southwest Quarter (SE/4, E/2 NE/4, S/2 SW/4).

Section 12: All.

Section 13: West Half, and South Half of the Southeast Quarter (W/2, S/2 SE/4).

Section 14: All.

Section 22: Northeast Quarter (NE/4).

Section 23: North Half, Southeast Quarter, East Half of the Southwest Quarter (N/2, SE, E/2 SW/4).

Section 24: North Half, Southeast Quarter, and East Half of the Southwest Quarter (N/2, SE/4, E/2 SW/4).

Section 25: Northwest Quarter of the Northeast Quarter, North Half of the Northwest Quarter, Southwest Quarter of the Northwest Quarter, and Northwest Quarter of the Southwest Quarter (NW/4 NE/4, N/2 NW/4, SW/4 NW/4, NW/4 SW/4.

Section 26: East Half of the Northeast Quarter, and Northeast Quarter of the Southeast Quarter (E/2 NE/4, NE/4 SE/4).

Section 30: West Half of the Northwest Quarter (W/4 NW/4).

Tract B

Township 44 North, Range 61 West, 6th P.M.

Section 24: Northeast Quarter of the Southeast Quarter (NE/4 SE/4).

Section 25: South Half of the Southwest Quarter, the Northeast Quarter of the Southwest Quarter, Northwest Quarter of the Southeast Quarter, East Half of the Northeast Quarter, Southwest Quarter of the Northeast Quarter, and Southeast Quarter of the Northwest Quarter (S/2 SW/4, NE/4 SW/4, NW/4 SE/4, E/4 NE/4, SW/4 NE/4, SE/4 NW/4).

Tract C

Township 44 North, Range 60 West, 6th P.M.

Section 18: North Half of the Southwest Quarter (N/2 SW/4).

Township 44 North, Range 61 West, 6th P.M.

Section 13: Northeast Quarter, and North Half of the Southeast Quarter (NE/4, N/2 SE/4).

Tract D

Township 44 North, Range 61 West, 6th P.M.

Section 22: Southeast Quarter (SE/4).

Section 27: West Half of the East Half (W/2 E/2).

Tract E

Township 44 North, Range 61 West, 6th P.M.

Section 24: Southeast Quarter of the Southeast Quarter (SE/4 SE/4).

Additional Property Leased from the Federal Government

Township 44 North, Range 61 West, 6th P.M.

Section 1: South Half of the Southwest Quarter (S/2 SW/4).

Section 10: Southwest Quarter of the Northeast Quarter (SW/4 NE/4).

Section 10: West Half of the Southeast Quarter, Southeast Quarter of the Southeast Quarter (W/2 SE/4, SE/4 SE/4), excluding 13.77 acres in railroad right-of-way WyW0119068.

Section 15: West Half of the East Half, East Half of the West Half, Northwest Quarter of the Northwest Quarter, Southwest Quarter of the Southwest Quarter (W/2 E/2, E/2 W/2, NW/4 NW/4, SW/4 SW/4).

Section 15: East Half of the East Half (E/2 E/2), excluding 23.78 acres in railroad right-of-way WyW0119068.

Section 25: Northeast Quarter of the Southeast Quarter, South half of the Southeast Quarter (NE/4 SE/4, S/2 SE/4).

Section 26: Northwest Quarter of the Northeast Quarter, East Half of the Southwest Quarter (NW/4 NE/4, E/2 SW/4).

Section 26: East Half of the Northwest Quarter, Southeast Quarter of the Southeast Quarter (E/2 NW/4, SE/4 SE/4), excluding 14.23 acres in railroad right-of-way WyW0119068.

2. Restriction as to depths

None

3. Parties To This Agreement With Addresses And Telephone Numbers For Notice Purposes:

Ivanhoe:	Ivanhoe Energy (USA) Inc. Mailing Address: P.O. Box 9279 Bakersfield, California 93389
Street Address:	5060 California Avenue Suite 400 Bakersfield, California 93309
Attention:	Robert Coffey
Telephone:	661-869-8332
Facsimile:	661-869-2820
e-mail:	rcoffey@ivanhoeenergy.com
Copy to:	Oscar Blake, General Counsel
Derek:	Derek Oil & Gas Corporation
Street Address:	Suite 1550, 355 Burrard Street Vancouver, B.C., Canada V6C 2GB
Attention:	Barry C.J. Ehrl
Telephone:	604-331-1757
Facsimile:	604-669-5193
e-mail:	info@derekresources.com

4. Percentages or fractional interest of parties to this Agreement

Pursuant to the LAK Ranch Farmin Agreement to which this Operating Agreement is attached.

5. **Oil and Gas Leases subject to this Agreement**

A. Oil and Gas Lease, dated January 10, 1981, by and between Donald B. Roberts, B.M. Stewart, and Sheri Vineyard, as Lessors, and Exoil Services, Inc., as Lessee, recorded in Book 87, Page 189, Official Records of Weston County, Wyoming.

B. Oil and Gas Lease, dated effective August 31, 1958, by and between the U.S. Department of Interior, Bureau of Land Management, as Lessor, and Edel P. Smith, as Lessee, being identified as Lease Serial Number W-022180-A.

C. Oil and Gas Lease, dated effective December 1, 1999, by and between the U.S. Department of Interior, Bureau of Land Management, as Lessor, and Cove Energy, as Lessee, being identified as Lease Serial Number W-149512.

6. **Burdens on Production**

Tract A 19.9501%
Tract B 19.9501%
Tract C 11.7875% (Based on 50% leased)
Tract D 11.7875% (Based on 50% leased)
Tract E 19.9501%

EXHIBIT "C"

Attached to and made a part of that certain Operating Agreement,dated January 15, 2004, by and between IVANHOE ENERGY (USA) INC., as Operator and DEREK RESOURCES (U.S.A.) INC. as Non-Operator

---

ACCOUNTING PROCEDURE

JOINT OPERATIONS

I. GENERAL PROVISIONS

1. DEFINITIONS

"Joint Property" shall mean the real and personal property subject to the agreement to which this Accounting Procedure is attached. "Joint Operations" shall mean all operations necessary or proper for the development, operation, protection and maintenance of the Joint Property.

"Joint Account" shall mean the account showing the charges paid and credits received in the conduct of the Joint Operations and which are to be shared by the Parties.

"Operator" shall mean the party designated to conduct the Joint Operations.

"Non-Operators" shall mean the Parties to this agreement other than the Operator.

"Parties" shall mean Operator and Non-Operators.

"First Level Supervisors" shall mean those employees whose primary function in Joint Operations is the direct supervision of other employees and/or contract labor directly employed on the Joint Property in a field operating capacity.

"Technical Employees" shall mean those employees having special and specific engineering, geological or other professional skills, and whose primary function in Joint Operations is the handling of specific operating conditions and problems for the benefit of the Joint Property.

"Personal Expenses" shall mean travel and other reasonable reimbursable expenses of Operator's employees.

"Material" shall mean personal property, equipment or supplies acquired or held for use on the Joint Property.

"Controllable Material" shall mean Material which at the time is so classified in the Material Classification Manual as most recently recommended by the Council or Petroleum Accountants Societies.

2. STATEMENT AND BILLINGS

Operator shall bill Non-Operators on or before the last day of each month for their proportionate share of the Joint Account for the preceding month. Such bills will be accompanied by statements which identify the authority for expenditure, lease or facility, and all charges and credits summarized by appropriate classifications of investment and expense except that items of Controllable Material and unusual charges and credits shall be separately identified and fully described in detail.

3. ADVANCES AND PAYMENTS BY NON-OPERATORS

A. Unless otherwise provided for in the agreement, the Operator may require the Non-Operators to advance their share of estimated cash outlay for the succeeding month's operation within fifteen (15) days after receipt of the billing or by the first day of the month for which the advance is required, whichever is later. Operator shall adjust each monthly billing to reflect advances received from the Non-Operators.

B. Each Non-Operator shall pay its proportion of all bills within fifteen (15) days after receipt. If payment is not made within such time, the unpaid balance shall bear interest monthly at the prime rate in effect at BANK OF AMERICA on the first day of the month in which delinquency occurs plus 1% or the maximum contract rate permitted by the applicable usury laws in the state in which the Joint Property is located, whichever is the lesser, plus attorney's fees, court costs, and other costs in connection with the collection of unpaid amounts.

4. ADJUSTMENTS

Payment of any such bills shall not prejudice the right of any Non-Operator to protest or question the correctness thereof; provided, however, all bills and statements rendered to Non-Operators by Operator during any calendar year shall conclusively be presumed to be true and correct after twenty-four (24) months following the end of any such calendar year, unless within the said twenty-four (24) month period a Non-Operator takes written exception thereto and makes claim on Operator for adjustment. No adjustment favorable to Operator shall be made unless it is made within the same prescribed period. The provisions of this paragraph shall not prevent adjustments resulting from a physical inventory of Controllable Material as provided for in Section V.

5. AUDITS

A. A Non-Operator, upon notice in writing to Operator and all other Non-Operators, shall have the right to audit Operator's accounts and records relating to the Joint Account for any calendar year within the twenty-four (24) month period following the end of such calendar year; provided, however, the making of an audit shall not extend the time for the taking of written exception to and the adjustments of accounts as provided for in Paragraph 4 of this Section I. Where there are two or more Non-Operators, the Non-Operators shall make every reasonable effort to conduct a joint audit in a manner which will result in a minimum of inconvenience to the Operator. Operator shall bear no portion of the Non-Operators' audit cost incurred under this paragraph unless agreed to by the Operator. The audits shall not be conducted more than once each year without prior approval of Operator, except upon the resignation or removal of the Operator, and shall be made at the expense of those Non-Operators approving such audit.

B. The Operator shall reply in writing to an audit report within 180 days after receipt of such report.

6. APPROVAL BY NON-OPERATORS

Where an approval or other agreement of the Parties or Non-Operators is expressly required under other sections of this Accounting Procedure and if the agreement to which this Accounting Procedure is attached contains no contrary provisions in regard thereto, Operator shall notify all Non-Operators of the Operator's proposal, and the agreement or approval of a majority in interest of the Non-Operators shall be controlling on all Non-Operators.

II. DIRECT CHARGES

Operator shall charge the Joint Account with the following items:

1. ECOLOGICAL AND ENVIRONMENTAL

Costs incurred for the benefit of the Joint Property as a result of governmental or regulatory requirements to satisfy environmental considerations applicable to the Joint Operations. Such costs may include surveys of an ecological or archaeological nature and pollution control procedures as required by applicable laws and regulations.

2. RENTALS AND ROYALTIES

Lease rentals and royalties paid by Operator for the Joint Operations.

3. LABOR

A. (1) Salaries and wages of Operator's field employees directly employed on the Joint Property in the conduct of Joint Operations.
   (2) Salaries of First level Supervisors in the field.
   (3) Salaries and wages of Technical Employees directly employed on the Joint Property if such charges are excluded from the overhead rates.
   (4) Salaries and wages of Technical Employees either temporarily or permanently assigned to and directly employed in the operation or the Joint Property if such charges are excluded from the overhead rates.

B. Operator's cost of holiday, vacation, sickness and disability benefits and other customary allowances paid to employees whose salaries and wages are chargeable to the Joint Account under Paragraph 3A of this Section II. Such costs under this Paragraph 3B may be charged on a "when and as paid basis" or by "percentage assessment" on the amount of salaries and wages chargeable to the Joint Account under Paragraph 3A of this Section II. If percentage assessment is used, the rate shall be based on the Operator's cost experience.

C. Expenditures or contributions made pursuant to assessments imposed by governmental authority which are applicable to Operator's costs chargeable to the Joint Account under Paragraphs 3A and 3B of this Section II.

D. Personal Expenses of those employees whose salaries and wages are chargeable to the Joint Account under Paragraphs 3A and 3B of this Section II.

4. EMPLOYEE BENEFITS

Operator's current costs or established plans for employees' group life insurance, hospitalization, pension, retirement, stock purchase, thrift, bonus, and other benefit plans of a like nature applicable to Operator's labor cost chargeable to the Joint Account under Paragraphs 3A and 3B of this Section II shall be Operator's actual cost not to exceed the percent most recently recommended by the Council of Petroleum Accountants Societies.

5. MATERIAL

Material purchased or furnished by Operator for use on the Joint Property as provided under Section IV. Only such Material shall be purchased for or transferred to the Joint Property as may be required for immediate use and is reasonably practical and consistent with efficient and economical operations. The accumulation of surplus stocks shall be avoided.

6. TRANSPORTATION

Transportation of employees and Material necessary for the Joint Operations but subject to the following limitations:

A. If Material is moved to the Joint Property from the Operator's warehouse or other properties, no charge shall be made to the Joint Account for a distance greater than the distance from the nearest reliable supply store where like material is normally available or railway receiving point nearest the Joint Property unless agreed to by the Parties.
B. If surplus Material is moved to Operator's warehouse or other storage point, no charge shall be made to the Joint Account for a distance greater than the distance to the nearest reliable supply store where like material is normally available, or railway receiving point nearest the Joint Property unless agreed to by the Parties. No charge shall be made to the Joint Account for moving Material to other properties belonging to Operator, unless agreed to by the Parties.
C. In the application of subparagraphs A and B above, the option to equalize or charge actual trucking cost is available when the actual charge is $400 or less excluding accessorial charges. The $400 will be adjusted to the amount most recently recommended by the Council of Petroleum Accountants Societies.

7. SERVICES

The cost of contract services, equipment and utilities provided by outside sources, except services excluded by Paragraph 10 of Section II and Paragraph i, ii, and iii, of Section III. The cost of professional consultant services and contract services of technical personnel directly engaged on the Joint Property if such charges are excluded from the overhead rates. The cost of professional consultant services or contract services of technical personnel not directly engaged on the Joint Property shall not be charged to the Joint Account unless previously agreed to by the Parties.

8. EQUIPMENT AND FACILITIES FURNISHED BY OPERATOR

A. Operator shall charge the Joint Account for use of Operator owned equipment and facilities at rates commensurate with costs of ownership and operation. Such rates shall include costs of maintenance, repairs, other operating expense, insurance, taxes, depreciation, and interest on gross investment less accumulated depreciation not to exceed TEN percent (10%) per annum. Such rates shall not exceed average commercial rates currently prevailing in the immediate area of the Joint Property.

B. In lieu of charges in Paragraph 8A above, Operator may elect to use average commercial rates prevailing in the immediate area of the Joint Property less 20%. For automotive equipment, Operator may elect to use rates published by the Petroleum Motor Transport Association.

9. DAMAGES AND LOSSES TO JOINT PROPERTY

All costs or expenses necessary for the repair or replacement of Joint Property made necessary because of damages or losses incurred by fire, flood, storm, theft, accident, or other cause, except those resulting from Operator's gross negligence or willful misconduct. Operator shall furnish Non-Operator written notice of damages or losses incurred as soon as practicable after a report thereof has been received by Operator.

10. LEGAL EXPENSE

Expense of handling, investigating and settling litigation or claims, discharging of liens, payment of judgments and amounts paid for settlement of claims incurred in or resulting from operations under the agreement or necessary to protect or recover the Joint Property, except that no charge for services of Operator's legal staff or fees or expense of outside attorneys shall be made unless previously agreed to by the Parties. All other legal expense is considered to be covered by the overhead provisions of Section III unless otherwise agreed to by the Parties, except as provided in Section I, Paragraph 3.

11. TAXES

All taxes of every kind and nature assessed or levied upon or in connection with the Joint Property, the operation thereof, or the production therefrom, and which taxes have been paid by the Operator for the benefit of the Parties. If the ad valorem taxes are based in whole or in part upon separate valuations of each party's working interest, then notwithstanding anything to the contrary herein, charges to the Joint Account shall be made and paid by the Parties hereto in accordance with the tax value generated by each party's working interest.

12. INSURANCE

Net premiums paid for insurance required to be carried for the Joint Operations for the protection of the Parties. In the event Joint Operations are conducted in a state in which Operator may act as self-insurer for Worker's Compensation and/or Employers Liability under the respective state's laws, Operator may, at its election, include the risk under its self-insurance program and in that event, Operator shall include a charge at Operator's cost not to exceed manual rates.

13. ABANDONMENT AND RECLAMATION

Costs incurred for abandonment of the Joint Property, including costs required by governmental or other regulatory authority.

14. COMMUNICATIONS

Cost of acquiring, leasing, installing, operating, repairing and maintaining communication systems, including radio and microwave facilities directly serving the Joint Property. In the event communication facilities/systems serving the Joint Property are Operator owned, charges to the Joint Account shall be made as provided in Paragraph 8 of this Section II.

15. OTHER EXPENDITURES

Any other expenditure not covered or dealt with in the foregoing provisions of this Section II, or in Section III and which is of direct benefit to the Joint Property and is incurred by the Operator in the necessary and proper conduct of the Joint Operations.

III. OVERHEAD

1. OVERHEAD - DRILLING AND PRODUCING OPERATIONS

(i) As compensation for administrative, supervision, office services and warehousing costs, Operator shall charge drilling and producing operations on either:

(X) Fixed Rate Basis, Paragraph lA, or
( ) Percentage Basis, Paragraph lB

Unless otherwise agreed to by the Parties, such charge shall be in lieu of costs and expenses of all offices and salaries or wages plus applicable burdens and expenses of all personnel, except those directly chargeable under Paragraph 3A, Section II. The cost and expense of services from outside sources in connection with matters of taxation, traffic, accounting or matters before or involving governmental agencies shall be considered as included in the overhead rates provided for in the above selected Paragraph of this Section III unless such cost and expense are agreed to by the Parties as a direct charge to the Joint Account.

(ii) The salaries, wages and Personal Expenses of Technical Employees and/or the cost of professional consultant services and contract services of technical personnel directly employed on the Joint Property:

( ) shall be covered by the overhead rates, or
(X) shall not be covered by the overhead rates.

(iii) The salaries, wages and Personal Expenses of Technical Employees and/or costs of professional consultant services and contract services of technical personnel either temporarily or permanently assigned to and directly employed in the operation of the Joint Property, irrespective of where such employee is located.

( ) shall be covered by the overhead rates, or
(X) shall not be covered by the overhead rates.

A. Overhead - Fixed Rate Basis

(1) Operator shall charge the Joint Account at the following rates per well per month:
Drilling Well Rate $10,000 – HORIZONTAL WELL; $6,000 -VERTICAL WELL (Prorated for less than a full month) Producing Well Rate $1,000 - HORIZONTAL WELL; $700 - VERTICAL INJECTOR OR PRODUCER.

(2) Application of Overhead – Fixed Rate Basis shall be as follows:

(a) Drilling Well Rate

(1) Charges for drilling wells shall begin on the date the well is spudded and terminate on the date the drilling rig, completion rig, or other units used in completion of the well is released, whichever is later, except that no charge shall be made during suspension of drilling or completion operations for fifteen (15) or more consecutive calendar days.

(2) Charges for wells undergoing any type of workover or recompletion for a period of five (5) consecutive work days or more shall be made at the drilling well rate. Such charges shall be applied for the period from date workover operations, with rig or other units used in workover, commence through date of rig or other unit release, except that no charge shall be made during suspension of operations for fifteen (15) or more consecutive calendar days.

(b) Producing Well Rates

(1) An active well either produced or injected into for any portion of the

month shall be considered as a one-well charge for the entire month.

(2) Each active completion in a multi-completed well in which production is not commingled down hole shall be considered as a one-well charge providing each completion is considered a separate well by the governing regulatory authority.

(3) An inactive gas well shut in because of overproduction or failure of purchaser to take the production shall be considered as a one-well charge providing the gas well is directly connected to a permanent sales outlet.

(4) A one-well charge shall be made for the month in which plugging and abandonment operations are completed on any well. This one-well charge shall be made whether or not the well has produced except when drilling well rate applies.

(5) All other inactive wells (including but not limited to inactive wells covered by unit allowable, lease allowable, transferred allowable, etc.) shall not qualify for an overhead charge. (3) The well rates shall be adjusted as of the first day of pril each year following the effective date of the agreement to which this Accounting Procedure is attached. The adjustment shall be computed by multiplying the rate urrently in use by the percentage increase or decrease in the average weekly earnings of Crude Petroleum and Gas Production Workers for the last calendar year compared to the calendar year preceding as shown by the index of average weekly earnings of Crude Petroleum and Gas Production Workers as published by the United States Department of Labor, Bureau of Labor Statistics, or the equivalent Canadian index as published by Statistics Canada, as applicable. The adjusted rates shall be the rates currently in use, plus or minus the computed adjustment.

2. OVERHEAD – MAJOR CONSTRUCTION

To compensate Operator for overhead costs incurred in the construction nd installation of fixed assets, the expansion of fixed assets, and ny other project clearly discernible as a fixed asset required for the evelopment and operation of the Joint Property, Operator shall either egotiate a rate prior to the beginning of construction, or shall harge the Joint Account for overhead based on the following rates for any Major onstruction project in excess of $____________________:

A. 5% of first $100,000 or total cost if less, plus
B. 3% of costs in excess of $100,000 but less than $1,000,000, plus
C. 2% of costs in excess of $1,000,000.

Total cost shall mean the gross cost of any one project. For the purpose of this paragraph, the component parts of a single project shall not be treated separately and the cost of drilling and workover wells and artificial lift equipment shall be excluded.

3. CATASTROPHE OVERHEAD

To compensate Operator for overhead costs incurred in the event of expenditures resulting from a single occurrence due to oil spill, blowout, explosion, fire, storm, hurricane, or other catastrophes as agreed to by the Parties, which are necessary to restore the Joint Property to the equivalent condition that existed prior to the event causing the expenditures, Operator shall either negotiate a rate prior to charging the Joint Account or shall charge the Joint Account for overhead based on the following rates:

A. 5% of total costs through $100,000; plus

B. 3% of total costs in excess of $100,000 but less than $1,000,000; plus

C. 2% of total costs in excess of $1,000,000.

Expenditures subject to the overheads above will not be reduced by insurance recoveries, and no other overhead provisions of this Section III shall apply.

4. AMENDMENT OF RATES

The overhead rates provided for in this Section III may be amended from time to time only by mutual agreement between the Parties hereto if, in practice, the rates are found to be insufficient or excessive.

IV. PRICING OF JOINT ACCOUNT MATERIAL PURCHASES, TRANSFERS AND DISPOSITIONS

Operator is responsible for Joint Account Material and shall make proper and timely charges and credits for all Material movements affecting the Joint Property. Operator shall provide all Material for use on the Joint Property; however, at Operator's option, such Material may be supplied by the Non-Operator. Operator shall make timely disposition of idle and/or surplus Material, such disposal being made either through sale to Operator or Non-Operator, division in kind, or sale to outsiders. Operator may purchase, but shall be under no obligation to purchase, interest of Non-Operators in surplus condition A or B Material. The disposal of surplus Controllable Material not purchased by the Operator shall be agreed to by the Parties.

1. PURCHASES

Material purchased shall be charged at the price paid by Operator after deduction of all discounts received. In case of Material found to be defective or returned to vendor for any other reasons, credit shall be passed to the Joint Account when adjustment has been received by the Operator.

2. TRANSFERS AND DISPOSITIONS

Material furnished to the Joint Property and Material transferred from the Joint Property or disposed of by the Operator, unless otherwise agreed to by the Parties, shall be priced on the following basis exclusive of cash discounts:

A. New Material (Condition A)

(1) Tubular Goods Other than Line Pipe

(a) Tubular goods, sized 2 3/8 inches OD and larger, except line pipe, shall be priced at Eastern mill published carload base prices effective as of date of movement plus transportation cost using the 80,000 pound carload weight basis to the railway receiving point nearest the Joint Property for which published rail rates for tubular goods exist. If the 80,000 pound rail rate is not offered, the 70,000 pound or 90,000 pound rail rate may be used. Freight charges for tubing will be calculated from Lorain, Ohio and casing from Youngstown, Ohio.

(b) For grades which are special to one mill only, prices shall be computed at the mill base of that mill plus transportation cost from that mill to the railway receiving point nearest the Joint Property as provided above in Paragraph 2.A.(1)(a). For transportation cost from points other than Eastern mills, the 30,000 pound Oil Field Haulers Association interstate truck rate shall be used.

(c) Special end finish tubular goods shall be priced at the lowest published out-of-stock price, f.o.b. Houston, Texas, plus transportation cost, using Oil Field Haulers Association interstate 30,000 pound truck rate, to the railway receiving point nearest the Joint Property.

(d) Macaroni tubing (size less than 2 3/8 inch OD) shall be priced at the lowest published out-of-stock prices f.o.b. the supplier plus transportation costs, using the Oil Field Haulers Association interstate truck rate per weight of tubing transferred, to the railway receiving point nearest the Joint Property.

(2) Line Pipe

(a) Line pipe movements (except size 24 inch OD and larger with walls 3/4 inch and over) 30,000 pounds or more shall be priced under provisions of tubular goods pricing in Paragraph A.(l)(a) as provided above. Freight charges shall be calculated from Lorain, Ohio.

(b) Line Pipe movements (except size 24 inch OD) and larger with walls 3/4 inch and over) less than 30,000 pounds shall be priced at Eastern mill published carload base prices effective as of date of shipment, plus 20 percent, plus transportation costs based on freight rates as set forth under provisions of tubular goods pricing in Paragraph A.(1)(a) as provided above. Freight charges shall be calculated from Lorain, Ohio.

(c) Line pipe 24 inch OD and over and 3/4 inch wall and larger shall be priced f.o.b. the point of manufacture at current new published prices plus transportation cost to the railway receiving point nearest the Joint Property.

(d) Line pipe, including fabricated line pipe, drive pipe and conduit not listed on published price lists shall be priced at quoted prices plus freight to the railway receiving point nearest the Joint Property or at prices agreed to by the Parties.

(3) Other Material shall be priced at the current new price, in effect at date of movement, as listed by a reliable supply store nearest the Joint Property, or point of manufacture, plus transportation costs, if applicable, to the railway receiving point nearest the Joint Property.

(4) Unused new Material, except tubular goods, moved from the Joint Property shall be priced at the current new price, in effect on date of movement, as listed by a reliable supply store nearest the Joint Property, or point of manufacture, plus transportation costs, if applicable, to the railway receiving point nearest the Joint Property. Unused new tubulars will be priced as provided above in Paragraph 2.A.(l) and (2).

B. Good Used Material (Condition B) Material in sound and serviceable condition and suitable for reuse without reconditioning:

(1) Material moved to the Joint Property
At seventy-five percent (75%) of current new price, as determined by Paragraph A.

(2) Material used on and moved from the Joint Property
   (a) At seventy-five percent (75%) of current new price, as determined by Paragraph A, if Material was originally charged to the Joint Account as new Material or
   (b) At sixty-five percent (65%) of current new price, as determined by Paragraph A, if Material was originally charged to the Joint Account as used Material

(3) Material not used on and moved from the Joint Property At seventy-five percent (75%) of current new price as determined by Paragraph A. The cost of reconditioning, if any, shall be absorbed by the transferring property.

C. Other Used Material

(1) Condition C
Material which is not in sound and serviceable condition and not suitable for its original function until after reconditioning shall be priced at fifty percent (50%) of current new price as determined by Paragraph A. The cost of reconditioning shall be charged to the receiving property, provided Condition C value plus cost of reconditioning does not exceed Condition B value.

(2) Condition D
Material, excluding junk, no longer suitable for its original purpose, but usable for some other purpose shall be priced on a basis commensurate with its use. Operator may dispose of Condition D Material under procedures normally used by Operator without prior approval of Non-Operators.
   (a) Casing, tubing, or drill pipe used as line pipe shall be priced as Grade A and B seamless line pipe of comparable size and weight.

Used casing, tubing or drill pipe utilized as line pipe shall be priced at used line pipe prices.

(b) Casing, tubing or drill pipe used as higher pressure service lines than standard line pipe, e.g. power oil lines, shall be priced under normal pricing procedures for casing, tubing, or drill pipe. Upset tubular goods shall be priced on a non upset basis.

(3) Condition E Junk shall be priced at prevailing prices. Operator may dispose of Condition E Material under procedures normally utilized by Operator without prior approval of Non-Operators.

D. Obsolete Material

Material which is serviceable and usable for its original function but condition and/or value of such Material is not equivalent to that which would justify a price as provided above may be specially priced as agreed to by the Parties. Such price should result in the Joint Account being charged with the value of the service rendered by such Material.

E. Pricing Conditions

(1) Loading or unloading costs may be charged to the Joint Account at the rate of twenty-five cents (25(cent)) per hundred weight on all tubular goods movements, in lieu of actual loading or unloading costs sustained at the stocking point. The above rate shall be adjusted as of the first day of April each year following January 1, 1985 by the same percentage increase or decrease used to adjust overhead rates in Section III, Paragraph 1.A.(3). Each year, the rate calculated shall be rounded to the nearest cent and shall be the rate in effect until the first day of April next year. Such rate shall be published each year by the Council of Petroleum Accountants Societies.

(2) Material involving erection costs shall be charged at applicable percentage of the current knocked-down price of new Material.

3. PREMIUM PRICES

Whenever Material is not readily obtainable at published or listed prices because of national emergencies. strikes or other unusual causes over which the Operator has no control, the Operator may charge the Joint Account for the required Material at the Operator's actual cost incurred in providing such Material, in making it suitable for use, and in moving it to the Joint Property; provided notice in writing is furnished to Non-Operators of the proposed charge prior to billing Non-Operators for such Material. Each Non-Operator shall have the right, by so electing and notifying Operator within ten days after receiving notice from Operator, to furnish in kind all or part of his share of such Material suitable for use and acceptable to Operator.

4. WARRANTY OF MATERIAL FURNISHED BY OPERATOR

Operator does not warrant the Material furnished. In case of defective Material, credit shall not be passed to the Joint Account until adjustment has been received by Operator from the manufacturers or their agents.

V. INVENTORIES

The Operator shall maintain detailed records of Controllable Material.

1. PERIODIC INVENTORIES, NOTICE AND REPRESENTATION

At reasonable intervals, inventories shall be taken by Operator of the Joint Account Controllable Material. Written notice of intention to take inventory shall be given by Operator at least thirty (30) days before any inventory is to begin so that Non-Operators may be represented when any inventory is taken. Failure of Non-Operators to be represented at an inventory shall bind Non-Operators to accept the inventory taken by Operator.

2. RECONCILIATION AND ADJUSTMENT OF INVENTORIES

Adjustments to the Joint Account resulting from the reconciliation of a physical inventory shall be made within six months following the taking of the inventory. Inventory adjustments shall be made by Operator to the Joint Account for overages and shortages, but, Operator shall be held accountable only for shortages due to lack of reasonable diligence.

3. SPECIAL INVENTORIES

Special inventories may be taken whenever there is any sale, change of interest, or change of Operator in the Joint Property. It shall be the duty of the party selling to notify all other Parties as quickly as possible after the transfer of interest takes place. In such cases, both the seller and the purchaser shall be governed by such inventory.

In cases involving a change of Operator, all Parties shall be governed by such inventory.

4. EXPENSE OF CONDUCTING INVENTORIES

A. The expense of conducting periodic inventories shall not be charged to the Joint Account unless agreed to by the Parties.
B. The expense of conducting special inventories shall be charged to the Parties requesting such inventories, except inventories required due to change of Operator shall be charged to the Joint Account.

EXHIBIT "D"

Attached to and made a part of that certain Operating Agreement dated January 15, 2004, by and between Ivanhoe Energy (USA) Inc., as Operator and Derek Resources (U.S.A.) Inc. as Non-Operator.

INSURANCE

1. At all times during the conduct of operations hereunder, Operator shall maintain in force the following minimum limits of insurance at the expense of, and for the benefit of the Joint Account:

Pursuant to Insurance Requirements, as provided under Article XIII of the Farmin Agreement, of which this Operating Agreement is attached to and made a part thereof.

EXHIBIT "E"

Attached to and made a part of that certain Operating Agreement dated January 15, 2004, by and between Ivanhoe Energy (USA) Inc., as Operator and Derek Resources (U.S.A.) Inc., as Non-Operator.

Gas Balancing Agreement Onshore

I. Definitions

A. "Agreement" shall mean this Gas Balancing Agreement.

B. "Balanced" is that condition which occurs when a party hereto has taken the same percentage of the cumulative volume of Gas production it is entitled to take pursuant to the terms of the Operating Agreement.

C. "Contract Area" is the area affected by the Operating Agreement.

D. "Gas" includes natural gas produced from a Well that produces Gas Well Gas, including all constituent parts of such natural gas, except liquid hydrocarbons and condensate recovered by primary separation equipment.

E. "Gas Well Gas" is gas produced from a Well classified as a gas well by the regulatory body having jurisdiction.

F. "Oil Well Gas" is gas produced from a Well classified as an oil well by the regulatory body having jurisdiction.

G. "Overproduced" is the status of a party when the percentage of the cumulative volume of Gas taken by that party exceeds that party's percentage interest of the volume of cumulative Gas production of all parties to the Operating Agreement under and pursuant to the terms of said Operating Agreement.

H. "Underproduced" is the status of a party when the percentage of cumulative volume of Gas taken by that party is less than that party's percentage interest of the volume of cumulative Gas production of all parties to the Operating Agreement under and pursuant to the terms of said Operating Agreement.

I. "Well" is defined as each well subject to the Operating Agreement that produces Gas. If a single Well is completed in two or more reservoirs, such Well shall be considered a separate Well with respect to, but only with respect to, each reservoir from which the Gas produced is not commingled in the wellbore.

II. Application of this Agreement

The parties to the Operating Agreement to which this Gas Balancing Agreement is attached, own the working or operating interests in the Gas underlying the Contract Area covered by such Agreement and are entitled to share in the percentages therein as stated in the Operating Agreement.

In accordance with the terms of the Operating Agreement, each party shall take its share of Gas produced from the Contract Area and market or otherwise dispose of same. In the event a party hereto does not take in kind or market its share of gas or has contracted to sell its share of Gas produced from the Contract Area to a purchaser which, at any time while this Agreement is in effect, fails to take the share of Gas attributable to the interest of such party, the terms of this Gas Balancing Agreement shall automatically become effective.

The Operator has the duty to control Gas Production and the responsibility of administering the provisions of this Gas Balancing Agreement. The Operator shall cause deliveries to be made to the Gas purchasers at such rates as may be required to give effect to the intent that the Gas Production accounts of all parties are, to the extent practicable, to be or become Balanced.

III. Storing and Making Up Gas Production

A. Right to Take and Market Gas

During any period or periods when any party hereto does not take, has no market for, or the market of a party is not sufficient to take that party's full share of the Gas produced from a Well located on the Contract Area, or such party's purchaser otherwise fails to take such party's share of Gas produced from any such Well in the contract Area, resulting in such party becoming Underproduced (such party being herein referred to as an "Underproduced party") the other party or parties shall be entitled, but not required, to produce from said Well in the Contract Area and take or deliver to their respective purchaser(s), each month, all or a part of that portion of the allowable Gas production assigned to such Well by the regulatory body having jurisdiction. Any party so taking or delivering Gas which results in such party becoming Overproduced is herein referred to as an "Overproduced party".

Those parties which are capable of taking and/or marketing quantities of Gas allocable to an Underproduced party, in the absence of any other agreement between them, shall each take a share of the Gas attributed to the Underproduced party or parties in the direct proportion that their respective interests bear to the total interest of all parties taking Gas who are also considered Overproduced.

All parties hereto shall share in and own the liquid hydrocarbons recovered from such Gas by primary separation equipment, in accordance with their respective interests and subject to the terms of the above-described Operating Agreement, whether or not such parties are actually taking and/or marketing Gas at such time.

B. Making Up Underproduction

Each party failing to market its share of the total volume of Gas produced or failing to take its full share of the total volume of Gas produced shall be considered Underproduced and shall be credited with Gas in storage equal to its percentage share of the total volume of Gas produced under this Agreement, less that portion of the Gas actually marketed or taken by such party, gas used in operations, vented, or lost. Any Underproduced party shall endeavor to bring its taking of Gas into a Balanced condition. Upon at least 30 days written notice to the Operator, any Underproduced party may thereafter begin taking or delivering to its purchaser its full share of the Gas produced from a Well (less any used in operations, vented or lost). To allow an Underproduced party to balance, upon 30 days written notice, an Underproduced party shall be entitled to take or deliver to a purchaser its full share of Gas produced from such Well (less any used in operations, vented, or lost) plus, (i) for the months of March, April, May, June, July, August, September and October only of any calendar year or years during which this agreement may be in place, an amount up to an additional fifty percent (50%) of the monthly quantity of gas attributable to the Overproduced party or parties, or (ii) for the months of November, December, January and February only of any calendar year or years during which this agreement may be in place, an amount up to an additional fifteen percent (15%) of the monthly quantity of Gas attributable to the Overproduced party or parties, beginning the first of the month immediately following the twenty (20) day written notice.

If more than one Underproduced party is entitled to take additional Gas, they shall divide the additional Gas in proportion to their respective Underproduced accounts. The first Gas made up shall be assumed to be the first Gas Underproduced.

C. Gas Balance Reporting

Each party taking Gas shall furnish or cause to be furnished to the Operator a monthly written statement of Gas volumes taken and the identity of its Gas purchaser, if any, no later than twenty (20) days after the production month. Operator shall not be required to adjust its Gas accounting statements reflecting a different Gas purchaser until the first day of the month following the month in which such notice is received by the Operator. The Operator will maintain appropriate accounting on a monthly and cumulative basis of the quantities of Gas each party is entitled to take and/or market and the quantities of Gas taken and/or marketed by each of the parties to their respective Gas purchasers. With respect to Gas purchased from or transported for more than one party by or through one pipeline connected to the Well, each party selling to or transporting through such one pipeline shall furnish to Operator or cause the pipeline owner to furnish to Operator monthly volume statements showing the split of ownership through such pipeline's sales or pipeline inlet meter during the preceding calendar month. Within ninety (90) days after the end of each producing calendar month, the Operator shall furnish each party a statement showing the status of the Overproduced and Underproduced accounts of all parties.

To determine respective volume of Gas taken by separate gas pipeline connected to the Well, measurement of Gas for overproduction and underproduction shall be accomplished by use of sales meters and well measurement equipment, which shall be in accordance with AGA requirements.

Each party to this agreement agrees that it will not utilize any information obtained hereunder for any purpose other than implementing or administering the terms of this Gas Balancing Agreement.

D. Royalty and Production Tax

At all times while Gas is produced from the Contract Area, unless otherwise required by any State or Federal law or regulations, each party shall pay or cause to be paid all royalty due and payable on its share of Gas production as if each party were taking or delivering to a Gas purchaser its share of Gas production. Each party agrees to hold each other party harmless from any and all claims for royalty payments asserted by its royalty owners. The term "royalty owner" shall include owners of royalty, overriding royalties, production payments, and similar interests payable out of production.

Each party producing and taking or delivering Gas to its Gas purchaser shall pay, or cause to be paid, all production and severance taxes due on all volumes of Gas actually taken or sold by such party.

IV. Cash Settlement

A. Final Cash Balancing

Should production of gas from the Contract Area be permanently discontinued before the gas accounts are balanced, the Operator shall make a final determination of the volume of the last accrued over- and underproduction, if any, as of the date of such permanent discontinuance and the identity of the party(s) who are over- or underproduced. A cash settlement will then be made between the Underproduced and Overproduced parties. In making such settlement, the Underproduced parties shall be paid a sum of money by the Overproduced parties equal to the value, computed as hereinafter set forth, of the unrecouped cumulative balance of overproduction, adjusted for applicable taxes and royalties theretofore paid.

In determining the value of the unrecouped cumulative balance of overproduction, beginning with the most recent month in which the Overproduced parties took a volume of gas in excess of the quantity to which such parties were entitled (hereafter called "overage") the volume of overage during such month shall be multiplied times the actual prices received for such overage during such month. The same calculation shall be made for the next preceding month in which an overage occurred and for each preceding month (progressing backward in time) in which an overage occurred until the total volume of the overages for these months equals the total volume of the unrecouped cumulative balance of overproduction for purposes of the cash settlement herein contemplated. Within ninety (90) days of permanent discontinuance, the Operator will supply to the Non-Operator final volume gas balancing statements. Within one hundred twenty (120) days of permanent discontinuance, the Overproduced parties will pay or cause to be paid to the Underproduced parties their share of overage. Operator assumes no liability with respect to any such third party payments due any Underproduced party. If refunds are later required by any governmental authority, each party shall be accountable for its respective share of such refunds as finally balanced hereunder.

B. Collection and Distribution

Operator shall provide within thirty (30) days of permanent termination of Gas production a final accounting of the Gas balance to all parties hereto. Overproduced parties, within thirty (30) days of receipt of the final accounting of the Gas balance, shall provide Operator with a monthly statement of volume for each month during which overproduction occurred that has not been made-up. Within thirty (30) days after the receipt of such monthly statements for Overproduced parties, Operator shall calculate and invoice each Overproduced party for its share of the cash settlement, based on said volume statements, and the Cash Out Value due each Underproduced party. Overproduced parties shall make settlement, based on the invoiced amount, to the Operator within thirty (30) days after receipt of said invoice. Such payment shall relieve an Overproduced party of liability to any other party for the sums paid. Operator shall promptly distribute the funds it receives to the Underproduced parties in that production that each Underproduced party's volume of gas in storage bears to the total of all Underproduced parties volumes for Gas in storage.

C. Responsibility and Liability for Collection

Operator shall not be liable to any Underproduced party for the failure of any Overproduced party to any amounts owned pursuant to the terms hereof. In the event that any party fails to pay any sum due under the terms hereof after demand therefore by the Operator, the Operator may turn responsibility for the collection of such sum to the party or parties to whom it is owned, and Operator shall have no further responsibility in the event that such sums are not paid. Any party shall have the right after expiration of thirty (30) days after the Operator shall have provided a final accounting of the Gas balance to all parties hereto to demand on thirty (30) days advance written notice to both Operator and all Overproduced parties that any payments due to such party for such party's Underproduced volumes shall be paid directly to such party by the Overproduced party(s), rather than being paid through Operator. In the event that any Overproduced party pays to Operator any sums due to an Underproduced party at any time after thirty (30) days following the receipt of such written notification of a demand that such Underproduced party receive such payment directly, the Overproduced party(s) shall continue to be liable to such Underproduced party for any sums so paid, until such payment is actually received by such Underproduced party. In no event shall Operator be liable or responsible for any amount of cash settlement based on a value asserted by an Underproduced party different than the value calculated based on the volume statements and Cash Out Value provided by the Overproduced party or parties.

D. Ownership Changes

In the event an Overproduced party intends to sell, assign, exchange or otherwise transfer any of its interest in a Well located on the Contract Area, such Overproduced party shall notify in writing the other working interest owners who are parties hereto in such Well of such fact within forty-five (45) days prior to closing the transaction. Any Underproduced party may demand of such Overproduced party in writing within twenty (20) days after receipt of the Overproduced party's notice of intent to sell, assign, exchange or otherwise transfer its interest in a Well, a cash settlement of its underproduction attributed to such Overproduced party's Overproduction in the Well. Any Underproduced party electing to cash settle with the Overproduced party shall thereby indemnify and hold the Overproduced party harmless against any causes of action, claims, losses or other actions which may be claimed by any third party, including, but not limited to, any purchaser of the Gas of the Underproduced party, as a result of the cash settlement. The Operator shall be notified of any such demand and of any cash settlement pursuant to this Paragraph IV.D., and the Gas balance accounts of the parties shall be adjusted accordingly. Any cash settlement pursuant to this Paragraph IV.D. shall be on the same basis as otherwise set forth in Paragraphs IV.A. through IV.C. hereof.

V. Miscellaneous

A. Term

This agreement shall remain in force and effect as long as the Operating Agreement to which it is attached remains in force and effect and thereafter until the Gas balance accounts between the parties are settled in full, and shall inure to the benefit of and be binding upon the parties hereto, their heirs, successors, legal representatives and assigns.

B. Expenses

Nothing herein shall change or affect each party's obligations to pay its proportionate share of all costs and liabilities incurred in operations on the Contract Area as its share thereof is set forth in the Operating Agreement to which this Agreement is attached.

C. Well Tests

Nothing herein shall be construed to deny any party the right, from time to time, to produce and take or deliver to its Gas purchaser up to one hundred percent (100%) of the entire well stream to meet the deliverability test required by its Gas purchaser, provided that such tests are reasonable in light of overall industry standards.

D. Monitoring of Takes of Production

Each party shall, at all times, use its best efforts to regulate its takes and deliveries from each well on said Contract Area so that no well will be shut-in for overproducing the allowable assigned thereto by the regulatory body having jurisdiction. Additionally, each party shall communicate, as necessary, the contents of this agreement to its respective Gas purchaser(s) or transporter(s) and shall monitor its deliveries to its respective Gas Purchaser(s) or transporter(s) so as to ensure to the greatest extent practicable that its Gas purchaser(s) or transporter(s) does not take Gas in excess of the quantities provided for herein.

E. Method of Tax Accounting

In the event Internal Revenue Service regulations require a uniform method of computing taxable income by all parties, each party agrees to compute and report income to the Internal Revenue Service based on the quantity of Gas taken for its account (the cumulative method) in accordance with such regulations.

EXHIBIT "F"

Attached to and made a part of that certain Operating Agreement dated January 15, 2004, by and between Ivanhoe Energy (USA) Inc., as Operator and Derek Resources (U.S.A.) Inc., as Non-Operator.

EQUAL OPPORTUNITY & NON-SEGREGATION OF FACILITIES CERTIFICATIONS AND AGREEMENTS

This contract shall be performed by Operator in compliance with all applicable laws, proclaimed, order, rules and regulations, including, without limitation, the following:

1. EQUAL EMPLOYMENT OPPORTUNITY
   A. Equal Opportunity Clause (41 CFR 60-1.4). (Applicable to all contracts for more than $10,000, individually; or if Operator has such contracts

or subcontracts with the Government in any 12-month period which have an aggregate total value (or can reasonable be expected to have an aggregate total value) exceeding \$10,000, the \$10,000 or under exemption does not apply, and the contracts are subject to the order and the regulations issued pursuant thereto regardless of whether any single contract exceeds \$10,000.)

The equal opportunity clause required by Executive Order 11246 of September 24, 1965, and prescribed in section 60-1.4 of Title 41 of the Code of Federal Regulations is incorporated by reference (as permitted by section 60-1.4(d) of said Regulations) as if set out in full at this point.

B. Certification of Nonsegregated Facilities (41 CFR 60-1.8). (Applicable only to contracts which are not exempt from the provisions of the Equal Opportunity Clause set out above.) Operator certifies that it does not, and will not, maintain or provide for its employees any segregated facilities at any of its establishments, and that it does not and will not permit its employees to perform their services at any location, under its control, where segregated facilities are maintained. Operator agrees that a breach of this certification is a violation of the Equal Opportunity Clause required by Executive Order 11246 of September 24, 1965.

As used in this certification, the term "segregated facilities" means any waiting rooms, work areas, rest rooms and wash room, restaurants and other eating areas, time clocks, locker rooms and other storage or dressing areas, parking lots, drinking fountains, recreation or entertainment areas, transportation, and housing facilities provided for employees which are segregated by explicit directive or are in fact segregated on the basis of race, color, religion, or national origin, because of habit, local custom, or otherwise.

Operator further agrees that (except where it has obtained identical certifications from proposed subcontractors for specific time periods), it will obtain identical certifications from proposed subcontractors prior to the award of subcontractors exceeding \$10,000 which are not exempt from the provisions of the Equal Opportunity Clause; that will forward that following notice to such proposed subcontractors (except where the proposed subcontractors have submitted identical certifications for specific time periods):

NOTICE TO PROSPECTIVE SUBCONTRACTORS OR REQUIREMENT FOR CERTIFICATIONS OF NONSEGREGATED FACILITIES

A certificate of Nonsegregated Facilities must be submitted prior to the award of a subcontract exceeding \$10,000 which is not exempt from the provisions of the Equal Opportunity Clause. The certification may be submitted either for each subcontract or for all subcontracts during a period (i.e., quarterly, semiannually, or annually).

C. Affirmative Action Compliance Program (41 CFR 60-1.4). (Applicable only if Operator (a) has 50 or more employees, and (b) has a contract for $50,000 or more). If required under Section 60-1.4 of Title 41 of Code of Federal Regulations, Operator certifies that it has developed, or agrees to develop, a written affirmative action program for each of its establishments within 120 days from the effectiveness of this contract or the first of the contracts of sale. Operator shall maintain such program until such time as it is no longer required by law or regulation. Operator shall maintain a copy of separate program for each establishment, including evaluation of utilization of minority group personnel and the job classifications tables, at each local office responsible for the personnel matters of such establishment.

D. Employer Information Report (41 CFR 60-1.7). (Applicable only if Operator (a) had 50 or more employees, (b) is non exempt pursuant to 41 CFR 60-1.5 from the requirement for filing Employer Information Report EEO-0, and (c) has a contract or subcontract amounting to $50,000 or more.)

If required under Section 60-1.7 of Title 41 of the Code of Federal Regulations to file, the Employer Information Report, Standard Form 100 (EEO-0), or such forms as may hereinafter be promulgated in its place, in accordance with the applicable instructions and will continue to file such report unless and until Operator is not required to so file by law or regulation.

2. AFFIRMATIVE ACTION FOR DISABLED VETERANS AND VETERANS OF THE VIETNAM ERA

A. Affirmative Action Clause (41 CFR 60-250.4). (Applicable only to contracts for $10,000 or more.)

The affirmative action clause prescribed in Section 60-250.4 of the Code of Federal Regulations is incorporated by reference (as permitted by Section 60-250.22 of said Regulations) as if set out in full at this point.

B. Affirmative Action Program (41 CFR 60-250.5). (Applicable to contracts for $10,000 or more only if Operator (a) has 50 or more employees and (b) holds a contract of $50,000 or more.)

The affirmation action program prescribed in sections 60-250.22 of said Regulations) as if set out in full at this point.

3. AFFIRMATIVE ACTION FOR HANDICAPPED WORKERS

A. Affirmative Action Clause (41 CFR 60-741.4). (Applicable to contracts for $2,500 or more.)

The affirmative action clause prescribed in Section 60-741.4 of Title 41 of the Code of Federal Regulations is incorporated herein by reference (as permitted by Section 60-741.22 of said Regulations) as if set out in full at this point.

B. Affirmative Action Program (41 CFR 60-741.5). (Applicable to contracts for $2,500 or more only if Operator (a) has 50 or more employees and (b) holds a contract of $50,000 or more.)

The affirmative action program prescribed in Sections 60-741.5 and 60-741.6 of Title 41 of the Code of Federal Regulations is incorporated by reference (as permitted by Section 60-741.22 of said Regulations) as if set out in full at this point.

4. MINORITY BUSINESS ENTERPRISES 41 CFR 1-1.13, FEDERAL PROCUREMENT REGULATIONS)

A. Utilization of Minority Business Enterprises (41 CFR 1-131310-2 (a)). (Applicable only to contracts which may exceed $10,000 except those, and all subcontracts thereunder, to be performed entirely outside the United States, its possessions, and Puerto Rico, and those of services of a personal nature.)

(1) It is the policy of the Government that minority business enterprises shall have the maximum practicable opportunity to participate in the performance of Government contracts.

(2) Operator agrees to use its best efforts to carry out this policy in the award of its subcontracts to the fullest extent consistent with the efficient performance of this contract. As used in this contract, the term "minority business enterprises" means a business, at least 50% of which is owned by minority group members or, in case of publicly owned businesses, at least 51 percent of the stock of which is owned by minority group members. For the purposes of this definition, minority group members are Negroes, Spanish speaking American persons, American-Oriental, American-Indians, American-Eskimos, and American-Aleut. Operator may rely on written representation by subcontractors regarding their status as minority business enterprises in lieu of an independent investigation.

B. Minority Business Enterprises Subcontracting Program (41 CFR 1-131310-2(b)). (Applicable to all contracts which may exceed $500,000 which contain the clause required by 41 CFR 1-1.1310-2(a) and which offer substantial subcontracting possibilities.)

(1) Operator agrees to establish and conduct a program which will enable minority business enterprises (as defined in the above clause entitled "Utilization of Minority business enterprises') to be considered fairly as subcontractors and suppliers under this contract. In this connection, Operator shall:

(a) Designate a liaison officer who will administer Operator's minority business enterprises program.

(b) Provide adequate and timely consideration of the potentialities of known minority business enterprises in all "make-or buy" decisions.

(c) Assure that known minority business enterprises will have an equitable opportunity to compete for subcontracts, particularly by arranging solicitations, time for the preparation of bids, quantities, specification, and delivery schedules so as to facilitate the participation of minority business enterprises.

(d) Maintain records showing (i) procedures which have been adopted to comply with the policies set forth in this clause, including the establishment of a source list of minority business enterprises, (ii) awards to minority business enterprises on the source list, and (iii) specific efforts to identify and award contracts to minority business enterprises.

(e) Include the Utilization of Minority Business Enterprises clause in subcontracts which offer substantial minority business enterprises subcontracting opportunities.

(f) Cooperate with the Contracting Officer in any studies and surveys of Operator's minority business enterprises procedures and practices that the Contracting Officer may from time to time conduct.

(g) Submit periodic reports of subcontracting to known minority business enterprises with respect to the records referred to in subparagraph (d), above, in such form and manner and at such time (not more often than quarterly) as the Contracting Officer may prescribe.

(2) Operator further agrees to insert, in any subcontract hereunder which may exceed $500,000, provisions which shall conform substantially to the language of this clause, including this Paragraph (2), and to notify the Contracting Officer of the names of such subcontractors.

EXHIBIT "H"

Attached to and made a part of that certain Operating Agreement dated January 15, 2004, by and between Ivanhoe Energy (USA) Inc., as Operator and Derek Resources (U.S.A.) Inc., as Non-Operator.

Model Form Recording Supplement to Operating Agreement and Financing Statement

THIS AGREEMENT, entered into by and between Ivanhoe Energy (USA) Inc. dba USA Ivanhoe Energy, Inc., in California (hereinafter referred to as "Operator") and the signatory party or parties other than Operator (hereinafter referred to individually as "Non-Operator", and collectively as "Non-Operators").

WHEREAS, the parties to this agreement are owners of Oil and Gas Leases and/or Oil and Gas Interests in the land identified in Exhibit "A" (said land, Leases and Interests being hereinafter called the "Contract Area"), and in any instance in which the Leases or Interests of a party are not of record, the record owner and the party hereto that owns the Interest or rights therein are reflected on Exhibit "A";

WHEREAS, the parties hereto have executed an Operating Agreement dated January 15, 2004 (herein the "Operating Agreement"), covering the

Contract Area for the purpose of exploring and developing such lands, Leases and Interests for Oil and Gas; and

WHEREAS, the parties hereto have executed this agreement for the purpose of imparting notice to all persons of the rights and obligations of the parties under the Operating Agreement and for the further purpose of perfecting those rights capable of perfection.

NOW, THEREFORE, in consideration of the mutual rights and obligations of the parties hereto, it is agreed as follows:

1. This agreement supplements the Operating Agreement, which Agreement in its entirety is incorporated herein by reference, and all terms used herein shall have the meaning ascribed to them in the Operating Agreement.
2. The parties do hereby agree that:
   A. The Oil and Gas Leases and/or Oil and Gas Interests of the parties comprising the Contract Area shall be subject to and burdened with the terms and provisions of this agreement and the Operating Agreement, and the parties do hereby commit such Leases and Interests to the performance thereof.
   B. The exploration and development of the Contract Area for Oil and Gas shall be governed by the terms and provisions of the Operating Agreement, as supplemented by this agreement.
   C. All costs and liabilities incurred in operations under this agreement and the Operating Agreement shall be borne and paid, and all equipment and materials acquired in operations on the Contract Area shall be owned, by the parties hereto, as provided in the Operating Agreement.
   D. Regardless of the record title ownership to the Oil and Gas Leases and/Oil and Gas Interests identified on Exhibit "A", all production of Oil and Gas from the Contract Area shall be owned by the parties as provided in the Operating Agreement; provided nothing contained in this agreement shall be deemed an assignment or cross-assignment of interests covered hereby.
   E. Each party shall pay or deliver, or cause to be paid or delivered, all burdens on its share of the production from the Contract Area as provided in the Operating Agreement.
   F. An overriding royalty, production payment, net profits interest or other burden payable out of production hereafter created, assignments of production given as security for the payment of money and those overriding royalties, production payments and other burdens payable out of production heretofore created and defined as Subsequently Created Interests in the Operating Agreement shall be (i) borne solely by the party whose interest is burdened therewith, (ii) subject to suspension if a party is required

to assign or relinquish to another party an interest which is subject to such burden, and (iii) subject to the lien and security interest hereafter provided if the party subject to such burden fails to pay its share of expenses chargeable hereunder and under the Operating Agreement, all upon the terms and provisions and in the times and manner provided by the Operating Agreement.

G. The Oil and Gas Leases and/or Oil and Gas Interests which are subject hereto may not be assigned or transferred except in accordance with those terms, provisions and restrictions in the Operating Agreement regulating such transfers. This agreement and the Operating Agreement shall be binding upon and shall inure to the benefit of the parties hereto, and their respective heirs, devisees, legal representatives, and assigns, and the terms hereof shall be deemed to run with the leases or interests included within the lease Contract Area.

H. The parties shall have the right to acquire an interest in renewal, extension and replacement leases, leases proposed to be surrendered, wells proposed to be abandoned, and interests to be relinquished as a result of non-participation in subsequent operations, all in accordance with the terms and provisions of the Operating Agreement.

I. The rights and obligations of the parties and the adjustment of interests among them in the event of a failure or loss of title, each party's right to propose operations, obligations with respect to participation in operations on the Contract Area and the consequences of a failure to participate in operations, the rights and obligations of the parties regarding the marketing of production, and the rights and remedies of the parties for failure to comply with financial obligations shall be as provided in the Operating Agreement.

J. Each party's interest under this agreement and under the Operating Agreement shall be subject to relinquishment for its failure to participate in subsequent operations and each party's share of production and costs shall be reallocated on the basis of such relinquishment, all upon the terms and provisions provided in the Operating Agreement.

K. All other matters with respect to exploration and development of the Contract Area and the ownership and transfer of the Oil and Gas Leases and/or Oil and Gas Interest therein shall be governed by the terms and provisions of the Operating Agreement.

3. The parties hereby grant reciprocal liens and security interests as follows:

A. Each party grants to the other parties hereto a lien upon any interest it now owns or hereafter acquires in Oil and Gas Leases and Oil and Gas Interests in the Contract Area, and a security interest and/or purchase money security interest in any interest it now owns or hereafter acquires in the personal property and fixtures on or used or obtained for use in connection therewith, to secure performance of all of its obligations under this agreement and the Operating Agreement including but not limited to payment of expense, interest and fees, the proper disbursement of all monies paid under this agreement and the Operating Agreement, the assignment or relinquishment of interest in Oil and Gas Leases as required under this agreement and the Operating Agreement, and the proper performance of operations under this agreement and the Operating Agreement. Such lien and security interest granted by each party hereto shall include such party's leasehold interests, working interests, operating rights, and royalty and overriding royalty interests in the Contract Area now owned or hereafter acquired and in lands pooled or unitized therewith or otherwise becoming subject to this agreement and the Operating Agreement, the Oil and Gas when extracted therefrom and equipment situated thereon or used or obtained for use in connection therewith (including, without limitation, all wells, tools, and tubular goods), and accounts (including, without limitation, accounts arising from the sale of production at the wellhead), contract rights, inventory and general intangibles relating thereto or arising therefrom, and all proceeds and products of the foregoing.

B. Each party represents and warrants to the other parties hereto that the lien and security interest granted by such party to the other parties shall be a first and prior lien, and each party hereby agrees to maintain the priority of said lien and security interest against all persons acquiring an interest in Oil and Gas Leases and Interests covered by this agreement and Operating Agreement by, through or under such party. All parties acquiring an interest in Oil and Gas Leases and Oil and Gas Interests covered by this agreement and the Operating Agreement, whether by assignment, merger, mortgage, operation of law, or otherwise, shall be deemed to have taken subject to the lien and security interest granted by the Operating Agreement and this instrument as to all obligations attributable to such interest under this agreement and the Operating Agreement whether or not such obligations arise before or after such interest is acquired.

C. To the extent that the parties have a security interest under the Uniform Commercial Code of the state in which the Contract Area is situated, they shall be entitled to exercise the rights and remedies of a secured party under the Code. The bringing of a suit and the obtaining of judgment by a party for the secured indebtedness shall not be deemed an election of remedies or otherwise affect the lien rights or security interest as security for the payment thereof. In addition, upon default by any party in the payment of its share of expenses, interest or fees, or upon the improper use of funds by the Operator, the other parties shall have the right, without prejudice to other rights or remedies, to collect from the purchaser the proceeds from the sale of such defaulting party's share of Oil and Gas until the amount owned by such party, plus interest, has been received, and shall have the right to offset the amount owned against the proceeds from the sale of such defaulting party or parties stating the amount due as a result of the default, and all parties waive any from the non-defaulting party or parties stating the amount due as a result of the default, and all parties waive any recourse available against purchasers for releasing production proceeds as provided in this paragraph.

D. If any party fails to pay its share of expenses within one hundred-twenty (120) days after rendition of a statement therefore by Operator the non-defaulting parties, including Operator, shall, upon request by Operator, pay the unpaid amount in the proportion that the interest of each such party bears to the interest of all such parties. The amount paid by each party so paying its share of the unpaid amount shall be secured by the liens and security rights described in this Paragraph 3 and in the Operating Agreement, and each paying party may independently pursue any remedy available under the Operating Agreement or otherwise.

E. If any party does not perform all of its obligations under this agreement of the Operating Agreement, and the failure to perform subjects such party to foreclosure or execution proceedings pursuant to the provisions of this agreement or the Operating Agreement, to the extent allowed by governing law, the defaulting party waives any available right of redemption from and after the date of judgment, any required valuation or appraisement of the mortgaged or secured property prior to sale, any available right to stay execution or to require a marshalling of assets and any required bond in the event a receiver is appointed. In addition, to the extent permitted by applicable law, each party hereby grants to the other parties a power of sale as to any property that is subject to the lien and security rights granted hereunder or under the Operating Agreement, such power to be exercised in the manner provided by applicable law or otherwise in a commercially reasonable manner and upon reasonable notice.

F. The lien and security interest granted in this Paragraph 3. supplements identical rights granted under the Operating Agreement.

G. To the extent permitted by applicable law, Non-Operators agree that Operator may invoke or utilize the mechanics' or materialmen's lien law of the state in which the Contract Area is situated in order to secure the payment to Operator of any sum due under this agreement and the Operating Agreement for services performed or materials supplied by Operator.

H. The above described security will be financed at the wellhead of the well or wells located on the Contract Area and this Recording Supplement may be filed in the land records in the County or Parish in which the Contract Area is located, and as a financing statement in all recording offices required under the Uniform Commercial Code or other applicable state statutes to perfect the above-described security interest, and any party hereto may file a continuation statement as necessary under the Uniform Commercial Code, or other state laws.

4. This agreement shall be effective as of the date of the Operating Agreement as above recited. Upon termination of this agreement and the Operating Agreement and the satisfaction of all obligations thereunder, Operator is authorized to file of record in all necessary recording offices a notice of termination, and each party hereto agrees to execute such a notice of termination as to Operator's interest, upon the request of Operator, if Operator has complied with all of its financial obligations.
5. This agreement and the Operating Agreement shall be binding upon and shall inure to the benefit of the parties hereto and their respective heirs, devisees, legal representatives, successors and assigns. No sale, encumbrance, transfer or other disposition shall be made by any party of any interest in the Leases or Interests subject hereto except as expressly permitted under the Operating Agreement and, if permitted, shall be made expressly subject to this agreement and the Operating Agreement and without prejudice to the rights of the other parties. If the transfer is permitted, the assignee of an ownership interest in any Oil and Gas Lease shall be deemed a party to this agreement and the Operating Agreement as to the interest assigned from and after the effective date of the transfer of ownership; provided, however, that the other parties shall not be required to recognize any such sale, encumbrance, transfer or other disposition for any purpose hereunder until thirty (30) days after they have received a copy of the instrument of transfer or other satisfactory evidence thereof in writing from the transferor or transferee. No assignment or other disposition of interest by a party shall relieve such party of obligations previously incurred by such party under this agreement or the Operating Agreement with respect to the interest transferred, including without limitation the obligation of a party to pay all costs attributable to an operation conducted under this agreement and the Operating Agreement in which such party has agreed to participate prior to making such assignment, and the lien and security interest granted by Article VII.B. of the Operating Agreement and hereby shall continue to burden the interest transferred to secure payment of any such obligations.
6. In the event of a conflict between the terms and provisions of this agreement and the terms and provisions of the Operating Agreement, then, as between the parties, the terms and provisions of the Operating Agreement shall control.

7. This agreement shall be binding upon each Non-Operator when this agreement or a counterpart thereof has been executed by such Non-Operator and Operator notwithstanding that this agreement is not then or thereafter executed by all of the parties to which it is tendered or which are listed on Exhibit "A" as owning an interest in the Contract Area or which own, in fact, an interest in the Contract Area. In the event that any provision herein is illegal or unenforceable, the remaining provisions shall not be affected, and shall be enforced as if the illegal or unenforceable provision did not appear herein. ______________ ______________________ who has prepared and circulated this form for execution, represents and warrants that the form was printed from and, with the exception(s) listed below, is identical to the AAPL Form 610RS-1989 Model Form Recording Supplement to Operating Agreement and Financing Statement, as published in computerized form by Forms On-A-Disk, Inc. No changes, alterations, or modifications, other than those made by strikethrough and/or insertion and that are clearly recognizable as changes in Articles ____________, have been made to the form.

IN WITNESS WHEREOF, this agreement shall be effective as of the 15th day of January 2004.

ATTEST OR WITNESS: OPERATOR

IVANHOE ENERGY (USA) INC.

______________________________ By: ______________________________

Title: ______________________________

______________________________ Date: ______________________________

NON-OPERATORS
DEREK RESOURCES (U.S.A.) INC.

______________________________ By: ______________________________

Title: ______________________________

______________________________ Date: ______________________________

5.4 TRANSPORTATION AGREEMENTS (PIPELINES)

In India demand for transporting petroleum products over long distances is expected to total 54.6 million tonnes in 2001-2002 and around 87.3 million tonnes in 2006-07. Pipeline transportation is the most efficient way of moving petroleum products across long distances. In India, barely 25 percent of long distance movement of petroleum products is by pipelines. In contrast, in the developed countries almost all such movement is through pipelines. Around Rs. 30,000 crore is expected to be invested over the next decade in setting up pipeline networks.

Globally, Pipeline transactions face unique and complex challenges that bring to bear an unusually broad array of legal specialties. In addition to the usual commercial and contractual issues that arise in virtually any transaction, pipeline deals often involve such distinct practice areas as the following:

1. Common carrier regulation in the countries
2. Economic regulation by state agencies in multiple states
3. Antitrust review by the government department
4. Pipeline safety regulation of the government
5. Right-of-way issues governed by state and International laws
6. Tax issues specifically applicable to the various kinds of pipeline entities
7. Negotiating transportation agreements for the transmission of crude oil, natural gas, refined petroleum products, and various petrochemicals
8. Handling contractual arrangements by pipeline, marine vessels, or rail cars
9. Rate development and rate design issues
10. Fuel requirements

International pipeline transactions raise a host of additional issues, by allowing unrestricted entry under a well-defined policy framework for the setting up and operation of pipelines, Set up new pipelines for products under the common carrier principle and place an even greater premium on experience and in-depth knowledge.

FORM OF FIRM TRANSPORTATION AGREEMENT
TRANSPORTATION AGREEMENT FOR FIRM TRANSPORTATION OF NATURAL GAS ALLIANCE PIPELINE L.P.

Firm Transportation Agreement No. _______

This TRANSPORTATION AGREEMENT FOR FIRM TRANSPORTATION OF NATURAL GAS ("**Firm Transportation Agreement" or "Agreement**") is made and entered into this ___ day of _____________, 20___, between:

ALLIANCE PIPELINE L.P., ("**Transporter**"), and _____________ _____________________________, (**"Shipper")**.

Witnesseth: That in consideration of the mutual covenants contained herein the parties agree as follows:

SECTION 1
SERVICE TO BE RENDERED

Transporter shall perform and Shipper shall receive Firm Transportation Service in accordance with the provisions of Transporter's effective Rate Schedule FT-1 and the applicable General Terms and Conditions (GTC) of Transporter's FERC Gas Tariff on file with the Federal Energy Regulatory Commission ("Commission") as the same may be amended or superseded in accordance with the Rules and Regulations of the Commission.

SECTION 2
TERM

This Agreement shall be effective as of the date first written above, for a term of ______ years. Shipper may extend the term of this Agreement for a minimum of ______________ upon __________ prior written notice of the extension. Pregranted abandonment of service shall apply upon termination of this agreement.

SECTION 3
RATES

[Shipper shall pay the currently effective Rate Schedule FT-1 Recourse Rates set forth at Sheet No. 10 of Transporter's Tariff, as such rates may be revised and superseded, subject to Commission approval, from time to time, unless Shipper has executed a Negotiated Rate Agreement in accordance with Section 39 of the GTC.]

OR

[Negotiated Rate]

SECTION 4
NOTICES

Notices to Transporter under this Agreement shall be addressed to:

Alliance Pipeline L.P.
6385 Old Shady Oak Road
Eden Prairie, MN 55344
Attention: Manager, Tariff Administration
Fax: (612) 944-9166

SECTION 5
SUPERSEDED AGREEMENTS

This Firm Transportation Agreement supersedes and cancels as of the effective date hereof the following agreements:

__________, __________.

IN WITNESS WHEREOF, the Parties have duly executed this Firm Transportation Agreement in several counterparts by their duly authorized officers with effect as of the day first above written.

ALLIANCE PIPELINE L.P. by its Managing General Partner, ALLIANCE PIPELINE INC.

[Date of Execution] Per: ______________________
Per: ______________________
[SHIPPER]
[Date of Execution] Per: ______________________
Per: ______________________

APPENDIX A

TO TRANSPORTATION AGREEMENT NO. [TA Contract No]

BETWEEN

ALLIANCE PIPELINE L.P.

and

[COMPANY FULL NAME]

CONTRACTED CAPACITY AND PRIMARY DELIVERY POINTS

Contracted Capacity ________ MMcf/day

Primary Delivery Point(s)	Primary Delivery Point Capacity (MMcf/day)
Total Primary Delivery Point Capacity (Not to exceed 150% of Contracted capacity)	

Agreement No. ________

APPENDIX B

RATE PRINCIPLES

(1) Subject to the incentive provisions, the reservation rates will be calculated on a per unit-of-capacity basis to provide for the recovery by the Transporter of all of the fixed costs of providing service. In addition, Shippers will pay a commodity or usage charge for volumes actually shipped, plus fuel.

(2) A deemed capital structure of 70% debt and 30% equity for the primary term and any extension of the primary term of the Firm Transportation Agreement.

(3) A cost of debt calculated using a rate of interest equal to the weighted average of the interest rates borne by Transporter's debt. Changes in Transporter's actual weighted average cost of debt will be reflected in Transporter's negotiated rates from time to time.

(4) Return on Equity

1. Base rate of return on equity of 12%.
2. Base rate of return on equity to be subject to an incentive adjustment. The resulting return on equity will apply for the primary term and any extension of the primary term of the Firm Transportation Agreement.
3. The base rate of return on equity will be increased or decreased inversely with increases or decreases in the actual capital cost versus the estimated capital cost of the Alliance Pipeline L.P. system. The adjustment formula will be linear so that any variations between actual and estimated capital cost will be reflected in the adjusted rate of return on equity. For example, a 10% increase in the actual capital cost, versus the estimated capital cost, would result in a decrease of 0.5% (50 basis points) in the base rate of return on equity. Similarly, a 20% decrease in the actual capital cost versus the estimated capital cost would result in an increase in the base rate of return on equity of 1.0% (100 basis points). The incentive rate of return increase or decrease will be limited to a maximum of 2% (200 basis points).
4. If (i) Transporter's facilities are not completed, tested and available to provide the service contemplated herein; or (ii) Transporter has not been granted permission to commence service by the Commission on or before November 1, 2001, then Transporter's return on equity shall be reduced by one percent (100 basis points) for the primary term set forth in section 2 of this Firm Transportation Agreement; provided, however, that Transporter's return on equity shall not be reduced if the delay beyond November 1, 2001 is attributable to (i) a Force Majeure event as defined in Transporter's FERC Gas Tariff; or (ii) a failure by Shipper to satisfy its obligations under this Firm Transportation Agreement.

(5) Income taxes will be calculated on a normalized basis, utilizing the federal and state corporate tax rates on income derived from pipeline operations for the primary term and any extension of the primary term of the Firm Transportation Agreement.

(6) The depreciation on transmission plant used for purposed of deriving rates will be calculated annually in accordance with Table 1. If, at any point, a Shipper elects not to exercise its rolling right to extend the primary term of its Firm Transportation Agreement, then during the final five (5) years of this Firm Transportation Agreement, that Shipper's rates will be adjusted so that the average depreciation rate over the term of the Firm Transportation Agreement is 4%.

(7) The rate base will include, among other things, actual capital costs.
(8) The actual reservation rates will be calculated based upon the higher of the sum of all of the Contracted Capacities or 1325 MMcfd, for the primary term and any extension of the primary term of the Firm Transportation Agreement.
(9) There will be a commodity or usage charge which will recover all of those costs that vary with volumes actually shipped, for the primary term and any extension of the primary term of the Firm Transportation Agreement.
(10) Fuel will be recovered on an actual tracked basis.
(11) The estimated capital cost of the Alliance Pipeline L.P. system, excluding AFUDC and at a system design of 1325 MMcfd, is U.S.$1,235,163,000 ($1235.2 million).
(12) Changes in Transporter's operating costs will be reflected in its rates from time to time, for the primary term and any extension of the primary term of the Firm Transportation Agreement.
(13) The rate for Authorized Overrun Service will be the negotiated commodity charge, plus fuel, for the primary term and any extension of the primary term of the Firm Transportation Agreement.

Table 1

Year	Depreciation Rate	Year	Depreciation Rate
1	0.290%	14	6.283%
2	0.751%	15	6.744%
3	1.212%	16	4.725%
4	1.673%	17	4.725%
5	2.134%	18	4.725%
6	2.595%	19	4.725%
7	3.056%	20	4.725%
8	3.517%	21	4.725%
9	3.978%	22	4.725%
10	4.439%	23	4.725%
11	4.900%	24	4.725%
12	5.361%	25	4.725%
13	5.822%		

CHAPTER 6

Negotiations

6.1 INTRODUCTION TO NEGOTIATIONS

What is Negotiation?

In simplest terms, negotiation is a discussion between two or more disputants who are trying to work out a solution to their problem. This interpersonal or inter-group process can occur at a personal level, as well as at a corporate or international (diplomatic) level. Negotiations typically take place because the parties wish to create something new that neither could do on his or her own, or to resolve a problem or dispute between them. The parties acknowledge that there is some conflict of interest between them and think they can use some form of influence to get a better deal, rather than simply taking what the other side will voluntarily give them. They prefer to search for agreement rather than fight openly, give in, or break off contact.

When parties negotiate, they usually expect give and take. While they have interlocking goals that they cannot accomplish independently, they usually do not want or need exactly the same thing. This interdependence can be either win-lose or win-win in nature, and the type of negotiation that is appropriate will vary accordingly. The disputants will either attempt to force the other side to comply with their demands, to modify the opposing position and move toward compromise, or to invent a solution that meets the objectives of all sides. The nature of their interdependence will have a major impact on the nature of their relationship, the way negotiations are conducted, and the outcomes of these negotiations.

Mutual adjustment is one of the key causes of the changes that occur during a negotiation. Both parties know that they can influence the other's outcomes and that the other side can influence theirs. The effective negotiator attempts to understand how people will adjust and readjust their positions during negotiations, based on what the other party does and is expected to do. The parties have to exchange information and make an effort to influence each other. As negotiations evolve, each side proposes changes to the other party's position and makes changes to its own. This process of give-and-take and making concessions is necessary if a settlement is to be reached. If one party makes several proposals that are rejected, and the other party makes no alternate proposal, the first party may break off negotiations. Parties typically will not want to concede too much if they do not sense that those with whom they are negotiating are willing to compromise.

The parties must work toward a solution that takes into account each person's requirements and hopefully optimizes the outcomes for both. As they try to find their way toward agreement, the parties focus on *interests, issues,* and *positions,* and use cooperative and/or competitive processes to come to an agreement.

6.2 APPROACHES TO NEGOTIATION

Negotiation theorists make several overlapping distinctions about approaches to negotiation. Fisher, Ury, and Patton distinguish between positional bargaining, which is competitive, and interest-based bargaining or principled negotiation, which is primarily cooperative. But they also make the distinction between soft, hard, and principled negotiation, the latter of which is neither soft, nor hard, but based on cooperative principles which look out for oneself as well as one's opponent.

Morton Deutsch also makes the distinction between competitive and cooperative approaches. According to Deutsch, the most important factors that determine whether an individual will approach a conflict cooperatively or competitively are the nature of the dispute and the goals each side seeks to achieve. Often the two sides' goals are linked together, or *interdependent.* The parties' interaction will be shaped by whether this interdependence is *positive* or *negative,* according to Deutsch:

Goals with *positive interdependence* are tied together in such a way that the chance of one side attaining its' goal is *increased* by the other side's attaining its goal. Positively interdependent goals normally result in cooperative approaches to negotiation, because any participant can "attain his goal if, and only if, the others with whom he is linked can attain their goals."

On the other hand, *negative interdependence* means the chance of one side attaining its goal is *decreased* by the other's success. Negatively interdependent goals force competitive situations, because the only way for one side to achieve its goals and "win" is for the other side to "lose."

Although Fisher, Ury, and Patton argue that almost any dispute can be resolved with interest-based bargaining (i.e., a cooperative approach), other theorists believe the two approaches should be used together. Lax and Sebenius, for example, argue that negotiations typically involve "creating" and "claiming" value. First, the negotiators work cooperatively to create value (that is, "enlarge the pie,") but then they must use competitive processes to claim value (that is, "divide up the pie").

However, a tension exists between creating and claiming value. This is because the competitive strategies used to claim value tend to undermine cooperation, while a cooperative approach makes one vulnerable to competitive bargaining tactics. The tension that exists between cooperation and competition in negotiation is known as "The Negotiator's Dilemma:"

If both sides cooperate, they will both have good outcomes. If one cooperates and the other competes, the cooperator will get a terrible outcome and the competitor will get a great outcome. If both compete, they will both have mediocre outcomes. In the face of uncertainty about what strategy the other side will adopt, each side's best choice is to compete. However, if they both compete, both sides end up worse off.

In real life, parties can communicate and commit themselves to a cooperative approach. They can also adopt norms of fair and cooperative behavior and focus on their future relationship. This fosters a cooperative approach between both parties and helps them to find joint gains.

Planning for Negotiations

Effective planning is crucial to meeting negotiation objectives. If the parties are to reach a stable agreement, specific events must take place before the parties ever come to the table.

1. Parties must frame the problem, and recognize that they have a common problem that they share an interest in solving. Frames are the conceptions that parties have of the situation and its risks. They allow the parties to begin to develop a shared definition of the issues involved, and the process needed to resolve them. When the frames of both parties match, they are more likely to focus on common issues and have a common definition of the situation. However, when the frames do not match, communication between the parties is likely to be more difficult. Unless the different outlooks on the problem begin to overlap, it is unlikely that negotiations will be successful. If negotiators understand what frame they are operating from and what frame the other is operating from, they may be able to shift the conversation and develop common definitions. The way in which parties define the problem can shape the rest of the planning process.
2. In the early stages of framing, negotiators must also determine their goals, anticipate what they want to achieve, and prepare for the negotiation process. They must define the issues to be discussed and analyze the conflict situation. In many cases, negotiators can appeal to research or consult with experts to help them develop a complete list of the issues at stake. Next, parties should assemble all the issues that have been defined into a comprehensive list. The combined list of issues and priorities from each side determines the negotiation agenda.
3. Negotiators often exchange and negotiate the list of issues to be discussed in advance. Consultation between negotiators prior to actual negotiation allows them to agree on the agenda of issues to be discussed, as well as the location of the negotiations, the time and duration of the sessions, the parties to be involved in the negotiations, and techniques to pursue if negotiation fails. Negotiators should also

agree on principles that will guide the drafting of a settlement, the procedures to be used in negotiations, and the formula by which a general agreement is to be reached. Discussions about these procedural issues are often crucial for the success of substantive negotiations. If parties cannot agree on negotiation procedures and proposed items for the agenda, they may very well decide to abandon the negotiations altogether.

4. After assembling issues on an agenda, the negotiators must prioritize their goals and evaluate the possible tradeoffs among them. Negotiators must be aware of their goals and positions and must identify the concerns, desires, and fears that underlie their substantive goals. They must determine which issues are most important, as well as whether the various issues are linked or separate. In addition, negotiators should be aware of the underlying interests and goals of the other side. Because the linkages between parties' goals often define the issue to be settled, these goals must be determined carefully. If one party wants more than the other party is capable or willing to give, the disputants must either change their goals or end the negotiation.
5. Once they have determined the relative importance of the issues, parties need to decide the order in which issues should be discussed. Many sequencing options are possible: going from easy to hard, hard to easy, or tackling everything together. Different situations suggest different answers to that question, and different negotiators and mediators prefer one approach over the others.
6. Negotiators that are operating on behalf of a constituency should consult with their constituents as well as with the other side to ensure that the constituents' needs and priorities are included in the negotiations.
7. The next step is for negotiators to define specific targets with respect to the key issues on the agenda. Parties should try to figure out the best resolution they can expect, what counts as a fair and reasonable deal, and what is a minimally acceptable deal. They should also be aware of the strongest points in their position and recognize the strongest points in the other side's position. This enables parties to become aware of the range of possible outcomes (see ZOPA) and to be flexible in what they will accept. It also improves the likelihood that they will arrive at a mutually satisfactory outcome.
8. Because negotiations typically involve more than one issue, it is helpful for negotiators to anticipate different ways of packaging issues. They can balance the issues they regard as most important by being more flexible about items they deem less important. They should also decide which items they can abandon and use as leverage to get what they really want with respect to the most important issues.
9. Planning for negotiation also involves the development of supporting arguments. Negotiators must be able to present supporting facts and arguments, anticipate how the other side will respond to these

arguments, and respond to the other party's claims with counter-arguments. This includes locating facts to support one's point of view, determining what sorts of arguments have been given in similar negotiations in the past, anticipating the arguments the other side is likely to make, and presenting facts in the most convincing way possible.

10. Finally, planning involves assessing the other party's priorities and interests and trying to get a better idea of what that party is likely to want. Negotiators should gather background information about the other party's current needs, resources, and interests. This can be done through preliminary interviews or consultations with those who have done business with the other party in the past. In addition, negotiators need to understand the other party's objectives. Professional negotiators will often exchange information about targets or initial proposals before negotiations begin. Third, negotiators should be aware of the other party's negotiation style, reputation, and the strategy and tactics they commonly use. They should investigate that party's past behavior in related settings, determine his or her organizational position, and find out whom he or she admires and whose advice carries weight. An individual's past negotiation behavior is a good indication of how he or she will behave in the future. Fourth, negotiators should understand the other party's alternatives. If the other negotiator has strong alternatives, he or she will probably be willing to set high objectives and be willing to push hard for these objectives during negotiation.

Sources of Power in Negotiation

Negotiation is a process of communication in which the parties aim to "send a message" to the other side and influence each other. Thus, power in negotiation lies in the ability to favorably affect someone else's decision. Some assume that because threats of physical force exert influence, the ability to make such threats is the essence of negotiating power. But making threats is a costly and dangerous way of trying to exert influence.

There is also a widespread belief that the best way to start a negotiation is with an extreme position. The idea is that negotiators should let others know they are in charge by taking a hard line, and should then soften their position later if appropriate. However, this may not be the most effective tactic. The more extreme the opening positions are, and the smaller the concessions, the more time and effort it often takes to move toward agreement. And as each side tries to use force to make the other change its position, anger and resentment result, putting a heavy strain on the relationship between the parties. Thus, it is a mistake to try to use force or threats before one has exhausted the other elements of negotiating power. Threatening to impose harsh consequences without having first made a firm and clear offer is actually likely to reduce a negotiator's level of power.

The ability to exert influence depends upon the combined total of a variety of factors. First, having a good alternative to negotiation contributes substantially to a negotiator's power. A negotiator with very strong alternatives does not need the negotiation in order to achieve at least a satisfactory outcome. In their 1981 bestseller, *Getting to Yes,* Roger Fisher and William Ury coined the term "BATNA" (best alternative to a negotiated agreement) to refer to this type of negotiating power. When parties have many options other than negotiation, they have more leverage in making demands. Therefore, parties should develop a strong understanding of their alternatives before participating in negotiations. Making one's BATNA as strong as possible, and then making that BATNA known to one's opponent, can strengthen one's negotiating position.

Second, a skilled negotiator who knows about the people and interests involved as well as the relevant facts is better able to influence the decisions of others. The abilities to listen, to empathize, and to communicate clearly and effectively are crucial in negotiating effective agreements. Likewise, an awareness of various negotiating styles and cultural differences can be a huge asset. A good working relationship also increases a negotiator's level of power. Such relationships are characterized by trust and the ability to communicate effectively and easily. If a negotiator establishes a reputation for candor and honesty, his or her capacity to exert influence is enhanced. A negotiator who understands the point of view from which the other party is operating is more likely to communicate persuasively, with minimal misunderstanding.

In addition, while facts and ideas are important in changing another person's opinions and perceptions, the effectiveness of persuasion depends on how these facts and ideas are selected and presented. To be persuasive, negotiators must consider the content of the message, the structure of the message, and the delivery style. First, negotiators must make the offer attractive to the other party and emphasize those features of the offer that the other party is likely to find appealing. Negotiators should also try to formulate a proposal, however minor, to which the other party can agree. This will put the other negotiator into the mindset of saying "yes" and will increase the likelihood that he or she will agree with a second, more significant proposal or statement. These yesable propositions can also help to reduce tension and hostility and create minor points of agreement.

In addition, negotiators should try to make their messages consistent with their opponent's values. Thus negotiators can enhance their chances of success by jointly developing objective criteria and standards of legitimacy, and then shaping proposed solutions so that they meet these joint standards, which may include appeals to principles of fairness and expert opinions. Negotiators should try to demonstrate by following the suggested course of action, that the other party will be acting in accordance with his own values or some higher code of conduct. Agreements about general principles can be an important first step in the negotiation process.

One final way to influence the other side is to invent a good solution to the problem. This typically involves devising an ingenious solution that addresses the interests of both sides. Brainstorming can be used to help generate such solutions in advance and increase a negotiator's chances of affecting the outcome in a favorable way. Introducing new opportunities for joint gain may also help to create a situation that is ripe for settlement. Each side recognizes that it has much to gain through collaboration, and that if both sides work together, they can reap rewards. Rather than seeking "power over" the other side, negotiators in this way exercise "power with" one another.

Obstacles to Negotiation

In intractable conflicts, removing the obstacles to negotiation is the critical first step in moving toward negotiated agreements. Sometimes people fail to negotiate because they do not recognize that they are in a bargaining position. They may fail to identify a good opportunity for negotiation, and may use other options that do not allow them to manage their problems as effectively. Or, they may recognize the need for bargaining but may bargain poorly because they do not fully understand the process and lack good negotiating skills.

In cases of intractable conflict, parties often will not recognize each other, talk with each other, or commit themselves to the process of negotiation. They may even feel committed, as a matter of principle, to not negotiate with an adversary. In such cases, getting parties to participate in negotiations is a very challenging process. In addition, both parties must be ready to negotiate if the process is to succeed. If efforts to negotiate are initiated too early, before both sides are ready, they are likely to fail. Then the conflict may not be open to negotiation again for a long time.

Before they will negotiate, parties must be aware of their alternatives to a negotiated settlement (their BATNA). They must believe that a negotiated solution would be preferable to continuing the current situation, that a fair settlement can be reached, and that the balance of forces permits such an agreement. William Zartman refers to this as the belief that there is a "Way Out." Weaker parties must feel assured that they will not be overpowered in a negotiation, and parties must trust that their needs and interests will be fairly considered in the negotiation process.

In many cases, conflicts become "ripe" for negotiation when both sides realize that they cannot get what they want through a power struggle and that they have reached a hurting stalemate. If the parties believe that their ideal solution is not available and that foreseeable settlement is better than the other available alternatives, the parties have a "Zone of Possible Agreement" (ZOPA). This means that a potential agreement exists that would benefit both sides more than their alternatives do.

However, it may take some time to determine whether a ZOPA exists. The parties must first explore their various interests, options, and alternatives. If the disputants can identify their ZOPA, there is a good chance that they will come to an agreement. But if they cannot, negotiation is very unlikely to succeed. In addition, each side must believe that the other side is willing to compromise. If the parties regard each other with suspicion and mistrust, they may conclude that the other side is not committed to the negotiation process and may withdraw.

When there is little trust between the negotiators, making concessions is not easy. First, there is the *dilemma of honesty*. On one hand, telling the other party everything about your situation may give that person an opportunity to take advantage of you. However, not telling the other person anything may lead to a stalemate. The *dilemma of trust* concerns how much you should believe of what the other party tells you. If you believe everything this person says, then he or she could take advantage of you. But if you believe nothing this other person says, then reaching an agreement will be very difficult. The search for an optimal solution is greatly aided if parties trust each other and believe that they are being treated honestly and fairly.

In many cases, the negotiators' relationship becomes entangled with the substantive issues under discussion. Any misunderstanding that arises between them will reinforce their prejudices and arouse their emotions. When conflict escalates, negotiations may take on an atmosphere of anger, frustration, distrust, and hostility. If parties believe that the fulfillment of their basic needs is threatened, they may begin to blame each other and may break off communication. As the issue becomes more personalized, perceived differences are magnified and cooperation becomes unlikely. If each side gets locked into its initial position and attempts to force the other side to comply with various demands, this hostility may prevent negotiators from reaching agreement or making headway toward a settlement. In addition, parties may maintain their commitment to a course of action even when that commitment constitutes irrational behavior on their part (see entrapment). Once they have adopted a confrontational approach, negotiators may seek confirming evidence for that choice and ignore contradictory evidence. In an effort to save face, they may refuse to go back on previous commitments or to revise their position.

To combat perceptual bias and hostility, negotiators should attempt to gain a better understanding of the other party's perspective and try to see the situation as the other side sees it. In some cases, parties can discuss each other's perceptions, making a point to refrain from blaming the other. In addition, they can look for opportunities to act in a manner that is inconsistent with the other side's perceptions. Such de-escalating gestures can help to combat the negative stereotypes that may interfere with fruitful negotiations. In ideal circumstances, negotiators also establish personal relationships that facilitate effective communication. This helps negotiators to focus on commonalities and find points of common interest.

Finally, if the "right" people are not involved in negotiations, the process is not likely to succeed. First, all of the interested and affected parties must be represented. Second, negotiators must truly represent and have the trust of those they are representing. If a party is left out of the process, they may become angry and argue that their interests have not been taken into account. Agreements can be successfully implemented only if the relevant parties and interests have been represented in the negotiations, in part because parties who participate in the negotiation process have a greater stake in the outcome. Similarly, if constituents do not recognize a negotiator as their legitimate representative, they may try to block implementation of the agreement. Negotiators must therefore be sure to consult with their constituents and to ensure that they adequately deal with constituents' concerns.

These concerns are related to what Guy and Heidi Burgess call the "scale-up" problem of getting constituency groups to embrace the agreements that negotiators create. In many cases, participation in the negotiation process helps negotiators to recognize the legitimacy of the other side's interests, positions, and needs. This transformative experience may lead negotiators to develop a sense of respect for the adversary, which their constituents do not share. As a result, negotiators may make concessions that their constituents do not approve of, and they may be unable to get the constituents to agree to the final settlement. This can lead to last-minute breakdown of negotiated agreements

Broadly speaking, **Negotiation** is an interaction of influences. Such interactions, for example, include the process of resolving disputes, agreeing upon courses of action, bargaining for individual or collective advantage, or crafting outcomes to satisfy various interests. Negotiation is thus a form of alternative dispute resolution.

Negotiation involves two basic elements: the process and the substance. The process refers to how the parties negotiate: the context of the negotiations, the parties to the negotiations, the relationships among these parties, the communication between these parties, the tactics used by the parties, and the sequence and stages in which all of these play out. The substance, however, refers to what the parties negotiate over: the agenda, the issues, the options, and the agreement(s) reached at the end.

Skilled negotiators may use a variety of tactics ranging from a straight forward presentation of demands or setting of preconditions to more deceptive approaches such as cherry picking. Intimidation and salami tactics may also play a part in swaying the outcome of negotiations.

Negotiation is one of three primary methods of alternative dispute resolution, typically evidenced by a trained negotiator acting on behalf of a particular organization or position. Compare this to mediation where a disinterested third party listens to each sides' arguments and attempts to help craft an agreement between the parties. Lastly, arbitration is similar to a legal proceeding, whereby both sides make an argument as to the merits of their "case" and then the arbitrator decides the outcome both parties should follow (non-binding arbitration) or must follow (binding arbitration).

6.3 NEGOTIATION THEORY

The foundations of **negotiation theory** are decision analysis, behavioral decision making, game theory, and negotiation analysis. Another classification of theories distinguishes between Structural Analysis, Strategic Analysis, Process Analysis, Integrative Analysis and behavioral analysis of negotiations.

Individuals should make separate, interactive decisions; and *negotiation analysis* considers how groups of reasonably bright individuals should and could make joint, collaborative decisions. These theories are interleaved and should be approached from the synthetic perspective.

Definitions of most concepts in Social Sciences are disputed, that holds also true for negotiations. There is no common and shared definition of what a negotiation is. Zartman defines negotiation as "a process of combining conflicting positions into a common position under a decision rule of unanimity, a phenomenon in which the outcome is determined by the process."

However, most theories of negotiations share the notion of negotiations as a process. Yet, they differ in their description of the process. Structural Analysis considers this process to be a power game. Strategic analysis thinks of it as a repetition of games (Game Theory). Integrative Analysis prefers the more intuitive notion of process, in which negotiations undergo successive stages, e.g. pre-negotiation, stalemate, settlement. Especially structural, strategic and processual analysis build on rational actors, who are able to prioritize clear goals, are able to make trade-offs between conflicting values, are consistent in their behavioural pattern, and are able to take uncertainty into account.

Negotiations differ from mere coercion, in that negotiating parties have the theoretic possibility to withdraw from negotiations. It is easier to study bi-lateral negotiations, as opposed to multilateral negotiations.

Structural Analysis

Structural Analysis is based on a distribution of empowering elements among two negotiating parties. Structural theory moves away from traditional Realist notions of power in that it does not only consider power to be a possession, manifested for example in economic or military resources, but also thinks of power as a relation.

Based on the distribution of elements, in structural analysis we find either power-symmetry between equally strong parties or power-asymmetry between a stronger and a weaker party. All elements from which the respective parties can draw power constitute structure. They may be of material nature, i.e. hard power, (such as weapons) or of social nature, i.e. soft power, (such as norms, contracts or precedents).

These instrumental elements of power, are either defined as parties' relative position (resources position) or as their relative ability to make their options prevail.

Structural analysis is easy to criticise, because it predicts that the strongest will always win. This, however, does not always hold true.

Strategic Alliance

According to structural analysis, negotiations can therefore be described with matrices, such as the Prisoner's Dilemma, a concept taken from Game Theory. Another common game is the Chicken Dilemma.

Strategic analysis starts with the assumption that both parties have a veto. Thus, in essence, negotiating parties can cooperate (C) or defect (D). Structural analysis then evaluates possible outcomes of negotiations (C, C; C, D; D, D; D, C), by assigning values to each of the possible outcomes. Often, co-operation of both sides yields the best outcome. The basic problem however is, that the parties can never be sure that the other is going to cooperate, mainly because of two reasons: first, decisions are made at the same time or, second, concessions of one side might not be returned. Therefore the parties have contradicting incentives to cooperate or defect. If one party cooperates or makes a concession and the other does not, the defecting party might relatively gain more.

Trust may be built only in repetitive games through the emergence of reliable patterns of behaviour such as tit-for-tat.

Process analysis

Process analysis is the theory closest to haggling. Parties start from two points and converge through a series of concessions. As in strategic analysis, both sides have a veto (e.g. sell, not sell; pay, not pay). Process analysis also features structural assumptions, because one side may be weaker or stronger (e.g. more eager to sell, not willing to pay a certain price). Process Analysis focuses on the study of the dynamics of processes. E.g. both Zeuthen and Cross tried to find a formula in order to predict the behaviour of the other party in finding a rate of concession, in order to predict the likely outcome.

The process of negotiation therefore is considered to unfold between fixed points: starting point of discord, end point of convergence. The so called security point, that is the result of optional withdrawal, is also taken into account.

Integrative Analysis

Integrative analysis divides the process into successive stages, rather than talking about fixed points. It extends analysis to pre-negotiations stages, in which parties make first contacts. The outcome is explained as the performance of the actors at different stages. Stages may include pre-negotiations, finding a formula of distribution, crest behaviour, settlement.

In negotiation theory, the **best alternative to a negotiated agreement** or **BATNA** is the course of action that will be taken by a party if the current negotiations fail and an agreement cannot be reached.

Alternative Dispute Resolution (ADR) includes dispute resolution processes and techniques that fall outside of the government judicial process. Despite historic resistance to ADR by both parties and their advocates, ADR has gained widespread acceptance among both the general public and the legal profession in recent years. In fact, some courts now require some parties to resort to ADR of some type, usually mediation, before permitting the parties' cases to be tried. The rising popularity of ADR can be explained by the increasing caseload of traditional courts, the perception that ADR imposes fewer costs than litigation, a preference for confidentiality, and the desire of some parties to have greater control over the selection of the individual or individuals who will decide their dispute.

ADR is generally classified into at least three subtypes: negotiation, mediation, and arbitration. (Sometimes a fourth type, conciliation, is included as well, but for present purposes it can be regarded as a form of mediation. See conciliation for further details.) The salient features of each type are as follows:

In negotiation, participation is voluntary and there is no third party who facilitates the resolution process or imposes a resolution.

In mediation, there is a third party, a mediator, who facilitates the resolution process (and may even suggest a resolution, typically known as a "mediator's proposal"), but does *not* impose a resolution on the parties. In some countries (for example, the United Kingdom), ADR is synonymous with what is generally referred to as mediation in other countries.

In arbitration, participation is typically voluntary, and there is a third party who, as a private judge, imposes a resolution. Arbitrations often occur because parties to contracts agree that any future dispute concerning the agreement will be resolved by arbitration. This is known as a 'Scott Avery Clause'. In recent years, the enforceability of arbitration clauses, particularly in the context of consumer agreements (e.g., credit card agreements), has drawn scrutiny from courts. Although parties may appeal arbitration outcomes to courts, such appeals face an exacting standard of review.

"Alternative" dispute resolution is usually considered to be alternative to litigation. It also can be used as a colloquialism for allowing a dispute to drop or as an alternative to violence.

ADR can increasingly be conducted online or by using technology. This branch of dispute resolution is known as online dispute resolution (ODR). It should be noted, however, that ODR services can be provided by government entities, and as such may form part of the litigation process. Moreover, they can be provided on a global scale, where no effective domestic remedies are available to disputing parties, as in the case of the UDRP and domain name disputes. In this respect, ODR might not satisfy the "alternative" element of ADR.

ADR in India

Alternative dispute resolution in India is not new and it was in existence even under the previous Arbitration Act, 1940. The Arbitration and Conciliation Act, 1996 has been enacted to accommodate the harmonisation mandates of UNCITRAL Model. To streamline the Indian legal system the traditional civil law known as Code of Civil Procedure, (CPC) 1908 has also been amended and section 89 has been introduced. Section 89(1) of CPC provides an option for the settlement of disputes outside the court. It provides that where it appears to the court that there exist elements, which may be acceptable to the parties, the court may formulate the terms of a possible settlement and refer the same for arbitration, conciliation, mediation or judicial settlement.

Approaches to Negotiation

Given the above definition, negotiation occurs in business, non-profit organizations, government branches, legal proceedings, among nations and in personal situations such as marriage, divorce and parenting. *See also negotiation theory.*

The Advocate's Approach

In the advocacy approach, a skilled negotiator usually serves as advocate for one party to the negotiation and attempts to obtain the most favorable outcomes possible for that party. In this process the negotiator attempts to determine the minimum outcome(s) the other party is (or parties are) willing to accept, then adjusts their demands accordingly. A "successful" negotiation in the advocacy approach is when the negotiator is able to obtain all or most of the outcomes their party desires, but without driving the other party to permanently break off negotiations, unless the BATNA is acceptable.

Traditional negotiating is sometimes called *win-lose* because of the assumption of a fixed "pie", that one person's gain results in another person's loss. This is only true, however, if only a single issue needs to be resolved, such as a price in a simple sales negotiation. If multiple issues are discussed, differences in the parties' preferences make win-win negotiation possible. For example, in a labor negotiation, the union might prefer job security over wage gains. If the employers have opposite preferences, a trade is possible that is beneficial to both parties. Such a negotiation is therefore not an adversarial zero-sum game.

The "win/win" Negotiator's Approach

During the early part of the twentieth century, academics such as Mary Parker Follett developed ideas suggesting that agreement often can be reached if parties look not at their stated positions but rather at their underlying interests and requirements. During the 1960s, Gerard I. Nierenberg recognized the role of negotiation in resolving disputes in personal, business and international relations. He published *The Art of Negotiating*, where he states that the philosophies of the negotiators determine the direction a negotiation takes. His *Everybody Wins* philosophy assures that all parties benefit from the negotiation process which also produces more successful outcomes than the adversarial "winner takes all" approach.

In the 1970s, practitioners and researchers began to develop *win-win* approaches to negotiation. *Getting to YES* was published by Roger Fisher and William Ury as part of the Harvard negotiation project. The book's approach, referred to as Principled Negotiation, is also sometimes called mutual gains bargaining. The mutual gains approach has been effectively applied in environmental situations (see Lawrence Susskind and Adil Najam) as well as labor relations where the parties (e.g. management and a labor union) frame the negotiation as "problem solving".

Do you feel that someone is continually taking advantage of you? Do you seem to have to fight your corner aggressively, or ally with others, to win the resources you need? Or do you struggle to get what you want from people whose help you need, but over whom you have little direct authority? If so, you may need to brush up your win-win negotiation skills. Effective negotiation helps you to resolve situations where what you want conflicts with what someone else wants. The aim of win-win negotiation is to find a solution that is acceptable to both parties, and leaves both parties feeling that they've won, in some way, after the event. There are different styles of negotiation, depending on circumstances.

Where you do not expect to deal with people ever again and you do not need their goodwill, then it may be appropriate to "play hardball", seeking to win a negotiation while the other person loses out. Many people go through this when they buy or sell a house – this is why house-buying can be such a confrontational and unpleasant experience. Similarly, where there is a great deal at stake in a negotiation, then it may be appropriate to prepare in detail and legitimate "gamesmanship" to gain advantage. Anyone who has been involved with large sales negotiations will be familiar with this. Neither of these approaches is usually much good for resolving disputes with people with whom you have an ongoing relationship: If one person plays hardball, then this disadvantages the other person – this may, quite fairly, lead to reprisal later. Similarly, using tricks and manipulation during a negotiation can undermine trust and damage teamwork. While a manipulative person may not get caught out if negotiation is infrequent, this is not the case when people work together routinely. Here, honesty and openness are almost always the best policies.

Preparing for a Successful Negotiation...

Depending on the scale of the disagreement, some preparation may be appropriate for conducting a successful negotiation.

For small disagreements, excessive preparation can be counter-productive because it takes time that is better used elsewhere. It can also be seen as manipulative because, just as it strengthens your position, it can weaken the other person's.

However, if you need to resolve a major disagreement, then make sure you prepare thoroughly. Using our free worksheet, think through the following points before you start negotiating:

- **Goals:** what do you want to get out of the negotiation? What do you think the other person wants?
- **Trades:** What do you and the other person have that you can trade? What do you each have that the other wants? What are you each comfortable giving away?
- **Alternatives:** if you don't reach agreement with the other person, what alternatives do you have? Are these good or bad? How much does it matter if you do not reach agreement? Does failure to reach an agreement cut you out of future opportunities? And what alternatives might the other person have?
- **Relationships:** what is the history of the relationship? Could or should this history impact the negotiation? Will there be any hidden issues that may influence the negotiation? How will you handle these?
- **Expected outcomes:** what outcome will people be expecting from this negotiation? What has the outcome been in the past, and what precedents have been set?
- **The consequences:** what are the consequences for you of winning or losing this negotiation? What are the consequences for the other person?
- **Power:** who has what power in the relationship? Who controls resources? Who stands to lose the most if agreement isn't reached? What power does the other person have to deliver what you hope for?

- **Possible solutions:** based on all of the considerations, what possible compromises might there be?

Negotiating successfully...

The negotiation itself is a careful exploration of your position and the other person's position, with the goal of finding a mutually acceptable compromise that gives you both as much of what you want as possible. People's positions are rarely as fundamentally opposed as they may initially appear – the other person may have very different goals from the ones you expect.

In an ideal situation, you will find that the other person wants what you are prepared to trade, and that you are prepared to give what the other person wants. If this is not the case and one person must give way, then it is fair for this person to try to negotiate some form of compensation for doing so – the scale of this compensation will often depend on the many of the factors we discussed above. Ultimately, both sides should feel comfortable with the final solution if the agreement is to be considered win-win. Only consider win-lose negotiation if you don't need to have an ongoing relationship with the other party as, having lost, they are unlikely to want to work with you again. Equally, you should expect that if they need to fulfill some part of a deal in which you have "won," they may be uncooperative and legalistic about the way they do this.

6.4 TYPES OF NEGOTIATION

Distributive Bargaining?

Distributive bargaining, also called "claiming value," "zero-sum," or "win-lose" bargaining, is a competitive negotiation strategy that is used to decide how to distribute a fixed resource, such as money. The parties assume that there is not enough to go around, and they cannot "expand the pie," so the more one side gets, the less the other side gets.

Why is Distributive Bargaining Important?

Distributive bargaining is important because there are some disputes that cannot be solved in any other way – they are inherently zero-sum. If the stakes are high, such conflicts can be very resistant to resolution. For example, if budgets in a government agency must be cut 30 percent, and people's jobs are at stake, a decision about what to cut is likely to be very difficult. If the cuts are small enough that the impact on employees will be minor, however, such distributive decisions can be made more easily.

Even in cooperative negotiations, distributive bargaining will come into play. Distributive bargaining and integrative bargaining are not mutually exclusive negotiation strategies. Integrative bargaining is a good way to make the pie (joint value) as large as it can possibly be, but ultimately the parties must distribute the value that was created. If they are able to expand the pie enough, distribution is easy. If there is still not enough to give each side what it wants, however, distributive negotiation will be more difficult.

The Pros and Cons of Distributive Bargaining

Some conflict resolution theorists believe that distributive bargaining is unnecessary. Any conflict, they argue, may be solved cooperatively through integrative bargaining. For example, in their book *Getting to Yes*, Fisher, Ury, and Patton argue that with creativity, disputants can almost always work together to "expand the pie" and create outcomes that benefit both sides. Even when budgets have to be cut, they would argue, the parties make the decisions together so that all sides get the best possible outcome. Distributive bargaining has also been criticized because it tends to lead to destructive actions and sometimes forces the involved parties to focus too much on their differences. If people want to maintain a good relationship with one another, it is argued, they should take an integrative approach to distribution as well as expansion of the pie.

However, in cases where the "negotiator wants to maximize the value obtained in a single deal and when the relationship with the other party is not important," distributive bargaining tactics may be very useful.

Process and Strategy in Distributive Negotiations

The process of distributive negotiation involves the interplay of one's walk away value – the minimum or maximum one can accept before "walking away" from the deal – and the adversary's walk away value. The trick is to get an idea of your opponent's walk away value and then try to negotiate an outcome that is closer to your own goals then theirs. Whether or not parties achieve their goals in distributive bargaining depends on the strategies and tactics they use.

Information is the key to gaining a strategic advantage in a distributive negotiation. You should do your best to guard your information carefully and also try to get information out of your opponent. To a large extent, your bargaining power depends on how clear you are about your goals, alternatives, and walk away values and how much you know about your opponents'. Once you know these values, you will be in a much stronger position to figure out when to concede and when to hold firm in order to best influence the response of the other side.

Integrative or Interest-Based Bargaining?

Integrative bargaining (also called "interest-based bargaining," "win-win bargaining") is a negotiation strategy in which parties collaborate to find a "win-win" solution to their dispute. This strategy focuses on developing mutually beneficial agreements based on the interests of the disputants. Interests include the needs, desires, concerns, and fears important to each side. They are the underlying reasons why people become involved in a conflict.

"Integrative refers to the potential for the parties' interests to be [combined] in ways that create joint value or enlarge the pie Potential for integration only exists when there are multiple issues involved in the negotiation. This is because the parties must be able to make trade-offs across issues in order for both sides to be satisfied with the outcome.

Why is Integrative Bargaining Important?

Integrative bargaining is important because it usually produces more satisfactory outcomes for the parties involved than does positional bargaining. Positional bargaining is based on fixed, opposing viewpoints (positions) and tends to result in compromise or no agreement at all. Oftentimes, compromises do not efficiently satisfy the true interests of the disputants. Instead, compromises simply split the difference between the two positions, giving each side half of what they want. ...

There are often many interests behind any one position. If parties focus on identifying those interests, they will increase their ability to develop win-win solutions. The classic example of interest-based bargaining and creating joint value is that of a dispute between two little girls over an orange. Both girls take the position that they want the whole orange. Their mother serves as the moderator of the dispute and based on their positions, cuts the orange in half and gives each girl one half. This outcome represents a compromise. However, if the mother had asked each of the girls why she wanted the orange – what her interests were – there could have been a different, win-win outcome. This is because one girl wanted to eat the meat of the orange, but the other just wanted the peel to use in baking some cookies. If their mother had known their interests, they could have both gotten all of what they wanted, rather than just half.

Integrative solutions are generally more gratifying for all involved in negotiation, as the true needs and concerns of both sides will be met to some degree. It is a collaborative process and therefore the parties actually end up helping each other. This prevents ongoing ill will after the negotiation concludes. Instead, interest-based bargaining facilitates constructive, positive relationships between previous adversaries.

Identifying Interests: The first step in integrative bargaining is identifying each side's interests. This will take some work by the negotiating parties, as interests are often less tangible than positions and are often not publicly revealed. A key approach to determining interests is asking "Why?" Why do you want that? Why do you need that? What are your concerns? Fears? Hopes? If you cannot ask these questions directly, get an intermediary to ask them.

The bottom line is you need to figure out *why* people feel the way they do, *why* they are demanding what they are demanding. Be sure to make it clear that you are asking these questions so you can understand their interests (needs, hopes, fears, or desires) better, not because you are challenging them or trying to figure out how to beat them.

Next you might ask yourself how the other side perceives your demands. What is standing in the way of them agreeing with you? Do they know your underlying interests? Do you know what your own underlying interests are? If you can figure out their interests as well as your own, you will be much more likely to find a solution that benefits both sides.

You must also analyze the potential consequences of an agreement you are advocating, as the other side would see them. This is essentially the process of weighing pros and cons, but you attempt to do it from the perspective of the other side. Carrying out an empathetic analysis will help you understand your adversary's interests. Then you will be better equipped to negotiate an agreement that will be acceptable to both of you.

There are a few other points to remember about identifying interests. First, you must realize that each side will probably have multiple interests it is trying to satisfy. Not only will a single person have multiple interests, but if you are negotiating with a group, you must remember that each individual in the group may have differing interests. Also important is the fact that the most powerful interests are basic human needs – security, economic well being, a sense of belonging, recognition, and control over one's life. If you can take care of the basic needs of both sides, then agreement will be easier. You should make a list of each side's interests as they become apparent. This way you will be able to remember them and also to evaluate their relative importance

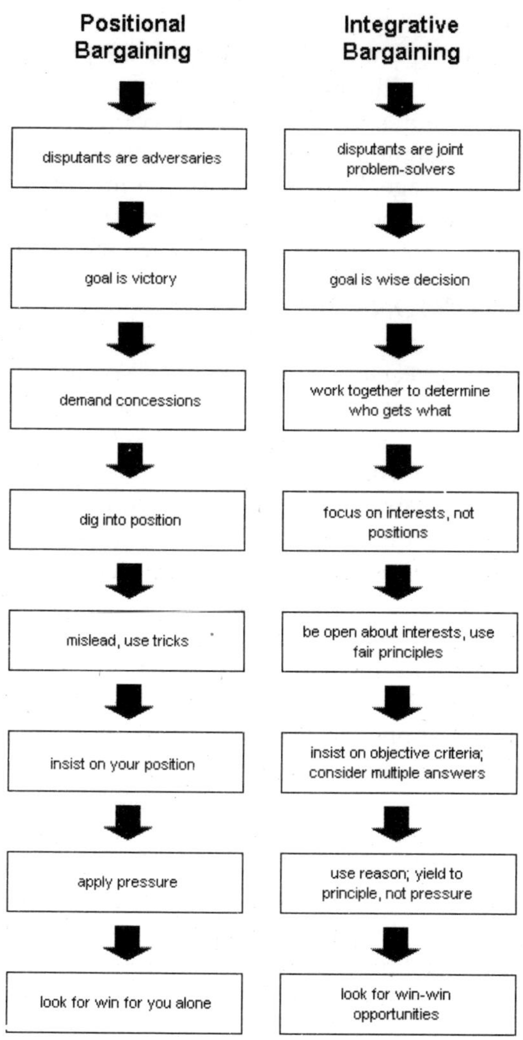

Creating Options

After interests are identified, the parties need to work together cooperatively to try to figure out the best ways to meet those interests. Often by "brainstorming" - listing all the options anyone can think of without criticizing or dismissing anything initially, parties can come up with creative new ideas for meeting interests and needs that had not occurred to anyone before. The goal is a win-win outcome, giving each side as much of their interests as possible, and enough, at a minimum that they see the outcome as a win, rather than a loss.

Using Integrative and Distributive Bargaining Together

Although distributive bargaining is frequently seen as the opposite of integrative bargaining, the two are not mutually exclusive. Distributive bargaining plays a role in integrative bargaining, because ultimately "the pie" has to be split up.

Integrative bargaining is a good way to make the pie (joint value) as large as it possibly can be, but ultimately the parties must distribute the value that was created through negotiation. They must agree on who gets what. The idea behind integrative bargaining is that this last step will not be difficult once the parties reach that stage. This is because the interest-based approach is supposed to help create a cooperative working relationship. Theoretically, the parties should know who wants what by the time they split the pie

6.5 THE NEGOTIATIONS PROCESS

The Difference Between Integrative and Distributive Negotiation

There are two types of negotiation: "integrative" and "distributive." Integrative negotiations are those typically referred to as "win/win" negotiations: all sides are looking for a solution that maximizes joint gain and allows everyone to walk away feeling like they won something. They involve looking at the issues being negotiated from multiple angles, considering multiple issues at once (thus allowing for trade-offs), and honestly trying to "expand the pie" rather than divide it. Anyone who imagines that they might see or do business with their fellow negotiator in the future should be attempting to negotiate in this way. Integrative negotiations foster trust and good working relationships, and leave *all* parties feeling good, not just one.

Distributive negotiations, on the other hand, are typically described as "win/lose" negotiations - one party gets what they want, and the other party gives something up. Think of negotiating for your car - you either get that extra $1000, or the dealership does. If you feel you got a good deal (and squeezed that salesman), you "won". If you walk away feeling like you paid too much money, you "lost." This type of negotiating does *not* lead to good long-term relationships. The parties' interests are often opposed (or *seem* to be opposed - this may not prove to be the case once you start getting creative), and usually good feelings are not plentiful when the negotiation is over.

Integrative vs. Distributive Negotiations

Characteristic	Integrative Negotiations	Distributive Negotiations
Outcome	Win / win	Win / lose
Motivation	Joint gain	Individual gain
Interests	Congruent	Opposed
Relationship	Long-term	Short-term
Issues	Multiple issues	Single Issue

Here's a story (probably familiar to many of you!) that demonstrates the difference between integrative and distributive negotiations: Two sisters were fighting over the last orange in the fruit bowl. They went back and forth, each girl insisting that she should get it and both refusing to give up. They were about to agree on cutting the orange in half when their aunt walked in and realized what was going on. She turned to the girls and asked them each *why* they wanted the orange. As it turns out, one wanted to eat the orange, and the other wanted the peel for an art project she was working on. Once they realized this, they were able to "split" the orange in such a way that both got exactly what they wanted. The girls were engaged in distributive negotiations (though it probably felt like bickering to them), while their aunt was suggesting an integrative solution.

As you might have guessed by now, partnership negotiations should *always* be integrative. The goal in any partnership should be maximizing all parties' abilities to advance their collective mission, and to enable both to better serve the community, deal effectively with their economic obligations and pressures, etc. One organization should not want to "win" at the expense of the other. You want joint gain, not just individual gain, and while your specific motivations for pursuing the partnership may be different, the spirit behind your interests should be congruent. Most importantly, you are entering into a long-term relationship with your prospective partner(s), and you want this to be a *good* relationship. The only way to build trust and expand the pie (or the orange) in a negotiation situation is be integrative. You must do this from the beginning. Once you start down the "distributive path" it is hard to change directions, as you establish a dynamic of competition and distrust. The moral of the story: be integrative!

Three Key Concepts of Negotiation

1. Know Your BATNA

In negotiation theory, the **best alternative to a negotiated agreement** or **BATNA** is the course of action that will be taken by a party if the current negotiations fail and an agreement cannot be reached.

If the current negotiations are giving you less value than your BATNA, there is no point in proceeding. Prior to the start of negotiations, the parties should have ascertained their own individual BATNAs.

BATNA was developed by negotiation researchers Roger Fisher and Bill

Ury of the Harvard Program on Negotiation (PON), in their series of books on Principled Negotiation that started with *Getting to YES*. Nobel Laureate John Forbes Nash has included such ideas in his early undergraduate research.

For example, if I have a written offer from a dealer to buy my car for $100 dollars, then my BATNA when dealing with other potential purchasers would be $100 since I can get $100 for my car even without reaching an agreement with such alternative purchaser.

A party should generally never accept a worse resolution than its BATNA. Care should be taken, however, to ensure that deals are accurately valued, taking into account all considerations (such as relationship value, time value of money, likelihood that the other party will live up to their side of the bargain, etc.) These other considerations are very difficult to value, since they are often based on uncertain considerations, rather than easily measurable and quantifiable factors.

Examples of other offers that might or might not be better than the BATNA in the example above might be:

An offer of $90 by a close relative (is the goodwill generated worth $10 or more?)

An offer of $125 in 45 days (what are the chances of this future commitment falling through, and would my prior BATNA ($100) still be available if it did?)

An offer from another dealer to offset $150 against the price of a new car (do I want to buy a new car right now, the offered car in particular? Also, is the probably minuscule reduction in monthly payments worth $100 to me today?)

BATNA is seen in negotiation fields as the single most important source of negotiation power. Negotiators don't use their BATNA merely as a safety net, but rather as a point of leverage in negotiations.

Consider the following business example: Company one can choose to buy from companies two, three and four - but companies two, three and four can only sell to company one. Company one can use their powerful BATNA position to leverage a better deal by playing companies two, three and four against each other. This is a common practice among purchasing and procurement managers in the business world.

Whilst your alternative options, and therefore your BATNA might be known to you, very often your BATNA may not be. So time and energy are demanded to figure out which options are *really* available to you and actionable. Even if your alternative options are known, they need to be real and actionable. A Project on Negotiation Executive Seminar experiment and other experiments have proven that most managers overestimate their BATNA whilst simultaneously investing too little time into researching their real options. This results in poor or faulty decision making from overconfidence and good choices being rejected.

What BATNAs Are

BATNA is a term coined by Roger Fisher and William Ury in their 1981 bestseller, *Getting to Yes: Negotiating Without Giving In.* It stands for "best alternative to a negotiated agreement." BATNAs are critical to negotiation because you cannot make a wise decision about whether to accept a negotiated agreement unless you know what your alternatives are. Your BATNA "is the only standard which can protect you both from accepting terms that are too unfavorable and from rejecting terms it would be in your interest to accept. In the simplest terms, if the proposed agreement is better than your BATNA, then you should accept it. If the agreement is not better than your BATNA, then you should reopen negotiations. If you cannot improve the agreement, then you should at least consider withdrawing from the negotiations and pursuing your alternative (though the costs of doing that must be considered as well).

Having a good BATNA increases your negotiating power. Therefore, it is important to improve your BATNA whenever possible. Good negotiators know when their opponent is desperate for an agreement. When that occurs, they will demand much more, knowing their opponent will have to give in. If the opponent apparently has many options outside of negotiation, however, they are likely to get many more concessions, in an effort to keep them at the negotiating table. Thus making your BATNA as strong as possible before negotiating, and then making that BATNA known to your opponent will strengthen your negotiating position.

Guy Burgess and Heidi Burgess have adapted the concept of BATNA slightly to emphasize what they call "EATNAs" estimated alternatives to a negotiated agreement" instead of "best alternatives." Even when disputants do not have good options outside of negotiations, they often think they do. (For example, both sides may *think* that they can prevail in a military struggle, even when one side is clearly weaker, or when the relative strengths are so balanced that the outcome is very uncertain.) Yet, perceptions are all that matter when it comes to deciding whether or not to accept an agreement. If a disputant thinks that he or she has a better option, she will, very often, pursue that option, even if it is not as good as she thinks it is.

BATNA and EATNAs also affect what William Zartman and may others have called "ripeness," the time at which a dispute is ready or "ripe" for settlement. When parties have similar ideas or "congruent images" about what BATNAs exist, then the negotiation is ripe for reaching agreement. Having congruent BATNA images means that both parties have similar views of how a dispute will turn out if they do not agree, but rather pursue their other rights-based or power-based options. In this situation, it is often smarter for them to negotiate an agreement without continuing the disputing process, thus saving the transaction costs. This is what happens when disputing parties who are involved in a lawsuit settle out of court, (which happens in the U.S. about 90 percent of the time). The reason the parties settle is that their lawyers have come to an understanding of the strength of each sides' case and how likely each is to prevail in court. They then can "cut to the chase," and get to the same result much more easily and more quickly through negotiation.

On the other hand, disputants may hold "dissimilar images" about what BATNAs exist, which can lead to a stalemate or even to intractability. For example, both sides may think they can win a dispute if they decide to pursue it in court or through force. If both sides' BATNAs tell them they can pursue the conflict and win, the likely result is a power contest. If one side's BATNA is indeed much better than the other's, the side with the better BATNA is likely to prevail. If the BATNAs are about equal, however, the parties may reach a stalemate. If the conflict is costly enough, eventually the parties may come to realize that their BATNAs were not as good as they thought they were. Then the dispute will again be "ripe" for negotiation.

The allure of the EATNA often leads to last-minute breakdowns in negotiations. Disputants can negotiate for months or even years, finally developing an agreement that they think is acceptable to all. But then at the end, all the parties must take a hard look at the final outcome and decide, "is this better than all of my alternatives?" Only if all the parties say "yes," can the agreement be finalized. If just one party changes his or her mind, the agreement may well break down. Thus, knowing one's own and one's opponent's BATNAs and EATNAs is critical to successful negotiation

Determining Your BATNA

BATNAs are not always readily apparent. Fisher and Ury outline a simple process for determining your BATNA:

1. develop a list of actions you might conceivably take if no agreement is reached;
2. improve some of the more promising ideas and convert them into practical options; and
3. select, tentatively, the one option that seems best.

BATNAs may be determined for any negotiation situation, whether it be a relatively simple task such as finding a job or a complex problem such as a heated environmental conflict or a protracted ethnic conflict.

Fisher and Ury offer a job search as a basic example of how to determine a BATNA. If you do not receive an attractive job offer by the end of the month from Company X, what will you do? Inventing options is the first step to determining your BATNA. Should you take a different job? Look in another city? Go back to school? If the offer you are waiting for is in New York, but you had also considered Denver, then try to turn that other interest into a job offer there, too. With a job offer on the table in Denver, you will be better equipped to assess the New York offer when it is made. Lastly, you must choose your best alternative option in case you do not reach an agreement with the New York company. Which of your realistic options would you really want to pursue if you do not get the job offer in New York?

More complex situations require the consideration of a broader range of factors and possibilities. For example, a community discovers that its water is being polluted by the discharges of a nearby factory. Community leaders first attempt to negotiate a cleanup plan with the company, but the business refuses to voluntarily agree on a plan of action that the community is satisfied with. In such a case, what are the community's options for trying to resolve this situation?

- They could possibly sue the business based on stipulations of the Clean Water Act.
- They could contact the Environmental Protection Agency and see what sort of authority that agency has over such a situation.
- They could lobby the state legislature to develop and implement more stringent regulations on polluting factories.
- The community could wage a public education campaign and inform citizens of the problem. Such education could lead voters to support more environmentally minded candidates in the future who would support new laws to correct problems like this one.

In weighing these various alternatives to see which is "best," the community members must consider a variety of factors.

- Which is most affordable and feasible?
- Which will have the most impact in the shortest amount of time?
- If they succeed in closing down the plant, how many people will lose their jobs?

These types of questions must be answered for each alternative before a BATNA can be determined in a complex environmental dispute such as this one.

BATNAs and the Other Side

At the same time you are determining your BATNA, you should also consider the alternatives available to the other side. Sometimes they may be overly optimistic about what their options are. The more you can learn about their options, the better prepared you will be for negotiation. You will be able to develop a more realistic view of what the outcomes may be and what offers are reasonable.

There are also a few things to keep in mind about revealing your BATNA to your adversary. Although Fisher and Ury do not advise secrecy in their discussions of BATNAs, according to McCarthy, "one should not reveal one's BATNA unless it is better than the other side thinks it is." But since you may not know what the other side thinks, you could reveal more than you should. If your BATNA turns out to be worse than the opponent thinks it is, then revealing it will weaken your stance.

BATNAs and the Role of Third Parties

Third parties can help disputants accurately assess their BATNAs through reality testing and costing. In reality testing, the third party helps clarify and ground each disputing party's alternatives to agreement. S/he may do this by asking hard questions about the asserted BATNA: "How could you do that? What would the outcome be? What would the other side do? How do you know?" Or the third party may simply insert new information into the discussion...illustrating that one side's assessment of its BATNA is likely incorrect. Costing is a more general approach to the same process...it is a systematic effort to determine the costs and benefits of all options. In so doing, parties will come to understand all their alternatives. If this is done together and the parties agree on the assessment, this provides a strong basis upon which to come up with a negotiated solution that is better than both sides' alternatives. But if the sides cannot come to such an agreement, then negotiations will break down, and both parties will pursue their BATNA instead of negotiation.

2. *ZOPA*

ZOPA is a UK-based company providing an online money exchange, allowing people who have money to lend to provide it directly to people wanting to borrow it without the need to go through banks, a process sometimes referred to as peer-to-peer lending.

The name, ZOPA, stands for Zone of Possible Agreement, a negotiating term identifying the bounds within which agreement can be reached between two parties. The idea for the service was identified following extensive socio-economic research which itself identified a group of individuals, coined Freeformers.

Freeformers were identified as displaying different attitudes towards many aspects of life, including their money, and ZOPA has developed a credit referencing process that goes beyond credit bureau information in order to establish consumers' attitudes towards money, as well as their credit-worthiness. This extends their ability to offer credit to this freeformer group, who may have irregular incomes or short-term cash-flow requirements, but who may not be offered credit through traditional methods of credit referencing.

The service matches people that want to offer to lend money with people that need to borrow money. Matching is done on a many-to-one basis, such that no one lender is exposed to any single borrower. Bad debt risk is taken by the lender, and priced into the rate offered to borrowers, although debt recovery processes are available.

When someone applies to ZOPA for a loan, the company's proprietary IT system carries out real-time authentication checks and accesses credit reporting data from Equifax. The customer is then put into one of four risk bands: A*, A, B or C.

Variously likened to eBay and Betfair in the UK press, ZOPA is an addition to the emerging group of peer-to-peer services enabled by the internet. Prosper is a similar service based in the U.S.Launched in 2005, ZOPA is led by a management team that comprises many of those that founded Egg in the UK.

The company is based in London and backed by Benchmark Capital and Wellington Partners. The Social Futures Observatory recently published a study titled Internet Based Social Lending, which seeks to understand the antecedents of Social Lending, drawing parallels with Friendly Societies, and using ZOPA as a major source of case study material.

Both borrowers and lenders are charged a 0.5% fee by ZOPA on the funds for each transaction, although founder members are exempt from this charge.

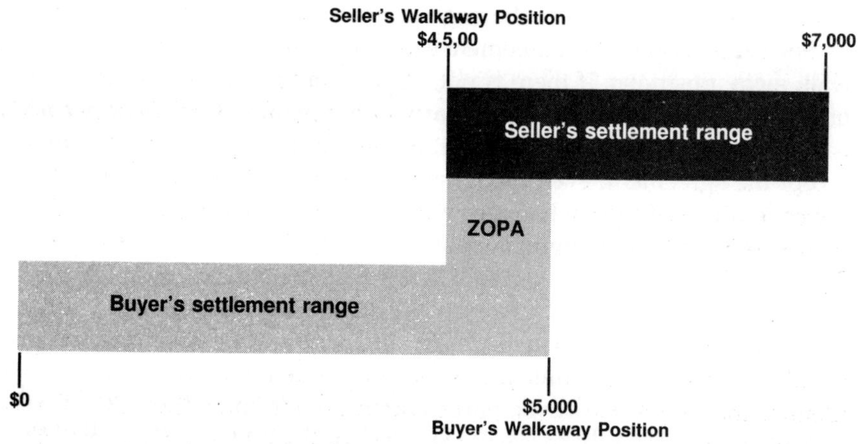

What is a ZOPA and Why Does it Matter?

A "Zone of Possible Agreement" (ZOPA) exists if there is a potential agreement that would benefit both sides more than their alternative options do. For example, if Fred wants to buy a used car for $5,000 or less, and Mary wants to sell one for $4,500, those two have a ZOPA. But if Mary will not go below $7,000 and Fred will not go above $5,000, they do not have a zone of possible agreement.

The ZOPA is critical to the successful outcome of negotiation, but it may take some time to determine whether a ZOPA exists. It may only become known once the parties explore their various interests and options. If the disputants can identify the ZOPA, there is a good chance that they will come to an agreement.

Foundations of ZOPA: BATNAs

In order for disputing parties to identify the ZOPA, they must first know their alternatives, and thus their "bottom line" or "walk away position."

Alternatives: Parties must determine what alternatives they have to any agreement. Roger Fisher and William Ury introduced the concept of "BATNA"

(Best Alternative To a Negotiated Agreement). This is the *best course of action* that a party can pursue if no agreement is reached.

For example, Mary might have two potential buyers. Georgio is willing to pay $6,950. Mary is now negotiating with Fred. If he will pay more than Georgio (Mary's BATNA), she'll sell to him. If he won't pay that much, she'll sell to Georgio. Likewise, if Fred has found another car he likes for $5,500, then he won't pay more for Mary's car than that...maybe even a bit less. Fred's BATNA is $5,500.

Bottom Lines or Walk-Away Positions: BATNAs determine each side's bottom lines. If you have an alternative car available for $5,000, $5,000 is your bottom line. If you can sell your car for $7,000, that is your bottom line. If you don't do better than that in the negotiation, you'll walk away.

So, a zone of possible agreement exists if there is an overlap between these walk away positions. If there is not, negotiation is very unlikely to succeed. In fact, it will only succeed if one party either realizes that his or her BATNA is not as good as he or she thought, or she decides for some other reason to accept the agreement, even though an alternative option might have yielded better results. (This often happens when parties do not explore or understand their BATNAs well enough.)

Identifying the ZOPA

If both sides know their BATNAs and walk away positions, the parties should be able to communicate, assess proposed agreements, and eventually identify the ZOPA. However, parties often do not know their own BATNAs, and are even less likely to know the other side's BATNA. Often parties may pretend they have a better alternative than they really do, as good alternatives usually translate into more power in the negotiations. This is explained more in the essay on BATNAs. The result of such deception, however, might be the apparent absence of a ZOPA, when one actually did exist. Shared uncertainties may also affect the parties' abilities to assess potential agreements because the parties may be unrealistically optimistic or pessimistic about the possibility of agreement or the value of alternative options

ZOPAs in Distributive and Integrative Negotiations

The nature of the ZOPA depends on the type of negotiation. In a distributive negotiation, in which the participants are trying to divide a "fixed pie," it is more difficult to find mutually acceptable solutions as both sides want to claim as much of the pie as possible. Distributive negotiations over a single issue tend to be zero-sum – there is a winner and a loser. There is no overlap of interests between the parties; therefore, no mutually beneficial agreement is possible.

For example, two people may be competing for one job. In the simplest case, there is no ZOPA because both people want the full-time job and are not willing to divide the job responsibilities and salary. One person must "win" and the other must "lose."

On the other hand, integrative negotiations involve creating value or "enlarging the pie." This is possible when parties have shared interests or are dealing with multiple issues. In an integrative negotiation, the parties can combine their interests to create joint value. To achieve integration, negotiators can deal with multiple issues at the same time and make trades between them. This is so that I might get more of something that I value while you get more of something that you value. That way both parties can "win," even though neither gets all that they originally thought they wanted. In the example above, if rewriting the job description could create an additional job then the distributive negotiation would change into an integrative negotiation between the employer and the two potential employees. If both applicants are qualified, now they may both get jobs. The ZOPA, in this case, exists when two jobs are created and each applicant prefers a different one of the two.

3. *Reservation Price*

In microeconomics, the **reservation price** or **reserve price** is the maximum price a buyer is willing to buy a good or service, or the minimum price a seller is willing to sell a good or service (e.g., in an auction). Reservation prices vary for the buyer according to their disposable income, their desire for the good, and the prices of, and their information about substitute goods. **Reservation demand** is a name for the schedule of reservation prices at which a seller would be willing to sell different quantities of the good in question.

The reservation price is used to help calculate the consumer surplus or the producer surplus with reference to the equilibrium price.

Go to a grocery or department store at any time and stand (discreetly) in one area for a while. Watch how buyers look at an item with interest and then check its price. Notice that some buyers will immediately put down the item and walk away. Others will ponder a while before deciding to keep it or leave it. While others will smile and immediately put it in their shopping cart. [If you see them put it in their pocket you should immediately contact store security!]

Why do different buyers have such different responses toward the same item?

When buyers think about paying for a good or service, they have a vague idea of the maximum they are willing to pay for it. The maximum price a buyer is willing to pay for one unit of a good or service, rather than do without, is called his *reservation price*. Buyer reservation prices vary due to differences in the strength of preference for the good, their knowledge of the price of substitute goods, their disposable income, and their expectations about what will happen to their income or the price of the good in the near future.

When sellers think about receiving payment for a good or service, they have a fairly clear idea of the minimum price they are willing to accept for it. The minimum price a seller is willing to accept in exchange for one unit of a good or service is called her *reservation price*. Seller reservation prices vary, in part, due to differences in the costs of production. Some firms may be subsidized or taxed, face different prices for inputs, have different expectations regarding how demand for their product will change, or use different production technology.

The 4 Key Steps in the Negotiating Process in Oil and Gas Business

Following are the negotiating issues at the time of license round stages:

1. **How do PSCs deal with the Exploration Phase?**
   - managing timescales, minimum work programmes, appraisal and relinquishment
   - how to set criteria, evaluate and agree when a discovery is "commercial"
2. **Negotiating Issues at the Development Stage**
   (a) Negotiating Financing and managing risk during development programmes
   - Where to find sources of funding from the public and private sector
   - Understanding the conditions and benefits

   (b) What are the PSC objectives in the Development & Production Phase?
   - Understanding the differences between host countries and contracting companies and how to bridge the gaps
   - check lists of the key elements for negotiation
   - bonuses, royalties, taxes and pricing issues
   - how to set up and manage cost recovery

   (c) How do we deal with "Non Financial Issues"?
   - how to meet national policies and contract commitments for employment, training, local purchasing, technology transfer, and health, safety and environmental compliance

   (d) Who gets the production?
   - how to work out the production splits
   - untangling the R factor

   (e) How to use economics in the negotiation of PSCs?
   - tools for assessing the value of success and the cost of failure
   - how do we set the right price?
   - special issues for marginal fields

   (f) Where is the balance of risk in the contract provisions?
   - how to use incentive structures to even out the upsides and downsides of the contract terms for each party
   - the challenge of abandonment obligations

   (g) In what ways are gas contracts different?
   - special PSC considerations for natural gas and LNG production, transportation, processing and sales agreements
3. **Bringing the PSC to a mutually satisfactory close out**
4. **Checklist of key negotiating steps for a PSC.**

Glossary

ABANDON – To allow an option to expire worthless.

ACCEPTANCE – Agreement of intention to carry out a transaction.

ACTUALS – The physical commodity.

AMERICAN STYLE – An option which can be exercised by the buyer (holder) at anytime during its life.

ARA – Antwerp, Rotterdam, Amsterdam.

ARBITRAGE – The simultaneous purchase of an instrument in one market and the sale of the same, or similar, instrument in another market in order to exploit price differentials.

ASK – An indication of willingness, to sell a specified amount of a commodity at a specific price (also known as offer).

ASK PRICE – The price at which a seller is prepared to trade.

AT-THE-MONEY – An option whose exercise price is equal, or close to, the current price in the underlying market.

BACKWARDATION – A market where the price of near delivery months is at a premium to more distant months.

BASIS – The difference between the futures price for a given commodity and the comparable cash or spot price for a commodity. Normally quoted as cash price minus futures price. i.e. a positive number indicates a futures discount; a negative number indicates a futures premium.

BBL – Barrel; also, less frequently, BL

BEARISH – The belief that the market price will fall.

BID – A proposal at a given time on the trading floor to buy a specific amount of a commodity at a specified price.

BID PRICE – The price at which a buyer is prepared to trade.

BULLISH – The belief that a market price is going to rise.

CALL OPTION – An option that gives the buyer (holder) the right but not the obligation to buy a specified quantity of the underlying instrument at a fixed price, on or before a specified date. The seller (writer) of the option has the obligation to deliver the underlying instrument if the holder exercises the option.

CARRYING CHARGES - The costs incurred in warehousing the physical commodity generally including interest, insurance and storage.

CASH MARKET - The physical market underlying a futures or options contract.

CASH & CARRY - An arbitrage transaction involving the simultaneous purchase of a cash commodity with borrowed money and the sale of the appropriate futures contract.

CASH SETTLEMENT - When the physical delivery of futures or options is cumbersome then contracts are settled by attaching a monetary value.

CLASS - All options of the same type pertaining to the same underlying commodity. Calls and puts form two separate classes.

CLEARING - The process of trade match, registration, settlement and provision of a guarantee of exchange traded contracts.

CLEARING FEE - A sum charged by the clearing house for clearing trades through its facilities.

CLEARING MEMBER - A member of a clearing house.

CLIENT ACCOUNT - Any individual or entity being serviced by an agent (broker) for a commission. Servicing generally includes advice, accounting, and order execution. A customer's business must be distinguished from the brokers principal in-house business and may be in a physically segregated account at the clients option.

CLOSE OUT - A finalising transaction by which an equal and opposite trade closes an open position.

CLOSING RANGE - A range of closely related prices at which transactions took place at the close of a market.

CODE OF CONDUCT - The trading procedures and etiquette of a market.

COMMISSION - The basic fee charged to a customer by a broker for a purchase or sale transaction. Commissions are negotiable.

COMPLIANCE- (1) Action to comply with rules and regulations.

(2) A department responsible for ensuring the maintenance and adherence by Members to the rules and regulations of an exchange.

CONFIRMATION - The process immediately is following a transaction whereby the traders confirm the details of the trade.

CONTANGO - Where the cash price of nearby future delivery months is at a discount to more distant months.

CONTRACT – A binding agreement between a buyer and a seller in a transaction or the expression used to relate to particular futures products i.e. Brent Crude futures contract.

CONTRA – A contra is a facility available on the TRS that is used to enable a trade or trades in any contract to be moved from one clearer to another subject to certain criteria. In certain circumstances the contra facility will also be used to delete ("bust") a trade.

CRACK SPREAD – The buying of crude oil whilst taking the reverse position in refined products (for example Brent Crude and Gas Oil). Any combination of energy futures can be used, provided that the number of crude contracts equals the number of product contracts.

CROSS/SELF TRADE – The simultaneous purchase and sale by the same broker in the pit for equal amounts of the same month of a contract at a single price.

DELIVERY – Settlement of a futures contract by delivery of the underlying physical product.

DELIVERY MONTH – The month specified in a given futures contract for delivery of the actual physical spot or cash commodity.

DISCOUNT – The amount by which a future or option is priced below its theoretical or "fair value" or below the price of the underlying instrument. This term is applied to both fair value and crude basis.

DISTANT MONTHS – The far dated delivery months in a futures or option contract.

EDSP – Exchange Delivery Settlement Price. The settlement price for physical delivery.

EFP/EFS – An EFP (EFS) is the exchange of a futures position for a physical (swap) position. With EFP (EFS) two agreeing parties switch cash (swap) and futures position through the Exchange.

EXECUTION – Refers to the transaction of an order.

EXERCISE – The procedure by which the option buyer (holder) takes up the rights to the contract and is delivered a long (call) or short (put) futures position by the seller (writer).

EXERCISE PRICE – The price at which an option buyer (holder) has the right to buy or sell the underlying commodity.

EXPIRY MONTH – Options terminology for delivery month.

FILL – Refers to the completed execution of an order.

FUTURES CONTRACT – A legally binding agreement on a recognised exchange e.g. IPE, to make or take delivery of a specified instrument, at a fixed date in the future, at a price agreed upon at the time of dealing.

HEDGE – The establishment of a position in one market which is equal and opposite to that held in another market. The reduction of risk.

HIGH – The highest price traded that day.

HOLDER – The person who buys an option contract to open a position.

HOUSE ORDER – An order for a Member's proprietary account.

IN-THE-MONEY – An option which has intrinsic value.

INITIAL MARGIN – The returnable collateral required to establish an options position.

INTRINSIC VALUE – The value to an option buyer (holder) if he were to exercise an option today.

LONG – The position established by the opening purchase of a contract.

LOW – The lowest price traded that day.

MARK-TO-MARKET – The process whereby positions are revalued daily using current prices to determine profit/loss and therefore variation margin.

MINIMUM PRICE FLUCTUATION – The minimum movement that a futures price can move. Otherwise known as a tick.

MNEMONIC – The abbreviation of a Member's company name or individual trader used on a trading badge.

OFFER – see ask price

OPEN INTEREST – The number of contracts left open in the market which need to be closed out or taken through to delivery.

OPEN POSITION – A long or short position that has not been offset by an equal and opposite position e.g. closed out.

OPENING PROCEDURE – The market procedure to establish the opening level of a particular contract.

ORDER – A buy or sell instruction, or combination thereof.

OUT-THE-MONEY – A call (put) option where the exercise price is above (below) the current price of the underlying.

PHYSICAL DELIVERY – The transfer of ownership of an underlying commodity between a buyer and seller party to a futures contract following expiry.

PREARRANGEMENT – The arrangement of a trade to be executed non-competitively in the pit. A pre-arranged trade represents a breach of the Rules (except in the case of cross trade).

PREMIUM – The price paid by the option holder (buyer) to the option writer (seller).

PUT OPTION – An option that gives the holder the right (but not the obligation) to sell a specified quantity of the underlying instrument at a fixed price, on or before a specified date. The writer of the option has the obligation to take delivery of the underlying instrument if the option is exercised by the holder.

QUOTATION – The method used to express the price of a contract for trading purposes.

RALLY – The rapid rise in a price following a fall.

ROLL OVER – The transfer of a position from one expiry month to another – always involving the purchase (sale) or the nearby month and simultaneous corresponding sale (purchase) of a longer dated expiry month.

ROUND TRIP – The opening purchase (sale) of a contract and the subsequent opposite and closing transaction in the same contract. Transaction costs are usually quoted on a "round trip" basis.

RISK MANAGEMENT – The application of financial analysis and diverse financial instruments to the control and the reduction of selected types of risk.

RULE BOOK – The Regulations of the Exchange.

SERIES – All options of the same class which have the same exercise price and expiry date.

SETTLEMENT PRICE – The official price established by the Exchange at the close of each trading day. This figure is used to calculate variation margin.

SHORT – A sale of a contract to open a position.

SPREAD – A trade in which two or more open positions are established in order to trade the differentials and offset risk. Option spreads may use different exercise prices and/or expiry dates.

STANDARD ORDER – The order form used as an official receipt of all incoming orders onto the market floor. These forms must be immediately completed and time-stamped before communication to the trading pit for execution.

STRIKE PRICE – See "exercise price".

TICK -The standardized minimum price movement of a contract's quotation.

TIME STAMP -The time logging machine which records the exact time an order is received and when executed.

TIME VALUE – That part of an options premium not accounted for by intrinsic value.

TIDYING UP PERIOD – Refers to the period of time at the clearing up process on the futures contracts.

TIME AND SALES – The bids, offers, traded prices and estimated volume as observed by Exchange Pit Observer and then recorded in the sequence in which they occurred.

TRADE DISPUTE – Where two traders or more disagree about a particular trade.

TRADE EXECUTION – The process by which orders are traded on the market floor via open outcry (except EFPs/EFSs).

TRADE PRICE – The price agreed between buyer and seller at which the transaction is executed.

TRADING HOURS – Refers to the hours when a particular contract is open for trading.

TRADING RANGE – The range of prices that have been traded over a particular period of time.

TRS – The Trade Registration System is a real-time dual sided matching and account assignment system which allows traders to actively confirm trades.

UNCOVERED / NAKED POSITION – An option position where the writer has no cover against a market move in the holder's favour. The writer of a naked call does not have the underlying commodity to deliver and the writer of a naked put has not sold the commodity short.

VARIATION MARGIN – Profits and losses on open positions which are calculated daily by the mark-to-market process, which are then paid or collected daily.

HISTORICAL VOLATILITY – The measurement given to reflect the degree to which a particular price has fluctuated in the past.

VOLUME – The number of transactions of a particular contract made during a specified period of time.

WRITER – The person who sells an option contract to open a position.

Index

D

E

F

P

R

S